ADVANCED ENGINEERING ANALYSIS

The Calculus of Variations and Functional Analysis with Applications in Mechanics

ADVANCED ENGINEERING ANALYSIS

The Calculus of Variations and Functional Analysis with Applications in Mechanics

Leonid P. Lebedev
Department of Mathematics,
National University of Colombia, Colombia

Michael J. Cloud
Department of Electrical and Computer Engineering,
Lawrence Technological University, USA

Victor A. Eremeyev
Institute of Mechanics, Otto von Guericke University Magdeburg, Germany
South Scientific Center of RASci
and South Federal University, Rostov on Don, Russia

NEW JERSEY • LONDON • SINGAPORE • BEIJING • SHANGHAI • HONG KONG • TAIPEI • CHENNAI

Published by

World Scientific Publishing Co. Pte. Ltd.
5 Toh Tuck Link, Singapore 596224
USA office: 27 Warren Street, Suite 401-402, Hackensack, NJ 07601
UK office: 57 Shelton Street, Covent Garden, London WC2H 9HE

British Library Cataloguing-in-Publication Data
A catalogue record for this book is available from the British Library.

ADVANCED ENGINEERING ANALYSIS
The Calculus of Variations and Functional Analysis with Applications in Mechanics

Desk Editor: Tjan Kwang Wei

ISBN-13 978-981-4390-47-7
ISBN-10 981-4390-47-X

Printed in Singapore.

Preface

A little over half a century ago, it was said that even an ingenious person could not be an engineer unless he had nearly perfect skills with the logarithmic slide rule. The advent of the computer changed this situation crucially; at present, many young engineers have never heard of the slide rule. The computer has profoundly changed the mathematical side of the engineering profession. Symbolic manipulation programs can calculate integrals and solve ordinary differential equations better and faster than professional mathematicians can. Computers also provide solutions to differential equations in numerical form. The easy availability of modern graphics packages means that many engineers prefer such approximate solutions even when exact analytical solutions are available.

Because engineering courses must provide an understanding of the fundamentals, they continue to focus on simple equations and formulas that are easy to explain and understand. Moreover, it is still true that students must develop some analytical abilities. But the practicing engineer, armed with a powerful computer and sophisticated canned programs, employs models of processes and objects that are mathematically well beyond the traditional engineering background. The mathematical methods used by engineers have become quite sophisticated. With insufficient base knowledge to understand these methods, engineers may come to believe that the computer is capable of solving any problem. Worse yet, they may decide to accept nearly any formal result provided by a computer as long as it was generated by a program of a known trademark.

But mathematical methods are restricted. Certain problems may appear to fall within the nominal solution capabilities of a computer program and yet lie well beyond those capabilities. Nowadays, the properties of sophisticated models and numerical methods are explained using terminology from functional analysis and the modern theory of differential equations. Without understanding terms such as "weak solution" and "Sobolev space", one cannot grasp a modern convergence proof or follow a rigorous discussion of the restrictions placed on a mathematical model. Unfortunately, the mathematical portion of the engineering curriculum remains preoccupied with 19th century topics, even omitting the calculus of variations and other classical subjects. It is, nevertheless, increasingly more important for the engineer to understand the theoretical underpinning of his instrumentation than to have an ability to calculate integrals or generate series solutions of differential equations.

The present text offers rigorous insight and will enable an engineer to communicate effectively with the mathematicians who develop models and methods for machine computation. It should prove useful to those who wish to employ modern mathematical methods with some depth of understanding.

The book constitutes a substantial revision and extension of the earlier book *The Calculus of Variations and Functional Analysis*, written by the first two authors. A new chapter (Chapter 2) provides applications of the calculus of variations to nonstandard problems in mechanics. Numerous exercises (most with extensive hints) have been added throughout.

The numbering system is as follows. All definitions, theorems, corollaries, lemmas, remarks, conventions, and examples are numbered consecutively by chapter (thus Definition 1.7 is followed by Lemma 1.8). Equations are numbered independently, again by chapter.

We would like to thank our World Scientific editor, Mr. Yeow-Hwa Quek.

Leonid P. Lebedev
Department of Mathematics, National University of Colombia, Colombia

Michael J. Cloud
Department of Electrical and Computer Engineering, Lawrence Technological University, USA

Victor A. Eremeyev
Institute of Mechanics, Otto von Guericke University Magdeburg, Germany
South Scientific Center of RASci and South Federal University, Rostov on Don, Russia

Contents

Preface v

1. Basic Calculus of Variations 1

1.1 Introduction 1
1.2 Euler's Equation for the Simplest Problem 15
1.3 Properties of Extremals of the Simplest Functional 21
1.4 Ritz's Method 23
1.5 Natural Boundary Conditions 31
1.6 Extensions to More General Functionals 34
1.7 Functionals Depending on Functions in Many Variables . 43
1.8 A Functional with Integrand Depending on Partial Derivatives of Higher Order 49
1.9 The First Variation 54
1.10 Isoperimetric Problems 65
1.11 General Form of the First Variation 72
1.12 Movable Ends of Extremals 76
1.13 Broken Extremals: Weierstrass–Erdmann Conditions and Related Problems 80
1.14 Sufficient Conditions for Minimum 85
1.15 Exercises 94

2. Applications of the Calculus of Variations in Mechanics 99

2.1 Elementary Problems for Elastic Structures 99
2.2 Some Extremal Principles of Mechanics 108
2.3 Conservation Laws 127
2.4 Conservation Laws and Noether's Theorem 131

2.5 Functionals Depending on Higher Derivatives of y 139
2.6 Noether's Theorem, General Case 143
2.7 Generalizations 147
2.8 Exercises 153

3. Elements of Optimal Control Theory 159

3.1 A Variational Problem as an Optimal Control Problem . . 159
3.2 General Problem of Optimal Control 161
3.3 Simplest Problem of Optimal Control 164
3.4 Fundamental Solution of a Linear Ordinary Differential Equation 170
3.5 The Simplest Problem, Continued 171
3.6 Pontryagin's Maximum Principle for the Simplest Problem 173
3.7 Some Mathematical Preliminaries 177
3.8 General Terminal Control Problem 189
3.9 Pontryagin's Maximum Principle for the Terminal Optimal Problem 195
3.10 Generalization of the Terminal Control Problem 198
3.11 Small Variations of Control Function for Terminal Control Problem 202
3.12 A Discrete Version of Small Variations of Control Function for Generalized Terminal Control Problem 205
3.13 Optimal Time Control Problems 208
3.14 Final Remarks on Control Problems 212
3.15 Exercises 214

4. Functional Analysis 215

4.1 A Normed Space as a Metric Space 217
4.2 Dimension of a Linear Space and Separability 223
4.3 Cauchy Sequences and Banach Spaces 227
4.4 The Completion Theorem 238
4.5 L^p Spaces and the Lebesgue Integral 242
4.6 Sobolev Spaces 248
4.7 Compactness 250
4.8 Inner Product Spaces, Hilbert Spaces 260
4.9 Operators and Functionals 264
4.10 Contraction Mapping Principle 269
4.11 Some Approximation Theory 276

4.12 Orthogonal Decomposition of a Hilbert Space and the Riesz Representation Theorem 280
4.13 Basis, Gram–Schmidt Procedure, and Fourier Series in Hilbert Space 284
4.14 Weak Convergence 291
4.15 Adjoint and Self-Adjoint Operators 298
4.16 Compact Operators 304
4.17 Closed Operators 311
4.18 On the Sobolev Imbedding Theorem 315
4.19 Some Energy Spaces in Mechanics 320
4.20 Introduction to Spectral Concepts 337
4.21 The Fredholm Theory in Hilbert Spaces 343
4.22 Exercises 352

5. Applications of Functional Analysis in Mechanics 359

5.1 Some Mechanics Problems from the Standpoint of the Calculus of Variations; the Virtual Work Principle 359
5.2 Generalized Solution of the Equilibrium Problem for a Clamped Rod with Springs 364
5.3 Equilibrium Problem for a Clamped Membrane and its Generalized Solution 367
5.4 Equilibrium of a Free Membrane 369
5.5 Some Other Equilibrium Problems of Linear Mechanics . 371
5.6 The Ritz and Bubnov–Galerkin Methods 379
5.7 The Hamilton–Ostrogradski Principle and Generalized Setup of Dynamical Problems in Classical Mechanics . . . 381
5.8 Generalized Setup of Dynamic Problem for Membrane . . 383
5.9 Other Dynamic Problems of Linear Mechanics 397
5.10 The Fourier Method 399
5.11 An Eigenfrequency Boundary Value Problem Arising in Linear Mechanics 400
5.12 The Spectral Theorem 404
5.13 The Fourier Method, Continued 410
5.14 Equilibrium of a von Kármán Plate 415
5.15 A Unilateral Problem 425
5.16 Exercises 431

Appendix A Hints for Selected Exercises 433

Bibliography 483

Index 485

Chapter 1

Basic Calculus of Variations

1.1 Introduction

Optimization is a universal goal. Students would like to learn more, receive better grades, and have more free time; professors (at least some of them) would like to give better lectures, see students learn more, receive higher pay, and have more free time. These are the optimization problems of real life. In mathematics, optimization makes sense only when formulated in terms of a function $f(x)$ or other expression. One then seeks the minimum value of the expression. (It suffices to discuss minimization because maximizing f is equivalent to minimizing $-f$.)

This book treats the minimization of *functionals*. The notion of functional generalizes that of function. Although the process of generalization does yield results of greater generality, as a rule the results are not sharper in particular cases. So to understand what can be expected from the calculus of variations, we should review the minimization of ordinary functions. All quantities will be assumed sufficiently differentiable for the purpose at hand. Let us recall some terminology for the one-variable case $y = f(x)$.

Definition 1.1. The function $f(x)$ has a *local minimum* at a point x_0 if there is a neighborhood $(x_0 - d, x_0 + d)$ in which $f(x) \geq f(x_0)$. We call x_0 the *global minimum* of $f(x)$ on $[a, b]$ if $f(x) \geq f(x_0)$ holds for all $x \in [a, b]$.

The necessary condition for a differentiable function $f(x)$ to have a local minimum at x_0 is

$$f'(x_0) = 0. \tag{1.1}$$

A simple and convenient sufficient condition is

$$f''(x_0) > 0. \tag{1.2}$$

Unfortunately, no available criterion for a local minimum is both sufficient and necessary. So the approach is to solve (1.1) for possible points of local minimum of $f(x)$ and then test these using an available sufficient condition.

The global minimum on $[a, b]$ can be attained at a point of local minimum. But there are two points, a and b, where (1.1) may not hold (because the corresponding neighborhoods are one-sided) but where the global minimum may still occur. Hence given a differentiable function $f(x)$ on $[a, b]$, we first find all x_k at which $f'(x_k) = 0$. We then calculate $f(a)$, $f(b)$, and $f(x_k)$ at the x_k, and choose the global minimum. Although this method can be arranged as an algorithm suitable for machine computation, it still cannot be reduced to the solution of an equation or system of equations.

These tools are extended to multivariable functions and to more complex objects called *functionals*. A simple example of a functional is an integral whose integrand depends on an unknown function and its derivative. Since the extension of ordinary minimization methods to functionals is not straightforward, we continue to examine some notions from calculus.

A continuously differentiable function $f(x)$ obeys Lagrange's formula

$$f(x+h) - f(x) = f'(x+\theta h)h \qquad (0 \le \theta \le 1). \tag{1.3}$$

Continuity of f' means that

$$f'(x+\theta h) - f'(x) = r_1(x,\theta,h) \to 0 \quad \text{as } h \to 0,$$

hence

$$f(x+h) = f(x) + f'(x)h + r_1(x,\theta,h)\,h$$

where $r_1(x,\theta,h) \to 0$ as $h \to 0$. The term $r_1(x,\theta,h)\,h$ is Lagrange's form of the remainder. There is also Peano's form

$$f(x+h) = f(x) + f'(x)h + o(h), \tag{1.4}$$

which means that

$$\lim_{h\to 0} \frac{f(x+h) - f(x) - f'(x)h}{h} = 0.$$

The principal (linear in h) part of the increment of f is the *first differential* of f at x. Writing $dx = h$ we have

$$df = f'(x)\,dx. \tag{1.5}$$

"Infinitely small" quantities are *not* implied by this notation; here dx is a finite increment of x (taken sufficiently small when used for approximation).

The first differential is invariant under the change of variable $x = \varphi(s)$:

$$df = f'(x)\,dx = \frac{df(\varphi(s))}{ds}\,ds, \quad \text{where} \quad dx = \varphi'(s)\,ds.$$

Lagrange's formula extends to functions having m continuous derivatives in some neighborhood of x. The extension for $x + h$ lying in the neighborhood is Taylor's formula:

$$f(x+h) = f(x) + f'(x)h + \frac{1}{2!}f''(x)h^2 + \cdots + \frac{1}{(m-1)!}f^{(m-1)}(x)h^{m-1}$$
$$+ \frac{1}{m!}f^{(m)}(x+\theta h)h^m \qquad (0 \le \theta \le 1). \tag{1.6}$$

Continuity of $f^{(m)}$ at x yields

$$f^{(m)}(x+\theta h) - f^{(m)}(x) = r_m(x,\theta,h) \to 0 \quad \text{as } h \to 0,$$

hence Taylor's formula becomes

$$f(x+h) = f(x) + f'(x)h + \frac{1}{2!}f''(x)h^2 + \cdots + \frac{1}{m!}f^{(m)}(x)h^m$$
$$+ \frac{1}{m!}r_m(x,\theta,h)h^m$$

with remainder in Lagrange form. The dependence of the remainder on the parameters is suppressed in Peano's form

$$f(x+h) = f(x) + f'(x)h + \frac{1}{2!}f''(x)h^2 + \cdots + \frac{1}{m!}f^{(m)}(x)h^m + o(h^m). \tag{1.7}$$

The conditions of minimum (1.1)–(1.2) can be derived via Taylor's formula for a twice continuously differentiable function having

$$f(x+h) - f(x) = f'(x)h + \frac{1}{2}f''(x)h^2 + o(h^2). \tag{1.8}$$

Indeed $f(x+h) - f(x) \ge 0$ if x is a local minimum. The right side has the form $ah + bh^2 + o(h^2)$. If $a = f'(x) \ne 0$, for example when $a < 0$, then for $h < h_0$ with sufficiently small h_0 the sign of $f(x+h) - f(x)$ is determined by that of ah; hence for $0 < h < h_0$ we have $f(x+h) - f(x) < 0$, which contradicts the assertion that x minimizes f. The case $a > 0$ is similar, resulting in the necessary condition (1.1). The increment formula gives

$$f(x+h) - f(x) = \frac{1}{2}f''(x)h^2 + o(h^2).$$

The term $f''(x)h^2$ defines the value of the right side when h is sufficiently close to 0, hence when $f''(x) > 0$ we see that for sufficiently small $|h| \ne 0$

$$f(x+h) - f(x) > 0.$$

So (1.2) is sufficient for x to be a minimum point of f.

A function in n variables

Consider the minimization of a function $y = f(\mathbf{x})$ with $\mathbf{x} = (x_1, \ldots, x_n)$. More cannot be expected from this theory than from the theory of functions in a single variable.

Definition 1.2. A function $f(\mathbf{x})$ has a *global minimum* at the point $\mathbf{x}^*$ if the inequality

$$f(\mathbf{x}^*) \le f(\mathbf{x}^* + \mathbf{h}) \tag{1.9}$$

holds for all nonzero $\mathbf{h} = (h_1, \ldots, h_n) \in \mathbb{R}^n$. The point $\mathbf{x}^*$ is a *local minimum* if there exists $\rho > 0$ such that (1.9) holds whenever $\|\mathbf{h}\| = (h_1^2 + \cdots + h_n^2)^{1/2} < \rho$.

Let $\mathbf{x}^*$ be a minimum point of a continuously differentiable function $f(\mathbf{x})$. Then $f(x_1, x_2^*, \ldots, x_n^*)$ is a function in one variable x_1 and takes its minimum at x_1^*. It follows that $\partial f / \partial x_1 = 0$ at $x_1 = x_1^*$. Similarly, the rest of the partial derivatives of f are zero at $\mathbf{x}^*$:

$$\left.\frac{\partial f}{\partial x_i}\right|_{\mathbf{x}=\mathbf{x}^*} = 0, \qquad i = 1, \ldots, n. \tag{1.10}$$

This is a necessary condition of minimum for a continuously differentiable function in n variables at the point $\mathbf{x}^*$.

To get sufficient conditions we must extend Taylor's formula. Let $f(\mathbf{x})$ possess all continuous derivatives up to order $m \ge 2$ in a ball centered at point $\mathbf{x}$, and suppose $\mathbf{x} + \mathbf{h}$ lies in this ball. Fixing these, we apply (1.7) to $f(\mathbf{x} + t\mathbf{h})$ and get Taylor's formula in the variable t:

$$\begin{aligned} f(\mathbf{x} + t\mathbf{h}) = f(\mathbf{x}) &+ \left.\frac{df(\mathbf{x} + t\mathbf{h})}{dt}\right|_{t=0} t + \frac{1}{2!} \left.\frac{d^2 f(\mathbf{x} + t\mathbf{h})}{dt^2}\right|_{t=0} t^2 \\ &+ \cdots + \frac{1}{m!} \left.\frac{d^m f(\mathbf{x} + t\mathbf{h})}{dt^m}\right|_{t=0} t^m + o(t^m). \end{aligned}$$

The remainder term is for the case when $t \to 0$. From this equality for sufficiently small t, the general Taylor formula can be derived.

The minimization problem for $f(\mathbf{x})$ is studied using only the first two terms of this formula:

$$f(\mathbf{x} + t\mathbf{h}) = f(\mathbf{x}) + \left.\frac{df(\mathbf{x} + t\mathbf{h})}{dt}\right|_{t=0} t + \frac{1}{2!} \left.\frac{d^2 f(\mathbf{x} + t\mathbf{h})}{dt^2}\right|_{t=0} t^2 + o(t^2). \tag{1.11}$$

We calculate $df(\mathbf{x} + t\mathbf{h})/dt$ as a derivative of a composite function:

$$\left.\frac{df(\mathbf{x} + t\mathbf{h})}{dt}\right|_{t=0} = \frac{\partial f(\mathbf{x})}{\partial x_1} h_1 + \frac{\partial f(\mathbf{x})}{\partial x_2} h_2 + \cdots + \frac{\partial f(\mathbf{x})}{\partial x_n} h_n.$$

The first differential is defined as

$$df = \frac{\partial f(\mathbf{x})}{\partial x_1}\, dx_1 + \frac{\partial f(\mathbf{x})}{\partial x_2}\, dx_2 + \cdots + \frac{\partial f(\mathbf{x})}{\partial x_n}\, dx_n. \tag{1.12}$$

The next term,

$$\left.\frac{d^2 f(\mathbf{x}+t\mathbf{h})}{dt^2}\right|_{t=0} = \sum_{i,j=1}^{n} \frac{\partial^2 f(\mathbf{x})}{\partial x_i\, \partial x_j} h_i h_j,$$

defines the second differential of f:

$$d^2 f = \sum_{i,j=1}^{n} \frac{\partial^2 f(\mathbf{x})}{\partial x_i\, \partial x_j}\, dx_i\, dx_j. \tag{1.13}$$

Taylor's formula of the second order becomes

$$f(\mathbf{x}+\mathbf{h}) = f(\mathbf{x}) + \sum_{i=1}^{n} \frac{\partial f(\mathbf{x})}{\partial x_i} h_i + \frac{1}{2!} \sum_{i,j=1}^{n} \frac{\partial^2 f(\mathbf{x})}{\partial x_i \partial x_j} h_i h_j + o(\|\mathbf{h}\|^2). \tag{1.14}$$

The necessary condition for a minimum, $df = 0$, follows from (1.11) or (1.10). By (1.11), the condition

$$\left.\frac{d^2 f(\mathbf{x}+t\mathbf{h})}{dt^2}\right|_{t=0} > 0 \text{ for any sufficiently small } \|\mathbf{h}\|$$

suffices for $\mathbf{x}$ to minimize f. The corresponding quadratic form in the variables h_i is

$$\frac{1}{2!} \sum_{i,j=1}^{n} \frac{\partial^2 f(\mathbf{x})}{\partial x_i \partial x_j} h_i h_j = \frac{1}{2} \begin{pmatrix} h_1 & \cdots & h_n \end{pmatrix} \begin{pmatrix} \frac{\partial^2 f(\mathbf{x})}{\partial x_1^2} & \cdots & \frac{\partial^2 f(\mathbf{x})}{\partial x_1 x_n} \\ \vdots & \ddots & \vdots \\ \frac{\partial^2 f(\mathbf{x})}{\partial x_n x_1} & \cdots & \frac{\partial^2 f(\mathbf{x})}{\partial x_n^2} \end{pmatrix} \begin{pmatrix} h_1 \\ \vdots \\ h_n \end{pmatrix}.$$

The $n \times n$ *Hessian matrix* is symmetric under our smoothness assumptions on f. Positive definiteness of the quadratic form can be verified via Sylvester's criterion.

The problem of global minimum for a function in many variables on a closed domain Ω is more complicated than the corresponding problem for a function in one variable. Indeed, the set of points satisfying (1.10) can be infinite for a multivariable function. Trouble also arises concerning the domain boundary $\partial\Omega$: since it is no longer a finite set (unlike $\{a, b\}$) we must also solve the problem of minimum on $\partial\Omega$, and the structure of such a set can be complicated. The algorithm for finding a point of global minimum of a function $f(\mathbf{x})$ cannot be described in several phrases; it depends on the structure of both the function and the domain.

Issues connected with the boundary can be avoided by considering the problem of global minimum of a function on an open domain. We will take this approach when treating the calculus of variations. Although analogous problems with closed domains arise in applications, the difficulties are so great that no general results are applicable to many problems. One must investigate each such problem separately.

Constraints of the form

$$g_i(\mathbf{x}) = 0, \qquad i = 1, \ldots, m, \tag{1.15}$$

permit reduction of constrained minimization to an unconstrained problem provided we can solve (1.15) and get

$$x_k = \psi_k(x_1, \ldots, x_{n-m}), \qquad k = n - m + 1, \ldots, n.$$

Substitution into $f(\mathbf{x})$ would yield an ordinary unconstrained minimization problem for a function in $n - m$ variables

$$f(x_1, \ldots, x_{n-m}, \ldots, \psi_n(x_1, \ldots, x_{n-m})).$$

The resulting system of equations is nonlinear in general. This situation can be circumvented by the use of Lagrange multipliers. The method proceeds with formation of the *Lagrangian function*

$$\mathcal{L}(x_1, \ldots, x_n, \lambda_1, \ldots, \lambda_m) = f(\mathbf{x}) + \sum_{j=1}^{m} \lambda_j g_j(\mathbf{x}), \tag{1.16}$$

by which the constraints g_j are adjoined to f. Then the x_i and λ_i are all treated as independent, unconstrained variables. The resulting necessary conditions form a system of $n + m$ equations in the $n + m$ unknowns x_i, λ_j:

$$\begin{aligned} &\frac{\partial f(\mathbf{x})}{\partial x_i} + \sum_{j=1}^{m} \lambda_j \frac{\partial g_j(\mathbf{x})}{\partial x_i} = 0, \qquad i = 1, \ldots, n, \\ &g_j(\mathbf{x}) = 0, \qquad j = 1, \ldots, m. \end{aligned} \tag{1.17}$$

Functionals

The kind of dependence in which a real number corresponds to another (or to a finite set) is not enough to describe many natural processes. Areas such as physics and biology spawn formulations not amenable to such description. Consider the deformations of an airplane in flight. At some point near an engine, the deformation is not merely a function of the force produced by the engine — it also depends on the other engines, air resistance, and passenger positions and movements (hence the admonition that everyone remain seated during potentially dangerous parts of the flight). In general, many real processes in a body are described by the dependence of the displacement field (e.g., the field of strains, stresses, heat, voltage) on other fields (e.g., loads, heat radiation) in the same body. Each field is described by one or more functions, so the dependence is that of a function uniquely defined by a set of other functions acting as whole objects (arguments). A dependence of this type, provided we specify the classes to which all functions belong, is called an *operator* (or *map*, or sometimes just a "function" again). Problems of finding such dependences are often formulated as boundary or initial-boundary value problems for partial differential equations. These and their analysis form the main content of any course in a particular science. Since a full description of any process is complex, we usually work with simplified models that retain only essential features. However, even these can be quite challenging when we seek solutions.

Humans often try to optimize their actions through an intuitive — not mathematical — approach to fuzzily-posed problems on minimization or maximization. This is because our nature reflects the laws of nature in total. In physics there are quantities, like energy and enthalpy, whose values in the state of equilibrium or real motion are minimal or maximal in comparison with other "nearby admissible" states. Younger sciences like mathematical biology attempt to follow suit: when possible they seek to describe system behavior through the states of certain fields of parameters, on which functions of energy type attain maxima or minima. The energy of a system (e.g., body or set of interacting bodies) is characterized by a number which depends on the fields of parameters inside the system. Thus the dependence described by quantities of energy type is such that *a numerical value E is uniquely defined by the distribution of fields of parameters characterizing the system.* We call this sort of dependence a *functional*. Of course, in mathematics we must also specify the classes to which the above fields may belong. The notion of functional generalizes that of function so that the minimization problem remains sensible. Hence we come to the object of investigation of our main subject: the calculus of variations. In actuality we shall consider a somewhat restricted class of functionals. (Optimization of general functionals belongs to *mathematical programming*, a younger science that contains the calculus of variations — a subject some 300 years old — as a special case.) In the calculus of variations we minimize functionals of integral type. A typical problem involves the total energy functional for an elastic membrane under load $F = F(x, y)$:

$$E(u) = \frac{1}{2}a \iint_S \left[\left(\frac{\partial u}{\partial x} \right)^2 + \left(\frac{\partial u}{\partial y} \right)^2 \right] dx\, dy - \iint_S Fu\, dx\, dy.$$

Here $u = u(x, y)$ is the deflection of a point (x, y) of the membrane, which occupies a domain S and has tension described by parameter a (we can put $a = 1$ without loss of generality). For a membrane with fixed edge, in equilibrium $E(u)$ takes its minimal value relative to all other *admissible* (or *virtual*) states. (An "admissible" function takes appointed boundary values and is sufficiently smooth, in this case having first and second continuous derivatives in S.) The equilibrium state is described by Poisson's equation

$$\Delta u = -F. \tag{1.18}$$

Let us also supply the boundary condition

$$u\big|_{\partial S} = \phi. \tag{1.19}$$

The problem of minimizing $E(u)$ over the set of smooth functions satisfying (1.19) is equivalent to the boundary value problem (1.18)–(1.19). Analogous situations arise in many other sciences. Eigenfrequency problems can also be formulated within the calculus of variations.

Other interesting problems come from geometry. Consider the following *isoperimetric problem:*

> *Of all possible smooth closed curves of unit length in the plane, find the equation of that curve L which encloses the greatest area.*

With $r = r(\phi)$ the polar equation of a curve, we seek to have

$$\int_0^{2\pi} \sqrt{r^2 + \left(\frac{dr}{d\phi} \right)^2}\, d\phi = 1, \qquad \frac{1}{2} \int_0^{2\pi} r^2\, d\phi \to \max.$$

Notice how we denoted the problem of maximization. Every high school student knows the answer, but certainly not the method of solution.

We cannot list all problems solvable by the calculus of variations. It is safe to say only that the relevant functionals possess an integral form, and that the integrands depend upon unknown functions and their derivatives.

Again, we can suppose that the theory for minimizing a functional should represent an extension of the theory for minimizing a multivariable function. As in the latter theory, we must appoint a domain on which the functional is determined. Even for a multivariable function, this is not always an easy task. For the functional it is much harder, as the arguments now belong to certain classes of functions, and the answer can depend on the class as well as the detailed calculations we perform. The study of function spaces falls under the heading of functional analysis, considered in Chapter 4. General description of the domains of functionals can be undertaken via normed spaces of functions. The classical calculus of variations arose long before functional analysis, and dealt with the classes of continuously differentiable (or n-times continuously differentiable) functions under certain conditions on the boundary of the integration domain.

We expect the notions of local minimum and global minimum to appear in the study of functionals. A definition of local minimum will require a precise notion of a neighborhood of the minimizing function. In this case functional analytic ideas are quite helpful. As we said, however, the calculus of variations predated functional analysis. The notion of a neighborhood of a function was developed in the calculus of variations and later inherited by functional analysis.

The necessary conditions (1.10) can be suitably extended to the problem of minimum for a functional. We will see this explicitly when we approximate the functional with a function in n variables. But for the complete treatment of a functional, the conditions should be given at any point along the minimizing function. These conditions are known as Euler equations or Euler–Lagrange equations. They are obtained when the minimizer lies *inside* the domain of the functional (i.e., the minimizer should lie some distance away from the boundary of the domain, and this will be assumed even if not stated).

Finally, the Euler equation for a functional represents only a necessary condition for a minimum. Sufficient conditions are more subtle and require separate investigation. However, in certain physical problems (such as those associated with linear models in continuum mechanics) where a point of minimum total potential energy is sought, we obtain a unique extremum that automatically turns out to be a minimum.

In the next section, we show how the problem of minimum for one special functional is related to the problem of minimum for a multivariable function.

Minimization of a simple functional using calculus

Consider a general functional of the form

$$F(y) = \int_a^b f(x, y, y')\,dx, \tag{1.20}$$

where $y = y(x)$ is smooth. (At this stage we do not stop to formulate strict conditions on the functions involved; we simply assume they have as many continuous derivatives as needed. Nor do we clearly specify the neighborhood of a function for which it is a local minimizer of a functional.)

From the time of Newton's *Principia*, mathematical physics has formulated and considered each problem so that it has a solution which, at least under certain conditions, is unique. Although the idea of determinism in nature was buried by quantum mechanics, it remained an important part of the older subject of the calculus of variations. We know that for the equilibrium problem for a membrane to have a unique solution, we must impose boundary conditions. So let us first understand whether the problem of minimum for (1.20) is well-posed; i.e., whether (at least for simple particular cases) a solution exists and is unique.

The particular form

$$\int_a^b \sqrt{1 + (y')^2}\,dx$$

yields the length of the plane curve $y = y(x)$ from $(a, y(a))$ to $(b, y(b))$. The obvious minimizer is a straight line $y = kx + d$. Without boundary conditions (i.e., with $y(a)$ or $y(b)$ unspecified), k and d are arbitrary and the solution is not unique. We can impose no more than two restrictions on $y(x)$ at the ends a and b, because $y = kx + d$ has only two indefinite constants. However, the problem without boundary conditions also makes sense; its solution is the set of horizontal segments $y = d$ starting at the vertical line $x = a$ and ending at $x = b$.

Problem setup is a tough yet important issue in mathematics. We shall eventually face the question of how to pose the main problems of the calculus of variations in a sensible fashion.

Let us consider the problem of minimum of (1.20) without additional restrictions, and attempt to solve it using calculus. Discretization, in this case the approximation of the integral by a Riemann sum, will reduce the functional to a multivariable function. In the calculus of variations other methods of investigation are customary; however, the current approach is instructive because it leads to some central results of the calculus of variations and shows that certain important ideas are extensions of ordinary calculus.

We begin by subdividing $[a, b]$ into n partitions each of length

$$h = \frac{b-a}{n}.$$

Denote $x_i = a + ih$ and $y_i = y(x_i)$, so $y_0 = y(a)$ and $y_n = y(b)$. Take an approximate value of $y'(x_i)$ as

$$y'(x_i) \approx \frac{y_{i+1} - y_i}{h}.$$

Approximation of (1.20) by the Riemann sum

$$\int_a^b f(x, y, y')\,dx \approx h \sum_{k=0}^{n-1} f(x_k, y_k, y'(x_k)) \tag{1.21}$$

gives

$$\begin{aligned} \int_a^b f(x, y, y')\,dx &\approx h \sum_{k=0}^{n-1} f(x_k, y_k, (y_{k+1} - y_k)/h) \\ &= \Phi(y_0, \ldots, y_n). \end{aligned} \tag{1.22}$$

Since $\Phi(y_0, \ldots, y_n)$ is an ordinary function in $n + 1$ independent variables, we set

$$\frac{\partial \Phi(y_0, y_1, \ldots, y_n)}{\partial y_i} = 0, \qquad i = 0, \ldots, n. \tag{1.23}$$

Again, any function f encountered is assumed to possess all needed derivatives. Henceforth we denote partial derivatives using

$$f_y = \frac{\partial f}{\partial y}, \qquad f_{y'} = \frac{\partial f}{\partial y'}, \qquad f_x = \frac{\partial f}{\partial x}, \tag{1.24}$$

and the total derivative using

$$\begin{aligned} \frac{df(x, y(x), y'(x))}{dx} &= f_x(x, y(x), y'(x)) + f_y(x, y(x), y'(x))\, y'(x) \\ &\quad + f_{y'}(x, y(x), y'(x))\, y''(x). \end{aligned} \tag{1.25}$$

Observe that in the notation $f_{y'}$ we regard y' as the name of a simple variable; we temporarily ignore its relation to y and even its status as a function in its own right.

Consider the structure of (1.23). The variable y_i appears in the sum (1.22) only once when $i = 0$ or $i = n$, twice otherwise. In the latter case (1.23) gives, using the chain rule and omitting the factor h,

$$\frac{f_{y'}\left(x_{i-1}, y_{i-1}, \dfrac{y_i - y_{i-1}}{h}\right)}{h} - \frac{f_{y'}\left(x_i, y_i, \dfrac{y_{i+1} - y_i}{h}\right)}{h} + f_y\left(x_i, y_i, \frac{y_{i+1} - y_i}{h}\right) = 0. \tag{1.26}$$

For $i = 0$ the result is

$$h\left[f_y\left(x_0, y_0, \frac{y_1 - y_0}{h}\right) - \frac{f_{y'}\left(x_0, y_0, \dfrac{y_1 - y_0}{h}\right)}{h}\right] = 0$$

or

$$f_{y'}\left(x_0, y_0, \frac{y_1 - y_0}{h}\right) - h\, f_y\left(x_0, y_0, \frac{y_1 - y_0}{h}\right) = 0. \tag{1.27}$$

For $i = n$ we obtain

$$f_{y'}\left(x_{n-1}, y_{n-1}, \frac{y_n - y_{n-1}}{h}\right) = 0. \tag{1.28}$$

In the limit as $h \to 0$, (1.27) and (1.28) give, respectively,

$$f_{y'}(x, y(x), y'(x))\big|_{x=a} = 0, \qquad f_{y'}(x, y(x), y'(x))\big|_{x=b} = 0.$$

Finally, considering the first two terms in (1.26) for $0 < i < n$,

$$-\frac{f_{y'}\left(x_i, y_i, \dfrac{y_{i+1} - y_i}{h}\right) - f_{y'}\left(x_{i-1}, y_{i-1}, \dfrac{y_i - y_{i-1}}{h}\right)}{h},$$

we recognize an approximation for the total derivative $-df_{y'}/dx$ at y_{i-1}. Hence (1.26), after $h \to 0$ in such a way that x_{i-1} remains a fixed value c, reduces to

$$f_y - \frac{d}{dx} f_{y'} = 0 \tag{1.29}$$

at $x = c$. A nonuniform partitioning will yield this equation similarly for any $x = c \in (a, b)$. In expanded form (1.29) is

$$f_y - f_{y'x} - f_{y'y}y' - f_{y'y'}y'' = 0, \qquad x \in (a, b). \tag{1.30}$$

The limit passage has given us this second-order ordinary differential equation and two boundary conditions

$$f_{y'}\Big|_{x=a} = 0, \qquad f_{y'}\Big|_{x=b} = 0. \tag{1.31}$$

Equations (1.29) and (1.31) play the same role for the functional (1.20) as equations (1.10) play for a function in many variables. In the absence of boundary conditions on $y(x)$, we get necessarily two boundary conditions for a function on which (1.20) attains a minimum.

Since the resulting equation is of second order, no more than two boundary conditions can be imposed on its solution (see, however, Remark 1.20). We could, say, fix the ends of the curve $y = y(x)$ by putting

$$y(a) = c_0, \qquad y(b) = c_1. \tag{1.32}$$

If we repeat the above process under this restriction we get (1.26) and correspondingly (1.29), whereas (1.31) is replaced by (1.32). We can consider the problem of minimum of this functional on the set of functions satisfying (1.32). Then the necessary condition which a minimizer should satisfy is the boundary value problem consisting of (1.29) and (1.32).

Conditions such as $y(a) = 0$ and $y'(a) = 0$ are normally posed for a Cauchy problem involving a second-order differential equation. In the present case, however, a repetition of the above steps implies the *additional* restriction $f_{y'}|_{x=b} = 0$. A problem for (1.29) with three boundary conditions is, in general, inconsistent.

We have obtained some possible ways to set up the problem of minimum of the functional (1.20).

Notation for various types of derivatives

It will be necessary to take derivatives of composite functions. When such functions are integrated by parts, we encounter "total derivatives" that must be distinguished from the usual partial derivatives. We denote total derivatives in the same way as ordinary derivatives, using the differential symbol d: therefore d/dx will denote a total derivative with respect to x. We often denote partial derivatives by subscripts so that $\partial(\cdot)/\partial x$ will be denoted by $(\cdot)_x$ or sometimes $(\cdot)_1$. Let us consider two common cases.

1. Suppose

$$f = f(x, y(x), y'(x))$$

so that f depends on x through (1) an independent variable x, and (2) the variables $p = y(x)$ and $q = y'(x)$ that are each functions of x as well. We will denote the partial derivative with respect to x as

$$f_x = \frac{\partial}{\partial x} f(x,p,q)\bigg|_{p=y(x),\, q=y'(x)}$$

where, during differentiation, we regard p and q as independent variables. Other partial derivatives are

$$f_y = \frac{\partial}{\partial p} f(x,p,q)\bigg|_{p=y(x),\, q=y'(x)}, \qquad f_{y'} = \frac{\partial}{\partial q} f(x,p,q)\bigg|_{p=y(x),\, q=y'(x)}.$$

The total derivative with respect to x, denoted d/dx, arises when we differentiate while considering $y(x)$ and $y'(x)$ to be functions of x. The total derivative of the partial derivative $f_{y'}$ is, by the chain rule,

$$\frac{d}{dx} f_{y'} \equiv \frac{d}{dx} f_{y'}(x, y(x), y'(x)) = f_{y'x} + f_{y'y} y' + f_{y'y'} y'',$$

where, for example,

$$f_{y'y} = \frac{\partial}{\partial p}\frac{\partial}{\partial q} f(x,p,q)\bigg|_{p=y(x),\, q=y'(x)}.$$

2. Consider the composite function

$$f = f(x, y, u(x,y), u_x(x,y), u_y(x,y))$$

depending on independent variables x, y and on a function u and its derivatives, which depend on x, y as well. Now we denote

$$p = u(x,y), \qquad q = u_x(x,y), \qquad r = u_y(x,y),$$

where u_x and u_y are partial derivatives with respect to x and y, respectively. Introducing variables p, q, r, we get a function $f = f(x, y, p, q, r)$ in five independent variables. The following notations are used for partial derivatives:

$$f_x = \frac{\partial}{\partial x} f(x,y,p,q,r)\bigg|_{p=u(x,y),\, q=u_x(x,y),\, r=u_y(x,y)},$$

$$f_y = \frac{\partial}{\partial y} f(x,y,p,q,r)\bigg|_{p=u(x,y),\, q=u_x(x,y),\, r=u_y(x,y)},$$

$$f_u = \frac{\partial}{\partial p} f(x,y,p,q,r)\bigg|_{p=u(x,y),\, q=u_x(x,y),\, r=u_y(x,y)},$$

$$f_{u_x} = \frac{\partial}{\partial q} f(x, y, p, q, r)\Big|_{p=u(x,y),\, q=u_x(x,y),\, r=u_y(x,y)},$$

and

$$f_{u_y} = \frac{\partial}{\partial r} f(x, y, p, q, r)\Big|_{p=u(x,y),\, q=u_x(x,y),\, r=u_y(x,y)}.$$

Finally, let us display the notation for the total derivative d/dx of f_{u_x}, where f denotes $f = f(x, y, p, q, r)$:

$$\frac{d}{dx} f_{u_x} = \left(f_{qx} + f_{qp} u_x + f_{qq} u_{xx} + f_{qr} u_{yx} \right)\Big|_{p=u(x,y),\, q=u_x(x,y),\, r=u_y(x,y)},$$

and a similar formula for the total derivative with respect to y:

$$\frac{d}{dy} f_{u_x} = \left(f_{qy} + f_{qp} u_y + f_{qq} u_{xy} + f_{qr} u_{yy} \right)\Big|_{p=u(x,y),\, q=u_x(x,y),\, r=u_y(x,y)}.$$

The formulas for higher derivatives are denoted similarly.

Brief summary of important terms

A *functional* is a correspondence assigning a real number to each function in some class of functions. The calculus of variations is concerned with *variational problems*: i.e., those in which we seek the *extrema* (maxima or minima) of functionals.

An *admissible function* for a given variational problem is a function that satisfies all the constraints of that problem.

A function is *sufficiently smooth* for a particular development if all required actions (e.g., differentiation, integration by parts) are possible and yield results having the properties needed for that development.

1.2 Euler's Equation for the Simplest Problem

We begin with the problem of local minimum of the functional

$$F(y) = \int_a^b f(x, y, y')\, dx \tag{1.33}$$

on the set of functions $y = y(x)$ that satisfy the boundary conditions

$$y(a) = c_0, \qquad y(b) = c_1. \tag{1.34}$$

The existence of a solution can depend on the properties of this set. We must compare the values of $F(y)$ on all functions y satisfying (1.34). In view of (1.29) it is reasonable to seek minimizers that have continuous first and second derivatives on $[a, b]$. How should we specify a neighborhood of a function $y(x)$? Since all admissible functions must satisfy (1.34), we can consider the set of functions of the form $y(x) + \varphi(x)$ where

$$\varphi(a) = \varphi(b) = 0. \tag{1.35}$$

With the intention of using tools close to those of classical calculus, we first introduce the idea of continuity of a functional with respect to an argument which, in turn, is a function on $[a, b]$. A suitably modified version of the classical definition of function continuity is as follows: given any small $\varepsilon > 0$, there exists a δ-neighborhood of $y(x)$ such that when $y(x) + \varphi(x)$ belongs to this neighborhood we have

$$|F(y + \varphi) - F(y)| < \varepsilon.$$

If the neighborhood of the zero function is specified by the inequality

$$\max_{x \in [a,b]} |\varphi(x)| + \max_{x \in [a,b]} |\varphi'(x)| < \delta, \tag{1.36}$$

the definition can become workable when $f(x, y, y')$ is continuous in the three independent variables x, y, y'. This is not the only possible definition of a neighborhood; later we shall discuss other possibilities. But one benefit is that the left side of (1.36) contains the expression usually used to define the norm on the set of all functions continuously differentiable on $[a, b]$:

$$\|\varphi(x)\|_{C^{(1)}(a,b)} = \max_{x \in [a,b]} |\varphi(x)| + \max_{x \in [a,b]} |\varphi'(x)|. \tag{1.37}$$

Definition 1.3. The space $C^{(1)}(a, b)$ is the normed space consisting of the set of all functions $\varphi(x)$ that are continuously differentiable on $[a, b]$, supplied with the norm (1.37). Its subspace of functions satisfying (1.35) is denoted $C_0^{(1)}(a, b)$. The set of all functions having k continuous derivatives on $[a, b]$ is denoted $C^{(k)}(a, b)$.

In many books these spaces are denoted by $C^{(k)}([a, b])$ to emphasize that $[a, b]$ is closed. To keep our notation reasonable throughout the book, we introduce

Convention 1.4. In cases where no ambiguity should arise, we typically abbreviate the space designation subscript on a norm symbol. □

For example, the notation $\|\cdot\|_{C^{(1)}(a,b)}$ (where the dot stands for the argument of the norm operation) is shortened to $\|\cdot\|$ in the present section. At times, only some aspect of the full label can be suppressed. For example, we may use the notation $\|\cdot\|_{C^{(1)}}$ if only the domain $[a, b]$ is understood. With this convention in mind let us proceed to

Definition 1.5. A δ-*neighborhood* of $y(x)$ of admissible functions is the set of all functions of the form $y(x) + \varphi(x)$ where $\varphi(x)$ is such that $\varphi(x) \in C_0^{(1)}(a, b)$ and $\|\varphi(x)\| < \delta$.

When no boundary conditions are imposed on y, then the definition of δ-neighborhood does not require φ to vanish at the endpoints.

Definition 1.6. A function $y(x)$ is a *point of local minimum* of $F(y)$ on the set satisfying (1.34) if there is a δ-neighborhood of $y(x)$, i.e., a set of functions $z(x)$ such that $z(x) - y(x) \in C_0^{(1)}(a, b)$ and $\|z(x) - y(x)\| < \delta$, in which $F(z) - F(y) \geq 0$. If in a δ-neighborhood the relation $F(z) - F(y) > 0$ holds for all $z(x) \neq y(x)$, then $y(x)$ is a *point of strict local minimum.*

We may speak of more than one type of local minimum. According to Definition 1.6, a function y is a minimum if there is a δ such that

$$F(y + \varphi) - F(y) \geq 0 \text{ whenever } \|\varphi\|_{C_0^{(1)}(a,b)} < \delta.$$

Historically this type of minimum is called "weak" and we shall use only this type and simply call it a minimum. Those who pioneered the calculus of variations also considered "strong" local minima, defining these as values of y for which there is a δ such that $F(y+\varphi) \geq F(y)$ whenever $\varphi(a) = \varphi(b) = 0$ and $\max|\varphi| < \delta$ on $[a, b]$. Here the modified condition on φ permits "strong variations" into consideration: i.e., functions φ for which φ' may be large even though φ itself is small. Note that when we "weaken" the condition on φ by changing the norm from the norm of $C_0^{(1)}(a, b)$ to the norm of $C_0(a, b)$ which contains only φ and not φ', we simultaneously strengthen the statement made regarding y when we assert the inequality $F(y+\varphi) \geq F(y)$.

Let us turn to a rigorous justification of (1.29). We restrict the class of possible integrands $f(x, y, z)$ of (1.33) to the set of functions that are continuous in (x, y, z) when $x \in [a, b]$ and $|y-y(x)|+|z-y'(x)| < \delta$. Suppose the existence of a minimizer $y(x)$ for $F(y)$ (see, however, Remark 1.13 on page 21). Consider $F(y + t\varphi)$ for an arbitrary but fixed $\varphi(x) \in C_0^{(1)}(a, b)$. It is a function in the single variable t, taking its minimum at $t = 0$. If it is differentiable then

$$\left.\frac{dF(y+t\varphi)}{dt}\right|_{t=0} = 0. \tag{1.38}$$

To justify differentiation under the integral sign, let $f(x, y, y')$ be continuously differentiable in the variables y and y'. But, since (1.30) shows that we shall need the existence of other derivatives of f as well, let us assume $f(x, y, y')$ is twice continuously differentiable, in any combination of its arguments, in the domain of interest. By the chain rule, (1.38) yields

$$\begin{aligned} 0 &= \frac{d}{dt}\int_a^b f(x, y+t\varphi, y'+t\varphi')\,dx\bigg|_{t=0} \\ &= \int_a^b [f_y(x,y,y')\varphi + f_{y'}(x,y,y')\varphi']\,dx. \end{aligned} \tag{1.39}$$

Definition 1.7. The right member of (1.39) is denoted $\delta F(y, \varphi)$ and called the *first variation* of the functional (1.33).

Integration by parts in the second term on the right in (1.39) gives

$$\int_a^b f_{y'}(x,y,y')\varphi'\,dx = -\int_a^b \varphi \frac{d}{dx} f_{y'}(x,y,y')\,dx$$

where the boundary terms vanish by (1.35). It follows that

$$\int_a^b \left[f_y(x,y,y') - \frac{d}{dx} f_{y'}(x,y,y')\right]\varphi\,dx = 0. \tag{1.40}$$

In the integrand we see the left side of (1.29). To deduce (1.29) from (1.40) we need the *fundamental lemma* of the calculus of variations.

Lemma 1.8. *Let $g(x)$ be continuous on $[a,b]$, and let*

$$\int_a^b g(x)\varphi(x)\,dx = 0 \tag{1.41}$$

hold for every function $\varphi(x)$ that is differentiable on $[a,b]$ and vanishes in some neighborhoods of a and b. Then $g(x) \equiv 0$.

Proof. Suppose to the contrary that (1.41) holds while $g(x_0) \neq 0$ for some $x_0 \in (a,b)$. Without loss of generality we may assume $g(x_0) > 0$. By continuity we have $g(x) > 0$ in a neighborhood $[x_0 - \varepsilon, x_0 + \varepsilon] \subset (a,b)$. It is easy to construct a nonnegative bell-shaped function $\varphi_0(x)$ such that $\varphi_0(x)$ is differentiable, $\varphi_0(x_0) > 0$, and $\varphi_0(x) = 0$ outside $(x_0 - \varepsilon, x_0 + \varepsilon)$:

$$\varphi(x) = \begin{cases} \exp\left(\dfrac{\varepsilon^2}{(x-x_0)^2 - \varepsilon^2}\right), & |x - x_0| < \varepsilon, \\ 0, & |x - x_0| \geq \varepsilon. \end{cases}$$

See Fig. 1.1. The product $g(x)\varphi_0(x)$ is nonnegative everywhere and positive near x_0. Hence $\int_a^b g(x)\varphi(x)\,dx > 0$, a contradiction. □

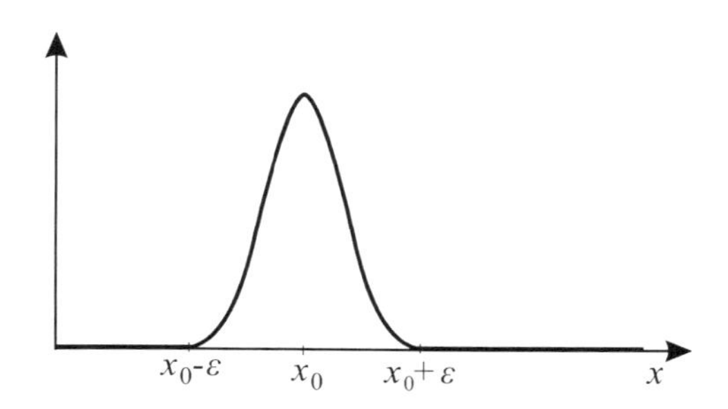

Fig. 1.1 Bell-shaped function for the proof of Lemma 1.8.

It is possible to further restrict the class of functions $\varphi(x)$ in Lemma 1.8.

Lemma 1.9. *Let $g(x)$ be continuous on $[a, b]$, and let (1.41) hold for any function $\varphi(x)$ that is infinitely differentiable on $[a, b]$ and vanishes in some neighborhoods of a and b. Then $g(x) \equiv 0$.*

The proof is the same as that for Lemma 1.8: it is necessary to construct the same bell-shaped function $\varphi(x)$ that is infinitely differentiable. This form of the fundamental lemma provides a basis for the theory of generalized functions or distributions. These are linear functionals on the sets of infinitely differentiable functions, and arise as elements of the Sobolev spaces to be discussed later.

Now we can formulate the main result of this section.

Theorem 1.10. *Suppose $y = y(x) \in C^{(2)}(a, b)$ locally minimizes the functional (1.33) on the subset of $C^{(1)}(a, b)$ consisting of those functions satisfying (1.34). Then $y(x)$ is a solution of the equation*

$$f_y - \frac{d}{dx} f_{y'} = 0. \tag{1.42}$$

Proof. Under the assumptions of this section (including that $f(x, y, y')$ is twice continuously differentiable in its arguments), the bracketed term in (1.40) is continuous on $[a, b]$. Since (1.40) holds for any $\varphi(x) \in C_0^{(1)}(a, b)$, Lemma 1.8 applies. □

Definition 1.11. Equation (1.42) is known as the *Euler equation*, and a solution $y = y(x)$ is called an *extremal* of (1.33). A functional is *stationary* if its first variation vanishes.

Taken together, (1.42) and (1.34) constitute a boundary value problem for the unknown $y(x)$.

Example 1.12. Find a function $\bar{y} = \bar{y}(x)$ that minimizes the functional

$$F(y) = \int_0^1 [y^2 + (y')^2 - 2y]\, dx$$

subject to the conditions $y(0) = 1$ and $y(1) = 0$.

Solution. Here $f(x, y, y') = y^2 + (y')^2 - 2y$, so we obtain

$$f_y = 2y - 2, \qquad f_{y'} = 2y',$$

and the Euler equation is

$$y'' - y + 1 = 0.$$

Subject to the given boundary conditions, the solution is

$$\bar{y}(x) = 1 - \frac{e^x - e^{-x}}{e - e^{-1}}.$$

We stress that this is an extremal: only supplementary investigation can determine whether it is an actual minimizer of $F(y)$. Consider the difference $F(\bar{y} + \varphi) - F(\bar{y})$ where $\varphi(x)$ vanishes at $x = 0, 1$. It is easily shown that

$$F(\bar{y} + \varphi) - F(\bar{y}) = \int_0^1 [\varphi^2 + (\varphi')^2]\, dx \geq 0,$$

so the global minimum of $F(y)$ really does occur at $\bar{y}(x)$. Although such direct verification is not always straightforward, a large class of important problems in mechanics (e.g., problems of equilibrium for linearly elastic structures under conservative loads) yield single extremals that minimize their corresponding total energy functionals. This happens because of the quadratic structure of the functional, as in the present example. □

Certain forms of f lead to simplification of the Euler equation:

(1) If f does not depend explicitly on y, then $f_{y'} = \text{constant}$.
(2) If f does not depend explicitly on x, then $f - f_{y'}y' = \text{constant}$.
(3) If f depends explicitly on y' only and $f_{y'y'} \neq 0$, then $y(x) = c_1 x + c_2$.

Remark 1.13. On page 17 we assumed the existence of a minimizer. This can lead to incorrect conclusions, and it is normally necessary to prove the existence of an object having needed properties. *Perron's paradox* illustrates the trouble we may encounter by supposing the existence of a nonexistent object. Suppose there exists a greatest positive integer N. Since N^2 is also a positive integer we must have $N^2 \leq N$, from which it follows that $N = 1$. If we knew nothing about the integers we might believe this result and attempt to base an entire theory on it. □

1.3 Properties of Extremals of the Simplest Functional

While attempting to seek a minimizer on a subset of $C^{(1)}(a,b)$, we imposed the illogical restriction that it must belong to $C^{(2)}(a,b)$ (note that f does not depend on y''). Let us consider how to circumvent this requirement.

Lemma 1.14. *Let $g(x)$ be a continuous function on $[a,b]$ for which the following equality holds for every $\varphi(x) \in C_0^{(1)}(a,b)$:*

$$\int_a^b g(x)\varphi'(x)\,dx = 0. \tag{1.43}$$

Then $g(x)$ is constant.

Proof. For a constant c it is clear that $\int_a^b c\varphi'(x)\,dx = 0$ whenever $\varphi(x) \in C_0^{(1)}(a,b)$. So $g(x)$ can be an arbitrary constant. We show that there are no other forms for g. From (1.43) it follows that

$$\int_a^b [g(x) - c]\varphi'(x)\,dx = 0. \tag{1.44}$$

Take $c = c_0 = (b-a)^{-1}\int_a^b g(x)\,dx$. The function $\varphi(x) = \int_a^x [g(s) - c_0]\,ds$ is continuously differentiable and satisfies $\varphi(a) = \varphi(b) = 0$. Hence we can put it into (1.44) and obtain

$$\int_a^b [g(x) - c_0]^2\,dx = 0,$$

from which $g(x) \equiv c$. □

Lemma 1.14 provides a necessary condition for a relative minimum.

Theorem 1.15. *Suppose* $y = y(x) \in C^{(1)}(a,b)$ *locally minimizes* (1.33) *on the subset of functions in* $C^{(1)}(a,b)$ *satisfying* (1.34). *Then* $y(x)$ *is a solution of the following equation, where* c *is a constant:*

$$\int_0^x f_y(s, y(s), y'(s))\, ds - f_{y'}(x, y(x), y'(x)) = c. \tag{1.45}$$

Proof. Let us return to the equality (1.39),

$$\int_a^b [f_y(x,y,y')\varphi + f_{y'}(x,y,y')\varphi']\, dx = 0,$$

which is valid here as well. Integration by parts gives

$$\int_a^b f_y(x, y(x), y'(x))\varphi(x)\, dx = -\int_a^b \int_a^x f_y(s, y(s), y'(s))\, ds\, \varphi'(x)\, dx.$$

The boundary terms were zero by (1.35). It follows that

$$\int_a^b \left[-\int_a^x f_y(s, y(s), y'(s))\, ds + f_{y'}(x, y(x), y'(x)) \right] \varphi'(x)\, dx = 0.$$

This holds for all $\varphi(x) \in C_0^{(1)}(a,b)$. So by Lemma 1.14 we have (1.45). □

The integro-differential equation (1.45) has been called the *Euler equation in integrated form.*

Corollary 1.16. *If*

$$f_{y'y'}(x, y(x), y'(x)) \neq 0$$

along a minimizer $y = y(x) \in C^{(1)}(a,b)$ *of* (1.33), *then* $y(x) \in C^{(2)}(a,b)$.

Proof. Rewrite (1.45) as

$$f_{y'}(x, y(x), y'(x)) = \int_0^x f_y(s, y(s), y'(s))\, ds - c.$$

The function on the right is continuously differentiable for any $y = y(x) \in C^{(1)}(a,b)$. Thus we can differentiate both sides of the last identity with respect to x and obtain

$$f_{y'x} + f_{y'y}y' + f_{y'y'}y'' = \text{a continuous function.}$$

Considering the term with $y''(x)$ on the left, we prove the claim. □

It follows that under the condition of the corollary equations (1.42) and (1.45) are equivalent; however, this is not the case when $f_{y'y'}(x, y(x), y'(x))$ can be equal to zero on a minimizer $y = y(x)$. Since $y''(x)$ does not appear in (1.45), it can be considered as defining a generalized solution of (1.42).

At times it becomes clear that we should change variables and consider a problem in another coordinate frame. For example, if we consider geodesic lines on a surface of revolution, then cylindrical coordinates may seem more appropriate than Cartesian coordinates. For the problem of minimum of a functional we have two objects: the functional itself, and the Euler equation for this functional. Let $y = y(x)$ satisfy the Euler equation in the original frame. Let us change variables, for example from (x, y) to (u, v):

$$x = x(u, v), \qquad y = y(u, v). \tag{1.46}$$

The forms of the functional and its Euler equation both change. Next we change variables for the extremal $y = y(x)$ and get a curve $v = v(u)$ in the new variables. Is $v = v(u)$ an extremal for the transformed functional? It is, provided the transformation does not degenerate in some neighborhood of the curve $y = y(x)$: that is, if the Jacobian

$$J = \begin{vmatrix} x_u & x_v \\ y_u & y_v \end{vmatrix} \neq 0$$

there. This property is called the *invariance* of the Euler equation. Roughly speaking, we can change all the variables of the problem at any stage of the solution and get the same solutions in the original coordinates. This invariance is frequently used in practice. We shall not stop to consider the issue of invariance for each type of functional we treat, but the results are roughly the same.

We have derived a necessary condition for a function to be a point of minimum or maximum of (1.33). Other functionals will be treated in the sequel. An Euler equation is the starting point for any variational investigation of a physical problem, and in practice its solution is often approached numerically. Let us consider some methods relevant to (1.33).

1.4 Ritz's Method

We now consider a numerical approach to minimizing the functional (1.33) with boundary conditions (1.34). Corresponding techniques for other problems will be presented later; we shall benefit from a consideration of this simple problem, however, since the main ideas will be the same.

In § 1.1 we obtained the Euler equation for (1.33). The intermediate equations (1.26) with boundary conditions (1.27)–(1.28), which for this case must be replaced by the Dirichlet conditions

$$y(a) = y_0 = d_0, \qquad y(b) = y_n = d_1,$$

present us with a finite difference variational method for solving the problem (1.42), (1.34), belonging to a class of numerical methods based on representing the derivatives of $y(x)$ in finite-difference form and the functional as a finite sum. These methods differ in how the functions and integrals are discretized. Despite widespread application of the finite element and boundary element methods, the finite-difference variational methods remain useful because of certain advantages they possess.

Other methods for minimizing a functional, and hence of solving certain boundary value problems, fall under the heading of *Ritz's method*. Included are modifications of the finite element method. Ritz's method was popular before the advent of the computer, and remains so, because it can yield accurate results for complex problems that are difficult to solve analytically.

The idea of Ritz's method is to reduce the problem of minimizing (1.33) on the space of all continuously differentiable functions satisfying (1.34) to the problem of minimizing the same functional on a finite dimensional subspace of functions that can approximate the solution. Formerly, the necessity of doing manual calculations forced engineers to choose such subspaces quite carefully, since it was important to get accurate results in as few calculations as possible. The choice of subspace remains an important issue because a bad choice can lead to computational instability.

In Ritz's method we seek a solution to the problem of minimization of the functional (1.33), with boundary conditions (1.34), in the form

$$y_n(x) = \varphi_0(x) + \sum_{k=1}^{n} c_k \varphi_k(x). \tag{1.47}$$

Here $\varphi_0(x)$ satisfies (1.34); a common choice is the linear function $\varphi_0(x) = \alpha x + \beta$ with

$$\alpha = \frac{d_1 - d_0}{b - a}, \qquad \beta = \frac{b d_0 - a d_1}{b - a}. \tag{1.48}$$

The remaining functions, called *basis functions*, satisfy the homogeneous conditions

$$\varphi_k(a) = \varphi_k(b) = 0, \qquad k = 1, \ldots, n.$$

The c_k are constants.

Definition 1.17. The function $y_n^*(x)$ that minimizes (1.33) on the set of all functions of the form (1.47) is called the *nth Ritz approximation.*

The Ritz approximations satisfy the boundary conditions (1.34) automatically. The above mentioned subspace is the space of functions of the form $\sum_{k=0}^{n} c_k \varphi_k(x)$. For a numerical solution it is necessary that the $\varphi_k(x)$ be linearly independent, which means that

$$\sum_{k=1}^{n} c_k \varphi_k(x) = 0 \quad \text{only if } c_k = 0 \text{ for } k = 1, \ldots, n.$$

For manual calculation this was supplemented by the requirement that a small value of n — say 1, 2, or 3 at most — would suffice. The requirement could be met since the corresponding boundary value problems described real objects, such as bent beams, whose shapes under load were understood. Now, to provide a theoretical justification of the method, we require that the system $\{\varphi_k(x)\}_{k=1}^{\infty}$ be *complete.* This means that given any $y = g(x) \in C_0^{(1)}(a, b)$ and $\varepsilon > 0$ we can find a finite sum $\sum_{k=1}^{n} c_k \varphi_k(x)$ such that

$$\left\| g(x) - \sum_{k=1}^{n} c_k \varphi_k(x) \right\| < \varepsilon.$$

Here the norm is defined by (1.37). It is sometimes required that $\{\varphi_k(x)\}_{k=1}^{\infty}$ be a basis of the corresponding space, but this is not needed for either the justification of the method or its numerical realization.

We therefore arrive at the problem of minimizing the functional

$$\int_a^b f(x, y_n, y_n')\, dx$$

where $y_n(x)$ is given by (1.47). The unknowns are the c_k, so the functional becomes a function in n real variables:

$$\Phi(c_1, \ldots, c_n) = \int_a^b f(x, y_n, y_n')\, dx.$$

To minimize this we solve the system

$$\frac{\partial \Phi(c_1, \ldots, c_n)}{\partial c_k} = 0, \qquad k = 1, \ldots, n. \tag{1.49}$$

Denoting $c_0 = 1$, we have

$$\begin{aligned}
\frac{\partial\Phi(c_1,\dots,c_n)}{\partial c_k} &= \frac{\partial}{\partial c_k}\int_a^b f(x, y_n, y_n')\,dx \\
&= \frac{\partial}{\partial c_k}\int_a^b f\left(x, \sum_{i=0}^n c_i\varphi_i(x), \sum_{i=0}^n c_i\varphi_i'(x)\right) dx \\
&= \int_a^b f_y\left(x, \sum_{i=0}^n c_i\varphi_i(x), \sum_{i=0}^n c_i\varphi_i'(x)\right) \varphi_k(x)\,dx \\
&\quad + \int_a^b f_{y'}\left(x, \sum_{i=0}^n c_i\varphi_i(x), \sum_{i=0}^n c_i\varphi_i'(x)\right) \varphi_k'(x)\,dx,
\end{aligned}$$

hence (1.49) becomes

$$\begin{aligned}
&\int_a^b f_y\left(x, \sum_{i=0}^n c_i\varphi_i(x), \sum_{i=0}^n c_i\varphi_i'(x)\right) \varphi_k(x)\,dx \\
&\quad + \int_a^b f_{y'}\left(x, \sum_{i=0}^n c_i\varphi_i(x), \sum_{i=0}^n c_i\varphi_i'(x)\right) \varphi_k'(x)\,dx = 0
\end{aligned} \tag{1.50}$$

for $k = 1, \dots, n$. This is a system of n simultaneous equations in the n variables $c_1, \dots, c_n$. It is linear only if f is a quadratic form in c_k; i.e., only if the Euler equation is linear in $y(x)$. For methods of solving simultaneous equations, the reader is referred to books on numerical analysis.

Note that (1.50) can be obtained in other ways. We could put $y = y_n$ and $\varphi = \varphi_k$ in (1.39), since while deriving (1.50) we used the same steps we used in deriving (1.39). Alternatively, we could put y_n into the left side of the Euler equation,

$$f_y(x, y_n, y_n') - \frac{d}{dx} f_{y'}(x, y_n, y_n'), \tag{1.51}$$

and then require it to be "orthogonal" to each φ_k. That is, we could multiply (1.51) by φ_k, integrate the result over $[a, b]$, use integration by parts on the term with the total derivative d/dx, and equate the result to zero. This is opposite the way we derived (1.50). This method of approximating the solution of the boundary value problem (1.42), (1.47) is *Galerkin's method.* In the Russian literature it is called the *Bubnov–Galerkin* method, because in 1915 I.G. Bubnov, who was reviewing a paper by S.P. Timoshenko on applications of Ritz's method to the solution of a problem for a bending beam, offered a brief remark on another method of obtaining the equations of Ritz's method. The journal in which Timoshenko's paper appeared happened to publish the comments of reviewers together with the papers (a nice way to hold reviewers responsible for their comments). Hence Bubnov became an originator of the method. Galerkin was Bubnov's successor, and his real achievement was the development of various forms and applications of the method. In particular, there is a modification wherein (1.51) is multiplied not by φ_k, the functions from the representation of y_n, but by other functions $\psi_1, \ldots, \psi_n$. This is sometimes a better way to minimize the *residual* (1.51).

Popular basis functions φ_k for one-dimensional problems include the trigonometric polynomials and functions of the form $(x-a)(x-b)P_k(x)$ where the $P_k(x)$ are polynomials. Here the factors $(x-a)$ and $(x-b)$ enforce the required homogeneous boundary conditions at $x=a$ and $x=b$.

When deriving the equations of the Ritz (or Bubnov–Galerkin) method, we imposed no special conditions on $\{\varphi_k\}$ other than linear independence and some smoothness: $\varphi_k(x) \in C_0^{(1)}(a,b)$. In general each of the equations (1.50) contains all of the c_k. By the integral nature of (1.50), if we select basis functions so that each $\varphi_k(x)$ is nonzero only on some small part of $[a,b]$, we get a system in which each equation involves only a subset of $\{\varphi_i\}$. This is the background for the finite element method based on Galerkin's method: depending on the problem each equation involves just a few of the c_k (typically three to five). Moreover, the derivation of Galerkin's equations suggests that it is not necessary to have basis functions with continuous derivatives — it suffices to take functions with piecewise continuous derivatives of higher order (first order for the problem under consideration) when it is possible to calculate the terms of (1.50).

Ritz's method can yield good results using low-order approximations. A disadvantage is that the calculations at a given step are almost independent from those of the previous step. The c_k do not change continuously from step to step; hence, although the next step gives a better approximation, the coefficients can change substantially. Accumulation of errors imposes limits on the number of basis functions in practical calculations.

Example 1.18. Consider the problem

$$\Psi(y) = \int_0^1 \{y'^2(x) + [1 + 0.1\sin(x)]y^2(x) - 2xy(x)\}\,dx \to \min$$

subject to $y(0) = 0$ and $y(1) = 10$. Find the Ritz approximations for $n = 1, 3, 5$ using $\varphi_0(x) = 10x$ and the following basis sets:

(a) $\varphi_k(x) = (1-x)x^k$, $k \geq 1$,
(b) $\varphi_k(x) = \sin k\pi x$, $k \geq 1$.

Solution. Note that $\varphi_0(x)$ was chosen to satisfy the given boundary conditions. We find the expansion coefficients c_k by solving the system

$$\frac{\partial}{\partial c_k}\Psi\left(\varphi_0(x) + \sum_{i=1}^{n} c_i\varphi_i(x)\right) = 0, \qquad i = 1, \ldots, n.$$

For brevity let us denote

$$\langle y, z\rangle = \int_0^1 \{y'(x)z'(x) + [1 + 0.1\sin(x)]y(x)z(x)\}\, dx$$

so that

$$\Psi(y) = \langle y, y\rangle - 2\int_0^1 xy(x)\, dx.$$

Using the symmetry of the form $\langle y, z\rangle$ we write out Ritz's equations:

$$\begin{aligned}
c_1\langle\varphi_1, \varphi_1\rangle + c_2\langle\varphi_2, \varphi_1\rangle + \cdots + c_n\langle\varphi_n, \varphi_1\rangle &= -\langle\varphi_0, \varphi_1\rangle + \int_0^1 x\varphi_1(x)\, dx,\\
c_1\langle\varphi_1, \varphi_2\rangle + c_2\langle\varphi_2, \varphi_2\rangle + \cdots + c_n\langle\varphi_n, \varphi_2\rangle &= -\langle\varphi_0, \varphi_2\rangle + \int_0^1 x\varphi_2(x)\, dx,\\
&\vdots\\
c_1\langle\varphi_1, \varphi_n\rangle + c_2\langle\varphi_2, \varphi_n\rangle + \cdots + c_n\langle\varphi_n, \varphi_n\rangle &= -\langle\varphi_0, \varphi_n\rangle + \int_0^1 x\varphi_n(x)\, dx.
\end{aligned} \tag{1.52}$$

For small n this system can be solved by hand, otherwise computer solution is required. In the present case we find that for the first basis set the Ritz approximations are

$$\begin{aligned}
y_1(x) &= 10x - 2.162x(1-x),\\
y_3(x) &= 10x + (-1.409x - 1.356x^2 - 0.246x^3)(1-x),\\
y_5(x) &= 10x + (-1.404x - 1.404x^2 - 0.140x^3 - 0.063x^4 - 0.007x^5)(1-x).
\end{aligned}$$

For the second basis set we obtain the Ritz approximations

$$\begin{aligned}
z_1(x) &= 10x - 0.289\sin\pi x,\\
z_3(x) &= 10x - 0.289\sin\pi x + 0.063\sin 2\pi x - 0.017\sin 3\pi x,\\
z_5(x) &= 10x - 0.289\sin\pi x + 0.063\sin 2\pi x - 0.017\sin 3\pi x\\
&\quad + 0.008\sin 4\pi x - 0.004\sin 5\pi x,
\end{aligned}$$

as required. □

In this example we employed the bilinear form $\langle y, z \rangle$. The symmetry of this form with respect to its arguments simplified the calculation. In the static problems of linear elasticity, such a form is naturally induced by the energy expression for an elastic body. Moreover, the form of the left sides of (1.52) is the same for all such problems, whether they are three-dimensional problems of elasticity, or problems describing elastic beams or shells.

In Ritz's time such approximate solutions were sought for problems describing elastic beams and plates. The resulting systems of equations were fairly hard to solve by hand. The method was justified by comparison with experimental data. A full justification of Ritz's and similar methods requires the tools of functional analysis, which forms the subject of Chapter 4. However, we would like to discuss some aspects of the method on an elementary level using Example 1.18 as a model.

Notes on basis functions

First let us comment on the approximations. The normal working viewpoint is that one compares each pair of successive approximations and terminates the calculation process upon reaching a pair whose difference is less than some predetermined tolerance ε.

For each type of approximation, if we appoint $\varepsilon = 0.01$ then we can stop at $k = 5$. Calculation out to $k = 10$ shows that the $k = 5$ approximations are both very good. However, they do differ from each other by a maximum of about 0.25. So which is "more" correct? We can answer this by substitution into the functional, which gives $\Psi(y_5) \approx 127.046$ and $\Psi(z_5) \approx 127.449$. This is evidence that polynomial approximation is preferable. It is not hard to see why: the true solution is not oscillatory, so the oscillatory behavior of the trigonometric polynomials is not helpful in this case. So the "practical" approach to terminating the numerical process may not work well for trigonometric approximation. In this particular example it can be shown that the trigonometric approximations do converge, but slowly.

We have selected the polynomial-type Ritz approximations. But our observation regarding trigonometric approximations is cause for concern since the situation with ordinary polynomials should not differ in principle from that with trigonometric polynomials. Let us further discuss the problem of basis functions.

In formulating Ritz's method we required completeness of the set of basis functions. Weierstrass's theorem of calculus states that any function $f(x)$ continuous on $[0,1]$ can be approximated uniformly by a polynomial to within any accuracy. In other words, given $\varepsilon > 0$ there exists an nth order polynomial $P_n(x)$ such that

$$\max_{x\in[0,1]} |f(x) - P_n(x)| < \varepsilon.$$

It follows that to within any accuracy we may use a polynomial to uniformly approximate a function $f(x)$ together with its continuous derivative. Indeed, given $\varepsilon > 0$, we begin with approximation of the derivative $f'(x)$ by a polynomial $Q_n(x)$:

$$\max_{x\in[0,1]} |f'(x) - Q_n(x)| < \varepsilon/2.$$

The polynomial

$$P_n(x) = f(0) + \int_0^x Q_n(t)\,dt$$

approximates $f(x)$:

$$\begin{aligned} |f(x) - P_n(x)| &= \left| f(0) + \int_0^x f'(t)\,dt - f(0) - \int_0^x Q_n(t)\,dt \right| \\ &\le \int_0^x |f'(t) - Q_n(t)|\,dt \\ &\le \varepsilon/2 \qquad \text{for } x \in [0,1]. \end{aligned}$$

In the same way it can be shown that a function n-times continuously differentiable on $[0,1]$ can be approximated to within any prescribed accuracy by a polynomial together with all n of its derivatives on $[0,1]$. The set of monomials $\{x^k\}$ constitutes a complete system of functions in $C^{(n)}[0,1]$ for any n.

Note that Weierstrass' theorem guarantees nothing more than the existence of an approximating polynomial. When we decrease ε we get a new polynomial where the coefficient standing at each term x^k may differ significantly from the corresponding coefficient of the previous approximating polynomial. This is because the set $\{x^k\}$ does not have the uniqueness property required of a true basis. Moreover, in mathematical analysis it is shown that we can arbitrarily remove infinitely many members of the family $\{x^k\}$ and still have a complete system $\{x^{k_r}\}$. It is necessary only to retain such members of the family that the series $\sum_{r=1}^{\infty} 1/k_r$ diverges. So the system $\{x^k\}$ contains more members than we need. Although any finite set of monomials x^k is linearly independent, as we take more and more elements the set gets closer to becoming linearly dependent; that is, given any $\varepsilon > 0$ we can find infinitely many polynomials approximating the zero function to within ε-accuracy on $[0, 1]$. This leads to numerical instability. The difficulty can be avoided by using other families of polynomials for approximation: namely, orthogonal polynomials for which numerical instability shows itself only in higher degrees of approximation.

As we know from the theory of Fourier expansion, the second system of functions $\{\sin k\pi x\}$ is orthonormal. It is, moreover, a basis (but not of $C_0^{(1)}(0, \pi)$) as we shall discuss later. This provides greater stability in calculations to within higher accuracy. However, in low-order Ritz approximations it can be worse than a polynomial approximation of the same problem, at least for many problems whose solutions do not oscillate.

One more aspect of the approximation is seen in the above results. For Ritz's approximations we compared their values. Comparing the values of their derivatives, we find that much better agreement is obtained for the values of the approximating functions than for the derivatives. It is obvious that the same holds for the difference between an exact solution and the approximating functions. This property is common to all projection methods. So, for example, in solving problems of elasticity we get comparatively good results in low-order approximations for the field of displacements, whereas the fields of stresses, which are expressed through the derivatives of the displacement fields, are approximated significantly worse.

1.5 Natural Boundary Conditions

In § 1.1 we found that by using discretization on the problem of minimum of the functional (1.33) without boundary conditions ("with free boundary") we obtain the Euler equation and some boundary conditions. We shall demonstrate that the same boundary conditions appear by the method of § 1.2. They are known as *natural boundary conditions*.

Consider the minimization of (1.33) when there are no restrictions on the boundary for $y = y(x)$.

Theorem 1.19. *Let $y = y(x) \in C^{(2)}(a, b)$ be a minimizer of the functional $\int_a^b f(x, y, y')\, dx$ over the space $C^{(1)}(a, b)$. Then for $y = y(x)$ the Euler equation*

$$f_y - \frac{d}{dx} f_{y'} = 0 \qquad \text{for all } x \in (a, b) \tag{1.53}$$

holds along with the natural boundary conditions

$$f_{y'}\Big|_{x=a} = 0, \qquad f_{y'}\Big|_{x=b} = 0. \tag{1.54}$$

Proof. We can repeat the initial steps of § 1.2. Namely, consider the values of the functional on the bundle of functions $y = y(x) + t\varphi(x)$ where $\varphi(x) \in C^{(1)}(a, b)$ is arbitrary but fixed. Here, however, there are no restrictions on $\varphi(x)$ at the endpoints of $[a, b]$.

For fixed $y(x)$ and $\varphi(x)$ the functional $\int_a^b f(x, y + t\varphi, y' + t\varphi')\,dx$ becomes a function of the real variable t, and attains its minimum at $t = 0$. Differentiating with respect to t we get

$$\int_a^b [f_y(x, y, y')\varphi + f_{y'}(x, y, y')\varphi']\,dx = 0.$$

Integration by parts gives

$$\int_a^b \left[f_y(x, y, y') - \frac{d}{dx} f_{y'}(x, y, y')\right] \varphi\,dx + f_{y'}(x, y(x), y'(x))\varphi(x)\Big|_{x=a}^{x=b} = 0. \tag{1.55}$$

From this we shall derive the Euler equation for $y(x)$ and the natural boundary conditions. The procedure is as follows. We limit the set of all continuously differentiable functions $\varphi(x)$ to those satisfying $\varphi(a) = \varphi(b) = 0$. For these functions we have

$$\int_a^b \left[f_y(x, y, y') - \frac{d}{dx} f_{y'}(x, y, y')\right] \varphi\,dx = 0. \tag{1.56}$$

This equation holds for all functions $\varphi(x)$ that participate in the formulation of Lemma 1.8. Hence the continuous multiplier of $\varphi(x)$ in the integrand of (1.56) is zero, and the Euler equation (1.53) holds in (a, b).

Now let us return to (1.55). The equality (1.56), because of the Euler equation, holds for all $\varphi(x)$. From (1.55) it follows that

$$f_{y'}(x, y(x), y'(x))\varphi(x)\Big|_{x=a}^{x=b} = 0 \tag{1.57}$$

for any $\varphi(x)$. Taking $\varphi(x) = x - b$ we find that $f_{y'}|_{x=a} = 0$; taking $\varphi(x) = x - a$ we find that $f_{y'}|_{x=b} = 0$. □

Let us call attention to the way this result was obtained. First we restricted the set of admissible functions to those for which we could get a certain intermediate result (the Euler equation); using this result, we obtained some simplification in the first variation. We finished the argument by considering the simplified first variation on *all* the admissible functions.

Natural boundary conditions are of great importance in mathematical physics. For some models of real bodies or processes it may be unclear which (and how many) boundary conditions are necessary for well-posedness of the problem. The variational approach usually clarifies the situation and provides natural boundary conditions dictated by the nature of the problem. The bending of a plate is a famous example. For her pioneering studies of this problem Sophie Germain received a prize from the French Academy of Sciences. She derived the biharmonic equation for the deflections of the midsurface of the plate, but with three boundary conditions as seemed to be in accordance with mechanical intuition; variational considerations later demonstrated that only two were independent.

It is worth noting that in mechanical problems, the natural boundary conditions are dual to kinematic conditions on the boundary. They do not arise at a boundary point when we "clamp" as fully as allowed by the model. Incomplete clamping at a point always results in a natural boundary condition of force type there. If no kinematic constraint prevails at a point, then the natural boundary conditions express the equilibrium of forces. A simple example is afforded by the stretched rod treated later on; application of a force F at the right end of the rod results in the natural boundary condition $ES(l)u'(l) = F$, which means that the cross section at point l is in mechanical equilibrium under F and the reaction of the remainder of the rod.

Remark 1.20. In § 1.1 we discussed the question of which boundary conditions can be imposed to get a well-posed boundary value problem for minimizing the functional (1.33). General considerations are nice; however, consider the minimization of

$$\int_0^1 (y'^2 + 2y)\, dx \tag{1.58}$$

on the set of continuously differentiable functions. Its Euler equation is $y'' = 1$, thus all the extremals take the form

$$y = \frac{1}{2}x^2 + kx + b.$$

The natural boundary conditions are $y'(0) = 0$, $y'(1) = 0$. These imply $k = 0$. So the problem of minimum of (1.58) (with natural boundary conditions) has a family of solutions $y = \frac{1}{2}x^2 + b$ with arbitrary constant b. Thus we may impose an additional condition, say $y(0) = 2$. But in general, such a third condition for an ordinary differential equation of second order can yield a boundary value problem that has no solution.

Although (1.58) is simple, the situation we just described is not unimportant. Indeed, the same situation holds for the whole class of functionals that govern the equilibrium states of linear elastic systems in terms of displacements. If we impose no geometrical restrictions on the position of an elastic body (it is normally the case of natural boundary conditions) we can always change the coordinate frame, and all the displacements can be changed in such a way that the body appears to be shifted as a whole (i.e., to move as a "rigid body"). Depending on the model of the body there are apparently one to six free constants describing such a motion — hence we can impose additional boundary conditions at some points and still preserve the well-posedness of the problem. In a one-dimensional problem (where the dimension is a spatial coordinate) the situation is exactly as it is for (1.58): it is possible to impose an additional boundary condition when considering the problem with "free" ends. Caution is often warranted when applying the outcomes of very general considerations. □

1.6 Extensions to More General Functionals

Let us consider two extensions of the above results.

The functional $\int_a^b f(x, \mathbf{y}, \mathbf{y}')\, dx$

Let us replace $y(x)$ in (1.33) by a vector function

$$\mathbf{y}(x) = (y_1(x), \ldots, y_n(x)).$$

We denote the integrand of the functional as

$$f(x, \mathbf{y}(x), \mathbf{y}'(x)) \quad \text{or} \quad f(x, y_1(x), \ldots, y_n(x), y_1'(x), \ldots, y_n'(x))$$

interchangeably. The task is to treat functionals of the form

$$F(\mathbf{y}) = \int_a^b f(x, \mathbf{y}, \mathbf{y}')\, dx. \tag{1.59}$$

First consider the problem of minimizing (1.59) when $\mathbf{y}(x)$ takes boundary values

$$\mathbf{y}(a) = \mathbf{c}_0, \qquad \mathbf{y}(b) = \mathbf{c}_1, \tag{1.60}$$

with vector constants $\mathbf{c}_0 = (c_{01}, c_{02}, \ldots, c_{0n})$, $\mathbf{c}_1 = (c_{11}, c_{12}, \ldots, c_{1n})$. We take $\mathbf{y}(x) \in C^{(k)}(a, b)$ to mean that each coordinate function $y_i(x) \in$

$C^{(k)}(a, b)$; that is, each $y_i(x)$ possesses all derivatives up to order k and these are all continuous on $[a, b]$. Imposing the norm

$$\|\mathbf{y}(x)\|_{C^{(k)}(a,b)} = \sum_{i=1}^{n} \|y_i(x)\|_{C^{(k)}(a,b)} \tag{1.61}$$

on $C^{(k)}(a, b)$, we can define ε-neighborhoods as needed to describe minimizers of (1.59). We seek a minimizer $\mathbf{y}(x)$ of (1.59) from among all vector functions belonging to $C^{(1)}(a, b)$ and satisfying (1.60).

Theorem 1.21. *Suppose $\mathbf{y}(x) \in C^{(2)}(a, b)$ locally minimizes the functional $\int_a^b f(x, \mathbf{y}, \mathbf{y}')\,dx$ on the subset of vector functions of $C^{(1)}(a, b)$ satisfying* (1.60). *Then $\mathbf{y}(x)$ satisfies*

$$\nabla_{\mathbf{y}} f - \frac{d}{dx} \nabla_{\mathbf{y}'} f = 0. \tag{1.62}$$

Here we use the gradient notation

$$\nabla_{\mathbf{y}} = \left(\frac{\partial}{\partial y_1}, \ldots, \frac{\partial}{\partial y_n} \right), \qquad \nabla_{\mathbf{y}'} = \left(\frac{\partial}{\partial y_1'}, \ldots, \frac{\partial}{\partial y_n'} \right).$$

The vector equation (1.62) *can be written as n scalar equations*

$$f_{y_i} - \frac{d}{dx} f_{y_i'} = 0, \qquad i = 1, \ldots, n, \tag{1.63}$$

each having the form of the Euler equation.

Proof. Over the same construction of admissible functions, $\mathbf{y}(x) + t\boldsymbol{\varphi}(x)$ where $\boldsymbol{\varphi}(a) = \boldsymbol{\varphi}(b) = 0$, we consider (1.59):

$$F(\mathbf{y}(x) + t\boldsymbol{\varphi}(x)) = \int_a^b f(x, \mathbf{y} + t\boldsymbol{\varphi}, \mathbf{y}' + t\boldsymbol{\varphi}')\,dx. \tag{1.64}$$

For fixed $\mathbf{y}(x)$ and $\boldsymbol{\varphi}(x)$ this becomes a function of the real variable t and takes its minimum at $t = 0$ for any $\boldsymbol{\varphi}(x)$. Take $\boldsymbol{\varphi}(x)$ of the special form $\boldsymbol{\varphi}_1(x) = (\varphi(x), 0, \ldots, 0)$ where the only nonzero component stands in the first position. Then (1.64) becomes

$$\begin{aligned} F(\mathbf{y}(x) + t\boldsymbol{\varphi}_1(x)) = \int_a^b f(x, y_1(x) + t\varphi(x), y_2(x), \ldots, y_n(x), \\ y_1'(x) + t\varphi'(x), y_2'(x), \ldots, y_n'(x))\,dx. \end{aligned} \tag{1.65}$$

Now the function of t becomes a particular case of the function of § 1.2, $F(y(x) + t\varphi(x))$, with the evident notational change $y \mapsto y_1$. A consequence of the minimum of (1.65) at $t = 0$ is the corresponding Euler equation

$$f_{y_1} - \frac{d}{dx} f_{y_1'} = 0.$$

This is the first equation of (1.63). Similarly, the ith equation of (1.63) is derived by taking $\boldsymbol{\varphi}(x)$ in the form $\boldsymbol{\varphi}_1(x) = (0, \ldots, \varphi_i(x), \ldots, 0)$, where the only nonzero component stands in the ith position. □

Let us derive the natural boundary conditions for (1.59). Now we should not impose any conditions for $\mathbf{y}$ at points $x = a$ and $x = b$ in advance, and thus it is the same for $\boldsymbol{\varphi}$ at these points. For a moment consider all components of the minimizer $\mathbf{y}(x)$ other than $y_i(x)$ to be given. Then (1.59) can be formally considered as a particular case of (1.33) with respect to the ordinary function $y = y_i(x)$. Admissible vector functions differ from $\mathbf{y}(x)$ only in the ith component: $\boldsymbol{\varphi}(x) = \boldsymbol{\varphi}_i(x) = (0, \ldots, \varphi(x), \ldots, 0)$. We can repeat the reasoning of § 1.3. Thus considering the problem of minimum of (1.59) without boundary restrictions, we get n pairs of boundary conditions:

$$f_{y_i'}\Big|_{x=a} = 0, \qquad f_{y_i'}\Big|_{x=b} = 0, \qquad\qquad i = 1, \ldots, n.$$

These are natural boundary conditions for a minimizer.

The functional $\int_a^b f(x, y, y', \ldots, y^{(n)})\, dx$

The functional

$$F_n(y) = \int_a^b f(x, y, y', \ldots, y^{(n)})\, dx \tag{1.66}$$

may be considered on the set of functions satisfying certain boundary conditions. Alternatively, we may impose no boundary conditions and seek natural boundary conditions.

First consider the problem with given boundary equations. The corresponding Euler equation will have order $2n$, hence we take n conditions at each endpoint:

$$\begin{aligned} y(a) &= c_0^*, & y(b) &= c_0^{**}, \\ y'(a) &= c_1^*, & y'(b) &= c_1^{**}, \\ &\vdots & &\vdots \\ y^{(n-1)}(a) &= c_{n-1}^*, & y^{(n-1)}(b) &= c_{n-1}^{**}. \end{aligned} \tag{1.67}$$

A sufficiently smooth integrand $f(x, y, y', \ldots, y^{(n)})$ belongs to $C^{(n)}$ on the domain of all of its variables, at least in some neighborhood of a minimizer.

Theorem 1.22. *Suppose $y(x) \in C^{(2n)}(a, b)$ locally minimizes $F_n(y)$ in* (1.66) *on the subset of vector functions of $C^{(n)}(a, b)$ satisfying* (1.67). *Then*

*$y(x)$ satisfies the **Euler–Lagrange equation***

$$f_y - \frac{d}{dx} f_{y'} + \frac{d^2}{dx^2} f_{y''} - \cdots + (-1)^n \frac{d^n}{dx^n} f_{y^{(n)}} = 0. \tag{1.68}$$

Proof. Let us recall what it means for $y(x)$ to be a local minimizer of $F_n(y)$. Consider the bundle of functions $y(x)+\varphi(x)$ where $\varphi(x)$ is arbitrary and belongs to $C^{(n)}(a,b)$. Because the bundle must satisfy (1.67) for any $\varphi(x)$, we see that $\varphi(x)$ must satisfy the homogeneous conditions

$$\begin{aligned} \varphi(a) &= 0, & \varphi(b) &= 0, \\ \varphi'(a) &= 0, & \varphi'(b) &= 0, \\ &\vdots & &\vdots \\ \varphi^{(n-1)}(a) &= 0, & \varphi^{(n-1)}(b) &= 0. \end{aligned} \tag{1.69}$$

Let $C_0^{(n)}(a,b)$ denote the subspace of $C^{(n)}(a,b)$ containing functions $\varphi(x)$ that satisfy (1.69). A function $y(x) \in C^{(n)}(a,b)$ satisfying (1.67) is a local minimizer of $F_n(y)$ if $F_n(y+\varphi) \geq F_n(y)$ for any $\varphi(x) \in C_0^{(n)}(a,b)$ such that $\|\varphi\|_{C^{(n)}(a,b)} < \varepsilon$ for some $\varepsilon > 0$.

As usual we introduce the parameter t and consider the values of $F_n(y)$ on the bundle $y(x)+t\varphi(x)$. Considering $F_n(y(x)+t\varphi(x))$ for a momentarily fixed $\varphi(x)$ as a function of t, we see that it takes its minimal value at $t=0$ and thus

$$\left.\frac{dF_n(y(x)+t\varphi(x))}{dt}\right|_{t=0} = 0.$$

In detail,

$$\begin{aligned} &\left.\frac{dF_n(y(x)+t\varphi(x))}{dt}\right|_{t=0} \\ &= \left.\frac{d}{dt}\int_a^b f(x, y+t\varphi, y'+t\varphi', y''+t\varphi'', \ldots, y^{(n)}+t\varphi^{(n)})\, dx\right|_{t=0} \\ &= \int_a^b \left(f_y\varphi + f_{y'}\varphi' + f_{y''}\varphi'' + \cdots + f_{y^{(n)}}\varphi^{(n)} \right) dx \end{aligned} \tag{1.70}$$

(in the last line of the formula the arguments are $f = f(x, y, y', \ldots, y^{(n)})$). Now we apply (multiple) integration by parts to each term containing derivatives of φ so that on the last step the integrand contains only φ. For the term $\int_a^b f_{y'}\varphi'\, dx$ we already have (1.55). For the term $\int_a^b f_{y''}\varphi''\, dx$

we produce

$$\int_a^b f_{y''}\varphi''\,dx = -\int_a^b \varphi'\frac{d}{dx}f_{y''}\,dx + \varphi' f_{y''}\Big|_{x=a}^{x=b}$$
$$= \int_a^b \varphi\frac{d^2}{dx^2}f_{y''}\,dx + \left(\varphi' f_{y''} - \varphi\frac{d}{dx}f_{y''}\right)\Big|_{x=a}^{x=b}.$$

Similarly

$$\int_a^b f_{y'''}\varphi'''\,dx = -\int_a^b \varphi\frac{d^3}{dx^3}f_{y'''}\,dx$$
$$+ \left(\varphi'' f_{y'''} - \varphi'\frac{d}{dx}f_{y'''} + \varphi\frac{d^2}{dx^2}f_{y'''}\right)\Big|_{x=a}^{x=b}$$

and, in general,

$$\int_a^b f_{y^{(n)}}\varphi^{(n)}\,dx = (-1)^n\int_a^b \varphi\frac{d^n}{dx^n}f_{y^{(n)}}\,dx$$
$$+ \left(\varphi^{(n-1)} f_{y^{(n)}} - \varphi^{(n-2)}\frac{d}{dx}f_{y^{(n)}} + \cdots + (-1)^{n-1}\varphi\frac{d^{n-1}}{dx^{n-1}}f_{y^{(n)}}\right)\Big|_{x=a}^{x=b}.$$

By (1.68) the boundary terms vanish, and collecting results we have

$$\int_a^b \left(f_y - \frac{d}{dx}f_{y'} + \frac{d^2}{dx^2}f_{y''} - \cdots + (-1)^n\frac{d^n}{dx^n}f_{y^{(n)}}\right)\varphi\,dx = 0. \qquad (1.71)$$

Since this holds for any $\varphi(x) \in C_0^{(n)}(a,b)$, we can quote the fundamental lemma to complete the proof. □

Let us investigate the natural boundary conditions for $F_n(y)$. Now $\varphi(x) \in C^{(n)}(a,b)$ with no boundary restrictions. The first steps of the previous discussion still apply; however, now there are the boundary terms in the expression for the first variation of $F_n(y)$ (the right side of (1.70)), so in obtaining the result analogous to (1.71) we should collect all terms including boundary terms. We rearrange the boundary terms, collecting coefficients of each $\varphi^{(i)}(x)$:

$$
\begin{aligned}
&\int_a^b \left(f_y - \frac{d}{dx} f_{y'} + \frac{d^2}{dx^2} f_{y''} - \cdots + (-1)^n \frac{d^n}{dx^n} f_{y^{(n)}} \right) \varphi \, dx \\
&\quad + f_{y^{(n)}} \varphi^{(n-1)} \Big|_{x=a}^{x=b} \\
&\quad + \left(f_{y^{(n-1)}} - \frac{d}{dx} f_{y^{(n)}} \right) \varphi^{(n-2)} \Big|_{x=a}^{x=b} \\
&\quad + \left(f_{y^{(n-2)}} - \frac{d}{dx} f_{y^{(n-1)}} + \frac{d^2}{dx^2} f_{y^{(n)}} \right) \varphi^{(n-3)} \Big|_{x=a}^{x=b} \\
&\quad \vdots \\
&\quad + \left(f_{y'} - \frac{d}{dx} f_{y''} + \cdots + (-1)^{n-1} \frac{d^{n-1}}{dx^{n-1}} f_{y^{(n)}} \right) \varphi \Big|_{x=a}^{x=b} = 0. \qquad (1.72)
\end{aligned}
$$

We now realize the common plan. First we consider (1.72) only on the subset $C_0^{(n)}(a,b)$ of all $\varphi(x) \in C^{(n)}(a,b)$. Then (1.72) reduces to (1.71), implying that (1.68) holds. Equation (1.72) becomes

$$
\begin{aligned}
&f_{y^{(n)}} \varphi^{(n-1)} \Big|_{x=a}^{x=b} \\
&\quad + \left(f_{y^{(n-1)}} - \frac{d}{dx} f_{y^{(n)}} \right) \varphi^{(n-2)} \Big|_{x=a}^{x=b} \\
&\quad + \left(f_{y^{(n-2)}} - \frac{d}{dx} f_{y^{(n-1)}} + \frac{d^2}{dx^2} f_{y^{(n)}} \right) \varphi^{(n-3)} \Big|_{x=a}^{x=b} \\
&\quad \vdots \\
&\quad + \left(f_{y'} - \frac{d}{dx} f_{y''} + \cdots + (-1)^{n-1} \frac{d^{n-1}}{dx^{n-1}} f_{y^{(n)}} \right) \varphi \Big|_{x=a}^{x=b} = 0. \qquad (1.73)
\end{aligned}
$$

It is easy to construct a set of polynomials $P_{ik}(x)$, for $k = 0, 1$ and $i = 0, \ldots, n-1$, with the following properties:

$$
\begin{aligned}
\left. \frac{d^j P_{i0}}{dx^j} \right|_{x=a} &= \delta_i^j, & \left. \frac{d^j P_{i0}}{dx^j} \right|_{x=b} &= 0, & j &= 0, 1, \ldots, n-1, \\
\left. \frac{d^j P_{i1}}{dx^j} \right|_{x=a} &= 0, & \left. \frac{d^j P_{i1}}{dx^j} \right|_{x=b} &= \delta_i^j, & j &= 0, 1, \ldots, n-1,
\end{aligned}
$$

where δ_i^j is the *Kronecker delta* symbol defined by $\delta_i^j = 1$ for $i = j$ and $\delta_i^j = 0$ otherwise. Substituting these polynomials into (1.73), we get the natural boundary conditions for a minimizer $y(x)$:

$$\left. f_{y^{(n)}} \right|_{x=a} = 0,$$
$$\left. f_{y^{(n)}} \right|_{x=b} = 0,$$
$$\left. \left(f_{y^{(n-1)}} - \frac{d}{dx} f_{y^{(n)}} \right) \right|_{x=a} = 0,$$
$$\left. \left(f_{y^{(n-1)}} - \frac{d}{dx} f_{y^{(n)}} \right) \right|_{x=b} = 0,$$
$$\left. \left(f_{y^{(n-2)}} - \frac{d}{dx} f_{y^{(n-1)}} + \frac{d^2}{dx^2} f_{y^{(n)}} \right) \right|_{x=a} = 0,$$
$$\left. \left(f_{y^{(n-2)}} - \frac{d}{dx} f_{y^{(n-1)}} + \frac{d^2}{dx^2} f_{y^{(n)}} \right) \right|_{x=b} = 0,$$
$$\vdots$$
$$\left. \left(f_{y'} - \frac{d}{dx} f_{y''} + \cdots + (-1)^{n-1} \frac{d^{n-1}}{dx^{n-1}} f_{y^{(n)}} \right) \right|_{x=a} = 0,$$
$$\left. \left(f_{y'} - \frac{d}{dx} f_{y''} + \cdots + (-1)^{n-1} \frac{d^{n-1}}{dx^{n-1}} f_{y^{(n)}} \right) \right|_{x=b} = 0.$$

Note that the last two conditions contain $y^{(2n-1)}(x)$. In general, the natural boundary conditions contain higher derivatives than the equations (1.67).

What if we appoint some of the boundary conditions (1.67)? For example, let $y(a) = c_1^*$ be the only boundary restriction for a minimizer. Then we need to require that $\varphi(a) = 0$, and we will get all the natural boundary conditions for $y(x)$ except the one whose expression is the multiplier of $\varphi(a)$ in the boundary sum (1.73). We must remove

$$\left. \left(f_{y'} - \frac{d}{dx} f_{y''} + \cdots + (-1)^{n-1} \frac{d^{n-1}}{dx^{n-1}} f_{y^{(n)}} \right) \right|_{x=a} = 0$$

from the list.

The reader should consider what happens to the natural boundary conditions in case the following apply (consider each case separately):

(1) $y(a) + ky'(a) = c$,
(2) $y(a) + ky(b) = c$.

Example 1.23. Derive the Euler–Lagrange equation and natural boundary conditions for the energy functional whose minimizer defines the equilibrium of a bent cantilever beam described by parameters E, I. The beam is subjected to a distributed load $q(x)$, as well as a shear force Q^* and torque M^* applied to the end $x = l$:

$$E(y) = \frac{1}{2}\int_0^l EI(y'')^2\,dx - \int_0^l qy\,dx - Q^*y(l) - M^*y'(l),$$
$$y(0) = y'(0) = 0.$$

Note that the natural boundary conditions now have mechanical meaning: they account for the given torque and shear force at the "free" end $x = l$.

Solution. In this case the energy functional involves terms outside an integral, so it makes sense to repeat the derivation of the Euler–Lagrange equation for the functional $\int_a^b f(x, y, y', \ldots, y^{(n)})\,dx$ to understand how M^* and Q^* enter the natural boundary conditions. Supposing y is a solution, we consider $E(y)$ on the bundle $y + t\varphi$ with arbitrary but fixed φ: that is, we consider $E(y + t\varphi)$ where $\varphi(0) = 0 = \varphi'(0)$. As a function of t this takes a minimum at $t = 0$, so its derivative at this point is zero:

$$\int_0^l EIy''\varphi''\,dx - \int_0^l q\varphi\,dx - Q^*\varphi(l) - M^*\varphi'(l) = 0.$$

Two integrations by parts in the first integral give

$$\int_0^l (EIy^{(4)} - q)\varphi\,dx + EIy''\varphi'\Big|_0^l - EIy'''\varphi\Big|_0^l - Q^*\varphi(l) - M^*\varphi'(l) = 0$$

and, because $\varphi(0) = 0 = \varphi'(0)$, we have

$$\int_0^l (EIy^{(4)} - q)\varphi\,dx + (EIy''(l) - M^*)\varphi'(l) - (EIy'''(l) + Q^*)\varphi(l) = 0.$$

Now we repeat the steps connected with the choice of φ. First we take those φ for which $\varphi(l) = 0 = \varphi'(l)$, which brings us to the equation

$$\int_0^l (EIy^{(4)} - q)\varphi\,dx = 0;$$

then, because of the arbitrariness of φ, we invoke the fundamental lemma to arrive at the Euler–Lagrange equation

$$EIy^{(4)} - q = 0 \quad \text{on } [0, l].$$

Hence for any φ that does not vanish at $x = l$ we have

$$(EIy''(l) - M^*)\varphi'(l) - (EIy'''(l) + Q^*)\varphi(l) = 0.$$

It follows that

$$EIy''(l) = M^*, \qquad EIy'''(l) = -Q^*,$$

which are the natural boundary conditions for the cantilever beam.

From the strength of materials we know the relations between the deflection y of the beam, the torque M, and the shear force Q:

$$M = EIy'', \qquad Q = -M' = -EIy'''.$$

We see that the natural boundary conditions really do represent the conditions on the torque and shear force given at the free end $x = l$. □

Let us discuss the example further. The solution of this simple boundary value problem constitutes a considerable part of any textbook on the strength of materials. At one time people relied on graphical approaches, although it is now easy to solve the problem analytically. In practice we encounter largely piecewise continuous load functions q displaying linear and parabolic-type dependences.

The example did force us to consider a case omitted by the general theory of this section: the integrand can have points of discontinuity. Essentially nothing happened though. The Euler–Lagrange equation holds everywhere except at a discontinuity of q, and at such a point a jump in q will give rise to a jump in $y^{(4)}$. The lower-order derivatives of y remain continuous.

In practice it is common to introduce external point torques and shear forces on the beam. What can we say in such cases? In the strength of materials, mechanical reasoning is used to show that at such points the moments and shear forces have corresponding jumps. Can we show this using the tools of the calculus of variations?

We consider a particular problem of the bending of a beam with fixed ends. The beam carries a distributed load q and is a subjected to a point torque M^* and shear force Q^* at some point c. The total energy functional, which takes its minimum value on a solution, has the form

$$\frac{1}{2}\int_0^l EI(y'')^2\,dx - \int_0^l qy\,dx - Q^*y(c) - M^*y'(c).$$

The hypothesis for the model of a beam requires continuity of y and y' at all points including $x = c$. Let us see what actually happens at this point. As in the example above, the energy functional is be considered on the bundle $y + t\varphi$ where φ, together with its first derivative, goes to zero at the endpoints of the segment $[0, l]$. Since we are unsure of what happens at $x = c$ it makes sense to split the integral into two parts: one over the domain $[0, c]$ and the other over the domain $[c, l]$. We shall use the notation $x = c - 0$ to denote a limit taken from the left, and $x = c + 0$ to denote a limit taken from the right. The approach taken in the example brings us to the following equation:

$$\begin{aligned}\int_0^c (EIy^{(4)} - q)\varphi\,dx &+ \int_c^l (EIy^{(4)} - q)\varphi\,dx \\ &+ EIy''(c-0)\varphi(c-0) - EIy''(c+0)\varphi(c+0) \\ &- EIy'''(c-0)\varphi(c-0) + EIy'''(c+0)\varphi(c+0) \\ &- M^*\varphi'(c) - Q^*\varphi(c) = 0.\end{aligned}$$

Supposing $\varphi(c) = 0 = \varphi'(c)$, we obtain the same equation $EIy^{(4)} - q = 0$ on both segments $[0, c)$ and $(c, l]$. Returning to the above equation with φ unrestricted at $x = c$, we see that the second and third derivatives of y do indeed have jumps at $x = c$ defined by M^* and Q^*, respectively:

$$EI(y''(c-0) - y''(c+0)) = M^*, \quad EI(y'''(c-0) - y'''(c+0)) = -Q^*.$$

The reader may treat the case in which the beam characteristic EI changes from EI_0 to EI_1 at $x = c$. He or she can derive the conditions for solving the equilibrium problem for a beam under load at point $x = c$. The solution is a point of minimum of the above total energy functional $E(y)$.

1.7 Functionals Depending on Functions in Many Variables

Although obtaining the Euler equation has become somewhat routine for us, we will not be fully prepared to treat practical problems until we can seek unknown minimizers in many variables.

The two variable case is the simplest; extension to three or more independent variables is straightforward. Consider a functional of the form

$$F(u) = \iint_S f(x, y, u(x, y), u_x(x, y), u_y(x, y))\,dx\,dy. \tag{1.74}$$

Here u_x and u_y denote the partial derivatives $\partial u/\partial x$ and $\partial u/\partial y$, respectively. We confine ourselves to cases where S is simple; practical problems normally involve such domains and much complexity is thereby avoided. Let S be a closed domain in $\mathbb{R}^2$ with a piecewise smooth boundary ∂S. (We do not elaborate on the meaning of "smooth." Our attitude toward this issue is common among practitioners: we simply require everything needed in intermediate calculations.)

We consider two main minimization problems for (1.74): the problem with the Dirichlet boundary condition

$$u(x,y)\Big|_{\partial S} = \psi(s), \tag{1.75}$$

and the problem "without" boundary conditions (i.e., the problem for which natural boundary conditions appear).

We first obtain the analogue to the Euler equation for (1.74). The general approach is to repeat the steps of § 1.2. Specifically we (1) introduce classes of functions over which we may consider the problem of minimum, (2) formulate the fundamental lemma for the two variable case, and (3) recall how to integrate by parts in the two variable case.

Let $C^{(k)}(S)$ denote the set of functions continuous on a compact domain S together with all their derivatives up to order k. The norm for defining a neighborhood of a function is

$$\|u\|_{C^{(k)}(S)} = \max_{\alpha+\beta\le k} \max_{(x,y)\in S} \left| \frac{\partial^{\alpha+\beta} u(x,y)}{\partial x^\alpha \partial y^\beta} \right|. \tag{1.76}$$

$C_0^{(k)}(S)$ is the subset of $C^{(k)}(S)$ consisting of functions which, together with all their derivatives up to order $k-1$, vanish on the boundary ∂S. We shall use the corresponding notations $C^{(\infty)}(S)$ and $C_0^{(\infty)}(S)$ for sets of functions infinitely differentiable on S.

Lemma 1.24. *Let $g(\mathbf{x})$ be continuous on S, and let*

$$\iint_S g(\mathbf{x})\varphi(\mathbf{x})\, dx\, dy = 0 \tag{1.77}$$

hold for any function $\varphi(\mathbf{x}) \in C_0^{(\infty)}(S)$. Then $g(\mathbf{x}) \equiv 0$.

Proof. We imitate the proof of Lemma 1.8. Suppose to the contrary that at some interior point $\mathbf{x}_0$ of S we have $g(\mathbf{x}_0) \neq 0$, say $g(\mathbf{x}_0) > 0$. Then $g(\mathbf{x}) > 0$ for all $\mathbf{x}$ in some disk C_ε having radius ε and center $\mathbf{x}_0$. It is easy to construct a bell-shaped surface of revolution centered at $\mathbf{x}_0$. The corresponding function $\varphi_0(\mathbf{x}) \in C_0^{(\infty)}(S)$ gives

$$\iint_S g(\mathbf{x})\varphi_0(\mathbf{x})\, dx\, dy = \iint_{C_\varepsilon} g(\mathbf{x})\varphi_0(\mathbf{x})\, dx\, dy > 0,$$

which contradicts (1.77). □

To integrate by parts we use

$$\iint_S u \frac{\partial v}{\partial x_i}\, dx\, dy = -\iint_S \frac{\partial u}{\partial x_i} v\, dx\, dy + \oint_{\partial S} uv\, n_i\, ds. \tag{1.78}$$

Here n_i is the cosine of the angle between the unit outward normal $\mathbf{n}$ and the unit vector along the x_i axis ($x_i = x, y$ for $i = 1, 2$, respectively). The length variable s parameterizes contour ∂S.

Remark 1.25. When applying integration by parts in this book, we encounter composite functions such as $f_{u_y}(x, y, u(x, y), u_x(x, y), u_y(x, y))$ which must be differentiated completely with respect to x via the chain rule, because u and its derivatives depend on x. Such derivatives are called *total derivatives*. The total derivatives with respect to the spatial variables x and y will be denoted by d/dx and d/dy. On the other hand, "ordinary" partial derivatives with respect to x and y will be denoted by f_x and f_y. Recall the discussion regarding notation, starting on page 13. □

The main result of this section is the following. Let $f(x, y, u, p, q)$ be a continuous function having continuous first partial derivatives with respect to all of its arguments.

Theorem 1.26. *Let $u = u(x, y) \in C^{(2)}(S)$ be a minimizer of the functional $\iint_S f(x, y, u, u_x, u_y)\, dx\, dy$ on the subset of $C^{(1)}(S)$ consisting of those functions satisfying* (1.75). *Then the Euler equation*

$$f_u - \left(\frac{df_{u_x}}{dx} + \frac{df_{u_y}}{dy} \right) = 0 \tag{1.79}$$

holds in S. Here d/dx and d/dy are total partial derivatives, analogous to the total derivative in the one-dimensional case, when the function $u = u(x, y)$ as well as its partial derivatives u_x and u_y are considered as depending on x and y respectively.

Proof. Consider the functional on the usual bundle $u = u(x, y) + t\varphi(x, y)$ where $\varphi(x, y)$ is a function from $C_0^{(1)}(S)$; that is, it has first derivatives continuous on S and satisfies

$$\varphi(x, y)\big|_{\partial S} = 0. \tag{1.80}$$

The functional $F(u + t\varphi)$ for a fixed $\varphi(x, y)$ becomes a function of the real variable t and takes its minimum at $t = 0$. Thus

$$\begin{aligned} 0 &= \left. \frac{dF(u+t\varphi)}{dt} \right|_{t=0} \\ &= \frac{d}{dt} \left(\iint_S f(x, y, u + t\varphi, u_x + t\varphi_x, u_y + t\varphi_y)\, dx\, dy \right) \Big|_{t=0} \\ &= \iint_S (f_u \varphi + f_{u_x} \varphi_x + f_{u_y} \varphi_y)\, dx\, dy. \end{aligned}$$

Integration by parts in the last two terms of the integrand gives

$$\iint_S \left[f_u - \left(\frac{df_{u_x}}{dx} + \frac{df_{u_y}}{dy} \right) \right] \varphi\, dx\, dy + \oint_{\partial S} (f_{u_x} n_x + f_{u_y} n_y)\, \varphi\, ds = 0. \tag{1.81}$$

Remembering that $\varphi(x, y)$ satisfies (1.80), we get

$$\iint_S \left[f_u - \left(\frac{df_{u_x}}{dx} + \frac{df_{u_y}}{dy} \right) \right] \varphi\, dx\, dy = 0. \tag{1.82}$$

Equation (1.79) follows from Lemma 1.24. □

Theorem 1.27. *Let $u = u(x, y) \in C^{(2)}(S)$ be a minimizer of the functional $\iint_S f(x, y, u, u_x, u_y)\, dx\, dy$ on $C^{(1)}(S)$ (without any boundary conditions). Then the Euler equation (1.79) holds in S, and $u(x, y)$ satisfies the natural boundary condition*

$$\left. (f_{u_x} n_x + f_{u_y} n_y) \right|_{\partial S} = 0. \tag{1.83}$$

Proof. Consider $F(u+t\varphi)$ on the bundle $u+t\varphi$ where $\varphi(x, y) \in C^{(1)}(S)$ is arbitrary but momentarily fixed. For all such functions we establish (1.81) using the same reasoning as above. Restriction of $\varphi(x, y)$ to the set $C_0^{(1)}(S)$ then shows that (1.74) holds in S. So (1.82) holds whether φ belongs to $C_0^{(1)}(S)$ or $C^{(1)}(S)$. Hence

$$\oint_S (f_{u_x} n_x + f_{u_y} n_y)\, \varphi\, ds = 0. \tag{1.84}$$

Now we use the fact that on S, $\varphi = \varphi(s)$ is an arbitrary differentiable function. We do not prove the corresponding fundamental lemma for such an integral, but it is clear that a proof could be patterned after that of Lemma 1.8. (We could use the function $\varphi_0(\mathbf{x})$ from the proof of Lemma 1.24; the point $\mathbf{x}_0$ would be a chosen point of the boundary where the corresponding multiplier $g(\mathbf{x})$ is not equal to zero, by the contrary assumption.) Hence (1.83) follows from (1.84). □

Example 1.28. Demonstrate that for the functional

$$\Psi(u) = \frac{1}{2} \iint_S (u_x^2 + u_y^2)\, dx\, dy - \iint_S F u\, dx\, dy \tag{1.85}$$

with $F = F(x, y)$ a given continuous function, the Euler equation and the natural boundary conditions are

$$\Delta u = -F \quad \text{in } S \tag{1.86}$$

and

$$\left. \frac{\partial u}{\partial n} \right|_{\partial S} = 0, \tag{1.87}$$

respectively. Show that on a solution u^* of the latter boundary value problem, if it exists, the functional $\Psi(u)$ attains a global minimum.

Solution. The derivation of (1.86) and (1.87) is straightforward. Denoting

$$f = \frac{1}{2}(u_x^2 + u_y^2) - Fu$$

we get

$$f_u - \left(\frac{df_{u_x}}{dx} + \frac{df_{u_y}}{dy} \right) = -F - \Delta u,$$

which leads to (1.86). The left-hand expression in (1.83) is

$$f_{u_x} n_x + f_{u_y} n_y = u_x n_x + u_y n_y,$$

which is $\partial u / \partial n$ on the boundary.

Before demonstrating the last statement in the example, we note that $\Psi(u)$ expresses the total energy of an elastic membrane. From physics we know that at points of minimum of a total energy functional for a mechanical system with conservative loads, the system is in equilibrium. In particle mechanics it is even shown that such an equilibrium state is stable at a point of strict minimum. Let us see what happens in this case of a spatially distributed object. We suppose that a solution u^* of the boundary value problem (1.86)–(1.87) exists. Consider the values of Ψ over the bundle $u^* + \varphi$, where φ is arbitrary:

$$\begin{aligned}\Psi(u^* + \varphi) &= \frac{1}{2}\iint_S \left((u_x^* + \varphi_x)^2 + (u_y^* + \varphi_y)^2\right) dx\,dy \\ &\quad - \iint_S F(u^* + \varphi)\,dx\,dy \\ &= \Psi(u^*) + \left[\iint_S (u_x^*\varphi_x + u_y^*\varphi_y)\,dx\,dy - \iint_S F\varphi\,dx\,dy\right] \\ &\quad + \frac{1}{2}\iint_S \left(\varphi_x^2 + \varphi_y^2\right) dx\,dy.\end{aligned}$$

Because of (1.86)–(1.87) (which, in the above theory, were derived as a direct consequence of the following equality and thus are equivalent to it when u^* is sufficiently smooth) we see that

$$\iint_S (u_x^*\varphi_x + u_y^*\varphi_y)\,dx\,dy - \iint_S F\varphi\,dx\,dy = 0.$$

So

$$\Psi(u^* + \varphi) - \Psi(u^*) = \frac{1}{2}\iint_S \left(\varphi_x^2 + \varphi_y^2\right) dx\,dy \geq 0,$$

which means that $\Psi(u)$ takes its global minimum at $u = u^*$. □

We are in the habit of supposing that a minimizer exists for each problem we encounter. But the problem of minimizing (1.85), which describes the equilibrium of a membrane, demonstrates that not every problem which seems sensible at first glance has a solution. Indeed, if we take $u = c$, a constant, then the first integral in (1.85) is zero. If $\iint_S F\,dx\,dy \neq 0$, then by changing c we make the value of the functional any large negative number. So the problem has no solution and (at least) the condition $\iint_S F\,dx\,dy = 0$ becomes necessary for the problem to be sensible. In fact, this has a clear mechanical sense: it is the condition of self-balance of the forces. A free membrane subjected to a load F can move as a whole in the direction normal to the membrane. In this model we neglect its inertia, so the problem of equilibrium of the membrane without the condition of self-balance of the load is senseless as we showed formally. Later we consider this question in more detail.

1.8 A Functional with Integrand Depending on Partial Derivatives of Higher Order

Now we derive the Euler equation for a minimizer $w = w(x, y)$ of a functional of the form

$$F(w) = \iint_S f(x, y, w, w_x, w_y, w_{xx}, w_{xy}, w_{yy})\,dx\,dy \tag{1.88}$$

on the functions of class $C^{(2)}(S)$ satisfying the boundary conditions

$$w\big|_{\partial S} = w_0(s), \qquad \left.\frac{\partial w}{\partial n}\right|_{\partial S} = w_1(s). \tag{1.89}$$

The steps are now routine. Assume a minimizer $w = w(x, y) \in C^{(4)}(S)$. Let $\varphi(x, y)$ be an arbitrary but fixed function from $C_0^{(2)}(S)$, which implies in particular that

$$\varphi\big|_{\partial S} = 0, \qquad \left.\frac{\partial \varphi}{\partial n}\right|_{\partial S} = 0. \tag{1.90}$$

$F(w + t\varphi)$ takes its minimum at $t = 0$ and thus $dF(w + t\varphi)/dt\big|_{t=0} = 0$. This equation takes the form

$$\iint_S (f_w\varphi + f_{w_x}\varphi_x + f_{w_y}\varphi_y + f_{w_{xx}}\varphi_{xx} + f_{w_{xy}}\varphi_{xy} + f_{w_{yy}}\varphi_{yy})\,dx\,dy = 0. \tag{1.91}$$

Supposing f has continuous derivatives of third order, we can integrate by parts in (1.91) and get

$$\iint_S \left(f_w - \frac{d}{dx} f_{w_x} - \frac{d}{dy} f_{w_y} + \frac{d^2}{dx^2} f_{w_{xx}} + \frac{d^2}{dxdy} f_{w_{xy}} + \frac{d^2}{dy^2} f_{w_{yy}} \right) \varphi\,dx\,dy = 0. \tag{1.92}$$

The boundary terms vanish by (1.90). By Lemma 1.24 we obtain the Euler equation for the functional (1.88):

$$f_w - \frac{d}{dx} f_{w_x} - \frac{d}{dy} f_{w_y} + \frac{d^2}{dx^2} f_{w_{xx}} + \frac{d^2}{dxdy} f_{w_{xy}} + \frac{d^2}{dy^2} f_{w_{yy}} = 0, \tag{1.93}$$

valid in S. Here d/dx and d/dy are total partial derivatives when $w = w(x, y)$ is considered as depending on its arguments x, y.

We could derive the form of the natural boundary conditions for (1.88), but this is cumbersome so we prefer to treat an illustrative case. We shall consider a problem of minimizing a total energy functional, whose solution describes the equilibrium of an elastic plate with free edge.

It is time to discuss how problems of minimization arose. Some came from geometrical considerations, like the isoperimetric problem mentioned in § 1.1; some were designed specifically as exercises, written out by analogy with other, more or less easily solved, problems. But for the most part the real problems of the calculus of variations came from physics — in particular, mechanics. There it was found that minimizers or maximizers of certain functionals describe important states of physical systems. It is interesting to note how this idea progressed in importance. Early in the development of classical mechanics, variational principles were derived using the "fundamental" equations of statics and mechanics; they were regarded as consequences, although in many circumstances they were actually equivalent. It was soon found that some problems were easier solved by variational methods, and the variational approach to mechanics gained a life of its own. In the theory of elasticity, for example, a great many variational principles have been derived; moreover, the name "variational principle" is applied not only to the minimization of functionals, but to any circumstance in which an important equation can be derived from an integro-differential equation having the form of the first variation of a functional being equal to zero, even if there is no functional for which it is the first variation. For example, the Virtual Work Principle arose as a consequence of the principle of minimum of potential energy of a mechanical system. But the former continues to hold in the case of nonconservative forces where it is impossible to compose the potential energy functional.

Early in the development of linear elasticity, an energy functional was derived whose minimizer describes the equilibrium of an elastic body. The procedure was to write out the equilibrium equations, multiply by appropriate components of the vector of displacements, and integrate over the region. Using integration by parts with regard for homogeneous Dirichlet boundary conditions, from the terms with second-order partial derivatives it was possible to get a symmetrical form (in the components of the strain tensor) for potential energy. The originators of this method were comforted by the fact that the associated natural boundary conditions coincided with the boundary conditions assigned to the same problem when considered as a problem of equilibrium with applied forces given on the boundary. This led to the idea that the Principle of Minimum Potential Energy (or, correspondingly, the Virtual Work Principle) could be used to derive boundary conditions for models of elastic plates and shells. Workers investigating such models had previously run into difficulty in posing appropriate boundary conditions: upon simplification from the three-dimensional case, uncertainties had arisen regarding precisely what force conditions should be appointed on the boundary of an object. The variational formalism provided the needed result in a simple fashion. Why are we taking the time to discuss this now? We are going to consider the problem of equilibrium of an elastic plate from the viewpoint of the calculus of variations. The first step is to formulate the energy functional. The left side of the equation describing a thin elastic plate bent under load contains a biharmonic operator. In this case there is no uniquely defined procedure to derive the energy functional. Moreover, integration by parts can yield several expressions for the energy of an elastic plate with homogeneous Dirichlet conditions (1.90). For each of these forms one can derive the natural boundary conditions, but only one form gives the conditions corresponding to mechanics. So to formulate the problem (i.e., the functional) properly, one should have some knowledge of mechanics — perhaps this is why so many pure mathematicians prefer to study only classical problems where everything is formulated in advance! To work purely mathematical exercises, one is seldom required to know the actual physical behavior of the object under consideration. But correct mathematical procedures often depend in large part on the details of a particular realm of application.

The energy functional of an isotropic homogeneous plate bending under load $F = F(x, y)$ is

$$\begin{aligned} E(w) = \frac{D}{2} \iint_S \left(w_{xx}^2 + w_{yy}^2 + 2\nu w_{xx} w_{yy} + 2(1-\nu) w_{xy}^2\right) dx\, dy \\ - \iint_S F w \, dx\, dy \end{aligned} \tag{1.94}$$

where D is the rigidity of the plate, ν is Poisson's ratio, and $w = w(x, y)$ is the deflection at point (x, y) of S, the compact domain occupied by the mid-surface of the plate. A minimizer of $E(w)$ describes the equilibrium deflection of the mid-surface. Using the standard method, we shall derive the Euler equation for the minimizer and the corresponding natural boundary conditions.

Let $w \in C^{(4)}(S)$ minimize the functional (1.94) over $C^{(2)}(S)$. Consider $E(w + t\varphi)$ at a fixed $\varphi \in C^{(2)}(S)$ as a function of the parameter t. It takes its minimum at $t = 0$, so as a consequence we have

$$\begin{aligned} D \iint_S [w_{xx}\varphi_{xx} + w_{yy}\varphi_{yy} + \nu(w_{xx}\varphi_{yy} + w_{yy}\varphi_{xx}) \\ + 2(1-\nu) w_{xy}\varphi_{xy}]\, dx\, dy - \iint_S F\varphi \, dx\, dy = 0 \end{aligned}$$

which is a particular case of (1.92). Now it is necessary to integrate by parts in the first integral on the left. We get

$$D\iint_S [(w_{xx}+\nu w_{yy})\varphi_{xx}+(w_{yy}+\nu w_{xx})\varphi_{yy}+2(1-\nu)w_{xy}\varphi_{xy}]\,dx\,dy$$

$$= -D\iint_S \left[\varphi_x \frac{\partial}{\partial x}(w_{xx}+\nu w_{yy})+\varphi_y\frac{\partial}{\partial y}(w_{yy}+\nu w_{xx}) + (1-\nu)w_{xyy}\varphi_x+(1-\nu)w_{xxy}\varphi_y\right] dx\,dy$$

$$+ D\oint_{\partial S} [(w_{xx}+\nu w_{yy})\varphi_x n_x+(w_{yy}+\nu w_{xx})\varphi_y n_y + (1-\nu)w_{xy}(\varphi_x n_y+\varphi_y n_x)]\,ds \qquad (1.95)$$

where $\mathbf{n}$, the unit normal to the boundary ∂S, has components (n_x, n_y). Note that we have preserved the symmetry of the expressions. Integrating by parts once more in the first integral on the right, denoted by A, we get

$$A = D\iint_S [(w_{xx}+\nu w_{yy})_{xx}+(w_{yy}+\nu w_{xx})_{yy}+2(1-\nu)w_{xxyy}]\varphi\,dx\,dy$$

$$- D\oint_S [(w_{xx}+\nu w_{yy})_x n_x+(w_{yy}+\nu w_{xx})_y n_y + (1-\nu)(w_{xyy}n_x+w_{xxy}n_y)]\varphi\,ds.$$

The first integral in A is

$$D\iint_S (w_{xxxx}+2w_{xxyy}+w_{yyyy})\varphi\,dx\,dy = D\iint_S \varphi\Delta^2 w\,dx\,dy.$$

Thus (1.95) takes the form

$$D\iint_S \varphi\Delta^2 w\,dx\,dy - \iint_S F\varphi\,dx\,dy$$

$$+ D\oint_S [(w_{xx}+\nu w_{yy})\varphi_x n_x+(w_{yy}+\nu w_{xx})\varphi_y n_y + (1-\nu)w_{xy}(\varphi_x n_y+\varphi_y n_x)]\,ds$$

$$- D\oint_S [(w_{xx}+\nu w_{yy})_x n_x+(w_{yy}+\nu w_{xx})_y n_y + (1-\nu)(w_{xyy}n_x+w_{xxy}n_y)]\varphi\,ds = 0. \qquad (1.96)$$

First consider the subset of admissible functions $\varphi(x,y)$ satisfying (1.90). Equation (1.96) reduces to

$$\iint_S (D\Delta^2 w - F)\varphi\,dx\,dy = 0. \qquad (1.97)$$

By the fundamental lemma we obtain the Euler equation

$$D\Delta^2 w - F = 0 \quad \text{in } S. \tag{1.98}$$

Because of (1.98) the equality (1.97) holds for any admissible $\varphi(x, y)$, thus the two first integrals over S disappear from (1.96). In equation (1.96) there remains the sum of two contour integrals that equals zero for any $\varphi \in C^{(2)}(S)$.

We might think that since we have three arbitrary functions φ, φ_x, φ_y on S, we could set their multipliers equal to zero and obtain three natural boundary conditions. But this is incorrect. We see this first on mechanical grounds: these "boundary conditions" would depend on x and y, hence would not be invariant under coordinate rotations. Mathematically, it appears that we cannot choose φ, φ_x, and φ_y independently on S. Indeed let us fix φ on S: then its derivative φ_τ in the tangential direction $\boldsymbol{\tau}$ is determined uniquely — only the derivative φ_n of φ in the normal direction is really independent of φ on the contour.

Thus we first need to introduce this change of coordinates, getting a local frame $(\boldsymbol{\tau}, \mathbf{n})$. The transformation formulas for derivatives are

$$\varphi_x = \varphi_n n_x - \varphi_s n_y, \qquad \varphi_y = \varphi_n n_y + \varphi_s n_x. \tag{1.99}$$

Let us put these into the integrand of the first contour integral:

$$\begin{aligned}
&(w_{xx} + \nu w_{yy})\varphi_x n_x + (w_{yy} + \nu w_{xx})\varphi_y n_y + (1-\nu)w_{xy}(\varphi_x n_y + \varphi_y n_x) \\
&= (w_{xx} + \nu w_{yy})(\varphi_n n_x - \varphi_s n_y)n_x + (w_{yy} + \nu w_{xx})(\varphi_n n_y + \varphi_s n_x)n_y \\
&\quad + (1-\nu)w_{xy}[(\varphi_n n_x - \varphi_s n_y)n_y + (\varphi_n n_y + \varphi_s n_x)n_x] \\
&= (1-\nu)\{(w_{yy} - w_{xx})n_x n_y + w_{xy}(n_x^2 - n_y^2)\}\varphi_s \\
&\quad + \{(w_{xx} + \nu w_{yy})n_x^2 + (w_{yy} + \nu w_{xx})n_y^2 + 2(1-\nu)w_{xy}n_x n_y\}\varphi_n \\
&= (1-\nu)\{(w_{yy} - w_{xx})n_x n_y + w_{xy}(n_x^2 - n_y^2)\}\varphi_s \\
&\quad + \{\nu\Delta w + (1-\nu)(w_{xx}n_x^2 + w_{yy}n_y^2 + 2w_{xy}n_x n_y)\}\varphi_n.
\end{aligned} \tag{1.100}$$

Change the integrand of the first contour integral in (1.96) by (1.100) and remember that $\varphi_s = \partial\varphi/\partial s$ and $\varphi_n = \partial\varphi/\partial n$:

$$\begin{aligned}
&D \oint_{\partial S} (1-\nu)\{(w_{yy} - w_{xx})n_x n_y + w_{xy}(n_x^2 - n_y^2)\}\frac{\partial\varphi}{\partial s}\, ds \\
&+ D \oint_{\partial S} \{\nu\Delta w + (1-\nu)(w_{xx}n_x^2 + w_{yy}n_y^2 + 2w_{xy}n_x n_y)\}\frac{\partial\varphi}{\partial n}\, ds \\
&- D \oint_{\partial S} [(w_{xx} + \nu w_{yy})_x n_x + (w_{yy} + \nu w_{xx})_y n_y \\
&\qquad\qquad + (1-\nu)(w_{xyy}n_x + w_{xxy}n_y)]\varphi\, ds = 0.
\end{aligned} \tag{1.101}$$

If S is smooth enough we can integrate by parts in the first integral with respect to s. This gives

$$D\oint_{\partial S}(1-\nu)\{(w_{yy}-w_{xx})n_xn_y+w_{xy}(n_x^2-n_y^2)\}\frac{\partial\varphi}{\partial s}\,ds$$
$$=-D(1-\nu)\oint_{\partial S}\varphi\frac{\partial}{\partial s}\{(w_{yy}-w_{xx})n_xn_y+w_{xy}(n_x^2-n_y^2)\}\,ds.$$

It follows that

$$\begin{aligned}-D\oint_{\partial S}&[(w_{xx}+\nu w_{yy})_xn_x+(w_{yy}+\nu w_{xx})_yn_y\\&+(1-\nu)(w_{xyy}n_x+w_{xxy}n_y)]\\&+(1-\nu)\frac{d}{ds}[(w_{yy}-w_{xx})n_xn_y+w_{xy}(n_x^2-n_y^2)]]\varphi\,ds\\+D\oint_{\partial S}&\{\nu\Delta w+(1-\nu)(w_{xx}n_x^2+w_{yy}n_y^2+2w_{xy}n_xn_y)\}\frac{\partial\varphi}{\partial n}\,ds=0.\end{aligned}$$

By independently choosing φ and $\partial\varphi/\partial n$, we get the following natural boundary conditions:

$$\nu\Delta w+(1-\nu)(w_{xx}n_x^2+w_{yy}n_y^2+2w_{xy}n_xn_y)\Big|_{\partial S}=0, \tag{1.102}$$

$$\begin{aligned}&[(w_{xx}+\nu w_{yy})_yn_x+(w_{yy}+\nu w_{xx})_yn_y+(1-\nu)(w_{xyy}n_x+w_{xxy}n_y)]\\&+(1-\nu)\frac{d}{ds}[(w_{yy}-w_{xx})n_xn_y+w_{xy}(n_x^2-n_y^2)]=0.\end{aligned}\tag{1.103}$$

The first means that the shear force on the lateral surface of the plate is zero, whereas the second means that the bending moment is zero.

We have assumed that ∂S is sufficiently smooth so we could integrate by parts in (1.101). At corner points (1.103) is not valid. The reader may wish to derive an appropriate corner condition.

1.9 The First Variation

This book is written for those who will use the calculus of variations. Although a simple exposition is the goal, continued exploitation of the same technique would prevent real progress. We need ideas applicable to more complex problems. As before, these will be extensions of elementary ideas from calculus. A principal analytical tool is the differential of a function. The first differential extracts the main part of the increment of the function when its argument changes by a small amount Δx. This main part is linear with respect to Δx. In this way, we approximate the change of a smooth function in some neighborhood of a point by an expression linear in Δx. The extension to functionals is called the first variation.

A few technical details

Definition 1.29. We say that $f(x) = o(g(x))$ when $x \to x_0$ if

$$\lim_{x \to x_0} \frac{f(x)}{g(x)} = 0. \tag{1.104}$$

Here x can be a real variable or an element of a more general metric or normed space; in the latter case, $x \to x_0$ refers to convergence in that space. We often use the abbreviated notation $f = o(g)$ and say that f is of a *higher order of smallness* than g.

So if the o relation holds then given any $\varepsilon > 0$ we can find $\delta > 0$ such that $|f(x)/g(x)| < \varepsilon$ whenever $\|x - x_0\| < \delta$.[1] Note the following.

(1) The functions $f(x)$ and $g(x)$ are not required to possess individual limits as $x \to x_0$; only the ratio must possess a limit.
(2) In practice, $g(x)$ will usually be some power of a simple real variable x.

The statement $f(x) = o(1)$ as $x \to x_0$, for example, means nothing more than $\lim_{x \to x_0} f(x) = 0$. If $f(x) = o(x - x_0)$ as $x \to x_0$, then $f(x)$ tends to zero even faster as $x \to x_0$ since the ratio $f(x)/(x - x_0)$ tends to zero even though its denominator tends to zero as $x \to x_0$.

Definition 1.30. We write $f(x) = O(g(x))$ as $x \to x_0$ if in some neighborhood of x_0 an inequality

$$\left| \frac{f(x)}{g(x)} \right| \le c \tag{1.105}$$

holds for some constant c. We often use the abbreviated notation $f = O(g)$ and say that f is of the *same order of smallness* as g.

The statement $f(x) = O(1)$ as $x \to 0$ means that in some neighborhood of 0 we have $|f(x)| < c$ (i.e., f is bounded in this neighborhood). If $f(x) = O(x)$ as $x \to 0$, then in some neighborhood of zero we have $|f(x)| < c|x|$.

[1] Here we refer to a more general vector norm. A reader unfamiliar with the subject of norms will find a more complete discussion in § 1.11. For now it is sufficient to think in terms of real numbers, where the role of norm is played by the absolute value.

This implies that $f(x) \to 0$ as $x \to 0$, hence that $f(x) = o(1)$. But $f(x) = O(x)$ tells *how fast* $f(x)$ tends to zero.

Let $f(x)$ and its first $n+1$ derivatives be continuous in an interval about $x = x_0$. Then according to Taylor's theorem

$$f(x) = f(x_0)+f'(x_0)(x-x_0)+\cdots+\frac{f^{(n)}(x_0)}{n!}(x-x_0)^n+\frac{f^{(n+1)}(\xi)}{(n+1)!}(x-x_0)^{n+1}$$

for some ξ between x_0 and x. The last term on the right is the Lagrange form of the remainder and is clearly $O(|x-x_0|^{n+1})$. Addition and subtraction of the term

$$\frac{f^{(n+1)}(x_0)}{(n+1)!}(x-x_0)^{n+1}$$

gives

$$\begin{aligned} f(x) = f(x_0) + f'(x_0)(x-x_0) + \cdots + \frac{f^{(n+1)}(x_0)}{(n+1)!}(x-x_0)^{n+1} \\ + \frac{[f^{(n+1)}(\xi) - f^{(n+1)}(x_0)]}{(n+1)!}(x-x_0)^{n+1}. \end{aligned}$$

This is a Taylor expansion with one more term and a new "remainder." Because $f^{n+1}(x)$ is continuous, the bracketed term $f^{(n+1)}(\xi) - f^{(n+1)}(x_0)$ tends to zero when $x \to x_0$ (recall that ξ is an intermediate point of (x, x_0)). Hence the ratio of the new remainder to the factor $|x-x_0|^{n+1}$ tends to zero as $x \to x_0$:

$$f(x) = f(x_0)+f'(x_0)(x-x_0)+\cdots+\frac{f^{(n+1)}(x_0)}{(n+1)!}(x-x_0)^{n+1}+o\left(|x-x_0|^{n+1}\right).$$

This is *Peano's form* of Taylor's theorem.

Theorem 1.31. *Let $f(x)$ and its first n derivatives be continuous in an interval about $x = x_0$. Then*

$$f(x) = f(x_0) + \frac{f'(x_0)}{1!}(x-x_0) + \cdots + \frac{f^{(n)}(x_0)}{n!}(x-x_0)^n + o\left(|x-x_0|^n\right).$$

With this we can say something about the behavior of the remainder term in the nth-order Taylor expansion even if we know nothing about continuity of the $(n+1)$th derivative.

Back to the first variation

In calculus, we consider the increment $f(x+\Delta x) - f(x)$ of a function $f(x)$ of a real variable x. If it is possible to represent it in the form

$$f(x+\Delta x) - f(x) = A\Delta x + \omega(\Delta x) \tag{1.106}$$

where $\omega(\Delta x) = o(\Delta x)$ as $\Delta x \to 0$, then

- $A\Delta x$ is called the *first differential of f at x*, and is denoted by $df(x)$,
- A is the derivative of f at x, denoted by $f'(x)$, and
- the increment Δx of the argument x is redenoted by dx and is called the differential of the argument.

We may therefore write

$$df(x) = A\Delta x = f'(x)\,dx.$$

In the mind of a calculus student the differential dx and its corresponding $df(x)$ are extremely small quantities. Let us now banish this misconception: both dx and $df(x)$ are finite. When dx is small then so is $df(x)$ and it approximates the difference $f(x+dx) - f(x)$: the smaller the value of dx, the better the relative approximation. However, neither dx nor $df(x)$ is small in general.

Let us repeat the same steps for a functional. This is especially easy to do for a quadratic functional. These arise in physics, corresponding to natural laws that are linear in form (of course, linearity is often a condition imposed rather artificially on models of real phenomena). Consider, for example,

$$F(u) = \frac{1}{2}\iint_S (u_x^2 + u_y^2)\,dx\,dy - \iint_S Fu\,dx\,dy. \tag{1.107}$$

We denote the "increment" of the argument $u = u(x)$ by $\varphi(x)$. Note that $\varphi(x)$ must have certain properties; it should be admissible in the sense of § 1.5. (Later we shall soften the smoothness conditions for this problem.) In mechanics φ is usually denoted by δu; this maintains a visual similarity between the two notions of increment dx and δu, and in this notation δu is called a *virtual displacement*. Now

$$\begin{aligned} F(u+\varphi) - F(u) &= \iint_S (u_x\varphi_x + u_y\varphi_y)\,dx\,dy - \iint_S F\varphi\,dx\,dy \\ &\quad + \frac{1}{2}\iint_S (\varphi_x^2 + \varphi_y^2)\,dx\,dy. \end{aligned} \tag{1.108}$$

The first two integrals on the right are linear in φ and pretend to analogy with the differential of calculus; together they are called the *first variation* of the functional $F(u)$ at u:

$$\iint_S (u_x\varphi_x + u_y\varphi_y)\,dx\,dy - \iint_S F\varphi\,dx\,dy. \tag{1.109}$$

The third integral in (1.108), quadratic in φ, is analogous to $\omega(\Delta x)$ in (1.106). We should introduce the smallness of the increment φ in such a way (and we did this in § 1.5) that this quadratic term becomes infinitely small in comparison with the linear terms.

In § 1.5 we found that if $u = u(x)$ is a minimizer of $F(u)$, then the expression (1.109) is zero for all admissible φ:

$$\iint_S (u_x\varphi_x + u_y\varphi_y)\,dx\,dy - \iint_S F\varphi\,dx\,dy = 0. \tag{1.110}$$

From this we derived the Euler equation (1.86) for the membrane. We now derive (1.110) in a different way. Let us suppose that $u = u(x,y)$ is a minimizer of $F(u)$; that is, $F(u+\varphi) - F(u) \geq 0$ for any admissible φ. Assume, contrary to (1.110), that

$$\iint_S (u_x\varphi_x^* + u_y\varphi_y^*)\,dx\,dy - \iint_S F\varphi^*\,dx\,dy \neq 0$$

for some admissible φ^*. Then putting another admissible function $t\varphi^*$ into the inequality $F(u+\varphi) - F(u) \geq 0$, we get

$$\begin{aligned}
0 &\leq F(u+t\varphi^*) - F(u) \\
&= \iint_S (u_x t\varphi_x^* + u_y t\varphi_y^*)\,dx\,dy - \iint_S Ft\varphi^*\,dx\,dy \\
&\qquad + \frac{1}{2}\iint_S t^2({\varphi_x^*}^2 + {\varphi_y^*}^2)\,dx\,dy \\
&= t\left[\iint_S (u_x\varphi_x^* + u_y\varphi_y^*)\,dx\,dy - \iint_S F\varphi^*\,dx\,dy\right] \\
&\qquad + \frac{t^2}{2}\iint_S ({\varphi_x^*}^2 + {\varphi_y^*}^2)\,dx\,dy.
\end{aligned} \tag{1.111}$$

Suppose the bracketed term differs from zero. If we take t such that it is sufficiently close to zero and the term $t[\cdots]$ is negative, then the term which is quadratic in t is much smaller than the term which is linear in t. Therefore $F(y+t\varphi) - F(y) < 0$, which contradicts the leftmost inequality of (1.111). So (1.110) holds for any admissible φ.

It is clear that we can repeat everything in terms of the plate problem of § 1.8. The differences are only technical.

We used the fact that at least for some (positive and negative) small t the function $t\varphi^*$ is admissible. In the membrane problem this is trivial. However, in some problems the set of admissible functions is restricted (e.g., it may be that $\varphi \geq 0$); free choice of t is thereby precluded. Such problems fall outside the scope of the classical theory, and in fact belong to the theory of variational inequalities.

We consider a general case of the simplest functional with respect to functions satisfying any of the types of boundary conditions we have discussed. Let us find its increment over the increment $\varphi(x)$ of the function $y(x)$. So we consider the increment of the functional

$$F(y) = \int_a^b f(x, y, y')\, dx$$

when the argument gets an admissible increment $\varphi = \varphi(x)$. Whether the boundary conditions are stipulated or not (free ends), we have

$$F(y + \varphi) - F(y) = \int_a^b [f(x, y + \varphi, y' + \varphi') - f(x, y, y')]\, dx.$$

Regarding the arguments of f as simple real variables, we can apply the Taylor expansion to f. If f has continuous second partial derivatives, then

$$f(x, y+\varphi, y'+\varphi') - f(x, y, y') = f_y(x, y, y')\varphi + f_{y'}(x, y, y')\varphi' + O(|\varphi|^2 + |\varphi'|^2).$$

Thus

$$\begin{aligned} F(y + \varphi) - F(y) = \int_a^b [f_y(x, y, y')\varphi + f_{y'}(x, y, y')\varphi']\, dx \\ + O\left(\int_a^b (|\varphi|^2 + |\varphi'|^2)\, dx\right). \end{aligned} \tag{1.112}$$

The last integral is of the order $O(\|\varphi\|^2_{C^{(1)}(a,b)})$ because

$$\begin{aligned} \int_a^b (|\varphi|^2 + |\varphi'|^2)\, dx &\leq \int_a^b (|\varphi| + |\varphi'|)^2\, dx \\ &\leq (b - a) \max_{x \in [a,b]} (|\varphi| + |\varphi'|)^2 \\ &\leq (b - a) \left[\max_{x \in [a,b]} (|\varphi| + |\varphi'|)\right]^2. \end{aligned}$$

For admissible functions φ that are small in the norm of $C^{(1)}(a,b)$, the last term on the right side of (1.112) has a higher order of smallness in φ than the integral term which is linear in φ. Thus we have a complete analogy with the first differential of a function.

Definition 1.32. The expression

$$\delta F(y,\varphi) \equiv \int_a^b [f_y(x,y,y')\varphi + f_{y'}(x,y,y')\varphi']\,dx, \tag{1.113}$$

often denoted simply by δF, is the *first variation* of $F(y)$.

Let $y = y(x)$ be a minimizer of $F(y)$ for some boundary conditions considered above. For any admissible function φ, the equation

$$\int_a^b [f_y(x,y,y')\varphi + f_{y'}(x,y,y')\varphi']\,dx = 0 \tag{1.114}$$

holds. Indeed, for any admissible φ we have $F(y+\varphi) - F(y) \geq 0$. Assume that (1.114) fails at some admissible φ^*. Suppose that $t\varphi^*$ for small t is also admissible so that

$$\begin{aligned} 0 &\leq F(y+t\varphi^*) - F(y) \\ &= t\int_a^b [f_y(x,y,y')\varphi^* + f_{y'}(x,y,y')\varphi^{*\prime}]\,dx + O(t^2\,\|\varphi^*\|^2_{C^{(1)}(a,b)}). \end{aligned} \tag{1.115}$$

Now the smallness of the increment of the argument is governed by t. For small t the sign of the right side of (1.115) is determined by the first integral term. Since we can choose t to be negative or positive and its coefficient is not zero, we can find a small t^* such that

$$t^*\int_a^b [f_y(x,y,y')\varphi^* + f_{y'}(x,y,y')\varphi^{*\prime}]\,dx + O(t^{*2}\,\|\varphi^*\|^2_{C^{(1)}(a,b)}) < 0.$$

This contradicts the leftmost inequality of (1.115).

Let us note that in $dF(y+t\varphi)/dt\big|_{t=0}$ we obtain the same expression (1.113), i.e., the first variation of the functional. The two methods of obtaining the first variation are equivalent if the integrand f is sufficiently smooth. But in the general theory of functionals our method of differentiation (i.e., the selection of the linear part of the difference $F(y+\varphi) - F(y)$) corresponds to the use of the *Fréchet derivative*, whereas the computation of $dF/dt|_{t=0}$ corresponds to the use of the *Gâteaux derivative*.

The reasoning of this section can be repeated for any of the functionals and their associated minimum problems we considered earlier. We leave this to the reader as a number of exercises.

Variational derivative

We have seen that the Euler equation is analogous to the equation $y'(x) = 0$ from elementary calculus. Let us consider another approach to deriving the Euler equation. This will provide a representation for the increment of a functional $F(y)$ under bell-shaped disturbances of $y(x)$. The resulting formula will be needed later for treatment of the isoperimetric problem.

Let us preview the approach before tackling the details. We first recall the proof of the fundamental lemma on page 18. The lemma states that $f(x)$ must vanish if it is continuous and if

$$\int_a^b f(x)g(x)\,dx = 0$$

for an arbitrary continuous function $g(x)$ that vanishes at the endpoints a, b. However, the proof required only a subset of such functions $g(x)$: those that were bell-shaped and whose supports were small enough. (The *support* of a function $g(x)$ is the closure of the set over which $g(x) \neq 0$.) Hence we can reframe the problem of minimizing a functional in terms of disturbance functions taken from this subset only. So let us consider what happens if take the set of bell functions of the form

$$\varphi_\varepsilon(x) = \begin{cases} \exp\left(\dfrac{\varepsilon^2}{x^2 - \varepsilon^2}\right), & |x| < \varepsilon, \\ 0, & |x| \geq \varepsilon, \end{cases} \tag{1.116}$$

which have supports of length 2ε and maximum values of unity. Clearly for the minimizer $y_0(x)$ of a functional F we also get

$$\frac{d}{dt}F(y_0(x) + t\varphi_\varepsilon(x - x_0))\bigg|_{t=0} = \int_{x_0-\varepsilon}^{x_0+\varepsilon} \left(f_y - \frac{d}{dx}f_{y'}\right)\varphi_\varepsilon(x - x_0)\,dx = 0.$$

From arbitrariness of x_0 and ε, and continuity of the parenthetical expression, it follows that the Euler equation holds at any $x_0 \in (a, b)$.

Let us use the smallness of ε. Recall the *second mean value theorem for integrals*:

Theorem 1.33. *Let $f(x)$ be continuous on $[a, b]$. If $g(x)$ is integrable and does not change sign in $[a, b]$, then*

$$\int_a^b f(x)g(x)\,dx = f(\xi)\int_a^b g(x)\,dx \tag{1.117}$$

for some $\xi \in [a, b]$.

Because $\varphi_\varepsilon(x)$ is nonnegative and $f_y - df_{y'}/dx$ is continuous, we get

$$\int_{x_0-\varepsilon}^{x_0+\varepsilon} \left(f_y - \frac{d}{dx}f_{y'}\right)\varphi_\varepsilon(x-x_0)\,dx = \left(f_y - \frac{d}{dx}f_{y'}\right)\Big|_{x=\xi}\sigma_\varepsilon$$
$$= \left[\left(f_y - \frac{d}{dx}f_{y'}\right)\Big|_{x=x_0} + \alpha(x,\varepsilon)\right]\sigma_\varepsilon$$

where

$$\sigma_\varepsilon = \int_{x_0-\varepsilon}^{x_0+\varepsilon} \varphi_\varepsilon(x-x_0)\,dx$$

is the area under the bell and $\alpha(x,\varepsilon) \to 0$ uniformly as $\varepsilon \to 0$. On the left we have $\delta F(y_0(x), \varphi_\varepsilon(x-x_0))$. If we divide both sides of this by σ_ε and let $\varepsilon \to 0$ (or equivalently $\sigma_\varepsilon \to 0$), we get

$$\lim_{\sigma_\varepsilon \to 0} \frac{\delta F(y_0(x), \varphi_\varepsilon(x-x_0))}{\sigma_\varepsilon} = \left(f_y - \frac{d}{dx}f_{y'}\right)\Big|_{x=x_0}.$$

The last equality holds for any smooth function y, not just for the minimizer, and can be rewritten as

$$\delta F(y(x), \varphi_\varepsilon(x-x_0)) = \left[\left(f_y - \frac{d}{dx}f_{y'}\right)\Big|_{x=x_0} + \alpha\right]\sigma_\varepsilon, \tag{1.118}$$

where $\alpha \to 0$ as $\sigma_\varepsilon \to 0$.

Since $\delta F(y(x), t\varphi_\varepsilon(x-x_0))$ is the principal linear part of the increment

$$\Delta F(y(x), t\varphi_\varepsilon(x-x_0)) = F(y(x), t\varphi_\varepsilon(x-x_0)) - F(y(x))$$

as $t \to 0$, we seek a similar relation for $\Delta F(y(x), t\varphi_\varepsilon(x-x_0))$, which is

$$\lim_{\sigma_{t,\varepsilon} \to 0} \frac{\Delta F(y_0(x), \varphi_\varepsilon(x-x_0))}{\sigma_{t,\varepsilon}} = \left(f_y - \frac{d}{dx}f_{y'}\right)\Big|_{x=x_0}, \tag{1.119}$$

where

$$\sigma_{t,\varepsilon} = \int_{x_0-\varepsilon}^{x_0+\varepsilon} t\varphi_\varepsilon(x-x_0)\,dx$$

is the area under the bell $t\varphi_\varepsilon(x-x_0)$. If (1.119) holds, then the limit on the left side is called the *variational derivative* of F at y and is denoted by

$$\frac{\delta F}{\delta y}\Big|_{x=x_0} = \lim_{\sigma_{t,\varepsilon} \to 0} \frac{\Delta F(y_0(x), \varphi_\varepsilon(x-x_0))}{\sigma_{t,\varepsilon}} = \left(f_y - \frac{d}{dx}f_{y'}\right)\Big|_{x=x_0}.$$

In this case, we obtain the relation for the increment

$$F(y(x) + t\varphi_\varepsilon(x-x_0)) - F(y(x)) = \left(\frac{\delta F}{\delta y}\Big|_{x=x_0} + \beta\right)\sigma_{t,\varepsilon} \tag{1.120}$$

where $\beta \to 0$ when $\sigma_{t,\varepsilon} \to 0$, or

$$\begin{aligned}\Delta F(y(x), t\varphi_\varepsilon(x-x_0)) &\equiv F(y(x)+t\varphi_\varepsilon(x-x_0)) - F(y(x)) \\ &= \left(f_y - \frac{d}{dx}f_{y'} + \beta\right)\bigg|_{x=x_0} \sigma_{t,\varepsilon} \end{aligned} \tag{1.121}$$

which will be of use in § 1.10.

Now we consider the question of when the variational derivative of F exists. It turns out that the limit exists only if, together with $\varepsilon \to 0$, we take $t \to 0$ in some specific relationship to the change in ε. In particular, we can take $t = \varepsilon^3$. Indeed, let us start with (1.112), which we rewrite for $\varphi = t\phi_\varepsilon(x - x_0)$ as follows:

$$\begin{aligned}&|\Delta F(y(x), t\varphi_\varepsilon(x-x_0)) - \delta F(y(x), t\varphi_\varepsilon(x-x_0))| \\ &\qquad < Ct^2 \int_{x_0-\varepsilon}^{x+\varepsilon} \left(\varphi_\varepsilon^2(x-x_0) + {\varphi_\varepsilon'}^2(x-x_0)\right) dx\end{aligned}$$

with some constant C. By (1.118),

$$\delta F(y(x), t\varphi_\varepsilon(x-x_0)) = \left[\left(f_y - \frac{d}{dx}f_{y'}\right)\bigg|_{x=x_0} + \alpha\right] t\sigma_\varepsilon$$

with $\alpha \to 0$ as $\varepsilon \to 0$. From these it is seen that, to prove (1.121), it suffices to find a dependence of t on ε such that

$$\frac{1}{t\sigma_\varepsilon} t^2 \int_{x_0-\varepsilon}^{x_0+\varepsilon} \left(\varphi_\varepsilon^2(x-x_0) + {\varphi_\varepsilon'}^2(x-x_0)\right) dx \to 0$$

as $\varepsilon \to 0$. To show that a workable dependence is $t = \varepsilon^3$, we must calculate a few integrals. It is sufficient to put $x_0 = 0$. The change of variables $x = \varepsilon u$ gives

$$t\sigma_\varepsilon = t\varepsilon K_1 \quad \text{where} \quad K_1 \equiv \int_{-1}^{1} \exp\left(\frac{1}{u^2-1}\right) du.$$

Observe that K_1 is a positive constant. Also

$${\varphi_\varepsilon'}^2(x) = \frac{4t^2\varepsilon^4x^2}{(x^2-\varepsilon^2)^4} \exp\left(\frac{2\varepsilon^2}{x^2-\varepsilon^2}\right)$$

and we obtain

$$\int_{x_0-\varepsilon}^{x_0+\varepsilon} (\varphi_\varepsilon^2(x) + {\varphi_\varepsilon'}^2(x))\, dx = K_2t^2\varepsilon + K_3\frac{t^2}{\varepsilon}$$

where K_2 and K_3 are the positive constants

$$K_2 = \int_{-1}^{1} \exp\left(\frac{2}{u^2-1}\right) du, \qquad K_3 = \int_{-1}^{1} \frac{4u^2}{(u^2-1)^4} \exp\left(\frac{2}{u^2-1}\right) du.$$

Hence when $t = \varepsilon^3$ we have

$$\frac{1}{t\sigma_\varepsilon}\int_{x_0-\varepsilon}^{x_0+\varepsilon}(\varphi_\varepsilon^2 + {\varphi'_\varepsilon}^2)\,dx = \frac{K_2}{K_1}t + \frac{K_3}{K_1}\frac{t}{\varepsilon^2} < K_4\varepsilon,$$

where K_4 is a constant which, for small ε, is less than $(K_2 + K_3)/K_1$. This completes the proof.

Brief review of important ideas

The *increment* $F(y + \varphi) - F(y)$ of the functional $F(y)$ can be written as

$$F(y + \varphi) - F(y) = \delta F(y, \varphi) + O(\|\varphi\|^2_{C^{(1)}(a,b)}) \tag{1.122}$$

where the *first variation*

$$\delta F(y, \varphi) = \int_a^b [f_y(x, y, y')\varphi + f_{y'}(x, y, y')\varphi']\,dx \tag{1.123}$$

is the principal part (i.e., the portion of the increment that is linear in φ).

We have

$$\delta F(y, \varphi) = 0 \tag{1.124}$$

when $y = y(x)$ is a minimizer of $F(y)$ for some given boundary conditions; this holds for any admissible increment φ of the function y. A functional is said to be *stationary* at y if its first variation vanishes.

The idea of the *variational derivative* is analogous to the idea of a partial derivative of a multivariable function. We define the variational derivative of a functional $F(y)$, at a point x_0, for a curve $y = y(x)$, as follows. We give $y(x)$ an increment which is nonzero only in a small neighborhood of x_0; we choose a small bell-shaped bump $t\varphi_\varepsilon(x - x_0)$, and denote the area between it and the x-axis by σ_ε. We then get the main linear (with respect to t) part δF of the increment ΔF under this special type of localized disturbance. By continuity of the Euler expression $f_y - \frac{d}{dx}f_{y'}$ we can approximate δF as the Euler expression times $t\sigma_\varepsilon$. Then we prove that for $t = \varepsilon^3$ and $t\sigma_\varepsilon \to 0$ the expression $\Delta F/t\sigma_\varepsilon S$ has the same limit as $\delta F/t\sigma_\varepsilon S$. In this way we define the variational derivative given by

$$\left.\frac{\delta F}{\delta y}\right|_{x=x_0} = \left.\left(f_y - \frac{d}{dx}f'_y\right)\right|_{x=x_0}. \tag{1.125}$$

If y is a minimizer of F, then

$$\left.\frac{\delta F}{\delta y}\right|_{x=x_0} = 0$$

for each $x_0 \in (a, b)$, which is the Euler equation.

1.10 Isoperimetric Problems

We have found a way (1.125) of obtaining the Euler equation by setting the variational derivative to zero. Let us apply this to the solution of an *isoperimetric problem.*

It is said that the first problem of this type was solved practically by Dido, legendary queen of ancient Carthage, who was offered as much land as she could surround with the skin of a bull. Using a fuzzy formulation of this "mathematical" problem, she cut the skin into thin bands, tied them end to end, and surrounded the town with this long "rope." Note that Dido's problem was quite hard; several issues had to be addressed, including (1) how to get the longest rope from the skin, (2) how to find the closed curve of a given length that would enclose the greatest planar area, and (3) how to choose the most desirable piece of land. We can only treat the second of these issues here. Let us begin by formulating the

Simplest Isoperimetric Problem. Find the minimum of the functional

$$F(y) = \int_a^b f(x, y, y')\, dx \tag{1.126}$$

from among the functions $y \in C^{(1)}(a, b)$ that satisfy

$$y(a) = c_0, \qquad y(b) = c_1, \tag{1.127}$$

and

$$G(y) = \int_a^b g(x, y, y')\, dx = l \tag{1.128}$$

where l is a given number.

Condition (1.128) is analogous to the condition that the length of a curve is given. We know a similar problem from calculus: given a restriction $g(x) = c$, find a minimum of $f(x)$. This is solved using Lagrange multipliers: there is a constant λ such that a minimizer of the problem is a stationary point of the function $f(x) + \lambda g(x)$ — that is, a solution of the equation $f'(x) + \lambda g'(x) = 0$. We correctly surmise that something similar should exist for the isoperimetric problem.

Note that our previous technique cannot be used because the restriction (1.128) has complicated the notion of the neighborhood of a function.

Indeed, if $g(x, y, y')$ is not linear in y and y' then we cannot expect that a sum of two admissible small increments of a minimizer is also admissible: condition (1.128) can fail for the sum. The same comment applies to increments of the form $t\varphi$ if φ is an admissible increment. However, the technique of § 1.9 does not depend on such transformations in the set of admissible increments, so we will try to use it.

Theorem 1.34. *Let $y = y(x)$ be a local solution of the Simplest Isoperimetric Problem, and suppose y is not an extremal of the functional $G(z)$. Then there is a real number λ such that $y = y(x)$ is an extremal of $F(z) + \lambda G(z)$ on the set of functions from $C^{(1)}(a, b)$ satisfying* (1.127).

The problem of finding this extremal is well defined in principle. A solution of the Euler equation for $F(z) + \lambda G(z)$ should have three independent constants: λ, and the two independent constants expected in the general solution of the (second-order) Euler equation. These can be determined from (1.128) and (1.127).

Proof. We will try the results of § 1.9. We must consider the set of small increments of the minimizer such that the incremented functions satisfy both (1.127) and (1.128). So we construct the set of increments by combining two bell-shaped functions of the class B_0 with centers of symmetry at x_1 and x_2, $x_1 < x_2$: that is, $A_i\varphi_{\varepsilon_i}(x - x_i)$, $|A_i| = \varepsilon_i^3$, $i = 1, 2$. Denote this increment by $\eta(x) = \sum_i A_i\varphi_{\varepsilon_i}(x - x_i)$. We can assume that $\varepsilon_i < (x_2 - x_1)/2$, so the two nonzero domains of such an increment do not intersect (or we could argue that we produced two bell-shaped increments of y at different points successively). Since the supports of the two bell-shaped functions do not intersect we can extend (1.120) to this case:

$$\Delta F(y, \eta) = \left[\left(f_y - \frac{d}{dx} f_{y'} \right) \bigg|_{x=x_1} + \alpha_1 \right] \sigma_{\varepsilon_1} + \left[\left(f_y - \frac{d}{dx} f_{y'} \right) \bigg|_{x=x_2} + \alpha_2 \right] \sigma_{\varepsilon_2} \tag{1.129}$$

where for $i = 1, 2$ we have

$$\sigma_{\varepsilon_i} = A_i \int_{x_i - \varepsilon}^{x_i + \varepsilon} \varphi_{\varepsilon_i}(x - x_i)\, dx, \qquad |A_i| = \varepsilon_i^3,$$

and $\alpha_i \to 0$ when $\sigma_{\varepsilon_i} \to 0$.

We must choose the increment η so that $y + \eta$ satisfies (1.128). Thus $G(y + \eta) - G(y) = 0$. This and the analogue of (1.121) for $G(y + \eta) - G(y)$

imply

$$\left[\left(g_y - \frac{d}{dx}g_{y'}\right)\Big|_{x=x_1} + \beta_1\right]\sigma_{\varepsilon_1} + \left[\left(g_y - \frac{d}{dx}g_{y'}\right)\Big|_{x=x_2} + \beta_2\right]\sigma_{\varepsilon_2} = 0$$

with the same σ_{ε_i} as in (1.129) and $\beta_i \to 0$ when $\sigma_{\varepsilon_i} \to 0$.

Since $y = y(x)$ is not an extremal of $G(z)$, there is a point $x_2 \in (a, b)$ where $g_y - \frac{d}{dx}g_{y'} \neq 0$. For sufficiently small ε_2 we get β_2 as small as desired, thus the second square bracket is nonzero in this case and so

$$\sigma_{\varepsilon_2} = -\frac{\left(g_y - \frac{d}{dx}g_{y'}\right)\Big|_{x=x_1} + \beta_1}{\left(g_y - \frac{d}{dx}g_{y'}\right)\Big|_{x=x_2} + \beta_2}\sigma_{\varepsilon_1}.$$

Then

$$\begin{aligned}\Delta F(y,\eta) = &\left[\left(f_y - \frac{d}{dx}f_{y'}\right)\Big|_{x=x_1} + \alpha_1\right]\sigma_{\varepsilon_1} \\ &- \left[\left(f_y - \frac{d}{dx}f_{y'}\right)\Big|_{x=x_2} + \alpha_2\right]\frac{\left(g_y - \frac{d}{dx}g_{y'}\right)\Big|_{x=x_1} + \beta_1}{\left(g_y - \frac{d}{dx}g_{y'}\right)\Big|_{x=x_2} + \beta_2}\sigma_{\varepsilon_1}.\end{aligned} \tag{1.130}$$

Denoting

$$\lambda = -\frac{\left(f_y - \frac{d}{dx}f_{y'}\right)\Big|_{x=x_2}}{\left(g_y - \frac{d}{dx}g_{y'}\right)\Big|_{x=x_2}}$$

we get from (1.130)

$$\Delta F(y,\eta) = \left[\left(f_y - \frac{d}{dx}f_{y'}\right)\Big|_{x=x_1} + \lambda\left(g_y - \frac{d}{dx}g_{y'}\right)\Big|_{x=x_1}\right]\sigma_{\varepsilon_1} + o(|\sigma_{\varepsilon_1}|).$$

The first variation of the functional that must vanish on the solution is

$$\delta F(y,\eta) = \left[\left(f_y - \frac{d}{dx}f_{y'}\right)\Big|_{x=x_1} + \lambda\left(g_y - \frac{d}{dx}g_{y'}\right)\Big|_{x=x_1}\right]\sigma_{\varepsilon_1} = 0.$$

Since we can choose σ_{ε_1} arbitrarily, it follows that for any $x_1 \in (a, b)$ we have

$$\left(f_y - \frac{d}{dx}f_{y'}\right)\Big|_{x=x_1} + \lambda\left(g_y - \frac{d}{dx}g_{y'}\right)\Big|_{x=x_1} = 0.$$

This means $y = y(x)$ is an extremal of $F + \lambda G$. □

For an isoperimetric problem where the functional F depends on a vector function $\mathbf{y} = (y_1, \ldots, y_n)$ and there are m restrictions of integral type

$$G_i = \int_a^b g_i(x, \mathbf{y}, \mathbf{y}')\, dx, \qquad i = 1, \ldots, k,$$

there is a corresponding statement. For this problem a minimizer $\mathbf{y}$ is an extremal of the functional $F + \sum_{i=1}^{k} \lambda_k G_i$. The reader can derive the corresponding Euler equations. It is clearly impossible to satisfy k integral restrictions for $\mathbf{y}$ considering only the two-belled increments, so here it is necessary to introduce increments composed of $k+1$ bell-shaped functions. This requires additional technical work.

Two problems

Let us consider two special problems. The first was mentioned in § 1.1: find the plane curve enclosing the maximum possible area for a given perimeter. One approach is to examine all curves $y(x)$ that, except for their endpoints, lie in the upper half of the xy-plane, and that have endpoints $(\pm a, 0)$ and a given length l. (Note that a is not specified in advance.) In the notation of Theorem 1.34 we have

$$F(y) = \int_{-a}^{a} y\, dx, \qquad G(y) = \int_{-a}^{a} \sqrt{1 + (y')^2}\, dx;$$

hence

$$f(x, y, y') = y, \qquad g(x, y, y') = \sqrt{1 + (y')^2},$$

and $f + \lambda g$ does not depend on x explicitly. So we can write

$$(f + \lambda g) - (f + \lambda g)_{y'} y' = y + \lambda \sqrt{1 + (y')^2} - \frac{\lambda (y')^2}{\sqrt{1 + (y')^2}} = c_1,$$

which simplifies to

$$y - c_1 = \frac{-\lambda}{\sqrt{1 + (y')^2}}.$$

Put

$$y' = \frac{dy}{dx} = \tan t \tag{1.131}$$

where t is a parameter; then

$$y - c_1 = \frac{-\lambda}{\sqrt{1 + \tan^2 t}} = \frac{-\lambda}{\sec t} = -\lambda \cos t. \tag{1.132}$$

Now from (1.131) and (1.132)

$$dx = \frac{1}{\tan t}\,dy = \frac{1}{\tan t}\frac{dy}{dt}\,dt = \frac{1}{\tan t}\lambda \sin t\,dt = \lambda \cos t\,dt$$

so that upon integration we have $x = \lambda \sin t + c_2$. From the equations

$$x - c_2 = \lambda \sin t, \qquad y - c_1 = -\lambda \cos t,$$

we may eliminate t to produce

$$(x - c_2)^2 + (y - c_1)^2 = \lambda^2.$$

Thus all extremals of $F(y) + \lambda G(y)$ are portions of a circle. The conditions

$$(-a - c_2)^2 + (0 - c_1)^2 = \lambda^2, \qquad (a - c_2)^2 + (0 - c_1)^2 = \lambda^2,$$

may be subtracted to show that $c_2 = 0$. The vertical shift c_1 of the center and the radius λ clearly depend on the given l. The reader can verify directly that a maximum has been obtained.

Another approach is to use polar coordinates. Calling these (r, ϕ) and placing the coordinate origin inside the desired closed curve $r = r(\phi)$, we have

$$f + \lambda g = \frac{1}{2}r^2 + \lambda\sqrt{r^2 + (r')^2}$$

and the corresponding Euler equation

$$r + \frac{\lambda r}{\sqrt{r^2 + (r')^2}} - \frac{d}{d\phi}\frac{\lambda r'}{\sqrt{r^2 + (r')^2}} = 0.$$

Differentiation and simplification give

$$\frac{1}{\lambda} = \frac{rr'' - 2(r')^2 - r^2}{[r^2 + (r')^2]^{3/2}},$$

which shows that the curvature of $r(\phi)$ is a constant $1/\lambda$ and yields a circle again.

It is worth noting that we formulated the problems for a minimum but solved for a maximum. This is analogous to the standard calculus trick of maximizing a function f by minimizing $-f$. Of even more interest is the idea of obtaining a *dual problem* by reversing the roles of the functionals F and G. For example, the maximum area that can be enclosed by a curve having length l is $l^2/4\pi$. The dual problem is to find a closed curve of minimum length that borders a flat domain with area $l^2/4\pi$. Of course, the solution is a circle having circumference l.

We now turn to another classical isoperimetric problem. Early in the development of mathematics people became curious about the precise form assumed by a chain hanging from both ends (such chains were used, for instance, as "fences" along the sides of bridges). This is a hard problem if one wishes to consider it in full detail (including friction, nonuniformities in the individual links, and so on); it is possible to show that many peculiarities arise, and even the full setup of the problem is quite cumbersome. A successful approach depended on the construction of a tractable model for the chain. First an ideal chain was introduced, consisting of extremely small elements that were all identical; this permitted the tools of calculus to be applied. An even simpler model was a uniform filamentary rope — heavy, flexible, and absolutely unstretchable. Unlike a chain, such an idealized rope could lie in a plane.

Let us therefore suppose that a uniform, flexible rope of a given fixed length hangs in equilibrium with its ends attached to two fixed points: what is the shape assumed by the rope? Denote by l the length of the rope, assume it has a unit mass density, and let the endpoints be (a, h_a) and (b, h_b). Clearly we need $b - a \leq l$. The y coordinate of the center of gravity is proportional to the integral $\int_a^b y(s)\,ds$ where s is arc length along the rope; since the center of gravity will find the lowest possible position, we are led to minimize the functional ($ds = \sqrt{1+(y')^2}\,dx$)

$$F(y) = \int_a^b y\sqrt{1+(y')^2}\,dx$$

subject to the side condition

$$G(y) = \int_a^b \sqrt{1+(y')^2}\,dx = l.$$

Accordingly we minimize

$$F(y) + \lambda G(y) = \int_a^b (y+\lambda)\sqrt{1+(y')^2}\,dx.$$

Since the integrand does not depend on x explicitly, we write out the first integral of the differential equation,

$$(y+\lambda)\sqrt{1+(y')^2} - \frac{(y+\lambda)(y')^2}{\sqrt{1+(y')^2}} = c_1,$$

and then simplify to obtain

$$y + \lambda = c_1\sqrt{1+(y')^2}.$$

We find a parametric representation of the solution, introducing a parameter t by the substitution $y' = \sinh t$. Then

$$y + \lambda = c_1 \cosh t$$

and the dependence of x on t is

$$dx = \frac{1}{\sinh t}\,dy = \frac{1}{\sinh t}\frac{dy}{dt}\,dt = \frac{1}{\sinh t}(c_1 \sinh t)\,dt = c_1 dt$$

so that

$$x - c_2 = c_1 t.$$

Elimination of t leads to the equation of a catenary:

$$y + \lambda = c_1 \cosh\left(\frac{x - c_2}{c_1}\right).$$

The given conditions can be used to determine c_1, c_2, and λ. (Of course $c_2 = 0$ if $b = -a$.)

Once again we do not provide formal verification that a minimum has actually been obtained. Indeed, with many problems that arise from geometry or physics it is intuitively clear whether we have the desired solution. For the hanging chain problem, we can assert on physical grounds that a solution exists; since the solution we obtained is unique, we can rest assured that it is the desired one.

It is possible to state other types of minimum problems with restrictions which, for their solution, require a technique similar to that of Lagrange multipliers. For example, it is possible to pose a problem of minimizing the functional $\int_{x_0}^{x_1} f(x, y, z, y', z')\,dx$ under some boundary conditions when there is a restriction $g(x, y, z) = 0$ (in more advanced books this is called minimizing a functional on a manifold). Here a minimizer is an extremal of a functional $\int_a^b [f - \lambda(x)g]\,dx$ without integral restrictions imposed by g, and $\lambda(x)$ is a new unknown function that is treated as given when we compose the Euler equations. Of course to define it one must use the equation $g(x, y, z) = 0$. Some problems in mechanics involve restrictions of even more general type; e.g., $g(x, y, z, y', z') = 0$.

Quick summary

We have concentrated on an isoperimetric problem of the following general form: find the minimizer of the simplest integral functional from among those functions y that satisfy

$$y(a) = c_0, \qquad y(b) = c_1, \qquad G(y) = \int_a^b g(x, y, y')\,dx = l$$

where $G(y)$ and l are given. A solution method is to introduce a real number λ (analogous to a Lagrange multiplier) and seek to minimize the functional $F + \lambda G$ subject to the given endpoint conditions on y.

1.11 General Form of the First Variation

We would like to consider the minimization problem for functionals of the form (1.33) when the endpoints of integration can change.

We have seen for various functionals that at a point of minimum the first variation is zero. Let us demonstrate this in general. First let us introduce some notions. In subsequent chapters we shall use the notion of a normed space; now we quote only the definition. A normed space is a linear space of elements x such that for each x a function called the *norm* $\|x\|$ is defined. The norm must possess the following three properties:

(i) for any x, $\|x\| \geq 0$; $\|x\| = 0$ if and only if $x = 0$;
(ii) $\|\lambda x\| = |\lambda|\ \|x\|$ for any real number λ;
(iii) $\|x + y\| \leq \|x\| + \|y\|$.

The third property is called the triangle inequality. For example, the norm (1.37) for functions in $C^{(1)}(a,b)$ satisfies the above properties.

We can define a functional on a general normed space. A functional on a normed space X is a function that takes values in $\mathbb{R}$; i.e., to any $x \in X$ there corresponds no more than one real number. A functional $\Phi(x)$ is *linear* if for any x, y belonging to its domain and any real λ, μ,

$$\Phi(\lambda x + \mu y) = \lambda \Phi(x) + \mu \Phi(y). \tag{1.133}$$

Finally, a linear functional $\Phi(x)$ is continuous in X if there is a constant c such that for any $x \in X$,

$$|\Phi(x)| \leq c \, \|x\| \, . \tag{1.134}$$

The infimum of all such c is called the *norm* of Φ and is denoted $\|\Phi\|$ (it is actually a norm according to the norm properties listed above).

Let $F(x)$ be a functional on X, and assume that in some ball about a point $x \in X$ (a ball is a set of elements $x + \delta x \in X$, where $\delta x \in X$, such that $\|\delta x\| \leq \varepsilon$ for some $\varepsilon > 0$) there is a representation

$$F(x + \delta x) - F(x) = \delta F(x, \delta x) + o(\|\delta x\|) \tag{1.135}$$

where $\delta F(x, \delta x)$ is a linear functional continuous in δx. We have called it the first variation of $F(x)$, but it also has another name: the *Fréchet differential* of $F(x)$ at x. Hence we have extended the definition of the first variation to abstract functionals.

Let x be a local minimizer of F: that is, $F(x + \delta x) - F(x) \geq 0$ for any $\|\delta x\| \leq \varepsilon$ with some $\varepsilon > 0$.

Theorem 1.35. *Let x be a minimizer of F on the set of elements $\{x+\delta x \mid \|\delta x\| \leq \varepsilon\}$, and suppose F has the first variation at x such that (1.135) holds on this set. Then $\delta F(x, \delta x) = 0$.*

Proof. Suppose to the contrary there exists an $x^* \in X$ such that $\delta F(x, x^*) \neq 0$. Then for small enough t we have

$$0 \leq F(x + tx^*) - F(x) = \delta F(x, tx^*) + o(t \|x^*\|) = t\, \delta F(x, x^*) + o(t).$$

For small $|t|$ the difference on the left is determined by the first term on the right. Choosing an appropriate t we get $t\, \delta F(x, x^*) < 0$, which contradicts the leftmost inequality. □

Thus for a problem of minimum of a functional, as a first step, we have to derive its first variation, equate it to zero, and then find solutions of this equation for any admissible disturbances (or virtual variations) δx.

We return to the beginning of this section and claim again that we would like to consider a minimization problem for a more general functional than (1.33), i.e., the functional

$$F(y) = \int_{x_0}^{x_1} f(x, y, y')\, dx \tag{1.136}$$

where the endpoints x_0 and x_1 can move. Thus we need the expression for the first variation in this case. To realize the above idea we must suppose that all changes are of the same order of smallness. Here we have not only a change φ in y to consider, but also changes δx_0 and δx_1 of the ends x_0 and x_1 respectively. Since δx_0 and δx_1 are arbitrary and we could have $\delta x_0 < 0$ or $\delta x_1 > 0$, we must agree on a way of extending a given function to points outside the segment $[x_0, x_1]$. We do this by linear extrapolation, using the tangent lines to $y = y(x)$ at x_0 and x_1 to define the values of the extension. The ends of the extended curve have coordinates $(x_0 + \delta x_0, y_0 + \delta y_0)$ and $(x_1 + \delta x_1, y_1 + \delta y_1)$.

Our problem is to derive the linear part of the increment for (1.136) when φ, φ', δx_0, δy_0, δx_1, and δy_1 have the same order of smallness; that is, to extract the part of the increment that is linear in each of these quantities. Denote

$$\varepsilon = \|\varphi\|_{C^{(1)}(x_0,x_1)} + |\delta x_0| + |\delta y_0| + |\delta x_1| + |\delta y_1|.$$

The increment is

$$\Delta F(y) = \int_{x_0+\delta x_0}^{x_1+\delta x_1} f(x, y+\varphi, y'+\varphi')\, dx - \int_{x_0}^{x_1} f(x, y, y')\, dx.$$

The first integral can be decomposed as

$$\int_{x_0+\delta x_0}^{x_1+\delta x_1} (\cdots)\, dx = \int_{x_0}^{x_1} (\cdots)\, dx + \int_{x_1}^{x_1+\delta x_1} (\cdots)\, dx - \int_{x_0}^{x_0+\delta x_0} (\cdots)\, dx.$$

Recall that all the functions $y = y(x)$, $\varphi = \varphi(x)$, are linearly extrapolated outside $[x_0, x_1]$, preserving continuity of the functions and their first derivatives. Thus

$$\begin{aligned}\Delta F(y) = &\int_{x_0}^{x_1} [f(x, y+\varphi, y'+\varphi') - f(x, y, y')]\, dx \\ &+ \int_{x_1}^{x_1+\delta x_1} f(x, y+\varphi, y'+\varphi')\, dx \\ &- \int_{x_0}^{x_0+\delta x_0} f(x, y+\varphi, y'+\varphi')\, dx. \qquad (1.137)\end{aligned}$$

The integral over $[x_0, x_1]$ can be transformed in the usual manner:

$$\begin{aligned}&\int_{x_0}^{x_1} [f(x, y+\varphi, y'+\varphi') - f(x, y, y')]\, dx \\ &\quad = \int_{x_0}^{x_1} \left[f_y(x, y, y') - \frac{d}{dx} f_{y'}(x, y, y') \right] \varphi\, dx \\ &\qquad + f_{y'}(x, y(x), y'(x))\varphi(x) \Big|_{x=x_0}^{x=x_1} + o(\varepsilon).\end{aligned}$$

Let us represent φ at the endpoints using δy_0 and δy_1.

Fig. 1.2 shows that

$$\varphi(x_1) = \delta y_1 - y'(x_1)\delta x_1 + o(\varepsilon). \qquad (1.138)$$

Similarly,

$$\varphi(x_0) = \delta y_0 - y'(x_0)\delta x_0 + o(\varepsilon). \qquad (1.139)$$

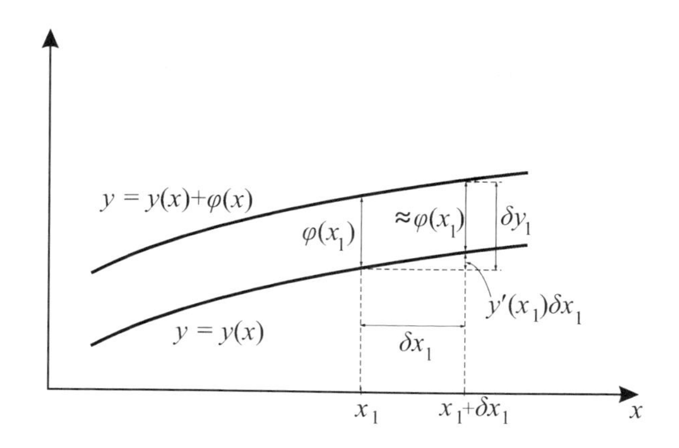

Fig. 1.2 Quantities appearing in equations (1.138) and (1.139).

Thus

$$\begin{aligned}
&\int_{x_0}^{x_1} [f(x, y+\varphi, y'+\varphi') - f(x, y, y')]\, dx \\
&\quad = \int_{x_0}^{x_1} \left[f_y(x, y, y') - \frac{d}{dx} f_{y'}(x, y, y') \right] \varphi \, dx \\
&\qquad + f_{y'}(x_1, y(x_1), y'(x_1))\delta y_1 - f_{y'}(x_0, y(x_0), y'(x_0))\delta y_0 \\
&\qquad - [f_{y'}(x_1, y(x_1), y'(x_1)) y'(x_1)\delta x_1 \\
&\qquad - f_{y'}(x_0, y(x_0), y'(x_0)) y'(x_0)\delta x_0] + o(\varepsilon).
\end{aligned}$$

Now consider the two other terms for ΔF in (1.137). Extracting the terms of the first order of smallness in ε we have

$$\begin{aligned}
\int_{x_1}^{x_1+\delta x_1} f(x, y+\varphi, y'+\varphi')\, dx &= \int_{x_1}^{x_1+\delta x_1} f(x, y, y')\, dx + o(\varepsilon) \\
&= f(x_1, y(x_1), y'(x_1))\delta x_1 + o(\varepsilon)
\end{aligned}$$

and similarly

$$\int_{x_0}^{x_0+\delta x_0} f(x, y+\varphi, y'+\varphi')\, dx = f(x_0, y(x_0), y'(x_0))\delta x_0 + o(\varepsilon).$$

Collecting terms we have

$$\begin{aligned}\Delta F = &\int_{x_0}^{x_1} \left[f_y(x,y,y') - \frac{d}{dx} f_{y'}(x,y,y') \right] \varphi\, dx \\ &+ f_{y'}(x_1, y(x_1), y'(x_1))\delta y_1 - f_{y'}(x_0, y(x_0), y'(x_0))\delta y_0 \\ &+ [f(x_1, y(x_1), y'(x_1)) - f_{y'}(x_1, y(x_1), y'(x_1))y'(x_1)]\delta x_1 \\ &- [f(x_0, y(x_0), y'(x_0)) - f_{y'}(x_0, y(x_0), y'(x_0))y'(x_0)]\delta x_0 + o(\varepsilon).\end{aligned}$$

So the following is the general form of the first variation of the functional when the ends of the curve can move:

$$\delta F = \int_{x_0}^{x_1} \left(f_y - \frac{d}{dx} f_{y'} \right) \varphi\, dx + f_{y'} \delta y \Big|_{x_0}^{x_1} + (f - y' f_{y'})\, \delta x \Big|_{x_0}^{x_1}. \qquad (1.140)$$

The reader can demonstrate that for a functional

$$F(\mathbf{y}) = \int_{x_0}^{x_1} f(x, \mathbf{y}, \mathbf{y}')\, dx$$

with movable boundaries, the general form of the first variation is

$$\delta F = \sum_{i=1}^{n} \int_{x_0}^{x_1} \left(f_{y_i} - \frac{d}{dx} f_{y_i'} \right) \varphi_i\, dx + \sum_{i=1}^{n} f_{y_i'} \delta y_i \Big|_{x_0}^{x_1} + \left(f - \sum_{i=1}^{n} y_i' f_{y_i'} \right) \delta x \Big|_{x_0}^{x_1}.$$

1.12 Movable Ends of Extremals

In the previous section we found the general form (1.140) of the first variation of a functional when the boundaries of integration can move. Note that when the boundaries are fixed then $\delta x_i = 0$ and (1.140) reduces to the left side of (1.55). Thus in this case the equation $\delta F = 0$ for a minimizer gives us the Euler equation and natural boundary conditions. The problem with natural boundary conditions can be reformulated as follows: given two vertical lines $x = a$ and $x = b$, find a minimizer of the functional (1.33) that starts on the line $x = a$ and ends on the line $x = b$ (or that connects these lines).

This formulation suggests that by using (1.140) it is possible to find equations to solve the following problem.

> Given two curves $y = \psi_0(x)$ and $y = \psi_1(x)$, find a minimizer of (1.33) that starts on $\psi_0(x)$ and ends on $\psi_1(x)$.

Let us call this the "problem with movable boundaries."

We assume any other functions of interest are defined (and twice continuously differentiable) wherever the boundary functions $\psi_i(x)$ are given. (If these latter functions are not defined on the same interval, we construct an interval that encompasses all points of interest and assume that everything is defined on this larger interval.) Moreover we assume the endpoints of the minimizer are not endpoints of the graph for the $\psi_i(x)$.

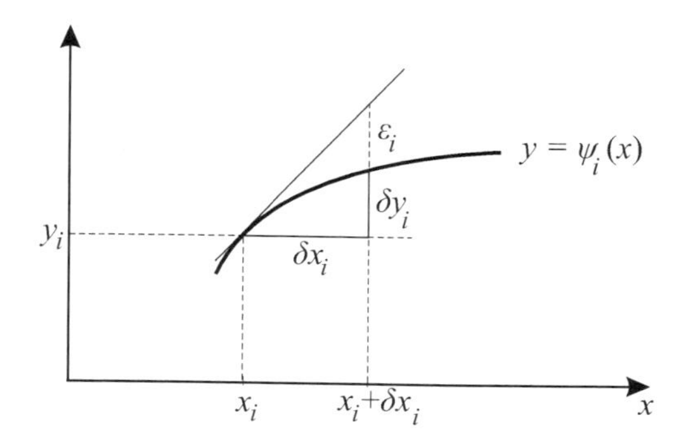

Fig. 1.3 Quantities near movable end of an extremal.

So we start with

$$\delta F = \int_{x_0}^{x_1} \left(f_y - \frac{d}{dx} f_{y'} \right) \varphi\, dx + f_{y'} \delta y_i \Big|_{x_0, i=0}^{x_1, i=1} + (f - y' f_{y'})\, \delta x_i \Big|_{x_0, i=0}^{x_1, i=1} . \tag{1.141}$$

For admissible increments φ of a minimizer $y = y(x)$, the first variation of the functional is equal to zero. Although the expression δF above contains all the terms of the increment of the first order of smallness, it is not the first variation in the present case. Admissible φ now are those that are continuously differentiable and such that both

$$(x_0, y(x_0)) \quad \text{and} \quad (x_0 + \delta x_0, y(x_0 + \delta x_0) + \varphi(x_0 + \delta x_0))$$

belong to the curve $y = \psi_0(x)$, and both

$$(x_1, y(x_1)) \quad \text{and} \quad (x_1 + \delta x_1, y(x_1 + \delta x_1) + \varphi(x_1 + \delta x_1))$$

belong to the curve $y = \psi_1(x)$.

Consider Fig. 1.3. Here each δy_i ($i = 0$ or 1) and its corresponding δx_i are no longer independent; it is clear that for small δx_i we have

$$\delta y_i = \psi_i'(x_i)\delta x_i + \varepsilon_i, \qquad i = 0, 1$$

where the ε_i are of a higher order of smallness than δx_i and δy_i. Substituting this into the right side of (1.141), we select only the terms of the first order of smallness and get

$$\int_{x_0}^{x_1} \left(f_y - \frac{d}{dx} f_{y'} \right) \varphi \, dx + f_{y'} \psi_i' \delta x_i \bigg|_{x_0, i=0}^{x_1, i=1} + (f - y' f_{y'}) \, \delta x_i \bigg|_{x_0, i=0}^{x_1, i=1} .$$

This is the first variation of the functional (note that it is equal to δF in (1.141) only up to terms of the first order of smallness in the norm of the increment). Thus

$$\int_{x_0}^{x_1} \left(f_y - \frac{d}{dx} f_{y'} \right) \varphi \, dx + f_{y'} \psi_i' \delta x_i \bigg|_{x_0, i=0}^{x_1, i=1} + (f - y' f_{y'}) \, \delta x_i \bigg|_{x_0, i=0}^{x_1, i=1} = 0 \tag{1.142}$$

for all admissible φ.

Let us derive the consequences of this equation. First, from among the admissible increments $y = \varphi(x)$ we take only those which satisfy the conditions $\varphi(x_0) = \varphi(x_1) = 0$. For any such φ we have

$$\int_{x_0}^{x_1} \left(f_y - \frac{d}{dx} f_{y'} \right) \varphi \, dx = 0$$

and thus by the fundamental lemma the Euler equation

$$f_y - \frac{d}{dx} f_{y'} = 0$$

is satisfied on (x_0, x_1). Hence the integral in (1.142) vanishes for any admissible φ, and it follows that

$$(f + (\psi_i' - y') f_{y'}) \, \delta x_i \bigg|_{x_0, i=0}^{x_1, i=1} = 0. \tag{1.143}$$

Because we can "move" the ends of the curve independently, (1.143) implies two boundary conditions for the minimizer:

$$(f + (\psi_1' - y') f_{y'}) \big|_{x_1} = 0, \qquad (f + (\psi_0' - y') f_{y'}) \big|_{x_0} = 0. \tag{1.144}$$

For the problem under consideration the minimizing curve $y = y(x)$ satisfies conditions (1.144) which are an extension of the natural boundary conditions. The way in which the minimizer intersects the boundary curves $y = \psi_i(x)$ has a special name:

Definition 1.36. The curve $y = y(x)$ is *transversal* to the curves $y = \psi_i(x)$, $i = 0, 1$.

Let us analyze the setting of the boundary value problem in this case. There is the Euler equation whose solution is determined up to two unknown constants (it is not always so; in nonlinear equations the situation with constants is sometimes much more complex, but when we analyze the problem qualitatively we keep in mind the terms of the linear case). The two conditions (1.144) could define those constants, but they contain unknown quantities x_0 and x_1 so we need to find two more equations. They are $y(x_0) = \psi_0(x_0)$ and $y(x_1) = \psi_1(x_1)$, and thus the setup of the necessary conditions for $y = y(x)$ to be a minimizer is completed.

Example 1.37. Show that for functionals of the form

$$\int_{x_0}^{x_1} q(x, y)\sqrt{1 + (y')^2}\, dx$$

where $q(x, y) \neq 0$ at the endpoints x_0 and x_1, conditions (1.144) imply orthogonal intersections between $y(x)$ and the curves $\psi_0(x)$ and $\psi_1(x)$ at the points x_0 and x_1, respectively.

Solution. Take, for example, the condition $\left(f + (\psi_1' - y')f_{y'}\right)\Big|_{x_1} = 0$. Direct substitution and a bit of simplification give

$$\left(q(x, y)\frac{1 + \psi_1' y'}{\sqrt{1 + (y')^2}}\right)\Bigg|_{x_1} = 0.$$

If $q(x, y)|_{x_1} \neq 0$, then $(1 + \psi_1' y')|_{x_1} = 0$; i.e.,

$$y'|_{x_1} = -\frac{1}{\psi_1'|_{x_1}}.$$

The slopes are negative reciprocals, so y is orthogonal to ψ_1 at $x = x_1$. □

Quick review

The problem with movable boundaries for the simplest integral functional involves finding a minimizer that connects two given curves $y = \psi_0(x)$ and $y = \psi_1(x)$. We first solve the Euler equation, obtaining a solution in terms of two unknown constants. We then impose the transversality conditions

$$\left(f + (\psi_1' - y')f_{y'}\right)\Big|_{x_1} = 0, \tag{1.145}$$

$$\left(f + (\psi_0' - y')f_{y'}\right)\Big|_{x_0} = 0, \tag{1.146}$$

where x_1 and x_0 are also unknowns. After the use of $y(x_0) = \psi(x_0)$ and $y(x_1) = \psi(x_1)$, all constants should be determined.

Special cases: (1) If one of the ψ_i is a horizontal line, say $\psi_1(x) = \text{constant}$, then $\psi_1' \equiv 0$ and the corresponding transversality condition becomes

$$(f - y' f_{y'})\Big|_{x_1} = 0.$$

(2) If ψ_1 is a vertical line ($x = \text{constant}$) then $f_{y'}\Big|_{x_1} = 0$.

1.13 Broken Extremals: Weierstrass–Erdmann Conditions and Related Problems

We have required a minimizer $y = y(x)$ of (1.33) to assume given values at the endpoints of $[a, b]$. Is it possible to retain these conditions and also require that $y(x)$ assume a third given value at an interior point of $[a, b]$? That is, can we impose three conditions of the form $y(a) = c_0$, $y(b) = c_1$, and $y(\alpha) = c_2$ where $\alpha \in (a, b)$? If we require the minimizer to be in $C^{(1)}(a, b)$, then the answer is, in general, no: a solution of the second-order Euler equation cannot be made to satisfy three conditions at once. If we omit the condition of continuity of the minimizer at $x = \alpha$, the problem can be solvable in principle. However, in this case we can consider two separate problems of minimizing two functionals, one of which is given on $[a, \alpha]$ and the other on $[\alpha, b]$. So in this case we reduce the three-point problem to the two-point problem already considered.

With some problems it makes sense to assume that a minimizing curve has a finite number of points at which continuity of its derivative fails. We cannot appoint the position of such points on (a, b) in advance. It happens that at such points the *Weierstrass–Erdmann* conditions must be satisfied. Let us derive these, assuming the existence of one point of discontinuity of the first derivative of the minimizer. They will hold at every such point.

Suppose $x = \alpha$ is a point at which the first derivative of a minimizer is not continuous.

Theorem 1.38. *Let $x = \alpha \in (a, b)$ be a point at which the tangent to a minimizer $y = y(x)$ of the functional $\int_a^b f(x, y, y')\,dx$ has a break. Then y satisfies the Euler equation on the intervals (a, α) and (α, b), and at $x = \alpha$ the* ***Weierstrass–Erdmann conditions***

$$f_{y'}\Big|_{x=\alpha-0} = f_{y'}\Big|_{x=\alpha+0}, \tag{1.147}$$

$$(f - y' f_{y'})\Big|_{x=\alpha-0} = (f - y' f_{y'})\Big|_{x=\alpha+0}, \tag{1.148}$$

hold.

Before giving the proof, let us discuss how to state the corresponding boundary value problem. On each of the intervals (a, α) and (α, b) the minimizer satisfies the Euler equation. So in general the minimizer is determined up to four unknown constants. Also unknown is α. There are five conditions to determine these constants: the two boundary conditions at a and b, the conditions (1.147)–(1.148), and the continuity condition $y(\alpha - 0) = y(\alpha + 0)$. Thus in principle the boundary value problem is formulated properly.

Proof. Consider for definiteness the boundary conditions $y(a) = c_0$ and $y(b) = c_1$ for a minimizer. The minimizer should be continuous at $x = \alpha$. Perturbing the minimizer by an admissible φ and supposing that the point $(\alpha, y(\alpha))$ gets the increments $(\delta x, \delta y)$, we apply the general formula for the first variation

$$\int_{x_0}^{x_1} \left(f_y - \frac{d}{dx} f_{y'} \right) \varphi \, dx + f_{y'} \delta y \Big|_{x_0}^{x_1} + (f - y' f_{y'}) \, \delta x \Big|_{x_0}^{x_1} \tag{1.149}$$

twice, on each of intervals (a, α) and (α, b) separately, taking into account that the increment $(\delta x, \delta y)$ at $(\alpha, y(\alpha))$ is the same on the left and the right of α. Remembering that δx and δy are zero at $x = a$ and $x = b$ for all admissible increments, we have

$$\begin{aligned} \delta F &= \delta \left(\int_a^\alpha f(x, y, y') \, dx + \int_\alpha^b f(x, y, y') \, dx \right) \\ &= \int_a^\alpha \left(f_y - \frac{d}{dx} f_{y'} \right) \varphi \, dx + f_{y'} \delta y \Big|_{x=\alpha-0} + (f - y' f_{y'}) \, \delta x \Big|_{x=\alpha-0} \\ &\quad + \int_\alpha^b \left(f_y - \frac{d}{dx} f_{y'} \right) \varphi \, dx - f_{y'} \delta y \Big|_{x=\alpha+0} - (f - y' f_{y'}) \, \delta x \Big|_{x=\alpha+0} . \end{aligned}$$

Thus for all admissible increments

$$\begin{aligned} &\int_a^\alpha \left(f_y - \frac{d}{dx} f_{y'} \right) \varphi \, dx + \int_\alpha^b \left(f_y - \frac{d}{dx} f_{y'} \right) \varphi \, dx \\ &\quad + \left[f_{y'} \big|_{x=\alpha-0} - f_{y'} \big|_{x=\alpha+0} \right] \delta y \\ &\quad + \left[(f - y' f_{y'}) \big|_{x=\alpha-0} - (f - y' f_{y'}) \big|_{x=\alpha+0} \right] \delta x = 0. \end{aligned} \tag{1.150}$$

Now we choose certain classes of admissible increments φ to show that each term summed in (1.150) is equal to zero separately. Let us take first those admissible φ that are zero on $[\alpha, b]$. Also take $\delta x = \delta y = 0$. All terms except the first integral on the left are equal to zero identically now. Thus

$$\int_a^\alpha \left(f_y - \frac{d}{dx} f_{y'}\right) \varphi \, dx = 0$$

for all differentiable functions φ that equal zero at a and α. By the fundamental lemma the Euler equation

$$f_y - \frac{d}{dx} f_{y'} = 0$$

holds on (a, α). Because of this the first integral is zero not only for those φ that satisfy $\varphi(\alpha) = 0$, but for all admissible increments. A similar choice of those φ that are zero on $[a, \alpha]$ together with the assumption $\delta x = \delta y = 0$ brings us to similar conclusions: the minimizer y satisfies the Euler equation on (α, b) and so for all admissible φ we have

$$\int_\alpha^b \left(f_y - \frac{d}{dx} f_{y'}\right) \varphi \, dx = 0.$$

It follows that

$$\left[f_{y'}\Big|_{x=\alpha-0} - f_{y'}\Big|_{x=\alpha+0} \right] \delta y + \left[(f - y' f_{y'})\Big|_{x=\alpha-0} - (f - y' f_{y'})\Big|_{x=\alpha+0} \right] \delta x = 0$$

for all admissible δx and δy, hence we obtain (1.147) and (1.148). □

For the functional $\int_a^b f(x, \mathbf{y}, \mathbf{y}') \, dx$ depending on a vector function, at a discontinuity of a component y_i there are the similar conditions

$$f_{y_i'}\Big|_{x=\alpha-0} = f_{y_i'}\Big|_{x=\alpha+0}, \qquad (f - y_i' f_{y_i'})\Big|_{x=\alpha-0} = (f - y_i' f_{y_i'})\Big|_{x=\alpha+0}.$$

Indeed, when deriving the corresponding equation for the first variation of the functional, we can appoint the increments of all the components except y_i to be zero, so formally the corresponding equation does not differ from (1.150).

The Weierstrass–Erdmann conditions are similar in form to the natural boundary conditions for a functional. The idea of the proof of Theorem 1.38 can be applied to other types of problems.

Example 1.39. Consider the problem of minimizing the functional

$$\int_a^\beta f(x, y, y') \, dx + \int_\beta^b g(x, y, y') \, dx \tag{1.151}$$

where β is a fixed point of (a, b), and y is continuous on $[a, b]$, twice continuously differentiable on (a, β) and (β, b), and satisfies $y(a) = c_0$ and

$y(b) = c_1$. Assume the integrand is discontinuous at $x = \beta$, hence y has no continuous derivative there.

Solution. Problems of this form are frequent in physics, arising from spatial discontinuities. A specific instance of this is when a ray of light crosses the interface between two media. We are interested in how to appoint the conditions at such points, since the equation of propagation is not valid there. Variational tools can often supply us with such conditions. Let us demonstrate how this can happen.

For the functional (1.151) we need to derive the expression for the first variation and set it to zero for admissible increment-functions. For this we use (1.149) as above, but should take into account that β is fixed so that $\delta x = 0$ at $x = \beta$. The changes are evident:

$$\int_a^\beta \left(f_y - \frac{d}{dx} f_{y'} \right) \varphi \, dx + \int_\beta^b \left(g_y - \frac{d}{dx} g_{y'} \right) \varphi \, dx \\ + \left[f_{y'} \big|_{x=\beta-0} - g_{y'} \big|_{x=\beta+0} \right] \delta y = 0.$$

Thus in a similar fashion at $x = \beta$, in addition to the continuity condition $y(\beta - 0) = y(\beta + 0)$ we get $f_{y'}\big|_{x=\beta-0} = g_{y'}\big|_{x=\beta+0}$. □

Let us now consider a particular problem of the same nature with another type of functional. We seek the deflections under transverse load $q(x)$ of a system consisting of a cantilever beam with parameters E and I and whose free end connects with a string as shown in Fig. 1.4.

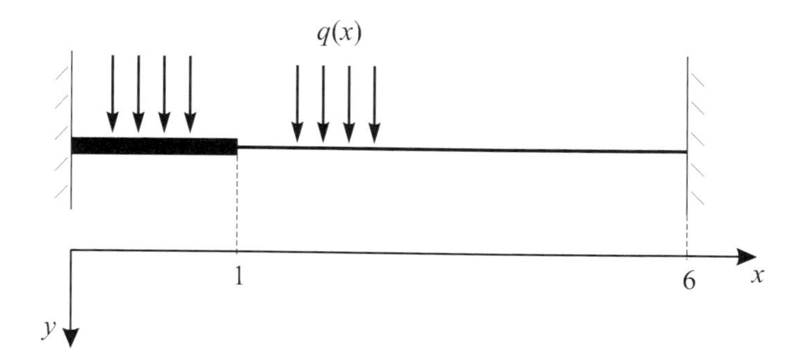

Fig. 1.4 A coupled mechanical system consisting of a beam and a string.

The models of a string and of a beam are of different natures; they are derived under different sets of assumptions, and the corresponding ordinary differential equations have different orders. It is clear that at the point of connection the function y describing the deflections must be continuous. However, we can imagine that the angles of inclination of the beam and the string can differ under certain loads; this means that we cannot require y' to be continuous at the point of coupling. What are the other conditions at this point? There are two ways to find them. One is to undertake a careful study of the theory of beams and strings and, understanding the mechanical meaning of each derivative at the point, to write out the conditions of equilibrium of the node (coupling unit). Another is to employ variational tools. Normally the latter is preferable, as it is less likely to yield incorrect conditions. We begin with the expression for total potential energy of the system: beam-string-load. We take the lengths of the beam and the string to be 1 m and 5 m, respectively. The stretching of the string is characterized by a parameter a:

$$E(y) = \frac{1}{2}\int_0^1 EI(y''(x))^2\,dx + \frac{a}{2}\int_1^6 (y'(x))^2\,dx - \int_0^6 q(x)y(x)\,dx.$$

We see from the figure that

$$y(0) = 0, \qquad y'(0) = 0, \qquad y(6) = 0.$$

Using tools developed earlier, we obtain the first variation

$$\delta E = \int_0^1 EIy''\varphi''\,dx + a\int_1^6 y'\varphi'\,dx - \int_0^6 q(x)\varphi(x)\,dx$$

of the energy functional. For all admissible functions that necessarily satisfy $\varphi(0) = 0$, $\varphi'(0) = 0$, and $\varphi(6) = 0$, we have

$$\delta E = 0.$$

Integration by parts gives

$$\int_0^1 EIy^{(4)}\varphi\,dx + EIy''\varphi'\Big|_{x=1-0} - EIy'''\varphi\Big|_{x=1-0}$$
$$- a\int_1^6 y''\varphi\,dx - ay'\varphi\Big|_{x=1+0} - \int_0^1 q\varphi\,dx - \int_1^6 q\varphi\,dx = 0.$$

We now reason as in the proof of Theorem 1.38. Putting $\varphi = 0$ on $[1, 6]$ and the "boundary" values $\varphi(1-0)$ and $\varphi'(1-0)$ equal to zero, we get

$$EIy^{(4)} - q = 0 \quad \text{on } (0, 1)$$

for the beam equation; similarly, we get

$$ay'' + q = 0 \quad \text{on } (1, 6)$$

for the string equation. Hence we deduce two additional boundary conditions at the point of connection:

$$EIy'''\big|_{x=1-0} = -ay'\big|_{x=1+0} \tag{1.152}$$

and

$$EIy''\big|_{x=1-0} = 0. \tag{1.153}$$

Condition (1.152) means that, at the connection point, the shear force in the beam is balanced by the vertical component of the tension force in the string. As the string cannot resist a torque, condition (1.153) states that the moment at this point of the beam is zero.

Such constructions consisting of elements of different natures are common in practice, and now the reader knows how to set up the corresponding boundary value problems.

Quick review

In some problems it becomes necessary to extend the class of admissible functions to include those that are piecewise smooth. Let $y(x)$ be a minimizer of the simplest integral functional, and suppose $y'(x)$ is continuous on the closed intervals $[a, \alpha]$ and $[\alpha, b]$ where $\alpha \in (a, b)$ is the sole corner point. The position of α cannot be determined in advance, but is subject to the Weierstrass–Erdmann conditions

$$f_{y'}\big|_{x=\alpha-0} = f_{y'}\big|_{x=\alpha+0}, \tag{1.154}$$

$$(f - y'f_{y'})\big|_{x=\alpha-0} = (f - y'f_{y'})\big|_{x=\alpha+0}. \tag{1.155}$$

In addition to the Euler equation on the intervals (a, α) and (α, b) then, y must satisfy (1) the Weierstrass–Erdmann conditions, (2) any given endpoint conditions on $y(a)$ and $y(b)$, and (3) the continuity condition $y(\alpha - 0) = y(\alpha + 0)$. A piecewise smooth extremal with a corner (or with multiple corners) is called a broken extremal.

1.14 Sufficient Conditions for Minimum

Thus far we have studied some of the techniques used to identify possible minimizers. It is also of interest to know how to solve the boundary value problems that yield corresponding extremals, although the treatment of this topic falls outside the scope of this book (and within the scope of books on ordinary and partial differential equations). But the solutions of these problems represent only the first step in a full solution of the problem of minimization; the next step is to learn whether an extremal is a minimizer. As we shall see, for many linear problems of mathematical physics an extremal satisfying boundary conditions is automatically a minimizer. Nonlinear problems, as a rule, need additional investigation. For this we need to derive sufficient conditions for an extremal to be a minimizer. First we shall derive conditions analogous to those found in the calculus of functions of many variables.

We reconsider the problem of minimum of the simplest functional

$$F(y) = \int_a^b f(x, y, y')\, dx$$

in the class $C^{(1)}(a, b)$ under the boundary conditions $y(a) = c_0$, $y(b) = c_1$. Let y be a minimizer of the problem under consideration and $\Delta y(x)$ an admissible increment of y. Consider the increment of F:

$$\begin{aligned} \Delta F &= F(y + \Delta y) - F(y) \\ &= \int_a^b [f(x, y + \Delta y, y' + \Delta y') - f(x, y, y')]\, dx. \end{aligned} \tag{1.156}$$

Denote $p = y(x)$, $q = y'(x)$, and $g(p, q) = f(x, p, q)$, and let Δp and Δq be the increments of p and q, respectively (in this case they are $\varphi(x)$ and $\varphi'(x)$ in our old notation). If in some small neighborhood of the point (p, q) the function g has continuous derivatives up to second order, then in this neighborhood we can write the Taylor expansion of g:

$$\begin{aligned} g(p + \Delta p, q + \Delta q) &= g(p, q) + [g_p(p, q)\Delta p + g_q(p, q)\Delta q] \\ &\quad + \frac{1}{2!}[g_{pp}(p, q)(\Delta p)^2 + 2g_{pq}(p, q)\Delta p \Delta q \\ &\quad + g_{qq}(p, q)(\Delta q)^2] + \beta(p, q, \Delta p, \Delta q)[(\Delta p)^2 + (\Delta q)^2] \end{aligned}$$

where $\beta(p, q, \Delta p, \Delta q) \to 0$ when $(\Delta p)^2 + (\Delta q)^2 \to 0$. We can write this expansion in terms of f, y, and Δy at each $x \in [a, b]$:

$$\begin{aligned} f(x, y + \Delta y, y' + \Delta y') &= f(x, y, y') + [f_y(x, y, y')\Delta y + f_{y'}(x, y, y')\Delta y'] \\ &\quad + \frac{1}{2!}[f_{yy}(x, y, y')(\Delta y)^2 + 2f_{yy'}(x, y, y')\Delta y \Delta y' \\ &\quad + f_{y'y'}(x, y, y')(\Delta y')^2] + \beta(x, y, y', \Delta y, \Delta y')[(\Delta y)^2 + (\Delta y')^2] \end{aligned} \tag{1.157}$$

(we keep the same notation β for the remainder function). Let us assume that for all $x \in [a,b]$ we have

$$|\beta(x,y,y',\Delta y,\Delta y')| \le \alpha(\Delta y,\Delta y')$$

where $\alpha(\Delta y,\Delta y') \to 0$ when $(\Delta y)^2 + (\Delta y')^2 \to 0$. This is an important assumption in what follows.

Let us return to the notation $\varphi = \Delta y$ and rewrite (1.157) as

$$\begin{aligned} f(x,y+\varphi,y'+\varphi') &= f(x,y,y') + [f_y(x,y,y')\varphi + f_{y'}(x,y,y')\varphi'] \\ &+ \frac{1}{2!}[f_{yy}(x,y,y')\varphi^2 + 2f_{yy'}(x,y,y')\varphi\varphi' \\ &+ f_{y'y'}(x,y,y')(\varphi')^2] + o(\varphi^2 + (\varphi')^2). \end{aligned} \tag{1.158}$$

Here $o(\varphi^2 + (\varphi')^2)$ indicates that the term which is uniform in x is small in comparison with $\varphi^2 + (\varphi')^2$. Now apply the expansion (1.158) to (1.156):

$$\begin{aligned} \Delta F = &\int_a^b [f_y(x,y,y')\varphi + f_{y'}(x,y,y')\varphi']\,dx \\ &+ \frac{1}{2!}\int_a^b [f_{yy}(x,y,y')\varphi^2 + 2f_{yy'}(x,y,y')\varphi\varphi' + f_{y'y'}(x,y,y')(\varphi')^2]\,dx \\ &+ o\left(\int_a^b (\varphi^2 + (\varphi')^2)\,dx\right). \end{aligned}$$

Since y is a minimizer of the problem we necessarily have

$$\int_a^b [f_y(x,y,y')\varphi + f_{y'}(x,y,y')\varphi']\,dx = 0$$

(cf., § 1.1) and thus

$$\Delta F = F(y+\varphi) - F(y) = \delta^2 F + o\left(\int_a^b (\varphi^2 + (\varphi')^2)\,dx\right)$$

where $\delta^2 F$ is the *second variation* defined by

$$\delta^2 F \equiv \frac{1}{2!}\int_a^b [f_{yy}(x,y,y')\varphi^2 + 2f_{yy'}(x,y,y')\varphi\varphi' + f_{y'y'}(x,y,y')(\varphi')^2]\,dx.$$

Integration by parts gives

$$\begin{aligned} \int_a^b 2f_{yy'}(x,y,y')\varphi\varphi'\,dx &= \int_a^b f_{yy'}(x,y,y')\frac{d}{dx}\varphi^2\,dx \\ &= -\int_a^b \varphi^2 \frac{d}{dx} f_{yy'}(x,y(x),y'(x))\,dx \end{aligned}$$

since $\varphi(a) = \varphi(b) = 0$. Then

$$\delta^2 F = \frac{1}{2!} \int_a^b \left\{ \left[f_{yy}(x, y, y') - \frac{d}{dx} f_{yy'}(x, y, y') \right] \varphi^2 + f_{y'y'}(x, y, y')(\varphi')^2 \right\} dx.$$

The quantity $\delta^2 F$ is quadratic in φ and φ'. Suppose it is bounded from below as follows:

$$\delta^2 F \geq m \int_a^b (\varphi^2 + (\varphi')^2)\, dx, \tag{1.159}$$

where the constant $m > 0$ does not depend on the choice of admissible increment φ (note that here we do not need assumptions on the smallness of φ). It then follows that

$$F(y + \varphi) - F(y) \geq 0$$

for all admissible increments φ (i.e., $\varphi \in C_0^{(1)}(a, b)$) with sufficiently small norm $\|\varphi\|_{C^{(1)}(a,b)}$. This means that (1.159) is sufficient for y to be a local minimizer of the problem under consideration.

Thus we seek conditions for (1.159) to hold. Let us denote

$$Q(x) = f_{yy}(x, y(x), y'(x)) - \frac{d}{dx} f_{yy'}(x, y(x), y'(x)),$$
$$P(x) = f_{y'y'}(x, y(x), y'(x)).$$

The functions $Q(x)$ and $P(x)$ can be regarded as momentarily given when we study whether $y = y(x)$ is a minimizer. So we must study the functional

$$\Phi(\varphi) = \int_a^b [P(x)\varphi'^2(x) + Q(x)\varphi^2(x)]\, dx$$

in the space $C_0^{(1)}(a, b)$.

It is easy to formulate the following restrictions:

$$P(x) \geq c \quad \text{and} \quad Q(x) \geq c > 0 \qquad \text{for all } x \in [a, b].$$

Under these the inequality (1.159) holds for all $\varphi \in C_0^{(1)}(a, b)$. Unfortunately these restrictions fail in many cases when $y = y(x)$ is really a minimizer, so we need more useful conditions.

Note that if $y = y(x)$ is a minimizer then $\Phi(\varphi) \geq 0$ at least. For if there were an admissible increment φ such that $\Phi(\varphi) < 0$ then we could find a t_0 so small that for all $0 < t < t_0$ we would have $F(y + t\varphi) - F(y) < 0$, and y would not be a minimizer. Let us suppose $\Phi(\varphi)$ is nonnegative.

Theorem 1.40. *Let $P(x)$ and $Q(x)$ be continuous on $[a, b]$ and $\Phi(\varphi) \geq 0$ for all $\varphi \in C_0^{(1)}(a, b)$. Then $P(x) \geq 0$ on $[a, b]$.*

Proof. Suppose to the contrary that $P(x_0) < 0$ for some x_0. Then $P(x) < \gamma < 0$ in some ε-neighborhood $[x_0 - \varepsilon, x_0 + \varepsilon]$ of x_0. Choose $\varphi(x) \in C^{(1)}(a, b)$ as the particular function

$$\varphi(x) = \begin{cases} \sin^2 \left[\dfrac{\pi(x - x_0)}{\varepsilon} \right], & x \in [x_0 - \varepsilon, x_0 + \varepsilon], \\ 0, & \text{otherwise.} \end{cases}$$

Then for $x \in [x_0 - \varepsilon, x_0 + \varepsilon]$ we have

$$\varphi'(x) = 2 \sin \left[\frac{\pi(x - x_0)}{\varepsilon} \right] \cos \left[\frac{\pi(x - x_0)}{\varepsilon} \right] \left(\frac{\pi}{\varepsilon} \right) = \frac{\pi}{\varepsilon} \sin \left[\frac{2\pi(x - x_0)}{\varepsilon} \right]$$

and therefore

$$\Phi(\varphi) = \left(\frac{\pi}{\varepsilon} \right)^2 \int_{x_0 - \varepsilon}^{x_0 + \varepsilon} P(x) \sin^2 \left[\frac{2\pi(x - x_0)}{\varepsilon} \right] dx + \int_{x_0 - \varepsilon}^{x_0 + \varepsilon} Q(x) \sin^4 \left[\frac{\pi(x - x_0)}{\varepsilon} \right] dx.$$

But

$$\int_{x_0 - \varepsilon}^{x_0 + \varepsilon} P(x) \sin^2 \left[\frac{2\pi(x - x_0)}{\varepsilon} \right] dx < \gamma \int_{x_0 - \varepsilon}^{x_0 + \varepsilon} \sin^2 \left[\frac{2\pi(x - x_0)}{\varepsilon} \right] dx = \gamma\varepsilon$$

and

$$\int_{x_0 - \varepsilon}^{x_0 + \varepsilon} Q(x) \sin^4 \left[\frac{\pi(x - x_0)}{\varepsilon} \right] dx \le M \int_{x_0 - \varepsilon}^{x_0 + \varepsilon} \sin^4 \left[\frac{\pi(x - x_0)}{\varepsilon} \right] dx = \frac{3M\varepsilon}{4}$$

where $M = \max_{x \in [a,b]} |Q(x)|$. Hence

$$\Phi(\varphi) < \left(\frac{\pi}{\varepsilon} \right)^2 \gamma\varepsilon + \frac{3M\varepsilon}{4} = \frac{\pi^2 \gamma}{\varepsilon} + \frac{3M\varepsilon}{4}.$$

Recall that $\gamma < 0$; for sufficiently small ε we can make $\Phi(\varphi) < 0$, a contradiction. □

Thus, besides the Euler equation we have established another necessary condition for y to be a minimizer of the problem under consideration:

$$f_{y'y'}(x, y(x), y'(x)) \ge 0 \quad \text{for all } x \in [a, b]. \tag{1.160}$$

This is *Legendre's condition.*

Legendre believed that satisfaction of the strict inequality $f_{y'y'} > 0$ for all $x \in [a, b]$ should be sufficient for y to be a minimizer, and even constructed a flawed proof. However, even the mistakes of great persons are useful — on the basis of this "proof" a useful sufficient condition was subsequently established. Jacobi proposed to study the functional $\Phi(\varphi)$ using the tools of the calculus of variations itself. The Euler equation for this functional is

$$[P(x)\varphi'(x)]' - Q(x)\varphi(x) = 0. \tag{1.161}$$

This clearly has the trivial solution $\varphi = 0$. Let $P(x)$ be continuously differentiable. Jacobi studied the zeros of a solution of (1.161) for the Cauchy problem $\varphi(0) = 0$, $\varphi'(0) = 1$. The nearest value $x_0 > a$ where $\varphi(x_0) = 0$ he called the point *conjugate* to a (with respect to the functional $\Phi(\varphi)$). This point is denoted a^* (we agree to call $a^* = \infty$ if $\varphi(x)$ has no zeros to the right of $x = a$). Jacobi established another necessary condition for y to be a minimizer: that the interval (a, b) does not contain a^*.

The following set of three conditions is sufficient for y to be a minimizer of the problem under consideration:

(1) y satisfies the Euler equation

$$f_y - \frac{d}{dx} f_{y'} = 0;$$

(2) $f_{y'y'}(x, y(x), y'(x)) > 0$ for all $x \in [a, b]$;

(3) $[a, b]$ does not contain points conjugate to a with respect to $\Phi(\varphi)$.

We shall not offer a proof of this, but do wish to note the following. The result is beautiful, but for many years it seemed impractical: the Jacobi condition (3) was quite difficult to check before the advent of the computer. Today, however, there are many good algorithms with which Cauchy problems for ordinary differential equations may be solved. Hence it is easy to check the Jacobi condition numerically.

Example 1.41. For which range of the constant c is an extremal of the functional

$$\int_0^1 (y'^2 - c^2 y^2 - 2y)\, dx, \qquad y(0) = 0, \quad y(1) = 1,$$

a minimizer?

Solution. The extremal exists, as the reader can verify. We suppose $c > 0$. Let us check the sufficiency conditions given above. Legendre's condition holds automatically. The Jacobi equation with initial conditions is

$$y'' + c^2 y = 0, \qquad y(0) = 0, \quad y'(0) = 1.$$

Its solution is $y = c^{-1} \sin cx$, hence the conjugate point occurs where $cx = \pi$. Thus, by sufficient conditions, the extremal really is a minimizer of the functional when $a^* = \pi/c > 1$, and by symmetry in c, the extremal is a minimizer when $|c| < \pi$. When $a^* < 1$, then extremal is not a minimizer and, moreover the functional has no minimizer at all (why?). □

The Jacobi theory of conjugate points and corresponding results can be established for a functional depending on an unknown vector-function.

Some field theory

We now turn to a brief, introductory discussion of certain concepts needed to express conditions sufficient for a strong minimum. The main idea is that of a *field of extremals.*

Let D be a domain in the xy-plane. Let

$$y = y(x; \alpha)$$

be a family of curves lying in D, a separate curve being generated by each choice of the parameter α. If a unique curve from the family passes through each point of D, then we call the family a *proper field* in D. A proper field can be regarded as a sort of cover for D, associating with each point $(x, y) \in D$ a unique slope $p(x, y)$ (i.e., the slope of the particular curve passing through that point). As a simple but standard example, let D be the unit disk

$$D = \{(x, y) \colon x^2 + y^2 < 1\}$$

and let $y = y(x; \alpha) = kx + \alpha$ where k is a fixed constant. This is a field of parallel straight lines with slopes $p(x, y) \equiv k$.

If all curves of a family $y = y(x; \alpha)$ pass through a certain point (x_0, y_0), then the family is known as a *pencil* of curves and (x_0, y_0) is called the *center* of the pencil. For example, the family $y = \alpha x$ is a pencil having center at the origin. Of course, a pencil of curves having center $(x_0, y_0) \in D$ cannot be a proper field of curves in D. However, if a pencil of curves assigns a unique slope $p(x, y)$ to all points in D *other than* (x_0, y_0), we speak of a *central field* of curves in D.

A *field of extremals* is a family of extremal curves (for some variational problem) that generates a proper or central field in a domain D. The Euler equation for the simplest functional

$$F(y) = \int_a^b f(x, y, y')\,dx \tag{1.162}$$

has solutions that form a two-parameter family of curves $y = y(x;\alpha;\beta)$. (Here α and β are the integration constants in the general solution of the Euler equation.) If one of the constants, say α, is determined by imposing a given fixed endpoint condition $y(a) = c_0$ on the general solution, then all the extremals in the resulting one-parameter family will issue from the same point (a, c_0). The resulting family $y = y(x;\beta)$ may be a field (proper or central) in some specified domain D. For example, consider the functional

$$\int_a^b [y^2 - (y')^2]\,dx$$

with $a = 0$ and $y(0) = 0$. The integrand does not depend explicitly on x, so $y^2 - (y')^2 - (-2y')y' = c_1$. It follows that the extremals have the form $y = c_2 \sin(x + c_3)$, which gives us a pencil having center $(0, 0)$. Another example we mention is for the functional

$$\int_a^b (y'^2 - 1)^2\,dx.$$

The extremals are straight lines. When suitably restricted, the two-parameter family of curves $y(x) = c_1 x + c_2$ can form a field in a couple of different ways: (1) when c_1 is fixed, we obtain a family $y = y(x;c_2)$ that can form a proper field in the unit disk D; (2) when $c_2 = 0$, the resulting pencil centered at the origin can form a central field in D.

Let $y = y(x;\alpha)$ generate a field of extremals (central or proper) in some domain D. Each choice of α then gives an extremal; by setting $\alpha = \alpha_0$, we select a particular extremal $y^*(x) = y(x;\alpha_0)$ from the field. If this extremal $y^*(x)$ has no common points with the boundary of D, it is said to be *admissible in the field*. Note that a given extremal may be admissible in more than one field covering a domain D. Returning to the example in which D is the unit circle, the two fields

$$y(x;\alpha) = c_1 x + \alpha, \qquad y(x;\alpha) = \alpha x,$$

mentioned above each admit the straight line extremal $y^*(x) = c_1 x$.

Armed with an understanding of the field concept, we proceed to the next step. Let D be a domain in which there is distributed a proper field of extremals for the simplest functional $F(y)$ of equation (1.162). Suppose further that this field admits the particular extremal $y = y^*(x)$ satisfying given endpoint conditions $y(a) = c_0$, $y(b) = c_1$. Now let $y = y(x)$ be *any* curve that lies in D and connects the desired endpoints (a, c_0) and (b, c_1). We also assume that the integral

$$H(y) = \int_a^b [f(x, y, p) + (y' - p) f_p(x, y, p)] \, dx \tag{1.163}$$

exists for $y = y(x)$, where $p = p(x, y)$ is the slope function (i.e., its value at (x, y) is the slope y' of the extremal through point (x, y)) of the field in D. This integral is extremely important for the theory.

When $y(x) = y^*(x)$, the integral (1.163) reduces to (1.162) because $y' \equiv p$ in that case. It can be shown that (1.163) is path independent in D. For this reason it is known as *Hilbert's invariant integral.*

We use these facts as follows. Defining

$$\Delta F = F(y) - F(y^*),$$

we have $\Delta F = F(y) - H(y^*) = F(y) - H(y)$ so that

$$\begin{aligned}\Delta F &= \int_a^b f(x, y, y') \, dx - \int_a^b [f(x, y, p) + (y' - p) f_p(x, y, p)] \, dx \\ &= \int_a^b [f(x, y, y') - f(x, y, p) - (y' - p) f_p(x, y, p)] \, dx.\end{aligned}$$

Thus

$$\Delta F = \int_a^b E(x, y, y', p) \, dx \tag{1.164}$$

where the integrand

$$E(x, y, y', p) = f(x, y, y') - f(x, y, p) - (y' - p) f_p(x, y, p) \tag{1.165}$$

is known as the *Weierstrass excess function.* The following conditions are sufficient for $y = y^*(x)$ to be a strong minimum of $F(y)$:

(1) The curve $y = y^*(x)$ is admissible in a field of extremals for $F(y)$, and
(2) $E(x, y, y', p) \geq 0$ for all points (x, y) lying sufficiently close to the curve $y = y^*(x)$ and for arbitrary values of y'.

These have been called the *Weierstrass conditions.* The proof is nearly obvious. Suppose condition (1) holds, and let $y = y(x)$ be any other curve lying in the domain covered by the field of extremals and connecting the desired endpoints. Then according to condition (2),

$$\Delta F = \int_a^b E(x, y, y', p) \, dx \geq 0$$

for all curves $y = y(x)$ that connect the endpoints and lie within some neighborhood of $y^*(x)$; moreover, the slope of y need not be close to that of y^* so the minimum is strong.

Although the Weierstrass conditions are attractive because of their simplicity, we can run into trouble when attempting to apply them to certain functionals. This happens, for example, with the problem of minimizing

$$\int_0^{3/2} \frac{y}{(y')^2}\, dx, \qquad y(0) = 1, \quad y(3/2) = 1/4.$$

The difficulty is related to the fact that the family of extremals has a so-called envelope.

Our treatment of sufficient conditions for the problem of minimum has been intentionally brief. We have formulated a couple of sets of such conditions; in fact, however, these are seldom used by practitioners. Rather, necessary conditions are usually applied to obtain extremals, and then various other methods are employed in place of sufficient conditions. For example, if a functional has a unique minimum residing in a class of functions, and if a unique extremal is found for the problem, then the desired minimum must be reached on the extremal found. If several extremals qualify as candidates for the minimum, it is often possible to test each one by calculating the corresponding values taken by the functional. The true minimum may then be identified and selected. Hence sufficient conditions may be viewed as largely of theoretical interest.

1.15 Exercises

1.1 Each functional below has the form (1.33). Write out the Euler equation and the natural boundary conditions,

$$f_y - \frac{d}{dx} f_{y'} = 0 \text{ in } (a, b), \qquad f_{y'}\big|_{x=a} = 0, \quad f_{y'}\big|_{x=b} = 0,$$

given in Theorem 1.19.

(a) $\displaystyle\int_0^1 \sqrt{1 + y'^2(x) + y^2(x)}\, dx.$

(b) $\displaystyle\int_{-1}^1 \left[y'^2(x) + (1 + x^2)\, y^2(x)\right] dx.$

(c) $\displaystyle\int_1^3 \left[\frac{1}{2} y'^2(x) - (1 + 2x^2)\, y^2(x)\right] dx.$

(d) $\displaystyle\int_a^b \left[7y'^2(x) - (1+x^2)\,y^2(x)\right] dx, \quad a < b.$

(e) $\displaystyle\int_a^b \left[y'^2(x) + (1+x^6)\,y^2(x)\right] dx + y(a)^2, \quad a < b.$

(f) $\displaystyle\int_1^3 \left[xy'^2(x) + (x^2-9)\,y^2(x)\right] dx + 5y^2(1) + y^2(3).$

(g) $\displaystyle\int_a^b \left[5y'^2(x) + \sqrt{x-a}\,y^2(x)\right] dx + y(a)^2, \quad 0 \le a < b.$

(h) $\displaystyle\int_1^4 \left[y'^2(x) + x^2y^2(x)\right] dx + y^2(2).$

(i) $\displaystyle\int_a^b \left[y'^2(x) + xy^2(x)\right] dx + y(c), \quad a < c < b.$

(j) $\displaystyle\int_0^\pi \left[3y'^2(x) - 2y^4(x)\right] dx + \left[y(\pi) - y(0)\right]^2.$

(k) $\displaystyle\int_0^\pi \left[yy'^2(x) - \cos(y(x))\right] dx + \left[y(\pi)\right]^2.$

(l) $\displaystyle\int_0^1 \left[(y'^2(x) - 1)^2 + y^2(x)\right] dx.$

1.2 For each functional below, write out the Euler–Lagrange equation (1.68) and the natural boundary conditions given on page 40.

(a) $\displaystyle\int_0^1 \left[y''^2(x) + 2y^2(x)\right] dx.$

(b) $\displaystyle\int_0^1 \left[y''^2(x) + 2y'^2(x)\right] dx.$

(c) $\displaystyle\int_0^1 \left[y''^2(x) + y'^2(x) + y^2(x)\right] dx.$

(d) $\displaystyle\int_0^1 \left[y'''^2(x) - y''^2(x) + 2(1-x^2)y^2(x)\right] dx.$

(e) $\displaystyle\int_a^b \left[y''''^2(x) - y'^2(x) + 2y(x)\right] dx, \quad a < b.$

1.3 The following functionals have the form (1.74). Write out the Euler equations and the natural boundary conditions given in Theorems 1.26 and 1.27:

$$f_u - \left(\frac{df_{u_x}}{dx} + \frac{df_{u_y}}{dy}\right) = 0 \quad \text{in } S, \qquad \left(f_{u_x}n_x + f_{u_y}n_y\right)\Big|_{\partial S} = 0.$$

(a) $\displaystyle\int_a^b \int_c^d \left[(u_x)^2 + 2(u_y)^2 + 3u^2 - 2u\right] dx\,dy, \quad a < b, \quad c < d.$

(b) $\displaystyle\int_a^b \int_c^d \left[(u_x)^2 + (u_y)^2 + u^2\right] dx\,dy + \int_a^b u^2(x,c)\,dx, \quad a < b, \quad c < d.$

(c) $\displaystyle\int_a^b \int_c^d \left[(u_x)^2 + (u_y)^2 - u^2\right] dx\,dy + \int_c^d u^2(b,y)\,dy, \quad a < b, \quad c < d.$

(d) $\displaystyle\int_a^b \int_c^d \left[(u_x)^2 - (u_y)^2 + 2u\right] dx\,dy, \quad a < b, \quad c < d.$

(e) $\displaystyle\int_a^b \int_c^d \left[(u_x)^n + (u_y)^n\right] dx\,dy, \quad a < b, \quad c < d, n \neq 1.$

(f) $\displaystyle\int_a^b \int_c^d \left[\sin(u_x) + \sin(u_y)\right] dx\,dy, \quad a < b, \quad c < d.$

(g) $\displaystyle\int_a^b \int_c^d \left[2 - \cos(u_x) - \cos(u_y)\right] dx\,dy, \quad a < b, \quad c < d.$

(h) $\displaystyle\int_0^1 \int_0^1 \sqrt{1 + (u_x)^2 + (u_y)^2}\,dx\,dy.$

(i) $\displaystyle\int_0^1 \int_0^1 (p(u_x) + q(u_y))\,dx\,dy.$

(j) $\displaystyle\int_0^1 \int_0^1 \left(1 + (u_x)^2 + (u_y)^2\right)^n dx\,dy, \quad n > 0.$

1.4 In the xy-plane, find the smooth curve between (a, y_0) and (b, y_1) which by revolution about the x-axis generates the surface of least area.

1.5 The *brachistochrone problem* is a famous classical problem in which one must find the equation of the plane curve down which a particle would slide from one given point to another in the least possible time when acted upon by gravity alone. Show that the required curve is a portion of an ordinary cycloid.

1.6 Show that if f in the simplest functional depends explicitly on y' only, then the extremals are straight lines.

1.7 During the time interval $[0, T]$ a particle having mass m is required to move along a straight line from the position $x(0) = x_0$ to the position $x(T) = x_1$. Determine the extremal for the problem of minimizing the particle's average kinetic energy. Explain your result physically.

1.8 Apply Ritz's method with basis functions of the form $\varphi_n(x) = x^2(1-x)^2 x^k$ to minimize the functional

$$\int_0^1 \{(y'')^2 + [1 + 0.1 \sin x](y')^2 + [1 + 0.1\cos(2x)]y^2 - 2\sin(2x)y\}\,dx.$$

The boundary conditions for the problem are $y(0) = y'(0) = y'(1) = 0$, $y(1) = 1$.

1.9 (a) Consider the problem of minimum for the simplest functional (1.20) with boundary condition $y(a) + y(b) = 1$. Find a supplementary natural boundary condition for this case. (b) Repeat for a condition of the more general form $\psi(y(a), y(b)) = 0$ where $\psi = \psi(\alpha, \beta)$ is a given function of two variables.

1.10 Find the equation of the plane curve down which a particle would slide from one given point (a, y_0) to cross the vertical line $x = b$ in the least possible time when acted upon by gravity alone.

1.11 Find the smooth curve of least length between two points on the surface of the cylinder of radius a.

1.12 For a functional of the form

$$F_2(y) = \int_a^b f(x, y, y', y'')\, dx,$$

find the Ritz system of equations corresponding to (1.50).

1.13 Find the first variation of the functional of the form

$$F(y) = \int_{x_0}^{x_1} f(x, y, y', y'')\, dx \tag{1.166}$$

where the endpoints x_0 and x_1 can move.

1.14 For problems of beam equilibrium posed as minimum energy problems, we know the extremals are given by continuous functions having continuous first derivatives; the second derivatives are continuous except at points where the beam parameters have jumps or point loads are applied. With this in mind, consider the extremals of the functional (1.166) that are continuous on (a, b), have continuous first derivatives, and have continuous second derivatives everywhere except at $x = c$ where y'' can have a jump. Find the differential equations for the extremals, the endpoint conditions, and an analogue of the Weierstrass–Erdmann conditions at point c.

1.15 What happens to the equations defining a "broken" extremal of the functional of Exercise 1.14 if the position c is known and fixed?

1.16 Consider the equilibrium problem for a plate when given forces f act on the edge. It is described as the minimization problem for the functional

$$\begin{aligned} F(w) = \frac{D}{2} \iint_S [w_{xx}^2 + w_{yy}^2 + 2\nu w_{xx} w_{yy} + 2(1-\nu) w_{xy}^2]\, dx\, dy \\ - \iint_S F w\, dx\, dy - \oint_{\partial S} f w\, ds. \end{aligned}$$

What is the form of the Euler equation? What are the natural boundary conditions for a minimizer?

1.17 Suppose a plate consists of two parts, with different constant rigidities D_1 and D_2, that join along a line Γ of the midplane (Fig. 1.5). Write out the conditions on the border line assuming the deflection w and its first derivatives are continuous over the whole domain. Note that these conditions have the same nature as the natural boundary conditions. They have a clear mechanical meaning.

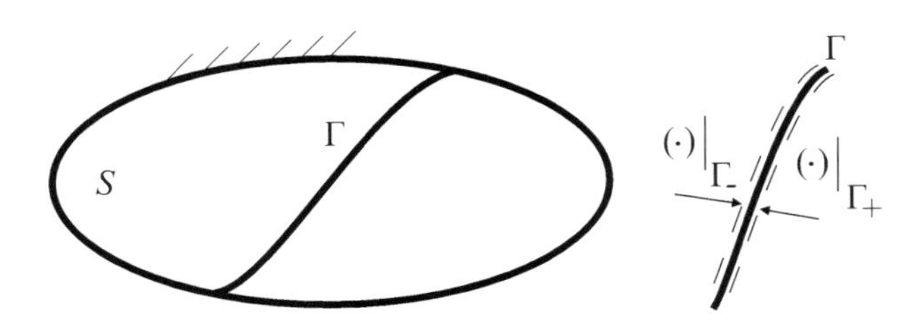

Fig. 1.5 Left: a compound plate. Right: calculation of one-sided limits.

1.18 Find the Euler equation and the natural boundary conditions for the functional $F(u) = E(u) - A(u)$, where

$$E(u) = \frac{1}{2}\iint_S (u_x^2 + u_y^2)\,dx\,dy + \frac{1}{2}\int_{\partial S} \alpha u_s^2\,ds, \quad A(u) = \iint_S fu\,dx\,dy + \int_{\partial S} gu\,ds.$$

Assume f is a given function on S, and g and α are given functions on the boundary ∂S of S.

1.19 Show that $E(u)$ from Problem 1.18 is unbounded from below if $\alpha < 0$ on any portion of ∂S.

Chapter 2

Applications of the Calculus of Variations in Mechanics

2.1 Elementary Problems for Elastic Structures

Now we consider some elementary problems in elasticity from the standpoint of the calculus of variations. In textbooks on the strength of materials, such problems are solved by elementary methods. Nonetheless, points of confusion often remain — typically in the problem setups. Our goal is to analyze proper variational setups and to show how these provide additional natural conditions describing the action of loads on the boundary. Along the way, we apply powerful variational methods to several interesting mechanics problems.

The structures considered in this section contain rods, beams, and springs connected at certain angles. The first question concerns the model to be employed for each structural element. The answer must come from engineering experience rather than pure mathematics. Model selection will provide a set of variables describing each structural element. Normally these sets are independent but must satisfy interrelationships along boundaries between the individual models. The length parameter along a beam is usually denoted by the same letter (such as x, s, or t) for all the elements. The displacement variables are denoted by different letters or by indices.

The general plan for posing minimization-type setups for such equilibrium problems involves the following steps.

(1) Establish notation for all variables in the structural description.

(2) Construct the strain energy functional W, which is the sum of the strain energies of all structural elements (rods, beams, cables, springs) described in local coordinates.

(3) Construct the work functional A for the external load over general

displacements.

(4) Write down the total potential energy functional $\mathcal{E} = W - A$ needed to pose the minimization problem.

To this list we should add

(5) Account for any geometrical boundary restrictions. These should be evident from a sketch of the structure proposed in the problem.

The leftmost portion of Fig. 2.1 indicates the only type of geometric restriction for a rod; this is rigid clamping, expressed as $u|_A = 0$ (or as $u|_A = c$ for a given constant c). For a beam, the same sketch implies that two con-

Fig. 2.1 Rigid clamping of rods and beams (A); hinged clamping (B_1–B_3).

ditions should be posed at A: $w = 0$ and $w' = 0$ (or possibly some given nonzero values). A beam can also be subjected to hinged supports as indicated in the rest of the figure. Note that the points B_k may be endpoints or intermediate points of the beam. A hinged connection restricts the displacement at a point (e.g., $w = 0$) but not the angle of rotation there (i.e., it does not restrict w'). We must also

(6) Consider compatibility restrictions, arising as mutual constraints on the displacements or rotation angles at points of coupling between structural elements.

Typical of these are rigid clamping. In the case of coupled beams (Fig. 2.2, left), both the displacement vectors of the coupled points and the rotation angles must be the same (hence, in the simple beam model, the values of w' must agree at the point of coupling). In the case of hinged beams (Fig. 2.2, right) the displacement vectors are equal but the angles of rotation (i.e., the values of w') are independent. Consideration of other modes of coupling are left to the exercises; they are usually clear from inspection of a diagram. For beams coupled under an angle, the equilibrium equations contain only lateral displacements as unknown functions; despite this fact, the conditions at the joint contain the full displacement vectors of the

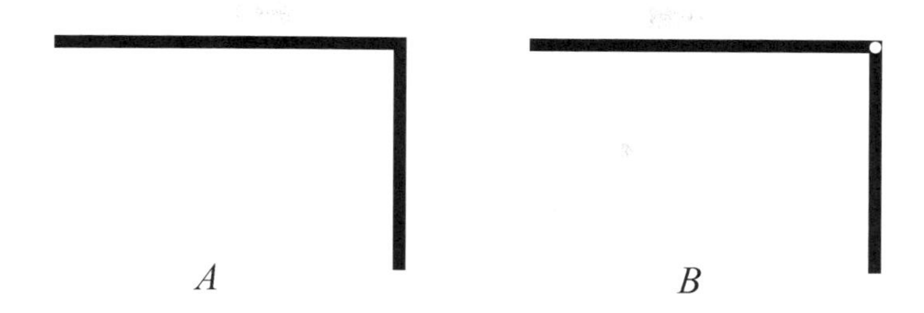

Fig. 2.2 Rigid joint of beams (A); hinged joint (B).

elements. Because the coordinate system used in the strength of materials is opposite that used in an ordinary calculus textbook, we will present the transformation formulas for the displacement vectors needed to formulate geometric compatibility conditions.

The coordinate unit vector $\mathbf{i}$ lies along the midline of a beam in the direction of increasing length coordinate x, and the unit vector $\mathbf{j}$ is orthogonal to $\mathbf{i}$ with the orientation shown in Fig. 2.3. For another beam, we introduce

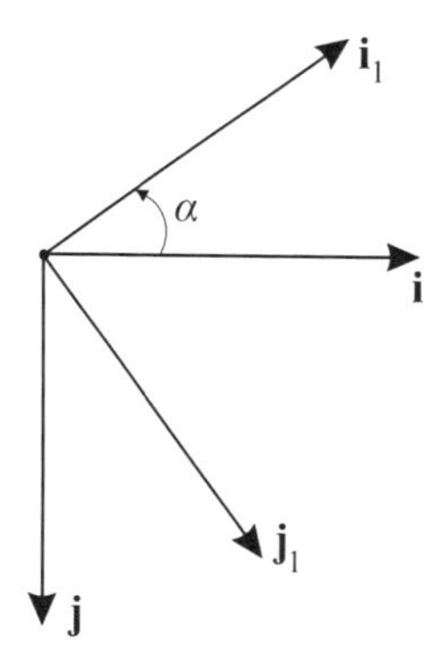

Fig. 2.3 Transformation of coordinate bases.

the respective unit basis vectors $\mathbf{i}_1$ and $\mathbf{j}_1$. The angle α between $\mathbf{i}$ and $\mathbf{i}_1$ is positive if the rotation from $\mathbf{i}$ to $\mathbf{i}_1$ is counterclockwise. The displacement vector $\mathbf{u}$ takes the form

$$\mathbf{u} = u\mathbf{i} + w\mathbf{j} = u_1\mathbf{i}_1 + w_1\mathbf{j}_1.$$

Dot multiplication yields

$$u_1 = u(\mathbf{i} \cdot \mathbf{i}_1) + w(\mathbf{j} \cdot \mathbf{i}_1), \qquad w_1 = u(\mathbf{i} \cdot \mathbf{j}_1) + w(\mathbf{j} \cdot \mathbf{j}_1),$$

and thus

$$
\begin{aligned}
u_1 &= u\cos\alpha - w\sin\alpha,\\
w_1 &= u\sin\alpha + w\cos\alpha.
\end{aligned}
\tag{2.1}
$$

When using these transformation formulas, one must remember that in the beam model the longitudinal displacement is uniform along the beam. In the rod model, the displacement normal to the rod axis takes the "rigid body motion" form $w = a + bx$.

We denote Young's modulus by E, the moment of inertia of the beam cross section by I, the cross-sectional area by S, and the length of the beam (or rod) by a.

Example 2.1. Consider the equilibrium of a cantilever beam of length a, under load $q(x)$, clamped at the left end and coupled with a spring at the right end (Fig. 2.4). The beam parameters are E and $I(x)$, and the spring coefficient is k. (1) Write down the total potential energy of the system, along with the boundary and compatibility equations. (2) Write down the functional that should be minimized in order to obtain the equilibrium equations and natural boundary conditions. (3) Applying the general procedure of the calculus of variations, derive the differential equation of equilibrium and the natural boundary conditions.

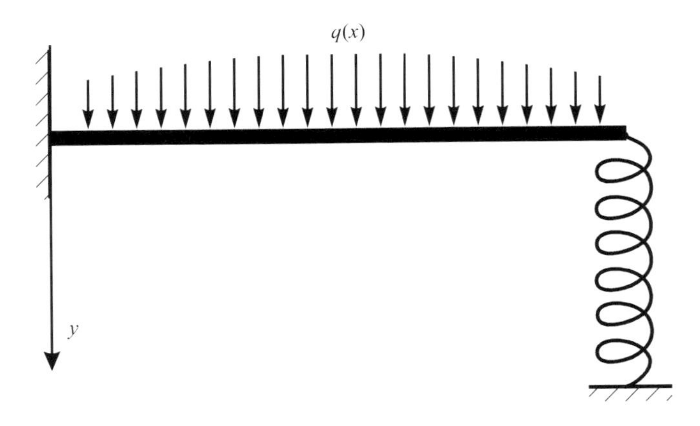

Fig. 2.4 System consisting of coupled beam and spring under load $q(x)$.

Solution. (1) Let $y(x)$ be the normal deflection of point x of the beam.

The system carries total potential energy

$$\mathcal{E}(y) = \frac{1}{2}\int_0^a EI(y'')^2\,dx - \int_0^a qy\,dx + \frac{1}{2}kz^2,$$

where z is the contraction suffered by the spring. The conditions at the left end are $y(0) = 0$ and $y'(0) = 0$. The compatibility condition between the beam and spring at $x = a$ is $z = y(a)$.

(2) The required functional is

$$\mathcal{E}(y) = \frac{1}{2}\int_0^a EI(y'')^2\,dx - \int_0^a qy\,dx + \frac{1}{2}ky^2(a).$$

(3) The equilibrium equation in integral form ($\delta\mathcal{E} = 0$) is

$$\int_0^a EIy''\varphi''\,dx - \int_0^a q\varphi\,dx + ky(a)\varphi(a) = 0.$$

Two integrations by parts in the first integral yield

$$\int_0^a (E(Iy'')'' - q)\varphi\,dx + EI(a)y''(a)\varphi'(a)$$
$$- E(Iy'')'\big|_{x=a}\varphi(a) + ky(a)\varphi(a) = 0,$$

since $y(0) = 0 = y'(0)$. Selecting the set of admissible φ such that $\varphi(a) = 0 = \varphi'(a)$ and using Lemma 1.8, we obtain the differential equation of equilibrium:

$$E(Iy'')'' - q = 0 \quad \text{on } (0, a).$$

Then, returning to the integral equation of equilibrium, we get

$$EI(a)y''(a)\varphi'(a) - E(I(x)y''(x))'\big|_{x=a}\varphi(a) + ky(a)\varphi(a) = 0$$

for all admissible φ. This yields

$$EI(a)y''(a) = 0, \qquad E(I(x)y''(x))'\big|_{x=a} = ky(a),$$

for the natural boundary conditions. □

Example 2.2. Two systems of coupled beams often encountered in the strength of materials are shown in Fig. 2.5. In the left portion of the figure the joints are hinged; in the right portion the beams are clamped. For both systems, construct (1) the models, (2) the energy functional to minimize, and (3) the kinematic restrictions. Finally, (4) derive the equilibrium differential equations and the natural boundary conditions.

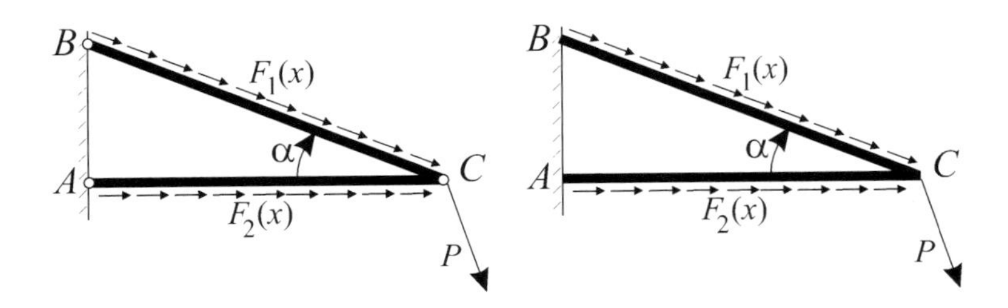

Fig. 2.5 Two beam systems.

Solution. The structural elements of the first system can be modeled as rods because they are in pure tension or compression. The elements of the second system, in contrast, are subject to bending as well. Hence we should consider them as rods in the axial direction and as the beams in the transverse direction.

(a) The system of coupled rods (left portion of the figure). The rods are described by parameters E and S. Suppose rod AC has length a, and denote its displacement components by (u, w) where u is the longitudinal component. The displacement components of the inclined rod BC are (u_1, w_1). Let the projections of the force P at the hinge onto the unit vectors $\mathbf{i}$ and $\mathbf{j}$ (recall Fig. 2.3) be P_1 and P_2, respectively. The total potential energy functional is

$$\mathcal{E}_L = \frac{1}{2}\int_0^a ESu'^2(x)\,dx - \int_0^a F_2(x)u(x)\,dx + \frac{1}{2}\int_0^a ESu'^2(x)\,dx$$
$$- \int_0^{a/\cos\alpha} F_1(x)u(x)\,dx - P_1u(a) - P_2w(a).$$

Note the presence of the normal displacement w. The geometrical boundary conditions are $u(0) = 0$ and $u_1(0) = 0$. Equations (2.1) apply at point C except that we must replace α by $-\alpha$ since the rotation from $\mathbf{i}$ to $\mathbf{i}_1$ is clockwise.

Next we derive the natural boundary conditions. The equilibrium equation, that the first variation of $\mathcal{E}_L$ must vanish, is

$$\int_0^a ESu'(x)\varphi'(x)\,dx - \int_0^{a/\cos\alpha} F_2(x)\varphi(x)\,dx + \int_0^a ESu'(x)\varphi_1(x)\,dx$$
$$- \int_0^{a/\cos\alpha} F_1(x)\varphi_1(x)\,dx - P_1\varphi(a) - P_2\psi(a) = 0, \tag{2.2}$$

where φ and φ_1 are independent admissible longitudinal displacements and $\psi(a)$ is the admissible normal (along $\mathbf{j}$) displacement at C. Taking $\varphi_1(x) = 0$ and an arbitrary φ that vanishes at the endpoints, we obtain the equilibrium equation for rod AC. Then, taking $\varphi = 0$ and $\varphi_1 \neq 0$, we get another equilibrium equation:

$$ESu''(x) + F_2(x) = 0 \text{ on } (0, a),$$
$$ESu_1''(x) + F_1(x) = 0 \text{ on } \left(0, \frac{a}{\cos\alpha}\right). \tag{2.3}$$

Returning to (2.2), we have

$$ESu'\varphi\big|_C + ESu_1'\varphi_1\big|_C - P_1\varphi\big|_C - P_2\psi\big|_C = 0. \tag{2.4}$$

The displacement vector at point C must be the same for both rods. Hence, by (2.1) (with $\alpha \to -\alpha$) the admissible displacements must satisfy

$$\varphi_1\big|_C = \varphi\big|_C \cos\alpha + \psi\big|_C \sin\alpha,$$
$$\psi_1\big|_C = -\varphi\big|_C \sin\alpha + \psi\big|_C \cos\alpha \tag{2.5}$$

(for this rod system we do not need the second transformation equation for the normal displacement). Substitution into (2.4) gives

$$ESu'\varphi\big|_C + ESu_1'(\varphi\big|_C \cos\alpha + \psi\big|_C \sin\alpha)\big|_C - P_1\varphi\big|_C - P_2\psi\big|_C = 0.$$

Using independence of $\varphi|_C$ and $\psi|_C$ we get two natural conditions

$$(ESu' + ESu_1' \cos\alpha)\big|_C = P_1, \qquad ESu_1' \sin\alpha\big|_C = P_2,$$

that express equilibrium at C.

(b) The rod–beam system (right portion of the figure). To account for the possibility of bending, we employ rod and beam models simultaneously. The total potential energy functional now includes the energy of bending for two beams:

$$\mathcal{E}_R = \mathcal{E}_L + \frac{1}{2}\int_0^a EIw''^2(x)\,dx + \frac{1}{2}\int_0^{a/\cos\alpha} EIw_1''^2(x)\,dx.$$

The boundary conditions are

$$u(0) = w(0) = w'(0) = 0, \qquad u_1(0) = w_1(0) = w_1'(0) = 0.$$

The beam-type joint conditions at C now include (2.1), which in this case become

$$u_1 = u\cos\alpha + w\sin\alpha,$$
$$w_1 = -u\sin\alpha + w\cos\alpha,$$

as well as the condition

$$w'\big|_C = w_1'\big|_C$$

expressing the requirement that the rotation angles for the deformations of both beams must match at C. To derive the natural conditions we will need both of (2.5) along with

$$\psi'\big|_C = \psi_1'\big|_C. \tag{2.6}$$

The rest of the derivation parallels that for part (a), modifications being required to treat the bending energy terms. We begin by writing $\delta\mathcal{E}_R = 0$. Then, selecting from the admissible virtual displacement functions $\varphi, \varphi_1, \psi, \psi_1$ that vanish on one of the beams, we derive four equations. Two of these are (2.3) for the displacements along the beams; the other two are for beam bending:

$$\begin{aligned} EIw^{(4)} &= 0 \text{ in } (0, a), \\ EIw_1^{(4)} &= 0 \text{ in } \left(0, \frac{a}{\cos\alpha}\right). \end{aligned}$$

Hence $\delta\mathcal{E}_R = 0$ yields

$$\begin{aligned} &ESu'\varphi\big|_C + ESu_1'\varphi_1\big|_C - P_1\varphi\big|_C - P_2\psi\big|_C \\ &\quad + EIw''\psi'\big|_C - EIw'''\psi\big|_C + EIw_1''\psi_1'\big|_C - EIw_1'''\psi_1\big|_C = 0. \end{aligned}$$

Using (2.5) and (2.6) and changing the set $\varphi_1, \psi_1, \psi_1'$ to φ, ψ, ψ' at point C, we get an equation that contains only φ, ψ, ψ' at C. As the values of these variables at C are arbitrary and independent, their coefficients must vanish. So we obtain three relations at C for u, w, u_1, w_1, which are the natural boundary conditions. These differ from the natural boundary conditions in part (a) but are still the equations of force balance of the section at C. Two of the equations contain all the terms seen in part (a); the third expresses the vanishing of the resultant couple at C. The details are left to the reader. □

Example 2.3. An elastic system consists of four identical beams rigidly clamped together (Fig. 2.6). Each beam has length a. Construct the model, write out the total potential energy functional and kinematic restrictions, and derive the equilibrium equations and natural boundary conditions.

Solution. Use the model of rigidly clamped elastic beams. Denote the deflection functions for the beams as follows: w_1 for AB, w_2 for BC, w_3 for BE, and w_4 for CD. For each beam, use an independent length parameter

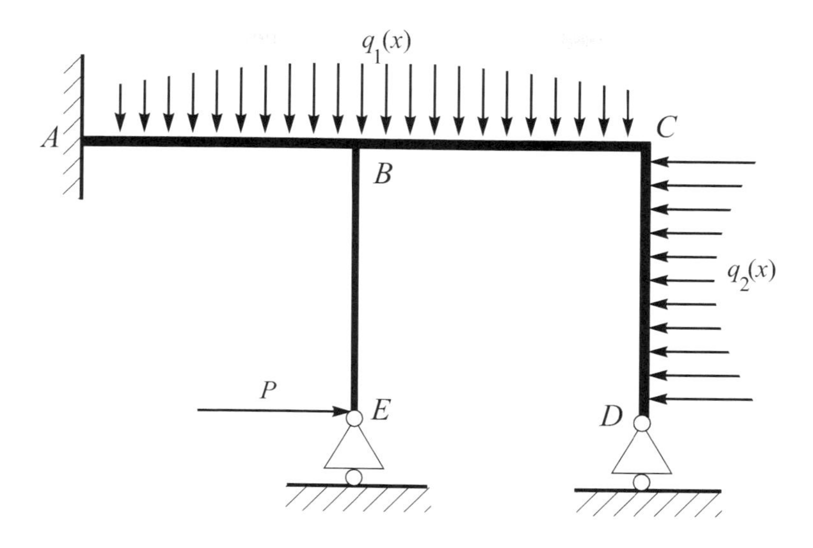

Fig. 2.6 Four beam system under load.

whose zero reference lies at the initial point of the beam. The total potential energy functional is

$$\mathcal{E} = \sum_{i=1}^{4} \frac{1}{2} \int_0^a EI w_i''^2(x)\, dx - \int_0^a q_1(x) w_1(x)\, dx$$
$$- \int_0^a q_1(x-a) w_2(x)\, dx - \int_0^a q_2(x) w_4(x)\, dx + P w_3(a).$$

Here we have considered that q_1 initially was a given function on the interval $[0, 2a]$. The kinematic boundary conditions are

$$w_1(0) = 0 = w_1'(0), \quad w_3(a) = 0, \quad w_1(a) = 0 = w_2(0), \quad w_2(a) = 0,$$
$$w_3(0) = 0, \quad w_4(a) = 0.$$

These take into account that the beams do not alter their lengths in the tangential directions. The joint constraints that define the additional natural boundary conditions are

$$w_1'(a) = w_2'(0) = w_3'(0), \qquad w_2'(a) = w_4'(0).$$

These define three additional natural restrictions; moreover, two natural restrictions also arise at the points E and D. A solution of the three fourth-order beam equilibrium equations will involve twelve integration constants. The total number of conditions at the points A, B, C, D is also twelve. □

2.2 Some Extremal Principles of Mechanics

Many physical — in particular, mechanical — problems drove the development of the calculus of variations. So it is not surprising that continuum mechanics and structural mechanics contain a host of variational principles. In this section we will consider some variational principles in the theory of elasticity and in linear plate theory, without penetrating too deeply into the mechanical details.

Elasticity

Describing small deformations of bodies under load, linear elasticity represents an extension of the ideas of § 2.1. But linear elasticity is only a first step toward describing spatial bodies made from various materials, elastic and non-elastic. Mechanicists employ more complex models related to heat transfer, viscous and plastic materials, problems of deterioration under load, etc.

Let us consider the mathematical formulation of boundary value problems in classical linear elasticity. (Henceforth the word "linear" will be omitted but understood.) These problems appear in many textbooks, e.g., [5; 15]. In Cartesian coordinates, the equations of motion are given by

$$\sigma_{ij,i} + f_j = \rho u_{j,tt} \qquad i,j = 1,2,3, \tag{2.7}$$

where σ_{ij} are the components of the stress tensor, u_i are the components of the displacement vector $\mathbf{u} = (u_1, u_2, u_3)$, f_i are the components of the volume force vector $\mathbf{f} = (f_1, f_2, f_3)$, and ρ is the material density. The stress components σ_{ij} describe the force interactions between portions of an elastic body. The following notation will be used for partial differentiation with respect to the spatial coordinates x_i and time t:

$$(\cdot)_{,i} = \frac{\partial(\cdot)}{\partial x_i}, \qquad (\cdot)_{,t} = \frac{\partial(\cdot)}{\partial t}.$$

In this chapter we modify our notation for partial derivatives in order to use Einstein's convention for repeated subscripts i, j, k, l, m, n. This will permit more concise expressions. Because the variable t is reserved for time and the subscripts x, y for partial derivatives with respect to the space variables, no summation over x, y, t is implied even when the symbols are repeated. For example, equation (2.7) is written instead of the more cumbersome form

$$\sum_{i=1}^{3} \frac{\partial \sigma_{ij}}{\partial x_i} + f_j = \rho \frac{\partial^2 u_j}{\partial t^2}, \qquad j = 1, 2, 3.$$

In elasticity it is shown that the matrix $[\sigma_{ij}]$ is symmetric so that $\sigma_{ij} = \sigma_{ji}$.

The Cartesian components ε_{sk} of the strain tensor are given by

$$\varepsilon_{sk} = \frac{1}{2}(u_{s,k} + u_{k,s}). \tag{2.8}$$

By definition, the matrix $[\varepsilon_{sk}]$ is also symmetric. The stress and strain tensors are related by the generalized version of Hooke's law. For an isotropic body this takes the form

$$\sigma_{ij} = \lambda \varepsilon_{kk} \delta_{ij} + 2\mu \varepsilon_{ij}, \tag{2.9}$$

where μ and λ are *Lamé's constants*. Relation (2.9) is an extension of Hooke's law for the rod: $\sigma = E\varepsilon$. Young's modulus E may be obtained from Lamé's constants as

$$E = \frac{\mu(3\lambda + 2\mu)}{\lambda + \mu}.$$

A more general form of Hooke's law for an anisotropic body is

$$\sigma_{ij} = C_{ijmn} \varepsilon_{mn}$$

where the *elastic moduli* C_{ijmn} (the components of a tensor of elastic moduli) satisfy

$$C_{ijmn} = C_{jimn} = C_{ijnm} = C_{mnij}.$$

As a result, the set C_{ijmn} consists of no more than 21 independent constants. For an isotropic body, the number of independent elastic constants is two; they can be chosen as the Lamé constants μ and λ so that

$$C_{ijmn} = \lambda \delta_{ij} \delta_{mn} + \mu(\delta_{im} \delta_{jn} + \delta_{in} \delta_{jm}),$$

where δ_{ij} is Kronecker's symbol.

Substitution of (2.9) and (2.8) into (2.7) yields the equations of motion in terms of the displacements:

$$(\lambda + \mu) u_{i,ki} + \mu u_{k,ii} + \rho f_k = \rho u_{k,tt} \qquad i, k = 1, 2, 3. \tag{2.10}$$

For equilibrium problems, the equations of motion reduce to

$$\sigma_{ij,i} + f_j = 0, \qquad j = 1, 2, 3 \tag{2.11}$$

or, in terms of displacements,

$$(\lambda + \mu)u_{i,ki} + \mu u_{k,ii} + f_k = 0, \qquad k = 1, 2, 3. \tag{2.12}$$

Two types of boundary conditions occur in the formulation of boundary value problems in elasticity. Suppose the boundary $S = \partial V$ of a body consists of two nonoverlapping portions S_1 and S_2 so that $S = S_1 \cup S_2$ and $S_1 \cap S_2 = \emptyset$. If the displacement vector is given on S_1, we have a boundary condition of the form

$$u_i\big|_{S_1} = u_i^0, \qquad i = 1, 2, 3, \tag{2.13}$$

where u_i^0 is a given function. If external forces (p_1, p_2, p_3) act over S_2, the condition is

$$n_i\sigma_{ij}\big|_{S_2} = p_j, \qquad j = 1, 2, 3, \tag{2.14}$$

where the p_j are given functions and n_i are the components of the outward unit normal to S. A *mixed* boundary value problem would involve both types of conditions. On the other hand, it is possible for a condition of the form (2.13) to prevail over all of S, or for a condition of the form (2.14) to prevail over all of S. The dynamic problems of elasticity also require initial conditions of the form

$$u_i\big|_{t=0} = \hat{u}_i, \quad u_{i,t}\big|_{t=0} = \hat{v}_i, \qquad i = 1, 2, 3.$$

In elasticity, the strain energy function W is introduced as a quadratic function of the ε_{mn}:

$$W(\varepsilon_{mn}) = \frac{1}{2}\varepsilon_{ij}C_{ijmn}\varepsilon_{mn}.$$

For an isotropic material this reduces to

$$W(\varepsilon_{mn}) = \frac{1}{2}\lambda\varepsilon_{ii}^2 + \mu\varepsilon_{ij}\varepsilon_{ij}. \tag{2.15}$$

From thermodynamic considerations it follows that W is positive definite:

$$W(\varepsilon_{mn}) > 0 \quad \text{whenever } \varepsilon_{mn} \neq 0. \tag{2.16}$$

This implies the following inequalities for the elastic moduli:

$$3\lambda + 2\mu > 0, \qquad \mu > 0. \tag{2.17}$$

It can be shown that W is the potential for stresses:

$$\sigma_{ij} = W_{,\varepsilon_{ij}}.$$

The existence of W allows us to formulate *Lagrange's variational principle for elasticity*:

Theorem 2.4. *A stationary point* $\mathbf{u} = (u_1, u_2, u_3)$ *of the total potential energy functional*

$$\mathcal{E}(\mathbf{u}) = \iiint_V W(\varepsilon_{mn})\, dV - \iiint_V f_i u_i \, dV - \iint_{S_2} p_i u_i \, dS$$

on the set of admissible displacements subject to (2.13) *satisfies the equilibrium equations* (2.11) *in the volume* V *and the boundary condition* (2.14). *The converse also holds. This stationary point is the minimum of* $\mathcal{E}$.

Proof. Using the formula

$$\delta\mathcal{E} = \frac{d}{d\tau}\mathcal{E}(\mathbf{u} + \tau\boldsymbol{\varphi})\Big|_{\tau=0}, \qquad \boldsymbol{\varphi} = (\varphi_1, \varphi_2, \varphi_3),$$

let us find the first variation of $\mathcal{E}$:

$$\begin{aligned}
\delta\mathcal{E} &= \iiint_V \frac{1}{2} W_{,\varepsilon_{ij}}(\varphi_{i,j} + \varphi_{j,i})\, dV - \iiint_V f_i \varphi_i \, dV - \iint_{S_2} p_i \varphi_i \, dS \\
&= \iiint_V W_{,\varepsilon_{ij}} \varphi_{j,i}\, dV - \iiint_V f_i \varphi_i \, dV - \iint_{S_2} p_i \varphi_i \, dS \\
&= \iiint_V \sigma_{ij} \varphi_{j,i}\, dV - \iiint_V f_i \varphi_i \, dV - \iint_{S_2} p_i \varphi_i \, dS.
\end{aligned}$$

We show that if $\delta\mathcal{E} = 0$ for all admissible φ_i, then (2.11) and (2.14) hold. The Gauss–Ostrogradski formula gives

$$\begin{aligned}
0 = \delta\mathcal{E} &= \iiint_V \sigma_{ij} \varphi_{j,i}\, dV - \iiint_V f_i \varphi_i \, dV - \iint_{S_2} p_i \varphi_i \, dS \\
&= - \iiint_V (\sigma_{ij,i} + f_j)\, \varphi_j \, dV + \iint_{S_1} n_k \sigma_{kj} \varphi_j \, dS \\
&\quad + \iint_{S_2} (n_k \sigma_{kj} - p_j) \varphi_j \, dS.
\end{aligned}$$

Recall that the φ_i satisfy the homogeneous version of (2.13), i.e., $\varphi_i|_{S_1} = 0$.

From the arbitrariness of φ_i, a two-step derivation (first for the volume integrals where we take $\varphi = 0$ on S_2, then for the surface integrals) yields the required equations

$$\sigma_{ij,i} + f_j = 0 \text{ in } V, \qquad n_k \sigma_{kj}\big|_{S_2} = p_j.$$

Conversely, on a solution $\mathbf{u}$ of the equilibrium problem we have $\delta\mathcal{E} = 0$ for any admissible φ_i that vanishes on S_1. Indeed, multiply the ith equation of (2.11) by φ_i, add the results, then integrate over V. Using a similar formula obtained by multiplying the ith equation of (2.14) by φ_i, summing, and integrating over S_2, we get

$$\begin{aligned}
0 &= \iiint_V (\sigma_{ki,k} + f_i)\,\varphi_i\,dV - \iint_{S_2} (n_k\sigma_{ki} - p_i)\varphi_i\,dS \\
&= -\iiint_V \sigma_{kj}\varphi_{j,k}\,dV + \iiint_V f_i\varphi_i\,dV \\
&\quad + \iint_S n_k\sigma_{ki}\varphi_i\,dS - \iint_{S_2} (n_k\sigma_{ki} - p_i)\varphi_i\,dS \\
&= -\iiint_V \sigma_{kj}\varphi_{j,k}\,dV + \iiint_V f_i\varphi_i\,dV + \iint_{S_2} p_i\varphi_i\,dS \\
&= -\delta\mathcal{E}.
\end{aligned}$$

Hence a stationary point of $\mathcal{E}$ is a solution to the equilibrium problem for the elastic body, and vice versa.

Finally we show that $\mathcal{E}$ attains its minimum at the stationary point. The proof uses the fact that W is a positive definite quadratic form in the strain components. Let $\tilde{\mathbf{u}} = (\tilde{u}_1, \tilde{u}_2, \tilde{u}_3)$ be another admissible vector function satisfying (2.13) and consider the difference

$$\Delta\mathcal{E} = \mathcal{E}(\tilde{\mathbf{u}}) - \mathcal{E}(\mathbf{u}).$$

We get

$$\begin{aligned}
\Delta\mathcal{E} &= \iiint_V W(\tilde{\varepsilon}_{mn})\,dV - \iiint_V f_i\tilde{u}_i\,dV - \iint_{S_2} p_i\tilde{u}_i\,dS \\
&\quad - \iiint_V W(\varepsilon_{mn})\,dV + \iiint_V f_i u_i\,dV + \iint_{S_2} p_i u_i\,dS \\
&= \iiint_V [W(\tilde{\varepsilon}_{mn}) - W(\varepsilon_{mn})]\,dV \\
&\quad - \iiint_V f_i(\tilde{u}_i - u_i)\,dV - \iint_{S_2} p_i(\tilde{u}_i - u_i)\,dS.
\end{aligned}$$

Let $\varphi_i = \tilde{u}_i - u_i$. Because $\tilde{u}_i$ and u_i coincide on S_1, we have $\varphi_i|_{S_1} = 0$. Next,

$$\begin{aligned}
2\,[W(\tilde{\varepsilon}_{mn}) - W(\varepsilon_{mn})] &= \lambda\tilde{\varepsilon}_{ii}^2 + 2\mu\tilde{\varepsilon}_{ij}\tilde{\varepsilon}_{ij} - \lambda\varepsilon_{ii}^2 - 2\mu\varepsilon_{ij}\varepsilon_{ij} \\
&= \lambda\tilde{\tilde{\varepsilon}}_{ii}^2 + 2\mu\tilde{\tilde{\varepsilon}}_{ij}\tilde{\tilde{\varepsilon}}_{ij} + 2\lambda\tilde{\tilde{\varepsilon}}_{ii}\tilde{\varepsilon}_{ii} + 4\mu\tilde{\tilde{\varepsilon}}_{ij}\tilde{\varepsilon}_{ij} \\
&= 2W(\tilde{\tilde{\varepsilon}}_{mn}) + 2\lambda\tilde{\tilde{\varepsilon}}_{ii}\tilde{\varepsilon}_{ii} + 4\mu\tilde{\tilde{\varepsilon}}_{ij}\tilde{\varepsilon}_{ij}
\end{aligned}$$

where

$$\tilde{\varepsilon}_{mn} = \frac{1}{2}(\tilde{u}_{m,n} + \tilde{u}_{n,m}), \qquad \tilde{\tilde{\varepsilon}}_{mn} = \frac{1}{2}(\varphi_{m,n} + \varphi_{n,m}).$$

Therefore

$$\begin{aligned}
\Delta\mathcal{E} &= \iiint_V W(\tilde{\tilde{\varepsilon}}_{mn})\,dV + \iiint_V (\lambda\varepsilon_{ii}\tilde{\tilde{\varepsilon}}_{ii} + 2\mu\varepsilon_{ij}\tilde{\tilde{\varepsilon}}_{ij})\,dV \\
&\quad - \iiint_V f_i\varphi_i\,dV - \iint_{S_2} p_i\varphi_i\,dS \\
&= \iiint_V W(\tilde{\tilde{\varepsilon}}_{mn})\,dV + \iiint_V \sigma_{ij}\varphi_{j,i}\,dV \\
&\quad - \iiint_V f_i\varphi_i\,dV - \iint_{S_2} p_i\varphi_i\,dS \\
&= \iiint_V W(\tilde{\tilde{\varepsilon}}_{mn})\,dV + \delta\mathcal{E}.
\end{aligned}$$

Because $\mathbf{u} = (u_1, u_2, u_3)$ is a solution, the first variation $\delta\mathcal{E} = 0$ for any admissible φ and we have

$$\Delta\mathcal{E} = \iiint_V W(\tilde{\tilde{\varepsilon}}_{mn})\,dV. \tag{2.18}$$

The positive definiteness of W means that $\Delta\mathcal{E} \geq 0$ for any admissible $\tilde{u}_i$. Hence the set of u_i are a global minimizer of $\mathcal{E}$. □

This proof also establishes the *virtual work principle*:

Theorem 2.5. *Sufficiently smooth functions u_i that vanish on S_1 are a solution to the boundary value problem* (2.11), (2.13), (2.14), *if and only if the equation*

$$\iiint_V \sigma_{ij}\varphi_{j,i}\,dV - \iiint_V f_i\varphi_i\,dV - \iint_{S_2} p_i\varphi_i\,dS = 0, \tag{2.19}$$

with σ_{ij} given by (2.9), *holds for any sufficiently smooth functions φ_i that also vanish on ∂S_1.*

The virtual work principle underlies the notion of weak solutions in elasticity (Chapter 5). It is more general than Lagrange's principle as it can be extended to nonconservative systems for which total potential energy functionals do not exist.

Hamilton's least action principle is the basis for variational formulations in dynamics. Let the kinetic energy density of a body be given by

$$K = \frac{1}{2}\rho(u_{1,t}^2 + u_{2,t}^2 + u_{3,t}^2).$$

In this case we say that a function is admissible if it (1) vanishes on S_1 and (2) takes the values of a solution to the dynamical problem at time instants t_1 and t_2. This means that we consider the admissible variations $\varphi_i(x_1, x_2, x_3, t)$ of the solution to the problem such that $\varphi_i\big|_{S_1} = 0$ and $\varphi_i\big|_{t=t_1} = 0 = \varphi_i\big|_{t=t_2}$. Hamilton's principle is formulated as follows.

Theorem 2.6. *A solution to a boundary value problem in the dynamics of elastic solids (i.e., a solution to (2.7), (2.13), and (2.14)) is a stationary point of the* ***action functional***

$$\mathcal{E}_A(\mathbf{u}) = \int_{t_1}^{t_2} \left(\iiint_V (K - W)\, dV + \iiint_V f_i u_i \, dV + \iint_{S_2} p_i u_i \, dS \right) dt$$

in the class of admissible functions that satisfy (2.13) and take prescribed values coincident with the solution at time instants t_1 and t_2. Conversely, a stationary point of $\mathcal{E}_A$ in the class of admissible functions is a solution to the dynamical boundary value problem for an elastic body.

Proof. The first variation of $\mathcal{E}_A$ is

$$\delta\mathcal{E}_A = \int_{t_1}^{t_2} \left(\iiint_V (\rho u_{i,t}\varphi_{i,t} - \sigma_{ij}\varphi_{j,i} + f_i\varphi_i)\, dV + \iint_{S_2} p_i\varphi_i \, dS \right) dt.$$

Integrating by parts, we have

$$\begin{aligned}\delta\mathcal{E}_A = \int_{t_1}^{t_2} &\left(\iiint_V (-\rho u_{i,tt}\varphi_i + \sigma_{ij,i}\varphi_j + f_i\varphi_i)\, dV - \iint_S n_k\sigma_{kj}\varphi_j \, dS \right. \\ &\left. + \iint_{S_2} p_i\varphi_i \, dS \right) dt + \iiint_V \rho u_{i,t}\varphi_i \, dV \Bigg|_{t=t_1}^{t=t_2}.\end{aligned}$$

But $\varphi_i\big|_{S_1} = 0$ and $\varphi_i\big|_{t=t_1} = 0 = \varphi_i\big|_{t=t_2}$, so

$$\begin{aligned}\delta\mathcal{E}_A = \int_{t_1}^{t_2} &\left(\iiint_V (-\rho u_{j,tt} + \sigma_{ij,i} + f_j)\varphi_j \, dV \right. \\ &\left. - \iint_S (n_k\sigma_{kj} + p_j)\varphi_j \, dS \right) dt.\end{aligned}$$

Hence if $\delta\mathcal{E}_A = 0$ for all admissible φ_i, the equations of motion (2.7) and the boundary conditions (2.14) follow.

Conversely, if $\mathbf{u}$ is a solution to the dynamic problem, the first variation of $\mathcal{E}_A$ is zero. The proof is similar to the proof of the corresponding part of Lagrange's principle. The difference lies in the sets of admissible functions and in the domain of integration, which for Hamilton's principle is $V \times [t_1, t_2]$. The details are left to the reader. □

We should note that Hamilton's principle is not minimal; it yields stationary points of the action functional.

Other variational principles in elasticity bear names such as Castigliano, Reissner, Washizu, Tonti, and Hashin-Strikman. Some are minimal or maximal like Lagrange's principle; others are stationary like Hamilton's principle. In addition to their roles in proving existence theorems, they form the basis for practical engineering approaches such as the finite element method. Moreover, extensions of variational methods turned out to be useful in the theory of more complex problems in nonlinear elasticity, plasticity, viscoelasticity, and so on.

Reissner–Mindlin plate theory

In Chapter 1 we examined the plate equations in the framework of Kirchhoff's theory. We used the energy functional (1.94) to derive the Euler–Lagrange equations, which are the equilibrium equations for the plate, and the natural boundary conditions (cf., equations (1.94)–(1.103) and Exercise 1.16). These results are revisited later in this section. Now we consider the more general plate theory of Reissner and Mindlin, also known as shear-deformable plate theory of first order.

In Reissner–Mindlin plate theory, the bending of an elastic plate is described by the equations

$$M_{11,1} + M_{21,2} - Q_1 = \rho J \vartheta_{1,tt}, \tag{2.20}$$

$$M_{12,1} + M_{22,2} - Q_2 = \rho J \vartheta_{2,tt}, \tag{2.21}$$

$$Q_{1,1} + Q_{2,2} + p = \rho h w_{,tt}, \tag{2.22}$$

where the $M_{\alpha\beta}$ are the bending and twisting moments ($\alpha, \beta = 1, 2$), the Q_α are the transverse shear forces, the ϑ_α are the averaged rotations of fibers normal to the plate midsurface before deformation, w is the deflection, ρ is the density, J is the moment of inertia, h is the plate thickness, and p is the transverse load. We recall that the partial derivatives of the components of vector functions are denoted by $(\cdot)_{,\alpha} = \partial(\cdot)/\partial x_\alpha$, where $x_1 = x$ and $x_2 = y$ are Cartesian coordinates in the midplane. Note that Greek letters are used for the subscripts. In shell theory, Greek indices usually range over the values 1 and 2. The Latin indices typically employed in the three-dimensional theory range over the values 1, 2, and 3.

The constitutive equations — i.e., the relations between the bending and twisting moments, the transverse shear forces, and the surface strain measures — are given by

$$M_{11} = D(\vartheta_{1,1} + \nu\vartheta_{2,2}), \qquad M_{22} = D(\vartheta_{2,2} + \nu\vartheta_{1,1}), \tag{2.23}$$

$$M_{12} = M_{21} = \frac{D(1-\nu)}{2}(\vartheta_{1,2} + \vartheta_{2,1}), \tag{2.24}$$

$$Q_1 = \Gamma(w_{,1} + \vartheta_1), \qquad Q_2 = \Gamma(w_{,2} + \vartheta_2), \tag{2.25}$$

$$D = \frac{Eh^3}{12(1-\nu^2)}, \qquad \Gamma = k\mu h, \tag{2.26}$$

where E is Young's modulus, μ is the shear modulus, ν is Poisson's ratio, D is the bending stiffness, Γ is the transverse shear stiffness, and k is the shear correction factor. For k, Reissner proposed $k = 5/6$ whereas Mindlin took $k = \pi^2/12$. Other values of k also appear in the literature.

In this theory, on the boundary contour ∂S or a portion ∂S_1, kinematic boundary conditions consist of given deflections and rotations:

$$w\big|_{\partial S_1} = w^0, \qquad \vartheta_\alpha\big|_{\partial S_1} = \vartheta_\alpha^0. \tag{2.27}$$

Static boundary conditions are

$$n_\alpha M_{\alpha\beta}\big|_{\partial S_2} = M_\beta^0, \qquad Q_\alpha n_\alpha\big|_{\partial S_2} = Q_n^0. \tag{2.28}$$

In (2.27)–(2.28), the quantities w^0, θ_α^0, M_β^0, and Q_n^0 are given functions of the arc-length parameter s. The quantities n_1 and n_2 are the components of the outward unit normal to ∂S.

In equilibrium, equations (2.20)–(2.22) reduce to

$$M_{11,1} + M_{21,2} - Q_1 = 0, \tag{2.29}$$

$$M_{12,1} + M_{22,2} - Q_2 = 0, \tag{2.30}$$

$$Q_{1,1} + Q_{2,2} + p = 0. \tag{2.31}$$

Solving (2.29) and (2.30) for Q_1 and Q_2, and substituting these into (2.31), we obtain

$$M_{11,11} + 2M_{12,12} + M_{22,22} + p = 0. \tag{2.32}$$

The strain energy density for plate bending is

$$\begin{aligned} W(\varkappa_{\alpha\beta}, \gamma_\alpha) &= \frac{1}{2}\left[M_{\alpha\beta}\varkappa_{\alpha\beta} + Q_\alpha\gamma_\alpha\right] \\ &= \frac{D}{2}\left[\varkappa_{11}^2 + \varkappa_{22}^2 + 2\nu\varkappa_{11}\varkappa_{22} + \frac{1-\nu}{2}(\varkappa_{12}^2 + 2\varkappa_{12}\varkappa_{21} + \varkappa_{21}^2)\right] \\ &\quad + \frac{\Gamma}{2}(\gamma_1^2 + \gamma_2^2), \end{aligned}$$

where $\varkappa_{\alpha\beta}$ are the components of the bending tensor (or tensor of change of curvature), and γ_α are the shear strain components defined by

$$\varkappa_{\alpha\beta} = \vartheta_{\alpha,\beta}, \quad \gamma_\alpha = w_{,\alpha} + \vartheta_\alpha.$$

It can be directly verified that

$$M_{\alpha\beta} = \frac{\partial W}{\partial \varkappa_{\alpha\beta}}, \qquad Q_\alpha = \frac{\partial W}{\partial \gamma_\alpha}. \tag{2.33}$$

In the Reissner–Mindlin plate theory, the kinetic energy density is

$$K = \frac{\rho h}{2}(w_{,t})^2 + \frac{\rho J}{2}\left[(\vartheta_{2,t})^2 + (\vartheta_{2,t})^2\right].$$

Plate theory features variational principles similar to those in linear elasticity. Lagrange's variational principle is exhibited in the following theorem.

Theorem 2.7. *A solution of boundary value problem* (2.29)–(2.31), (2.27), (2.28) *is a stationary point of the energy functional*

$$\mathcal{E}(w, \vartheta_1, \vartheta_2) = \iint_S W\, dS - \iint_S pw\, dS - \int_{\partial S_2} (Q_n^0 w + M_\beta^0 \vartheta_\beta)\, ds. \tag{2.34}$$

Conversely, sufficiently smooth functions ϑ_α and w that constitute a stationary point of $\mathcal{E}$ in the class of all admissible functions (i.e., satisfying the kinematic boundary conditions (2.27)*), satisfy the equilibrium equations* (2.29)–(2.31) *and boundary conditions* (2.28)*. Moreover, at a stationary point $\mathcal{E}$ takes its global minimum value.*

Proof. If $\delta\mathcal{E} = 0$ for all admissible variations, then (2.29)–(2.31) and (2.28) hold. Indeed, let φ_0, φ_1, and φ_2 be any three continuously differentiable functions that vanish on ∂S_1. Consider $\mathcal{E}(w+\tau\varphi_0, \vartheta_1+\tau\varphi_1, \vartheta_2+\tau\varphi_1)$ and calculate its derivative with respect to τ at $\tau = 0$. Using (2.33) and integration by parts, we get

$$\begin{aligned}
\delta\mathcal{E} &= \frac{d}{d\tau}\mathcal{E}(w + \tau\varphi_0, \vartheta_1 + \tau\varphi_1, \vartheta_2 + \tau\varphi_1)\Big|_{\tau=0} \\
&= \iint_S (M_{\alpha\beta}\varphi_{\beta,\alpha} + Q_\alpha(\varphi_{0,\alpha} + \varphi_\alpha))\, dS - \iint_S p\varphi_0\, dS \\
&\quad - \int_{\partial S_2} (Q_n^0 \varphi_0 + M_\beta^0 \varphi_\beta)\, ds \\
&= - \iint_S (M_{\alpha\beta,\alpha} - Q_\beta)\varphi_\beta + (Q_{\alpha,\alpha} + p)\varphi_0\, dS \\
&\quad + \int_{\partial S_2} \left[(n_\alpha Q_\alpha - Q_n^0)\varphi_0 + (n_\alpha M_{\alpha\beta} - M_\beta^0)\varphi_\beta\right] ds.
\end{aligned}$$

Since $\varphi_0, \varphi_1, \varphi_2$ are arbitrary, from $\delta\mathcal{E} = 0$ it follows that (2.29)–(2.31) are the Euler–Lagrange equations and (2.28) are the natural boundary conditions for the energy functional $\mathcal{E}$ from the Reissner–Mindlin theory.

Conversely, let w and ϑ_α constitute a solution to the boundary value problem (2.29)–(2.31), (2.27), (2.28). We show that the first variation of $\mathcal{E}$ vanishes on this solution. Again let $\varphi_0, \varphi_1, \varphi_2$ be any smooth functions that vanish on ∂S_1. Multiply (2.29) and (2.30) by φ_1 and φ_2, respectively; multiply (2.31) by φ_0. Then add the results and integrate over S. We perform similar operations with the boundary conditions: multiply $(2.28)_1$ by φ_β and $(2.28)_2$ by φ_0, add the results, and integrate over ∂S_2. We get

$$
\begin{aligned}
0 &= \iint_S (M_{\alpha\beta,\alpha} - Q_\beta)\varphi_\beta + (Q_{\alpha,\alpha} + p)\varphi_0 \, dS \\
&\quad - \int_{\partial S_2} \left[(n_\alpha Q_\alpha - Q_n^0)\varphi_0 + (n_\alpha M_{\alpha\beta} - M_\beta^0)\varphi_\beta\right] ds \\
&= -\iint_S (M_{\alpha\beta}\varphi_{\beta,\alpha} - p\varphi_0 + Q_\alpha(\varphi_{0,\alpha} + \varphi_\alpha)) \, dS \\
&\quad + \int_{\partial S_2} (Q_n^0\varphi_0 + M_\beta^0\varphi_\beta) \, ds \\
&= -\delta\mathcal{E},
\end{aligned}
$$

which proves the assertion.

To show that a stationary point of $\mathcal{E}$ is its global minimum, we use the positive definiteness of the quadratic form representing the strain energy W. Let w and ϑ_α constitute a stationary point of $\mathcal{E}$. Suppose $\tilde{w}$ and $\tilde{\vartheta}_\alpha$ also satisfy (2.27). Consider the difference between the values of the strain energy functional for these two sets:

$$\Delta\mathcal{E} = \mathcal{E}(\tilde{w}, \tilde{\vartheta}_\alpha) - \mathcal{E}(w, \vartheta_\alpha).$$

It can be shown that

$$\Delta\mathcal{E} = \iint_S W(\Delta\varkappa_{\alpha\beta}, \Delta\gamma_\alpha) \, dS,$$

where

$$\Delta\varkappa_{\alpha\beta} = \tilde{\vartheta}_{\alpha,\beta} - \vartheta_{\alpha,\beta}, \quad \Delta\gamma_\alpha = \tilde{w}_{,\alpha} + \tilde{\vartheta}_\alpha - w_{,\alpha} - \vartheta_\alpha.$$

Hence $\mathcal{E}$ takes its global minimum value at w, ϑ_α. □

This proof also establishes the virtual work principle in plate theory:

Theorem 2.8. *Sufficiently smooth functions w, ϑ_α that vanish on ∂S_1 constitute a solution of the boundary value problem (2.29)–(2.31), (2.27), (2.28) if and only if the equation*

$$\iint_S (M_{\alpha\beta}\varphi_{\beta,\alpha} + Q_\alpha\varphi_{0,\alpha} + Q_\alpha\varphi_\alpha - p\varphi_0)\,dS - \int_{\partial S_2} (Q_n^0\varphi_0 + M_\beta^0\varphi_\beta)\,ds = 0, \tag{2.35}$$

with $M_{\alpha\beta}$ and Q_α given by (2.23)–(2.25), holds for any sufficiently smooth functions $\varphi_0, \varphi_1, \varphi_2$ that also vanish on ∂S_1.

Equation (2.35) forms the basis for various versions of the finite element method in plate theory.

Hamilton's variational principle holds for dynamic problems in plate theory:

Theorem 2.9. *A solution to the dynamical boundary value problem (2.20)–(2.22), (2.27), (2.28) is a stationary point of the action functional*

$$\mathcal{E}_A = \int_{t_1}^{t_2} \left(\iint_S (K - W)\,dS + \iint_S pw\,dS + \int_{\partial S_2} (Q_n^0 w + M_\beta^0\vartheta_\beta)\,ds \right) dt.$$

in the class of admissible functions (i.e., satisfying (2.27) and taking prescribed values coincident with the solution at times t_1 and t_2). Conversely, a stationary point of $\mathcal{E}_A$ in the class of admissible functions is a solution of the dynamical boundary value problem for the plate.

The proof mimics the proof of Hamilton's principle in elasticity and is left to the reader. As in elasticity, Hamilton's principle for plates is not a minimal principle; it is only a stationary principle.

Kirchhoff plate theory

The classical Kirchhoff theory is easily derived from the Reissner–Mindlin theory. In the former, the rotations ϑ_α and the deflection w are related by

$$\vartheta_1 = -w_{,1}, \qquad \vartheta_2 = -w_{,2}. \tag{2.36}$$

So in Kirchhoff theory, bending of the plate is described by one function: the deflection $w(x, y, t)$. This allows us to return to a simpler notation for the partial derivatives of w. We shall write $w_{,1} = w_x$, $w_{,12} = w_{xy}$, etc. The constitutive equations for the moments now take the form

$$M_{11} = -D(w_{xx} + \nu w_{yy}), \qquad M_{22} = -D(w_{yy} + \nu w_{xx}), \tag{2.37}$$

$$M_{12} = M_{21} = -D(1 - \nu)w_{xy}. \tag{2.38}$$

Equilibrium equation (2.32) reduces to an equation in w that we saw in Chapter 1:

$$D\Delta w + p = 0. \tag{2.39}$$

The strain energy functional for a Kirchhoff plate is (1.94):

$$\mathcal{E}(w) = \iint_S W\, dS - \iint_S pw\, dS, \tag{2.40}$$

$$W = \frac{D}{2}\left[w_{xx}^2 + w_{yy}^2 + 2\nu w_{xx}w_{yy} + 2(1-\nu)w_{xy}^2\right]. \tag{2.41}$$

To avoid awkward formulas we assumed here an absence of boundary loads, i.e., $Q_n^0 = M_\beta^0 = 0$. See also the derivation of the natural boundary conditions (1.102) and (1.103) in Chapter 1. In Kirchhoff's plate theory, the rotational inertia is usually neglected, so the kinetic energy becomes

$$K = \frac{\rho h}{2}(w_t)^2.$$

Hamilton's variational principle reduces to finding stationary points of the functional

$$\mathcal{E}_A(w) = \int_{t_1}^{t_2} (K - W + pw)\, dS\, dt$$

on the class of admissible functions $w(x, y, t)$ that take prescribed values at times t_1 and t_2. The main results of the Kirchhoff theory parallel the corresponding theorems formulated above in the Reissner–Mindlin theory. Detailed formulations and proofs are left to the reader.

Interaction of a plate with elastic beams

In engineering, plates are sometimes reinforced with elastic beams. Deduction of compatibility conditions for deformation of a plate-beam system is not a trivial problem. Mathematically, we must seek compatibility equations for a system of partial differential equations for the plate and a system of ordinary differential equations for the beams. Physically, it is important to analyze the deformation and tension fields in the neighborhood of the joints between the plate and the beams.

Here we will consider the variational deduction of the equilibrium equations for a plate connected with an elastic beam over a portion of its boundary contour (Fig. 2.7). The approach is to represent the potential energy functional as a sum of the energy functionals for the coupled plate and

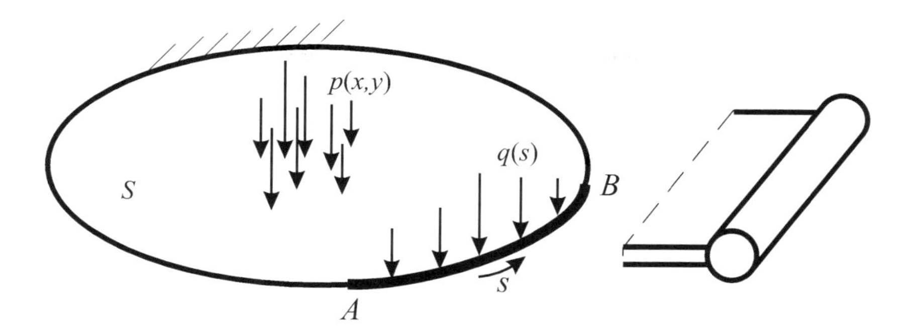

Fig. 2.7 Left: plate with part of its boundary contour supported by beam AB. Right: detail of the beam support.

beam. Formulating kinematic compatibility conditions for the displacements and rotations for the plate and the beam, we then derive static compatibility conditions for the plate and the beam. These are the natural boundary conditions for the energy functional.

To understand what is happening in this problem, we will treat a simplified problem that has its own significance. Consider a rectangular plate of dimensions a and b, supported by two straight beams as shown in Fig. 2.8. Suppose the edge AD is rigidly clamped and the edge BC is free from kinematic restrictions. The beams are clamped along AB and CD. Physically, bending of the plate implies rotation of the beam cross sections. Therefore in describing the deformations of the coupled system we must account for torsion as well as bending. Earlier we considered the bending equations for a beam. The energy functionals for beam bending for AB and CD are

$$\mathcal{E}_{b1}(u_1) = \frac{1}{2}\int_0^b E_1 I_1 (u_1''(y))^2\,dy - \int_0^b q_1(y)u_1(y)\,dy,$$
$$\mathcal{E}_{b2}(u_2) = \frac{1}{2}\int_0^b E_2 I_2 (u_2''(y))^2\,dy - \int_0^b q_2(y)u_2(y)\,dy,$$

where the E_α are Young's moduli, I_α are the moments of inertia of the beams, and $u_\alpha(y)$ are the vertical beam deflections.

Torsion in a beam is a classical problem in the strength of materials [29; 28]. The energy functionals for torsion in the beams AB and CD are

$$\mathcal{E}_{t1}(\psi_1) = \frac{1}{2}\int_0^b D_{T1}(\psi_1'(y))^2\,dy, \quad \mathcal{E}_{t2}(\psi_2) = \frac{1}{2}\int_0^b D_{T2}(\psi_2'(y))^2\,dy.$$

Here ψ_α denotes the beam twisting angles per unit length, and $D_{T\alpha}$ is the torsional rigidity.

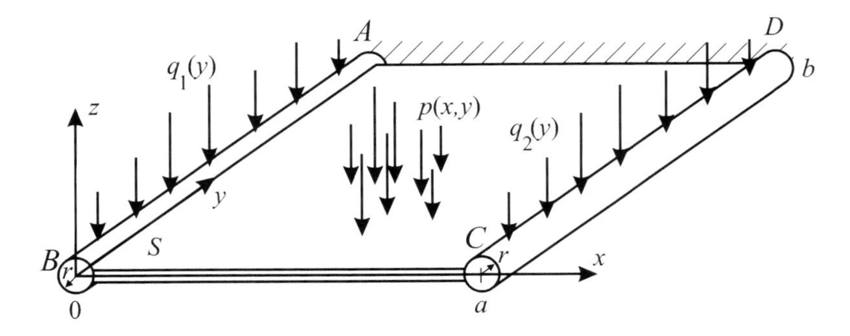

Fig. 2.8 A cantilever plate supported by two straight beams.

Thus the total potential energy functional of the plate with two reinforcement beams is

$$\begin{aligned}\mathcal{E}(w, \vartheta_1, \vartheta_2, u_1, u_2, \psi_1, \psi_2) &= \iint_S W\,dx\,dy - \iint_S pw\,dx\,dy \\ &+ \mathcal{E}_{b1}(u_1) + \mathcal{E}_{b2}(u_2) + \mathcal{E}_{t1}(\psi_1) + \mathcal{E}_{t2}(\psi_2).\end{aligned} \tag{2.42}$$

Kinematic boundary conditions are the equations that describe rigid clamping of the plate along AD, clamping of the beams at points A and D, and the equality of the twisting angle to zero at A and D:

$$\begin{aligned}&w\Big|_{y=b} = 0 = \vartheta_1\Big|_{y=b} = \vartheta_1\Big|_{y=b}, \\ &u_1\Big|_{y=b} = u_2\Big|_{y=b} = 0 = \psi_1\Big|_{y=b} = \psi_2\Big|_{y=b}.\end{aligned} \tag{2.43}$$

Kinematic compatibility of deformation for the plate and beams requires equality between the defections of the plate edges and the beams,

$$u_1(y) = w(0, y) - r\psi_1(y), \quad u_2(y) = w(a, y) + r\psi_2(y), \tag{2.44}$$

and equality of the corresponding rotation angles:

$$\psi_1(y) = \vartheta_1(0, y), \quad \psi_2(y) = \vartheta_1(a, y). \tag{2.45}$$

The kinematic compatibility conditions (2.44) describe coupling between a plate and a pair of beams having circular cross sections of radius r as in Fig. 2.8. Clearly this is not the only way to fix beams to a plate. For beams of more complicated cross section, the kinematic compatibility conditions can differ from (2.44); however, the analysis will be similar.

By (2.44) and (2.45), the energy functional takes the form

$$\begin{aligned}\mathcal{E}(w,\vartheta_1,\vartheta_2) &= \iint_S W\,dx\,dy - \iint_S pw\,dx\,dy \\ &\quad + \mathcal{E}_{b1}(w(0,y) - r\vartheta_1(0,y)) + \mathcal{E}_{b2}(w(a,y) + r\vartheta_1(a,y)) \\ &\quad + \mathcal{E}_{t1}(\vartheta_1(0,y)) + \mathcal{E}_{t2}(\vartheta_1(a,y)). \end{aligned} \tag{2.46}$$

Natural boundary conditions for the plate follow from the condition $\delta\mathcal{E} = 0$. We have

$$\begin{aligned} 0 = \delta\mathcal{E} = &-\int_0^a\int_0^b \left[(M_{\alpha\beta,\alpha} - Q_\beta)\varphi_\beta + (Q_{\alpha,\alpha} + p)\varphi_0\right] dx\,dy \\ &+ \int_{\partial S_2} (n_\alpha M_{\alpha\beta}\varphi_\alpha + n_\alpha Q_\alpha\varphi_0)\,ds \\ &+ \int_0^b \left[E_1I_1(w_{yy} - r\vartheta_{1yy})(\varphi_{0yy} - r\varphi_{1yy}) - q_1(\varphi_0 - r\varphi_1)\right]\big|_{x=0}\,dy \\ &+ \int_0^b \left[E_2I_2(w_{yy} + r\vartheta_{1yy})(\varphi_{0yy} + r\varphi_{1yy}) - q_2(\varphi_0 + r\varphi_1)\right]\big|_{x=a}\,dy \\ &+ \int_0^b D_{T1}\vartheta_{1y}\varphi_{1y}\big|_{x=0}\,dy + \int_0^b D_{T2}\vartheta_{1y}\varphi_{1y}\big|_{x=a}\,dy \end{aligned}$$

after use of integration by parts.

In this problem, ∂S_2 is the contour $ABCD$. On side AB we have $n_1 = -1$ and $n_2 = 0$. On side BC we have $n_1 = 0$ and $n_2 = -1$. On side CD we have $n_1 = 1$ and $n_2 = 0$. In the equation $\delta\mathcal{E} = 0$, the integral over S and the contour integral are zero independently. This is achieved by appropriate selection of admissible variations. Vanishing of the surface integral yields the equilibrium equations (2.29)–(2.31). Vanishing of the contour integral yields

$$\begin{aligned} &\int_0^b (M_{1\beta}\varphi_\beta + Q_1\varphi_0)\big|_{x=0}\,dy - \int_0^a (M_{2\beta}\varphi_\beta + Q_2\varphi_0)\big|_{y=0}\,dx \\ &\quad + \int_0^b (M_{1\beta}\varphi_\beta + Q_1\varphi_0)\big|_{x=a}\,dx \\ &\quad + \int_0^b \left[E_1I_1(w_{yy} - r\vartheta_{1yy})(\varphi_{0yy} - r\varphi_{1yy}) - q_1(\varphi_0 - r\varphi_1)\right]\big|_{x=0}\,dy \\ &\quad + \int_0^b \left[E_2I_2(w_{yy} + r\vartheta_{1yy})(\varphi_{0yy} + r\varphi_{1yy}) - q_2(\varphi_0 + r\varphi_1)\right]\big|_{x=a}\,dy \\ &\quad + \int_0^b D_{T1}\vartheta_{1y}\varphi_{1y}\big|_{x=0}\,dy + \int_0^b D_{T2}\vartheta_{1y}\varphi_{1y}\big|_{x=a}\,dy = 0. \end{aligned}$$

Integration by parts reduces this to the form

$$\begin{aligned}
&\int_0^b \Big(M_{1\beta}\varphi_\beta - D_{T1}\vartheta_{1yy}\varphi_1 + Q_1\varphi_0 \\
&\quad + E_1I_1(w_{yyyy} - r\vartheta_{1yyyy})(\varphi_0 - r\varphi_1) - q_1(\varphi_0 - r\varphi_1)\Big)\Big|_{x=0}\,dy \\
&- \int_0^a (M_{2\beta}\varphi_\beta + Q_2\varphi_0)\Big|_{y=0}\,dx \\
&+ \int_0^b \Big(M_{1\beta}\varphi_\beta - D_{T2}\vartheta_{1yy}\varphi_1 + Q_1\varphi_0 \\
&\quad + E_2I_2(w_{yyyy} + r\vartheta_{1yyyy})(\varphi_0 + r\varphi_1) - q_2(\varphi_0 + r\varphi_1)\Big)\Big|_{x=a}\,dx \\
&+ \Big[E_1I_1(w_{yy} - r\vartheta_{1yy})(\varphi_{0y} - r\varphi_{1y})\big|_{x=0} \\
&\qquad -E_1I_1(w_{yyy} - r\vartheta_{1yyy})(\varphi_0 - r\varphi_1)\big|_{x=0}\Big]\Big|_{y=0}^{y=b} \\
&+ \Big[E_2I_2(w_{yy} + r\vartheta_{1yy})(\varphi_{0y} + r\varphi_{1y})\big|_{x=a} \\
&\qquad -E_2I_2(w_{yyy} + r\vartheta_{1yyy})(\varphi_0 + r\varphi_1)\big|_{x=a}\Big]\Big|_{y=0}^{y=b} \\
&+ \Big(D_{T1}\vartheta_{1y}\varphi_1\big|_{x=0}\Big)\Big|_{y=0}^{y=b} + \Big(D_{T2}\vartheta_{1y}\varphi_1\big|_{x=a}\Big)\Big|_{y=0}^{y=b} = 0.
\end{aligned}$$

The functions φ_0 and φ_α are zero on AD, i.e., when $y = b$. As they are arbitrary, we get the following set of natural boundary conditions:

$$\begin{aligned}
AB:&\quad M_{11} - D_{T1}\vartheta_{1yy} + rQ_1 = 0, \quad M_{12} = 0, \\
&\quad Q_1 + E_1I_1(w_{yyyy} - r\vartheta_{1yyyy}) - q_1 = 0, \\
BC:&\quad M_{21} = 0, \quad M_{22} = 0, \quad Q_2 = 0, \\
CD:&\quad M_{11} - D_{T2}\vartheta_{1yy} - rQ_1 = 0, \quad M_{12} = 0, \\
&\quad Q_1 + E_2I_2(w_{yyyy} + r\vartheta_{1yyyy}) - q_2 = 0.
\end{aligned}$$

At the corner $B = (0, 0)$ the conditions

$$w_{yy} - r\vartheta_{1yy} = w_{yyy} - r\vartheta_{1yyy} = 0, \quad \vartheta_{1y} = 0$$

hold, at the corner $C = (a, 0)$

$$w_{yy} + r\vartheta_{1yy} = w_{yyy} + r\vartheta_{1yyy} = 0, \quad \vartheta_{1y} = 0$$

hold, while at $A = (0, b)$ and $D = (a, b)$ we have

$$w_{yy} - r\vartheta_{1yy} = 0 \quad \text{and} \quad w_{yy} + r\vartheta_{1yy} = 0.$$

We note the following.

(1) On a plate edge supported by a beam, the boundary conditions contain the functions and their normal derivatives as for an unsupported edge. But they also contain derivatives of the functions in the direction tangential to the boundary contour. Moreover, the tangential derivatives are of higher (namely, second and fourth) order.

(2) At the beam endpoints we also see natural boundary conditions. These correspond to the conditions given at corresponding points of the plate edge. Mathematically, such conditions can lead to singularities in the solution.

(3) In this elementary problem we have shown how to obtain the compatibility conditions for a Reissner–Mindlin plate with a classical beam of symmetric cross section clamped to the plate edge. For other ways of establishing the compatibility equations for coupled beams, plates, and shells, see, e.g., [22; 4].

Let us return to the more general problem of Fig. 2.7. The kinematic conditions for coupling between the plate and a beam along AB take the form

$$u = (w - r\vartheta_n)\big|_{AB}, \quad \psi(s) = \vartheta_n \equiv (n_1\vartheta_1 + n_2\vartheta_2)\big|_{AB}.$$

Here $u(s)$ is the vertical deflection and $\psi(s)$ is the twisting angle of the beam. It follows that the energy functional of a plate with edge reinforced by a beam along a portion of its contour is

$$\begin{aligned}\mathcal{E}(w, \vartheta_1, \vartheta_2) = \iint_S W\,dS - \iint_S pw\,dS - \int_{\partial S_2} (Q_n^0 w + M_\beta \vartheta_\beta)\,ds \\ + \frac{1}{2}\int_A^B [EI(w'' - r\vartheta_n'')^2 + D_T(\vartheta_n')^2]\,ds - \int_A^B qw\,ds,\end{aligned} \tag{2.47}$$

where $(\cdot)' = \frac{\partial}{\partial s}(\cdot)$. Skipping some technical details, we present the final form of the conditions on the reinforced edge:

$$M_n - D_T\vartheta_1'' + rQ_n = 0, \quad M_\tau = 0, \quad Q_n + EI(w'''' - r\vartheta_n'''') - q = 0,$$

where

$$M_n = n_\alpha M_{\alpha\beta} n_\beta, \quad M_\tau = -n_\alpha M_{\alpha 1} n_2 + n_\alpha M_{\alpha 2} n_1, \quad Q_n = n_\alpha Q_\alpha.$$

It is worth noting that in many cases, the variational derivation allows us to formulate correct natural boundary conditions. However, a purely formal application of this method can lead to errors. Sometimes an understanding of the physical features of a model are more important than mathematical rigor. This can be seen in the following example.

Example 2.10. Consider a horizontal square plate with a vertical rod of length a attached at point (x_0, y_0) and loaded with a force P and a twisting moment M at the endpoint (Fig. 2.9). Find the natural boundary condition

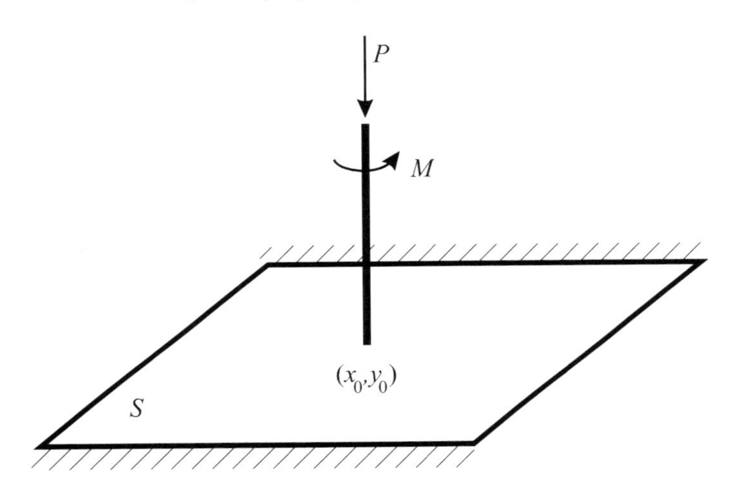

Fig. 2.9 Plate with a vertical rod.

related to contact between the rod and plate.

Solution. Clearly the rod does not bend, hence the rod model is justified (instead of a beam model). The energy functional is

$$\begin{aligned}\mathcal{E}(w, \vartheta_1, \vartheta_2, y, \psi) = \iint_S W\,dS + \frac{1}{2}\int_0^a (EA(u')^2 + D_T(\psi')^2)\,ds \\ - Pu(a) - M\psi(a),\end{aligned} \tag{2.48}$$

where $u(s)$ is the longitudinal displacement along the rod and $\psi(s)$ is the twisting angle. Taking into account the kinematic compatibility conditions $u(0) = w(x_0, y_0)$, we get a natural boundary condition that corresponds to the action of a point force at (x_0, y_0):

$$Q_{1,1} + Q_{2,2} + EAu'(0)\,\delta(x - x_0, y - y_0) = 0.$$

There is no problem with this physically. But if we consider the influence of the drilling moment M, we see that there is no kinematic compatibility condition relating ψ, w, and ϑ_α. Correspondingly, the torsion problem for the rod yields the natural conditions

$$D_T\psi'(0) = 0, \qquad D_T\psi'(a) = M.$$

Hence the torsion in the rod does not appear to be affected by clamping to the plate: the lower end of the rod appears to be free. Clearly, this strange conclusion must come from physical assumptions hidden in the model. The Reissner–Mindlin plate theory is derived under assumptions in which the drilling moment does not enter as a load. In this situation we could use a more general plate or shell theory (see [17; 4]). Alternatively, we could consider the three-dimensional tension-deformation fields near the coupling points, using three-dimensional elasticity for both the plate and rod. □

2.3 Conservation Laws

The conservation laws (conservation of energy, momentum, etc.) play a central role in physics. They are all statements of a similar nature, but exhibit very different external forms. A united mathematical presentation of the conservations laws related to the calculus of variations is the goal of this section.

Let u_i, $i = 1, \ldots, m$, be functions depending on the variables x_j, $j = 1, \ldots, n$. Functions u_i describing a physical object are defined by some simultaneous differential equations

$$l_p(x_j, u_i, u_{i,j}) = 0, \qquad p = 1, \ldots, k, \tag{2.49}$$

where for brevity we have denoted $u_{i,j} = \partial u_i/\partial x_j$ as in elasticity. Let $\mathbf{P} = (P_1, \ldots, P_n)$ be a vector function with components $P_q = P_q(x_j, u_i, u_{i,j})$, $q = 1, \ldots, n$. The equation

$$\frac{dP_1}{dx_1} + \cdots + \frac{dP_n}{dx_n} = 0, \tag{2.50}$$

which holds for all solutions of the system (2.49), is called a *conservation law* for the physical object. Here dP_q/dx_j denotes the *complete derivative* of P_q with respect to x_j; for a function $g(x_j, u_i, u_{i,j})$, we have

$$\frac{dg}{dx_k} = \frac{\partial g}{\partial x_k} + \sum_i \frac{\partial g}{\partial u_i}\frac{\partial u_i}{\partial x_k} + \sum_{i,j} \frac{\partial g}{\partial u_{i,j}}\frac{\partial u_{i,j}}{\partial x_k}. \tag{2.51}$$

In the following part of the chapter, Einstein's summation convention will be in force.

Using the n-dimensional divergence operator, we can rewrite (2.50) in the form

$$\operatorname{div} \mathbf{P} = 0. \tag{2.52}$$

Let us relate the definition (2.50) of a conservation law to some known physical conservation laws.

Let $n = 1$; that is, consider a system described by ordinary differential equations with respect to some unknown functions $u_i(x)$ of the variable x:

$$l_p(x, u_i, u_i') = 0, \qquad i = 1, \ldots, m. \tag{2.53}$$

Equation (2.50) reduces to

$$\frac{dP}{dx} = 0 \tag{2.54}$$

and it follows that any solution of (2.53) takes the form $P =$ constant. This is the typical form of a conservation law in physics: over any solution of (2.53), the value of P is preserved. In the theory of ordinary differential equations, an equality of the form $P =$ constant valid for any solution of a system is called a *first integral.* First integrals play an important role, as they provide general information about solutions in the absence of the solutions themselves. A set of m independent first integrals is equivalent to the solution of (2.53).

The familiar law of energy conservation for a particle in the gravitational field can be broadly extended to particle systems, to rigid body dynamics, and to other objects in the same general form: the sum of the kinetic energy K and the potential energy W is constant with respect to time t. This relation $K + W =$ constant can be written in the above form of a conservation law by time differentiation:

$$\frac{d}{dt}(K + W) = 0.$$

This equality can also be regarded as the result of minimizing $K + W$ over the solutions of the system of equations governing some object; it is a functional dependent on the functions that describe the object. This is another reason why conservation laws are discussed in the calculus of variations.

Consider the more complex case of a system of equations with u_i depending on two variables (x_1, x_2). With $n = 2$, (2.50) takes the form

$$\frac{dP_1}{dx_1} + \frac{dP_2}{dx_2} = 0. \tag{2.55}$$

Integration over an arbitrary domain S and application of the divergence theorem give

$$0 = \iint_S \operatorname{div} \mathbf{P} \, dx_1 \, dx_2 = \int_{\partial S} (P_1 n_1 + P_2 n_2) \, ds,$$

where ∂S is the boundary contour of S (Fig. 2.10) and n_1, n_2 are the components of the outward unit normal $\mathbf{n}$ to ∂S. This is the integral form of (2.55). From a physical standpoint the quantity $\mathbf{P} \cdot \mathbf{n} = P_1 n_1 + P_2 n_2$ is the flux of $\mathbf{P}$ through some portion of ∂S over a unit time interval. Thus, for solutions of (2.49), the flux $\mathbf{P}$ through ∂S is zero.

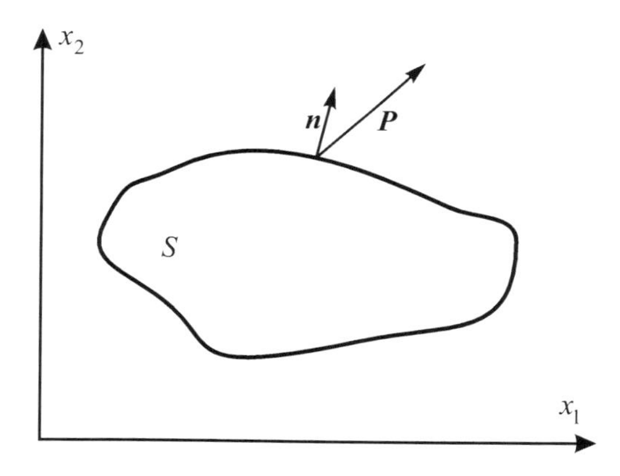

Fig. 2.10 A two-dimensional domain.

In the three-dimensional case, the conservation laws for the mass of a liquid or for electric charge distributed in space also fall under the above definition of a conservation law. Let $\rho = \rho(x_1, x_2, x_3, t)$ be the density of a liquid that depends on the Cartesian space coordinates (x_1, x_2, x_3) and time t. The mass of liquid contained in an arbitrary volume V at time t is

$$\iiint_V \rho(x_1, x_2, x_3, t) \, dV.$$

During the time interval $[t_1, t_2]$, the change of mass within V is

$$\iiint_V \rho(x_1, x_2, x_3, t_2) \, dV - \iiint_V \rho(x_1, x_2, x_3, t_1) \, dV.$$

Assume this change of mass is due purely to flux of liquid through the boundary ∂V of V over the same time interval. The flux is given by

$$\int_{t_1}^{t_2} \left(\iint_{\partial V} \rho \mathbf{v} \cdot \mathbf{n} \, dS \right) dt,$$

where $\mathbf{v}$ is the velocity of the particle having coordinates (x_1, x_2, x_3) and $\mathbf{n}$ is the outward unit normal from ∂V. The divergence theorem permits us to rewrite this as

$$\int_{t_1}^{t_2} \left(\iiint_V \operatorname{div}(\rho \mathbf{v})\, dV \right) dt.$$

So in integral form, mass conservation is expressed by the equality

$$\iiint_V \rho(x_1, x_2, x_3, t_2)\, dV - \iiint_V \rho(x_1, x_2, x_3, t_1)\, dV$$
$$= \int_{t_1}^{t_2} \left(\iiint_V \operatorname{div}(\rho \mathbf{v})\, dV \right) dt.$$

Dividing through by $t_2 - t_1$, letting $t_2 \to t_1$, and rearranging slightly, we find that at $t = t_1$ the integral equation

$$\iiint_V \left(\frac{\partial \rho}{\partial t} - \operatorname{div}(\rho \mathbf{v}) \right) dV = 0$$

holds for any volume V. Provided the integrand is a continuous function, we see that at any interior point of the liquid

$$\frac{\partial \rho}{\partial t} - \operatorname{div}(\rho \mathbf{v}) = 0.$$

The law of charge conservation may be obtained by simply changing the interpretation of ρ to electric charge density. Conservation laws for other quantities, whose values in a volume can change only via flux through the boundary surface, follow similarly.

We see that mass conservation takes the general form

$$\frac{dP_1}{dx_1} + \frac{dP_2}{dx_2} + \frac{dP_3}{dx_3} + \frac{dP_4}{dt} = 0 \tag{2.56}$$

with $P_4 = \rho$ and $P_k = -\rho v_k$ for $k = 1, 2, 3$. Clearly (2.56) can describe more than just the conservation of mass or charge. By reversing the steps taken above, we can obtain the integral form of this conservation law:

$$\frac{d}{dt} \iiint_V P_4\, dV + \iint_{\partial V} (P_1 n_1 + P_2 n_2 + P_2 n_3)\, dS = 0 \tag{2.57}$$

where $\mathbf{n} = (n_1, n_2, n_3)$. Physically, this states that the change of quantity P_4 in volume V is determined by the flux $P_1 n_1 + P_2 n_2 + P_2 n_3$ over the boundary ∂V.

Many physical conservation laws take the form (2.57). Again, conservation laws play a pivotal role in physics. In § 2.4 we show how they are obtained for a general system of equations representing the Euler–Lagrange equations for a functional.

2.4 Conservation Laws and Noether's Theorem

We would like to derive the conservation laws for systems of differential equations representing the Euler–Lagrange equations for certain functionals. One approach was proposed by Amalie Emmy Noether (1882–1935). It employs a type of invariance of the functional under *infinitesimal transformations.*

The simplest case

Recall that the simplest functional from the calculus of variations,

$$F(y) = \int_a^b f(x, y(x), y'(x))\, dx, \tag{2.58}$$

depends on an unknown function $y(x)$ of the variable x and has Euler equation

$$f_y - \frac{d}{dx} f_{y'} = 0. \tag{2.59}$$

We seek the quantity P in the corresponding conservation law

$$\frac{d}{dx} P(x, y(x), y'(x)) = 0.$$

Consider the question of infinitesimal invariance of F under transformations of the form

$$x \to x^* = x + \varepsilon\xi(x, y), \tag{2.60}$$

$$y \to y^* = y + \varepsilon\phi(x, y), \tag{2.61}$$

where ε is a small parameter and $\xi(x, y)$ and $\phi(x, y)$ are given functions. Denoting the value of F under this change of variables by F^*, we have

$$F^*(y^*) = \int_{a^*}^{b^*} f\left(x^*, y^*, \frac{dy^*}{dx^*}\right) dx^*$$

where $a^* = a + \varepsilon\xi(a, y(a))$ and $b^* = b + \varepsilon\xi(b, y(b))$.

Definition 2.11. The functional F is *infinitesimally invariant under the transformation* (2.60)–(2.61), for some fixed functions ξ and ϕ, if the equality $F^* = F$ holds in the asymptotic sense up to linear terms in ε as $\varepsilon \to 0$ for an extremal $y = y(x)$ of F; this means that

$$\lim_{\varepsilon \to 0} F^* = F \quad \text{and} \quad \lim_{\varepsilon \to 0} \frac{F^* - F}{\varepsilon} = 0. \tag{2.62}$$

We may also say that F has *variational symmetry* with respect to this infinitesimal transformation.

Later on, we will see how one may relate the variational symmetry of F with conservative laws for solutions of the corresponding Euler equation for F.

Very provisionally, the geometrical relation between F and F^* may be envisioned as in Fig. 2.11. In the terminology of asymptotic analysis, F

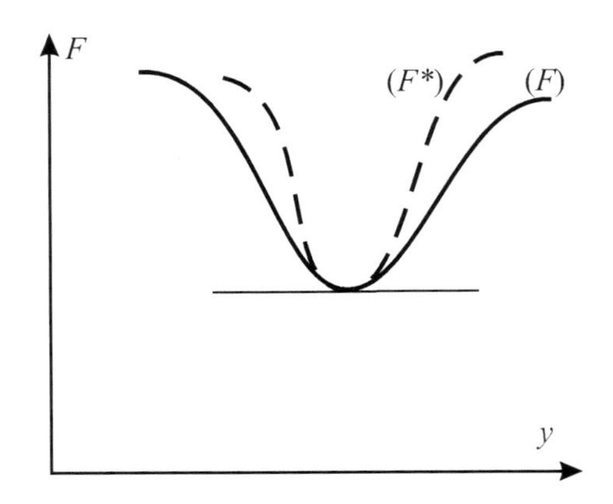

Fig. 2.11 The values of F and F^* coincide on an extremal with respect to which we apply Definition 2.11, and the influence of the transformation is absent in the terms that are linear in ε.

(which does not depend on ε) and F^* are asymptotically equal up to linear terms in a neighborhood of $\varepsilon = 0$.

With some smoothness assumed for F, relations (2.62) can be rewritten as

$$F^*|_{\varepsilon=0} = F, \qquad \left.\frac{dF^*}{d\varepsilon}\right|_{\varepsilon=0} = 0.$$

Let us express the variational invariance property of F in terms of ϕ, ξ, and f.

Theorem 2.12. *The functional F is infinitesimally invariant under the transformation* (2.60)–(2.61) *if*

$$\left[\xi\frac{\partial}{\partial x} + \phi\frac{\partial}{\partial y} + \left(\frac{d\phi}{dx} - y'\frac{d\xi}{dx}\right)\frac{\partial}{\partial y'} + \frac{d\xi}{dx}\right] f = 0 \tag{2.63}$$

or, equivalently,

$$\frac{d}{dx}\left[(\phi - y'\xi)f_{y'} + \xi f\right] + (\phi - \xi y')\left[f_y - \frac{d}{dx}f_{y'}\right] = 0. \tag{2.64}$$

Proof. To investigate some consequences of the variational invariance of F, let us express F^* in terms of x, y, and y'. First we derive dy^*/dx^*. It is clear that

$$\frac{d}{dx^*} = \frac{dx}{dx^*}\frac{d}{dx} = \left(\frac{dx^*}{dx}\right)^{-1}\frac{d}{dx} = \left(1 + \varepsilon\frac{d\xi}{dx}\right)^{-1}\frac{d}{dx}$$

where we have used (2.51). So

$$\frac{dy^*}{dx^*} = \left(1 + \varepsilon\frac{d\xi}{dx}\right)^{-1}\left(y' + \varepsilon\frac{d\phi}{dx}\right)$$

and therefore

$$F^* = \int_a^b f\left(x + \varepsilon\xi(x,y),\, y + \varepsilon\phi(x,y),\, \left(1 + \varepsilon\frac{d\xi}{dx}\right)^{-1}\left(y' + \varepsilon\frac{d\phi}{dx}\right)\right) \times \left(1 + \varepsilon\frac{d\xi}{dx}\right) dx.$$

Let us expand F^* in a series with respect to ε at zero, explicitly showing the linear part of the expansion. Using

$$\left(1 + \varepsilon\frac{d\xi}{dx}\right)^{-1} = 1 - \varepsilon\frac{d\xi}{dx} + O(\varepsilon^2)$$

we get

$$\frac{dy^*}{dx^*} = y' + \varepsilon\left(\frac{d\phi}{dx} - y'\frac{d\xi}{dx}\right) + O(\varepsilon^2).$$

In a neighborhood of $\varepsilon = 0$ we expand the integrand in a Taylor series with respect to ε, keeping only terms of the first order of smallness in ε:

$$\begin{aligned} F^* &= \int_a^b f(x,y,y')\,dx \\ &\quad + \varepsilon\int_a^b \left[f_x\xi + f_y\phi + f_{y'}\left(\frac{d\phi}{dx} - y'\frac{d\xi}{dx}\right) + f\frac{d\xi}{dx}\right] dx + O(\varepsilon^2) \\ &= F + \varepsilon\int_a^b \left[\xi\frac{\partial}{\partial x} + \phi\frac{\partial}{\partial y} + \left(\frac{d\phi}{dx} - y'\frac{d\xi}{dx}\right)\frac{\partial}{\partial y'} + \frac{d\xi}{dx}\right] f\,dx + O(\varepsilon^2). \end{aligned}$$

Clearly F is invariant in the above sense if the integrand of the integral coefficient of ε in the expansion vanishes:

$$\left[\xi\frac{\partial}{\partial x}+\phi\frac{\partial}{\partial y}+\left(\frac{d\phi}{dx}-y'\frac{d\xi}{dx}\right)\frac{\partial}{\partial y'}+\frac{d\xi}{dx}\right]f=0$$

for some functions ξ and ϕ. This proves (2.63). Next, in (2.63) we select the first terms with the derivative d/dx that we see in (2.64). We have

$$\begin{aligned}\left(\frac{d\phi}{dx}-y'\frac{d\xi}{dx}\right)&f_{y'}+\frac{d\xi}{dx}f\\
&=\frac{d}{dx}\left[(\phi-y'\xi)f_{y'}+\xi f\right]-\phi\frac{d}{dx}f_{y'}+y''\xi f'_y+y'\xi\frac{d}{dx}f_{y'}-\xi\frac{df}{dx}\\
&=\frac{d}{dx}\left[(\phi-y'\xi)f_{y'}+\xi f\right]-\phi\frac{d}{dx}f_{y'}+y''\xi f'_y+y'\xi\frac{d}{dx}f_{y'}\\
&\quad-\xi(f_x+y'f_y+y''f_{y'})\\
&=\frac{d}{dx}\left[(\phi-y'\xi)f_{y'}+\xi f\right]-(\phi-\xi y')\frac{d}{dx}f_{y'}-(\xi f_x+y'\xi f_y)\end{aligned}$$

so that

$$\begin{aligned}0&=\left[\xi\frac{\partial}{\partial x}+\phi\frac{\partial}{\partial y}+\left(\frac{d\phi}{dx}-y'\frac{d\xi}{dx}\right)\frac{\partial}{\partial y'}+\frac{d\xi}{dx}\right]f\\
&=\frac{d}{dx}\left[(\phi-y'\xi)f_{y'}+\xi f\right]+(\phi-\xi y')\left[f_y-\frac{d}{dx}f_{y'}\right].\end{aligned}$$

This completes the proof. □

A consequence of the variational invariance of F under an infinitesimal transformation is *Noether's theorem.*

Theorem 2.13. *Suppose that for an extremal y, the functional F is infinitesimally invariant under (2.60)–(2.61). Then a conservation law of the form*

$$\frac{d}{dx}P=0 \tag{2.65}$$

holds, where

$$P=\phi f_{y'}+\xi(f-y'f_{y'}). \tag{2.66}$$

In this case, P is termed a ***flux***.

A more general version of Noether's theorem was published in 1918 (see, e.g., [23]). The above formulation is commonly used in physics and engineering.

Proof. In the proof of Theorem 2.12, we saw that variational invariance yields the equation used to find the transformation functions ξ and ϕ:

$$\frac{d}{dx}\left[(\phi - y'\xi)f_{y'} + \xi f\right] + (\phi - \xi y')\left[f_y - \frac{d}{dx}f_{y'}\right] = 0.$$

Since $y(x)$ is an extremal of F, it satisfies the Euler–Lagrange equation (2.59). Hence the second term is zero and (2.64) takes the form of (2.65):

$$\frac{d}{dx}\left[(\phi - y'\xi)f_{y'} + \xi f\right] = 0.$$

Because

$$(\phi - y'\xi)f_{y'} + \xi f = \phi f_{y'} + \xi(f - y'f_{y'}),$$

we obtain the needed expression for P in (2.66). □

Thus we have established that the infinitesimal invariance of F involves the conservation law in one dimension

$$\frac{d}{dx}P = 0.$$

Under this condition, the transformed functional F^* takes a simple form:

$$\begin{aligned} F^* &= F + \varepsilon \int_a^b \frac{dP}{dx}\, dx + O(\varepsilon^2) \\ &= F + \varepsilon \left(P\big|_{x=b} - P\big|_{x=a}\right) + O(\varepsilon^2). \end{aligned} \tag{2.67}$$

Let us pause for a brief overview. The condition for invariance (or variational symmetry) of the functional

$$F = \int_a^b f(x, y, y')\, dx$$

under the transformation

$$x \to x^* = x + \varepsilon\xi(x, y), \qquad y \to y^* = y + \varepsilon\phi(x, y), \tag{2.68}$$

is

$$\left[\xi\frac{\partial}{\partial x} + \phi\frac{\partial}{\partial y} + \left(\frac{d\phi}{dx} - y'\frac{d\xi}{dx}\right)\frac{\partial}{\partial y'} + \frac{d\xi}{dx}\right] f = 0. \tag{2.69}$$

The functions $\xi(x, y)$ and $\phi(x, y)$ should be found from equation (2.69). We should note that finding variational symmetry is a very nontrivial problem. It consists of finding two unknown functions $\xi(x, y)$ and $\phi(x, y)$ such that (2.69) holds for all $y(x)$ satisfying the Euler–Lagrange equation

$$f_y - \frac{d}{dx}f_{y'} = 0.$$

Certain variational symmetries, i.e., the functions $\xi(x,y)$ and $\phi(x,y)$, can be suggested by the form taken by the integrand of the functional — when it is independent of some of the variables, for example, as will be shown below. In the general case, some nontrivial and less obvious symmetries can be identified using (2.69) (and similar equations for other cases), e.g., via symbolic machine computation. Nontrivial conservation laws have been regularly discovered up to the present time. On the other hand, in physics (and particularly in mechanics), there are known systems of equations possessing "poor" sets of symmetries; in these cases the conservation laws were established by other methods.

If $\xi(x,y)$ and $\phi(x,y)$ are known and (2.69) holds, then the conservation law is easily obtained from

$$\frac{d}{dx}P = 0, \quad \text{with} \quad P = \phi f_{y'} + \xi\,(f - y' f_{y'})\,. \tag{2.70}$$

Let us consider a particular case.

Example 2.14. Find a conservation law for the functional

$$F = \int_a^b f(y, y')\,dx.$$

Solution. Consider the transformation (which we can write out because we already know the answer to the problem)

$$x \to x^* = x + \varepsilon, \qquad y \to y^* = y.$$

In other words, we took $\xi = 1$ and $\phi = 0$. Equation (2.69) reduces to $f_x = 0$, which evidently holds. Then from (2.70) we get the expression

$$P = f - y' f_{y'}$$

for the flux. □

Functional depending on a vector function

The above considerations can be easily extended to the functional

$$F(\mathbf{y}) = \int_a^b f(x, \mathbf{y}, \mathbf{y}')\,dx, \tag{2.71}$$

where $\mathbf{y} = (y_1(x), \ldots, y_m(x))$. We will consider the invariance of F under a transformation of the form

$$x \to x^* = x + \varepsilon\xi(x,\mathbf{y}), \quad y_i \to y_i^* = y_i + \varepsilon\phi_i(x,\mathbf{y}), \quad i = 1, \ldots, m \tag{2.72}$$

which can be rewritten in vector notation as

$$x \to x^* = x + \varepsilon\xi(x, \mathbf{y}), \qquad \mathbf{y} \to \mathbf{y}^* = \mathbf{y} + \varepsilon\boldsymbol{\phi}(x, \mathbf{y}).$$

With suitable changes in the meanings of F^* and F, Definition 2.11 continues to apply. We present only the steps that involve more than trivial modifications from the case where F was given by (2.58). Expanding the new F^* in the vicinity of zero out to linear terms, we get the following relation between F^* and F:

$$F^* = F + \varepsilon \int_a^b \left[\xi \frac{\partial}{\partial x} + \phi_1 \frac{\partial}{\partial y_1} + \cdots + \phi_n \frac{\partial}{\partial y_n} \right.$$
$$\left. + \left(\frac{d\phi_1}{dx} - y_1' \frac{d\xi}{dx} \right) \frac{\partial}{\partial y_1'} + \cdots + \left(\frac{d\phi_n}{dx} - y_n' \frac{d\xi}{dx} \right) \frac{\partial}{\partial y_n'} + \frac{d\xi}{dx} \right] f \, dx.$$

We conclude that a sufficient condition for invariance of F under the transformation is

$$\left[\xi \frac{\partial}{\partial x} + \phi_i \frac{\partial}{\partial y_i} + \left(\frac{d\phi_i}{dx} - y_i' \frac{d\xi}{dx} \right) \frac{\partial}{\partial y_i'} + \frac{d\xi}{dx} \right] f = 0, \tag{2.73}$$

where summation over i is implied. In vector notation this is

$$\left[\xi \frac{\partial}{\partial x} + \boldsymbol{\phi} \cdot \nabla_{\mathbf{y}} + \left(\frac{d\boldsymbol{\phi}}{dx} - \frac{d\xi}{dx} \mathbf{y}' \right) \cdot \nabla_{\mathbf{y}'} + \frac{d\xi}{dx} \right] f = 0. \tag{2.74}$$

As in the case of the simplest functional, (2.73) or (2.74) can be reduced to

$$\frac{d}{dx} \left[\phi_i f_{y_i'} + \xi \left(f - y_i' f_{y_i'} \right) \right] + \left(\phi_i - \xi y_i' \right) \left(f_{y_i} - \frac{d}{dx} f_{y_i'} \right)$$
$$= \frac{d}{dx} \left[\boldsymbol{\phi} \cdot \nabla_{\mathbf{y}'} f + \xi \left(f - \mathbf{y}' \cdot \nabla_{\mathbf{y}'} f \right) \right] + \left(\boldsymbol{\phi} - \xi \mathbf{y}' \right) \cdot \left(\nabla_{\mathbf{y}} f - \frac{d}{dx} \nabla_{\mathbf{y}'} f \right) = 0.$$

Since $\mathbf{y}$ satisfies the Euler–Lagrange equations (1.62), we arrive at the conservation law

$$\frac{d}{dx} P = 0 \quad \text{with} \quad P = \boldsymbol{\phi} \cdot \nabla_{\mathbf{y}'} f + \xi \left(f - \mathbf{y}' \cdot \nabla_{\mathbf{y}'} f \right). \tag{2.75}$$

A reformulation of Theorem 2.12 for (2.71) is left to the reader. As an example, we derive the conservation laws for particle motion in a central force field.

Example 2.15. Obtain conservation laws for the functional

$$F = \int_0^1 \left[\frac{1}{2} \left({y_1'}^2 + {y_2'}^2 \right) - W(r) \right] dx, \qquad r = r(x) = \sqrt{y_1^2 + y_2^2},$$

where $W(r)$ is the potential of central forces acting on the particle with coordinates y_1 and y_2, which is a differentiable function of r, and x plays the role of the time variable.

Solution. In this example, f is the Lagrangian of the unit point mass in a central force field; it equals the difference between the kinetic energy $K = ({y'_1}^2 + {y'_2}^2)/2$ and the potential energy W of the particle. The Euler–Lagrange equations are the equations of motion of the particle:

$$y''_1 = \frac{dW}{dr}\frac{y_1}{\sqrt{y_1^2 + y_2^2}}, \qquad y''_2 = \frac{dW}{dr}\frac{y_2}{\sqrt{y_1^2 + y_2^2}}.$$

It is clear that F possesses two symmetries. The first is with respect to translation along the x-direction by a distance ε, which we express as the transformation

$$x \to x^* = x + \varepsilon, \qquad \mathbf{y} \to \mathbf{y}^* = \mathbf{y}.$$

The second is with respect to rotation of the vector $\mathbf{y}$ through an infinitesimal angle ε, which is described by the transformation

$$x \to x^* = x, \qquad y_1 \to y_1^* = y_1 + \varepsilon y_2, \qquad y_2 \to y_2^* = -\varepsilon y_1 + y_2.$$

We see that F does not change when $\mathbf{y}$ rotates through any finite angle ε, i.e., with respect to the finite transformation

$$y_1 \to y_1^* = y_1 \cos\varepsilon + y_2 \sin\varepsilon, \qquad y_2 \to y_2^* = -y_1 \sin\varepsilon + y_2 \cos\varepsilon,$$

but at present it will suffice for us to consider the conservation of F over infinitesimal angles ε. Invariance with respect to the first symmetry transformation is similar to that treated in Example 2.14; it yields the conservation flux law $P_1 =$ constant where

$$P_1 = f - y'_1 f_{y'_1} - y'_2 f_{y'_2} = -\frac{1}{2}\left({y'_1}^2 + {y'_2}^2\right) - W(r).$$

This is the energy conservation law: the sum of the kinetic and potential energies is constant, $K + W =$ constant. For the rotational transformation, in general terms we have $\xi = 0$, $\phi_1 = y_2$, $\phi_2 = -y_1$. Then (2.75) reduces to $P_2 =$ constant with

$$P_2 = y_2 f_{y'_1} - y_1 f_{y'_2} = y_2 y'_1 - y_1 y'_2.$$

In celestial mechanics this law is known as the conservation of kinetic momentum; it was published by Johannes Kepler in 1609. □

2.5 Functionals Depending on Higher Derivatives of y

Functional depending on y''

Let us extend the above considerations, regarding infinitesimal invariance with respect to transformations, to functionals dependent on higher derivatives. We start with a dependence F on y'':

$$F = \int_a^b f(x, y, y', y'')\, dx.$$

Consider the change of F under infinitesimal transformations of the form

$$x \to x^* = x + \varepsilon\xi(x, y), \qquad y \to y^* = y + \varepsilon\phi(x, y). \tag{2.76}$$

Now F^* denotes

$$F^*(y^*) = \int_{a^*}^{b^*} f\left(x^*, y^*, \frac{dy^*}{dx^*}, \frac{d^2y^*}{dx^{*2}}\right) dx^*.$$

To define infinitesimal invariance of this F with respect to the transformation (2.76), we again use Definition 2.11 with the new F and F^*.

Theorem 2.16. *For an extremal y, let F be infinitesimally invariant under the transformation* (2.76). *Then a conservation law of the form*

$$\frac{d}{dx}P = \frac{d}{dx}\left[\xi f + (\phi - \xi y')\left(f_{y'} - \frac{d}{dx}f_{y''}\right) + f_{y''}\frac{d}{dx}(\phi - \xi y')\right] = 0 \tag{2.77}$$

holds.

Proof. We express F^* in the initial variables x and y, first deriving d^2y^*/dx^{*2} in these terms. Using the formula

$$\frac{d}{dx^*} = \left(1 + \varepsilon\frac{d\xi}{dx}\right)^{-1}\frac{d}{dx},$$

we get

$$\frac{dy^*}{dx^*} = \left(1 + \varepsilon\frac{d\xi}{dx}\right)^{-1}\left(y' + \varepsilon\frac{d\phi}{dx}\right),$$

and so

$$\frac{d^2y^*}{dx^{*2}} = \frac{d}{dx^*}\frac{dy^*}{dx^*} = \left(1 + \varepsilon\frac{d\xi}{dx}\right)^{-1}\frac{d}{dx}\left[\left(1 + \varepsilon\frac{d\xi}{dx}\right)^{-1}\left(y' + \varepsilon\frac{d\phi}{dx}\right)\right]$$
$$= \left(1 + \varepsilon\frac{d\xi}{dx}\right)^{-2}\left(y'' + \varepsilon\frac{d^2\phi}{dx^2}\right) - \left(1 + \varepsilon\frac{d\xi}{dx}\right)^{-3}\varepsilon\frac{d^2\xi}{dx^2}\left(y' + \varepsilon\frac{d\phi}{dx}\right).$$

Taylor's expansion of d^2y^*/dx^{*2} with respect to ε at zero gives

$$\frac{d^2y^*}{dx^{*2}} = y'' + \varepsilon\left(\frac{d^2\phi}{dx^2} - 2y''\frac{d\xi}{dx} - y'\frac{d^2\xi}{dx^2}\right) + O(\varepsilon^2).$$

Thus we get

$$\begin{aligned} F^* &= \int_a^b f(x,y,y',y'')\,dx + \varepsilon\int_a^b \left[f_x\xi + f_y\phi + f_{y'}\left(\frac{d\phi}{dx} - y'\frac{d\xi}{dx}\right)\right. \\ &\quad \left. + f_{y''}\left(\frac{d^2\phi}{dx^2} - 2y''\frac{d\xi}{dx} - y'\frac{d^2\xi}{dx^2}\right) + f\frac{d\xi}{dx}\right] dx + O(\varepsilon^2) \\ &= F + \varepsilon\int_a^b \left[\xi\frac{\partial}{\partial x} + \phi\frac{\partial}{\partial y} + \left(\frac{d\phi}{dx} - y'\frac{d\xi}{dx}\right)\frac{\partial}{\partial y'}\right. \\ &\quad \left. + \left(\frac{d^2\phi}{dx^2} - 2y''\frac{d\xi}{dx} - y'\frac{d^2\xi}{dx^2}\right)\frac{\partial}{\partial y''} + \frac{d\xi}{dx}\right] f\,dx + O(\varepsilon^2). \end{aligned}$$

As in the case of the simplest functional, we obtain the sufficient condition for infinitesimal invariance of F as the equality to zero of the integrand of the integral coefficient of ε:

$$\begin{aligned} &\left[\xi\frac{\partial}{\partial x} + \phi\frac{\partial}{\partial y} + \left(\frac{d\phi}{dx} - y'\frac{d\xi}{dx}\right)\frac{\partial}{\partial y'}\right. \\ &\quad \left. + \left(\frac{d^2\phi}{dx^2} - 2y''\frac{d\xi}{dx} - y'\frac{d^2\xi}{dx^2}\right)\frac{\partial}{\partial y''} + \frac{d\xi}{dx}\right] f = 0. \end{aligned} \tag{2.78}$$

This condition can be presented in another form:

$$\begin{aligned} &\frac{d}{dx}\left[\phi f_{y'} + \xi(f - y'f_{y'}) - \frac{d(\xi y')}{dx}f_{y''} + \xi y'\frac{d}{dx}f_{y''} + \frac{d\phi}{dx}f_{y''} - \phi\frac{d}{dx}f_{y''}\right] \\ &\quad + (\phi - \xi y')\left[f_y - \frac{d}{dx}f_{y'} + \frac{d^2}{dx^2}f_{y''}\right] = 0. \end{aligned} \tag{2.79}$$

The second term vanishes on solutions of the Euler–Lagrange equation, hence the infinitesimal invariance condition for F reduces to $P =$ constant with

$$\begin{aligned} P &= \phi f_{y'} + \xi(f - y'f_{y'}) - \frac{d(\xi y')}{dx}f_{y''} + \xi y'\frac{d}{dx}f_{y''} + \frac{d\phi}{dx}f_{y''} - \phi\frac{d}{dx}f_{y''} \\ &= \xi f + (\phi - \xi y')\left[f_{y'} - \frac{d}{dx}f_{y''}\right] + f_{y''}\frac{d}{dx}(\phi - \xi y') \end{aligned} \tag{2.80}$$

as stated in the theorem. □

To illustrate, let us consider a functional dependent on y'' only.

Example 2.17. Find a conservation law for the functional

$$F = \int_a^b f(y'')\,dx.$$

With $f = EIy''^2/2$, this becomes the strain energy functional for a beam; use this fact to illustrate the results.

Solution. Clearly, F possesses a few types of symmetry.

(1) It is invariant with respect to translation along the x-direction by a distance ε, i.e., with respect to the transformation

$$x \to x^* = x + \varepsilon, \qquad y \to y^* = y.$$

With $\xi = 1$ and $\phi = 0$, equation (2.80) yields the following expression for the flux:

$$P = f + y'\frac{d}{dx}f_{y''} - y''f_{y''}.$$

For the beam this law reads

$$EI(y'y''' - y''^2/2) = \text{constant}.$$

(2) It is invariant with respect to translation along the y-direction by a distance ε:

$$x \to x^* = x, \qquad y \to y^* = y + \varepsilon.$$

So with the corresponding $\xi = 0$ and $\phi = 1$, equation (2.80) gives

$$P = -\frac{d}{dx}f_{y''}.$$

For the beam this takes the form

$$-EIy''' = \text{constant}$$

and expresses constancy of the shear force along the beam. Indeed there is no load acting on the beam.

(3) It is invariant with respect to translation along the y-direction by a distance εx:

$$x \to x^* = x, \qquad y \to y^* = y + \varepsilon x.$$

With $\xi = 0$ and $\phi = x$, equation (2.80) gives

$$P = -x\frac{d}{dx}f_{y''} + f_{y''}.$$

For the beam this is

$$EI(y'' - xy''') = \text{constant}.$$

Differentiating this, we get $EIxy^{(4)} = 0$. In terms of the shear force $Q = -EIy'''$ we get $Q' = 0$ or $Q =$ constant, which is mechanically evident because the beam is loaded at its ends. □

Because $f = EIy''^2/2$ is a quadratic form, beam theory provides additional symmetries as shown in the next example.

Example 2.18. Find an additional conservation law for the functional

$$F = \frac{1}{2}\int_a^b EIy''^2\, dx.$$

Solution. The quadratic nature of F implies variational symmetry with respect to coordinate scaling of the form $y \to k^\alpha y$, where k is the scale parameter and α is a value to be determined. The corresponding transformation

$$x \to x^* = x + \varepsilon x, \qquad y \to y^* = y + \alpha\varepsilon y,$$

is obtained by changing k to $1 + \varepsilon$ and dropping terms with ε^r for $r > 1$. Thus we take $\xi = x$ and $\phi = \alpha y$. Substituting ξ and ϕ into (2.78), we find that $\alpha = 3/2$, with which (2.78) holds. Hence F has a symmetry with respect to the above transformation if $\alpha = 3/2$. The respective flux is

$$P = EI\left[-\frac{xy''^2}{2} - \frac{3}{2}yy''' + xy'y''' + \frac{1}{2}y'y''\right].$$

The reader can verify that $P' = 0$ when $y^{(4)}(x) = 0$, the Euler–Lagrange equation for the functional, holds. □

Conditions for variational symmetry of functionals depending on higher derivatives of y, along with their respective conservation laws, can be found in [23]. See also Exercise 2.12.

2.6 Noether's Theorem, General Case

Functional depending on a function in n variables and its first derivatives

Let us consider Noether's theorem for the functional

$$F(u) = \int_V f(x_i, u, u_{,i})\, dV = \int_V f(\mathbf{x}, u, \nabla u)\, dV, \tag{2.81}$$

where $\mathbf{x} = (x_1, \ldots, x_n)$ and $\nabla u = (u_{,1}, \ldots, u_{,n})$ is the gradient of u. As before, we consider the infinitesimal transformations of the form

$$\begin{aligned} x_i \to x_i^* &= x_i + \varepsilon \xi_i(\mathbf{x}, u), \quad i = 1, \ldots, n, \\ u \to u^* &= u + \varepsilon \phi(\mathbf{x}, u). \end{aligned} \tag{2.82}$$

So the transformation is defined by a set of $n+1$ functions ξ_i, $i = 1, \ldots, n$, and ϕ. The vector form of (2.82) is

$$\begin{aligned} \mathbf{x} \to \mathbf{x}^* &= \mathbf{x} + \varepsilon \boldsymbol{\xi}(\mathbf{x}, u), \\ u \to u^* &= u + \varepsilon \boldsymbol{\phi}(\mathbf{x}, u). \end{aligned}$$

Variational symmetry of F is again expressed by the formulas

$$\lim_{\varepsilon \to 0} F^* = F \quad \text{and} \quad \lim_{\varepsilon \to 0} \frac{F^* - F}{\varepsilon} = 0,$$

where

$$F^* = \int_{V^*} f\left(x_i^*, u^*, \frac{\partial u^*}{\partial x_i^*}\right) dV^* = \int_{V^*} f(\mathbf{x}^*, u^*, \nabla^* u^*)\, dV^*,$$

and V^* is the new integration domain corresponding to the change $\mathbf{x} \to \mathbf{x}^*$.

The existence and form of the conservation laws under infinitesimal invariance is given by the following version of Noether's theorem. In this case the flux is a vector function $\mathbf{P}$.

Theorem 2.19. *For an extremal u, let the functional F be infinitesimally invariant under transformation* (2.82), *i.e.*,

$$\left[\xi_i \frac{\partial}{\partial x_i} + \phi \frac{\partial}{\partial u} + \left(\frac{d\phi}{dx_i} - u_{,p} \frac{d\xi_p}{dx_i}\right) \frac{\partial}{\partial u_{,i}} + \xi_{i,i}\right] f = 0. \tag{2.83}$$

Then a conservation law of the form

$$\operatorname{div} \mathbf{P} = 0 \tag{2.84}$$

holds, where

$$\mathbf{P} = (P_1, \ldots, P_n), \quad P_i = \phi \frac{\partial f}{\partial u_{,i}} + \xi_k \left(f \delta_{ik} - u_{,k} \frac{\partial f}{\partial u_{,i}}\right). \tag{2.85}$$

Proof. We proceed as in the proofs of Theorems 2.12 and 2.13. We find the linear approximation of F^* with respect to ε, neglecting higher-order terms, and obtain the variational symmetry condition in terms of $\boldsymbol{\xi}$ and ϕ. Then, remembering that the extremals of F satisfy the Euler–Lagrange equation, we cast the variational symmetry condition in divergence form.

To deduce the form of F^*, we derive expressions for the partial derivatives of u^*. The formulas for a change of variables in partial differentiation are

$$\frac{\partial}{\partial x_i^*} = \frac{\partial x_k}{\partial x_i^*}\frac{\partial}{\partial x_k}, \qquad \frac{\partial}{\partial x_i} = \frac{\partial x_k^*}{\partial x_i}\frac{\partial}{\partial x_k^*},$$

where we sum over the repeated index k. The transformation matrix takes the form

$$\frac{\partial x_k^*}{\partial x_i} = \delta_{ik} + \varepsilon\frac{d\xi_k}{dx_i}.$$

The portion of the inverse of this matrix that is linear in ε is

$$\frac{\partial x_k}{\partial x_i^*} = \delta_{ik} - \varepsilon\frac{d\xi_k}{dx_i} + O(\varepsilon^2).$$

So we have

$$\frac{\partial u^*}{\partial x_i^*} = \frac{\partial x_k}{\partial x_i^*}\frac{\partial u^*}{\partial x_k} = \frac{\partial x_k}{\partial x_i^*}\left(u_{,k} + \varepsilon\frac{d\phi}{dx_k}\right) = u_{,i} + \varepsilon\left(\frac{d\phi}{dx_i} - u_{,k}\frac{d\xi_k}{dx_i}\right) + O(\varepsilon^2).$$

Expanding F^* in a Taylor series with respect to ε and dropping higher-order terms, we get

$$\begin{aligned}
F^* &= \int_{V^*} f(\mathbf{x}^*, u^*, \nabla^* u^*)\, dV^* \\
&= \int_V f\left[\mathbf{x} + \varepsilon\boldsymbol{\xi}, u + \varepsilon\phi, u_{,i} + \varepsilon\left(\frac{d\phi}{dx_i} - u_{,k}\frac{d\xi_k}{dx_i}\right) + O(\varepsilon^2)\right]\frac{dV^*}{dV}\, dV \\
&= F + \varepsilon\int_V \left[f_{x_i}\xi_i + f_u\phi + f_{u,i}\left(\frac{d\phi}{dx_i} - u_{,k}\frac{d\xi_k}{dx_i}\right) + f\operatorname{div}\boldsymbol{\xi}\right] dV + O(\varepsilon^2)
\end{aligned}$$

after using the formula

$$\left.\frac{d}{d\varepsilon}\frac{dV^*}{dV}\right|_{\varepsilon=0} = \frac{d\xi}{dx_i} \equiv \operatorname{div}\boldsymbol{\xi}$$

for differentiating the Jacobian (see, e.g., [6]).

Thus, when F has variational symmetry, the following integrand is zero:

$$f_{x_i}\xi_i + f_u\phi + f_{u,i}\left(\frac{d\phi}{dx_i} - u_{,k}\frac{d\xi_k}{dx_i}\right) + f\operatorname{div}\boldsymbol{\xi} = 0.$$

Formulation (2.83) for variational symmetry is proved.

By technically cumbersome transformations it can be shown that (2.83) takes the form

$$\begin{aligned} f_{x_i}\xi_i + f_u\phi + f_{u_{,i}}\left(\frac{d\phi}{dx_i} - u_{,k}\frac{d\xi_k}{dx_i}\right) + f\,\mathrm{div}\,\boldsymbol{\xi} \\ = \frac{d}{dx_i}\left[\phi\frac{\partial f}{\partial u_{,i}} + \xi_k\left(f\delta_{ik} - u_{,k}\frac{\partial f}{\partial u_{,i}}\right)\right] \\ + (\phi - \xi_i u_{,i})\left(f_u - \frac{d}{dx_i}f_{u_{,i}}\right) = 0. \end{aligned}$$

The last parenthetical term vanishes on extremals u, i.e., on solutions of the Euler–Lagrange equation. So (2.83) can be reduced to divergence form, and a conservation law (2.84) holds with flux given by (2.85). □

Let us consider some special cases.

Example 2.20. Find a conservation law for the functional

$$F = \int_V f(\nabla u)\,dV. \tag{2.86}$$

Solution. Since $f_u = 0$, the functional has the symmetry with respect to the transformation

$$\mathbf{x} \to \mathbf{x}^* = \mathbf{x}, \qquad u \to u^* = u + \varepsilon.$$

Correspondingly we have $\boldsymbol{\xi} = \mathbf{0}$, $\phi = 1$ and the conservation law

$$\frac{d}{dx_i}f_{u_{,i}} = 0.$$

This law is uninteresting, as it coincides with the Euler–Lagrange equation for F, so we proceed to another possibility. Because $\partial f/\partial x_k = 0$ for $k = 1, \ldots, n$, the functional has symmetry with respect to the transformation

$$\mathbf{x} \to \mathbf{x}^* = \mathbf{x} + \varepsilon\mathbf{i}_k, \quad \mathbf{i}_k = (0, 0, \ldots, 1, \ldots, 0), \qquad u \to u^* = u,$$

where $\mathbf{i}_k$ is the kth Cartesian basis vector. This corresponds to $\boldsymbol{\xi} = \mathbf{i}_k$ and $\phi = 0$. As k runs from 1 to n, we obtain the n conservation laws

$$\mathrm{div}\,\mathbf{P}_k = 0, \qquad \mathbf{P}_k = f\mathbf{i}_k - (\mathbf{i}_k \cdot \nabla u)\frac{\partial f}{\partial \nabla u},$$

where we denote

$$\frac{\partial f}{\partial \nabla u} = \left(\frac{\partial f}{\partial u_{,1}}, \ldots, \frac{\partial f}{\partial u_{,n}}\right).$$

In component form,

$$\mathbf{P}_k = \left(-u_{,k}\frac{\partial f}{\partial u_{,1}}, \ldots, f - u_{,k}\frac{\partial f}{\partial u_{,k}}, \ldots, -u_{,k}\frac{\partial f}{\partial u_{,n}}\right).$$

□

Example 2.21. Treat the functional

$$F = \frac{1}{2}\iint_S (u_x^2 - u_y^2)\,dx\,dy,$$

noting that its Euler–Lagrange equation is the wave equation $u_{xx} - u_{yy} = 0$.

Solution. We use the solution to Example 2.20 with $f(\nabla u) = (u_x^2 - u_y^2)/2$ and $x_1 = x$, $x_2 = y$. Symmetry with respect to the translations in the u-direction gives

$$\operatorname{div}\mathbf{P} = 0, \qquad \mathbf{P} = (u_x, -u_y).$$

Symmetry with respect to translations in the x- and y-directions gives

$$\operatorname{div}\mathbf{P}_1 = 0, \qquad \mathbf{P}_1 = \left(-\frac{1}{2}u_x^2 - \frac{1}{2}u_y^2,\ u_x u_y\right),$$

$$\operatorname{div}\mathbf{P}_2 = 0, \qquad \mathbf{P}_2 = \left(-u_x u_y,\ \frac{1}{2}u_x^2 + \frac{1}{2}u_y^2\right).$$

See [23] for other conservation laws associated with the wave equation. □

Functional depending on vector function in several variables

In physics, and in mechanics in particular, we encounter functionals more general than those treated above. Let us consider a functional of the form

$$F(\mathbf{u}) = \int_V f(x_i, u_j, u_{j,i})\,dV = \int_V f(\mathbf{x}, \mathbf{u}, \nabla\mathbf{u})\,dV, \tag{2.87}$$

where $\mathbf{x} = (x_1, \ldots, x_n)$, $\mathbf{u} = (u_1, \ldots, u_m)$, and $\nabla\mathbf{u}$ denotes the matrix of the first partial derivatives $\partial u_j/\partial x_i = u_{j,i}$ for $j = 1, \ldots, m$ and $i = 1, \ldots, n$.

By analogy with previous cases, we consider infinitesimal transformations of the form

$$x_i \to x_i^* = x_i + \varepsilon\xi_i(\mathbf{x}, \mathbf{u}), \qquad i = 1, \ldots, n, \tag{2.88}$$

$$u_j \to u_j^* = u_j + \varepsilon\phi_j(\mathbf{x}, \mathbf{u}), \qquad j = 1, \ldots, m. \tag{2.89}$$

So the transformation is defined by $m + n$ functions ξ_i and ϕ_j. In vector form, (2.88) and (2.89) are

$$\mathbf{x} \to \mathbf{x}^* = \mathbf{x} + \varepsilon\boldsymbol{\xi}(\mathbf{x}, \mathbf{u}), \qquad \mathbf{u} \to \mathbf{u}^* = \mathbf{u} + \varepsilon\boldsymbol{\phi}(\mathbf{x}, \mathbf{u}). \tag{2.90}$$

The conditions for variational symmetry of (2.87) in terms of $\boldsymbol{\xi}$ and $\boldsymbol{\phi}$ are given, along with the formulation of the conservation law, in

Theorem 2.22. *For an extremal* $\mathbf{u}$*, let* F *be infinitesimally invariant under the transformation* (2.88)–(2.89)*, i.e.,*

$$\left[\xi_i \frac{\partial}{\partial x_i} + \phi_j \frac{\partial}{\partial u_j} + \left(\frac{d\phi_j}{dx_i} - u_{j,p}\frac{d\xi_p}{dx_i}\right)\frac{\partial}{\partial u_{j,i}} + \xi_{i,i}\right] f = 0. \tag{2.91}$$

Then the conservation law $\operatorname{div}\mathbf{P} = 0$ *holds, where* $\mathbf{P} = (P_1, P_2, \ldots, P_n)$ *and*

$$P_i = \phi_j \frac{\partial f}{\partial u_{j,i}} + \xi_k \left(f\delta_{ik} - u_{p,k}\frac{\partial f}{\partial u_{p,i}}\right). \tag{2.92}$$

Proof. The proof mostly follows the proofs of the simpler versions of Noether's theorem. As in those cases, we expand F^* in a Taylor series in ε and retain only first-order terms:

$$F^* = F + \varepsilon \int_V \left[f_{x_i}\xi_i + f_{u_j}\phi_j + f_{u_{j,i}}\left(\frac{d\phi_j}{dx_i} - u_{j,k}\frac{d\xi_k}{dx_i}\right) + f\operatorname{div}\boldsymbol{\xi}\right] dV$$
$$+ O(\varepsilon^2).$$

It can be shown that

$$\begin{aligned} f_{x_i}\xi_i + f_{u_j}\phi_j + f_{u_{j,i}}&\left(\frac{d\phi_j}{dx_i} - u_{j,k}\frac{d\xi_k}{dx_i}\right) + f\operatorname{div}\boldsymbol{\xi} \\ &= \frac{d}{dx_i}\left[\phi_j \frac{\partial f}{\partial u_{j,i}} + \xi_k\left(f\delta_{ik} - u_{p,k}\frac{\partial f}{\partial u_{p,i}}\right)\right] \\ &\qquad +(\phi_j - \xi_i u_{j,i})\left(f_{u_j} - \frac{d}{dx_i}f_{u_{j,i}}\right) = 0. \end{aligned}$$

The last parenthetical expression is the left side of the Euler–Lagrange equation. The details are left to the reader, including the derivation of the Euler–Lagrange equation itself. □

2.7 Generalizations

Divergence invariance

Noether's theorem is not the only way to establish the conservation laws. One of its extensions, obtained in 1921, is referred to as the *Bessel–Hagen extension.*

Definition 2.23. Consider the infinitesimal transformation

$$\mathbf{x} \to \mathbf{x}^* = \mathbf{x} + \varepsilon\boldsymbol{\xi}(\mathbf{x}, \mathbf{u}), \qquad \mathbf{u} \to \mathbf{u}^* = \mathbf{u} + \varepsilon\boldsymbol{\phi}(\mathbf{x}, \mathbf{u}).$$

The functional F is *infinitesimally divergence invariant* under this transformation if, in an asymptotic sense, the difference $F^* - F$ is an integral of the divergence of some vector field $\mathbf{K}$. This means that

$$\lim_{\varepsilon \to 0} F^* = F \quad \text{and} \quad \lim_{\varepsilon \to 0} \frac{F^* - F}{\varepsilon} = \int_V \operatorname{div} \mathbf{K} \, dV \tag{2.93}$$

where $\mathbf{K} = (K_1, \ldots, K_n)$ is a vector function dependent on $\mathbf{x}$, $\mathbf{u}$, and $\nabla \mathbf{u}$. We also say that F has *divergence symmetry* with respect to the infinitesimal transformation.

If $\mathbf{K} = \mathbf{0}$, the definition reduces to that of variational symmetry. The function $\mathbf{K}$ must be found together with the other unknowns $\boldsymbol{\xi}$ and $\boldsymbol{\phi}$. In terms of $\mathbf{K}$, $\boldsymbol{\xi}$, and $\boldsymbol{\phi}$, the condition for infinitesimal divergence invariance and the consequent form of the conservation law are given by

Theorem 2.24. *For an extremal* $\mathbf{u}$, *let* F *be infinitesimally invariant under the transformation* (2.88)–(2.89), *i.e.,*

$$\left[\xi_i \frac{\partial}{\partial x_i} + \phi_j \frac{\partial}{\partial u_j} + \left(\frac{d\phi_j}{dx_i} - u_{j,p} \frac{d\xi_p}{dx_i} \right) \frac{\partial}{\partial u_{j,i}} + \xi_{i,i} \right] f = K_{i,i}. \tag{2.94}$$

Then a conservation law of the form

$$\operatorname{div}(\mathbf{P} - \mathbf{K}) = 0 \tag{2.95}$$

holds, where $\mathbf{P} = (P_1, \ldots, P_n)$ *and*

$$P_i = \phi_j \frac{\partial f}{\partial u_{j,i}} + \xi_k \left(f \delta_{ik} - u_{p,k} \frac{\partial f}{\partial u_{p,i}} \right). \tag{2.96}$$

In advance of the proof let us note that this extension of Noether's theorem is quite natural. Indeed, Noether's theorem requires asymptotic equality of F^* and F to within quadratic precision with respect to ε. On the extremals this condition of coincidence reduces to the divergence form, which yields a conservation law. However, to get a conservation law, it is sufficient that F^* and F coincide asymptotically up to an integral whose integrand is the divergence of some field. Such an integrand is called a *null Lagrangian*.

If we can find a field $\mathbf{K}$, a new conservation law is formulated for the vector field $\mathbf{P} - \mathbf{K}$.

Proof. As in Theorem 2.22, expansion of F^* into a Taylor series in ε yields

$$\lim_{\varepsilon \to 0} \frac{F^* - F}{\varepsilon} = \int_V \left[f_{x_i} \xi_i + f_{u_j} \phi_j + f_{u_{j,i}} \left(\frac{d\phi_j}{dx_i} - u_{j,k} \frac{d\xi_k}{dx_i} \right) + f \operatorname{div} \boldsymbol{\xi} \right] dV.$$

So the condition for infinitesimal divergence invariance is

$$f_{x_i}\xi_i + f_{u_j}\phi_j + f_{u_{j,i}}\left(\frac{d\phi_j}{dx_i} - u_{j,k}\frac{d\xi_k}{dx_i}\right) + f\operatorname{div}\boldsymbol{\xi} = \operatorname{div}\mathbf{K}.$$

Transforming as in the proof of Theorem 2.22, we get

$$\frac{d}{dx_i}\left[\phi_j\frac{\partial f}{\partial u_{j,i}} + \xi_k\left(f\delta_{ik} - u_{p,k}\frac{\partial f}{\partial u_{p,i}}\right) - K_i\right]$$
$$+(\phi_j - \xi_i u_{j,i})\left(f_{u_j} - \frac{d}{dx_i}f_{u_{j,i}}\right) = 0.$$

On the solutions of the Euler–Lagrange equations where the last parenthetical term vanishes, this reduces to the conservation law for $\mathbf{P} - \mathbf{K}$. □

This fruitful extension of Noether's theorem can provide additional conservation laws. We consider an example from classical mechanics.

Example 2.25. Find the conservation laws for a system of particles having masses M_k and position vectors $\mathbf{y}_k \in \mathbb{R}^3$, $k = 1,\ldots,N$. The variable x plays the role of time. The Lagrangian for a system of N particles is the difference between the kinetic energy K and the potential energy W:

$$F = \int_0^1 \left[\frac{1}{2}\sum_{k=1}^{N} M_k\mathbf{y}'_k\cdot\mathbf{y}'_k - W(\mathbf{y}_1,\ldots,\mathbf{y}_N)\right]dx.$$

The Euler–Lagrange equations for the functional having this Lagrangian are the equations of motion

$$M_k\mathbf{y}''_k = W_{\mathbf{y}_k}, \qquad k = 1,\ldots,N. \tag{2.97}$$

Solution. Let us return to our earlier notation for Noether's theorem by introducing a "long vector" $\mathbf{y} = (\mathbf{y}_1,\ldots,\mathbf{y}_N) \in \mathbb{R}^{3N}$.

From physical considerations it is clear that the potential energy of the system does not change if we shift the coordinate origin by a vector $\mathbf{a} = (a_1, a_2, a_3) \in \mathbb{R}^3$; this just means that the whole set of particles has undergone the same parallel translation. Hence F is invariant under a transformation of the form

$$x \to x^* = x, \qquad \mathbf{y}_k \to \mathbf{y}_k^* = \mathbf{y}_k + \varepsilon\mathbf{a}.$$

This can also be represented in the "long vector" form $\mathbf{y} \to \mathbf{y}^* = \mathbf{y} + \varepsilon\tilde{\mathbf{a}}$ where $\tilde{\mathbf{a}} = (\mathbf{a},\ldots,\mathbf{a}) = (a_1, a_2, a_3,\ldots,a_1,a_2,a_3)$. The functions defining the transformation are

$$\xi = 1, \qquad \boldsymbol{\phi}_k = \mathbf{a} \quad \text{or} \quad \boldsymbol{\phi} = \hat{\mathbf{a}}.$$

For this transformation the conservation law (2.75) reduces to

$$\sum_{k=1}^{N} M_k \mathbf{y}'_k \cdot \mathbf{a} = \text{constant.}$$

By arbitrariness of $\mathbf{a}$, the quantity

$$\sum_{k=1}^{N} M_k \mathbf{y}'_k = \text{constant.}$$

This is the conservation of linear momentum.

The given F also has variational symmetry with respect to time shifts, i.e., with respect to the transformation

$$x \to x^* = x + \varepsilon, \qquad \mathbf{y}_k \to \mathbf{y}^*_k = \mathbf{y}_k.$$

Here $\boldsymbol{\phi} = \mathbf{0}$ and $\xi = 1$. The corresponding conservation law (2.75) is the law of energy conservation

$$K + W = \text{constant}, \qquad K = \frac{1}{2}\sum_{k=1}^{N} M_k \mathbf{y}'_k \cdot \mathbf{y}'_k.$$

Finally, consider the Galilean boost

$$x \to x^* = x, \qquad \mathbf{y}_k \to \mathbf{y}^*_k = \mathbf{y}_k + \varepsilon x \mathbf{a}. \tag{2.98}$$

Here $\xi = 0$, and $\boldsymbol{\phi}_k = x\mathbf{a}$ or $\boldsymbol{\phi} = x\tilde{\mathbf{a}}$. Unlike the two cases considered above, the conditions of infinitesimal invariance do not hold for the Galilean boost. Indeed, substitution of ξ, $\boldsymbol{\phi}$, and f into (2.73) or into (2.74) gives

$$\sum_{k=1}^{N} \left[\boldsymbol{\phi}_k \cdot \nabla_{\mathbf{y}_k} f + \frac{d\boldsymbol{\phi}_k}{dx} \cdot \nabla_{\mathbf{y}'_k} f \right] = \sum_{k=1}^{N} [-x\mathbf{a} \cdot \nabla_{\mathbf{y}_k} W + M_k \mathbf{a} \cdot \mathbf{y}'_k] = 0.$$

This holds only when $\mathbf{a} = \mathbf{0}$. In other words, the Galilean boost does not correspond to the variational symmetry of F. However, F does have infinitesimal divergence invariance. Let us verify that (2.94) holds for (2.98). For convenience we introduce $K = \sum_{k=1}^{N} K_k$. Equation (2.94) reduces to

$$\sum_{k=1}^{N} [-x\mathbf{a} \cdot \nabla_{\mathbf{y}_k} W + M_k \mathbf{a} \cdot \mathbf{y}'_k - K'_k] = 0. \tag{2.99}$$

Changing $\nabla_{\mathbf{y}_k} W$ to $M_k \mathbf{y}''_k$ via the equations of motion (2.97), we reduce (2.99) to the divergence form

$$\sum_{k=1}^{N} [-x\mathbf{a} \cdot M_k \mathbf{y}'_k + 2M_k \mathbf{a} \cdot \mathbf{y}_k - K_k]' = 0.$$

Let us take

$$K_k = -M_k x\mathbf{a} \cdot \mathbf{y}'_k + 2M_k \mathbf{a} \cdot \mathbf{y}_k.$$

The corresponding conservation law $P - K =$ constant takes the form

$$\sum_{k=1}^{N} (xM_k \mathbf{a} \cdot \mathbf{y}'_k - M_k \mathbf{a} \cdot \mathbf{y}_k) = \text{constant},$$

and since $\mathbf{a}$ is constant,

$$\sum_{k=1}^{N} (xM_k \mathbf{y}'_k - M_k \mathbf{y}_k) = \text{constant}.$$

Dividing through by the total mass of the system $M = \sum_{k=1}^{N} M_k$, and recalling that the momentum $M_k \mathbf{y}_k =$ constant, we obtain a familiar result from classical mechanics that the center of mass of the particle system undergoes uniform, rectilinear motion:

$$\bar{\mathbf{y}} \equiv \frac{1}{M} \sum_{k=1}^{N} M_k \mathbf{y}_k = x\mathbf{C}_1 + \mathbf{C}_2,$$

where $\mathbf{C}_1$ and $\mathbf{C}_2$ are constant vectors. □

These examples show how divergence symmetry can provide additional conservation laws. As an example, for $f = y'^2/2$, infinitesimal divergence invariance extends the number of conservation laws from three to five (Exercise 2.13).

Other generalizations

There are still other ways to obtain conservation laws. For example, it is possible to study the infinitesimal symmetries of equations (2.49) without restricting oneself to the case where the equations are the Euler–Lagrange equations for some functional (see [23]). Let us consider one such method that uses the notion of null Lagrangian and its properties.

Definition 2.26. A function $f(\mathbf{x}, \mathbf{u}, \nabla\mathbf{u})$ is a *null Lagrangian* if the Euler–Lagrange equations for the functional with integrand $f(\mathbf{x}, \mathbf{u}, \nabla\mathbf{u})$ vanish identically for all $\mathbf{x}$ and $\mathbf{u}$.

An example is the function $f = yy'$. Indeed, its Euler equation reduces to the identity

$$f_y - \frac{d}{dx} f_{y'} = y' - \frac{d}{dx} y = 0.$$

Similar cases exist for f depending on a function in many variables: $f = uu_x + uu_y$, for instance.

Note that

$$yy' = \frac{d}{dx}\left(\frac{y^2}{2}\right), \qquad uu_x + uu_y = \frac{d}{dx}\left(\frac{u^2}{2}\right) + \frac{d}{dy}\left(\frac{u^2}{2}\right).$$

We see that an expression that happens to be the divergence of some vector field can be a null Lagrangian. This idea is valid, and so is the converse idea:

Theorem 2.27. *A function $f(\mathbf{x}, \mathbf{u}, \nabla \mathbf{u})$ is a null Lagrangian if and only if it is equal to the divergence of some field $\mathbf{P}$:*

$$f = \operatorname{div} \mathbf{P}$$

where $\mathbf{P} = \mathbf{P}(\mathbf{x}, \mathbf{u}, \nabla \mathbf{u})$.

See [23] for a proof. It follows that two functionals have the same Euler–Lagrange equations if and only if their integrands differ by the divergence of some vector field. These facts serve as a basis for constructing conservation laws for the system (2.49). The idea of the method is as follows.

(1) Multiply each equation from (2.49) by a function $q_p(x_j, u_i, u_{i,j})$, $p = 1, \ldots, k$ and add the resulting equations to get

$$\sum_{p=1}^{k} q_p(x_j, u_i, u_{i,j}) l_p(x_j, u_i, u_{i,j}) = 0. \tag{2.100}$$

(2) By Theorem 2.27, the left side of the last expression is the divergence of some vector field if and only if the Euler–Lagrange equations for the functional having integrand

$$\tilde{f} = \sum_{p=1}^{k} q_p(x_j, u_i, u_{i,j}) l_p(x_j, u_i, u_{i,j})$$

are satisfied identically. So the functions q_p should be selected according to the condition that the Euler–Lagrange equations for the functional having integrand $\tilde{f}$ become identities.

This *neutral action method* leads to overdetermined systems of equations in the q_p. On the other hand, it applies to systems of partial differential equations that are not the Euler–Lagrange equations of some functional. Applications to mechanical problems can be found in [12].

2.8 Exercises

2.1 Let the structure of Example 2.1 be strengthened with another spring of rigidity k_2 as shown in Fig. 2.12. (a) Write down the total potential energy of the system and the boundary and compatibility equations. (b) Write down the functional that should be minimized to get the equilibrium equations and natural boundary conditions. (c) Applying the general procedure of the calculus of variations, derive the differential equation of equilibrium and the natural boundary conditions.

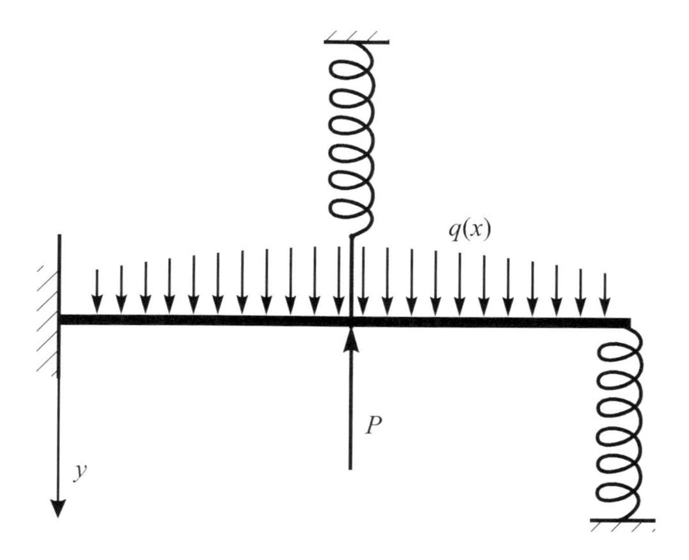

Fig. 2.12 Beam with two springs under load $q(x)$ and P.

2.2 A cantilever beam having length $2a$ and parameters E, I is supported with a spring of rigidity k at point a and a clock spring of rigidity c at point $2a$ (Fig. 2.13). Choose a mathematical model, write down the total potential energy functional and kinematic restrictions, and derive the equilibrium equations and natural boundary conditions.

2.3 Consider the system shown in Fig. 2.14. Using Example 2.2 as a guide, write down the energy functional, kinematic restrictions, and first variation of the energy functional. Then derive the equilibrium equation and the natural boundary conditions. Repeat for the system of Fig. 2.15; also consider the case when both beam and rod models are employed simultaneously.

2.4 Fig. 2.16 shows a system of three rigidly coupled beams having parameters E, I and respective lengths a_1, a_2, a_3. Construct a mathematical model of the

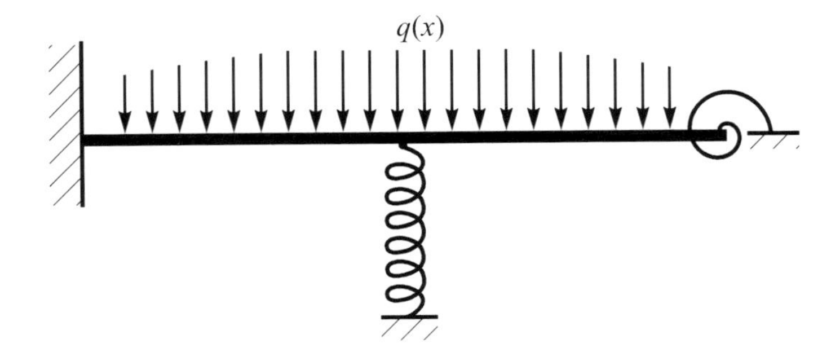

Fig. 2.13 Cantilever beam supported with springs under load.

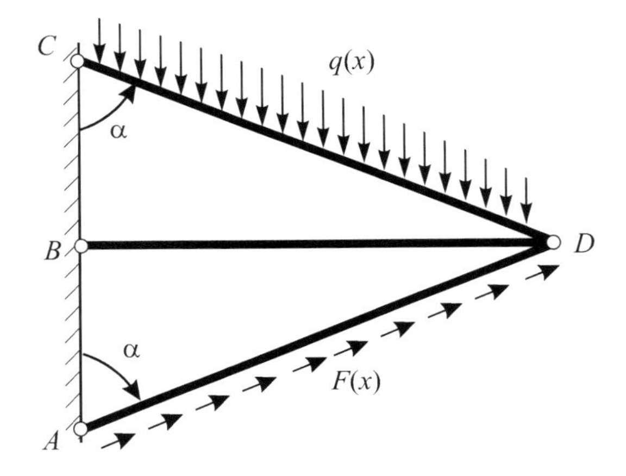

Fig. 2.14 System consisting of three rods.

structure, write out the energy functional and kinematic restrictions, and derive the equilibrium equations and natural boundary conditions.

2.5 Two beams having parameters E, I, a are related elastically. Using Winkler's model of the elastic junction with parameter k (Fig. 2.17), construct a model for the system, write down the energy functional and kinematic restrictions, and derive the equilibrium equations and natural boundary conditions.

2.6 A square is composed of equal beams having parameters E, I, a (Fig. 2.18). Construct the model, write out the energy functional and kinematic restrictions, and derive the equilibrium equations and natural boundary conditions. Find the conditions under which the equilibrium problem makes sense (has a solution).

2.7 A system of coupled beams with a supporting clock spring appears in

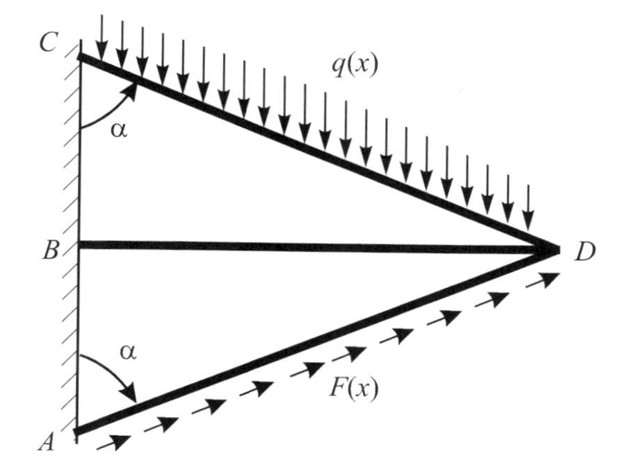

Fig. 2.15 System consisting of three beams.

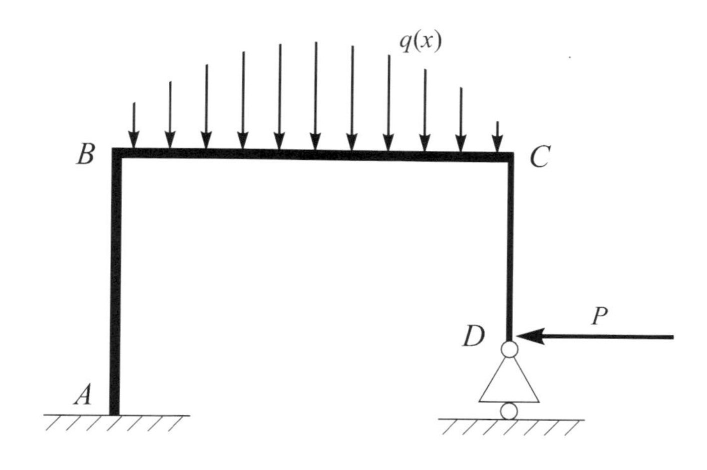

Fig. 2.16 Three beam system under load.

Fig. 2.19. Construct the mathematical model, write out the total potential energy functional and kinematic restrictions, and derive the equilibrium equations and natural boundary conditions.

2.8 A system of coupled beams with a supporting spring is shown in Fig. 2.20. (Also known as *von Mises truss*, the system is used to study the stability of elastic systems.) Construct a mathematical model, write out the energy functional and kinematic restrictions, and derive the equilibrium equations and natural boundary conditions.

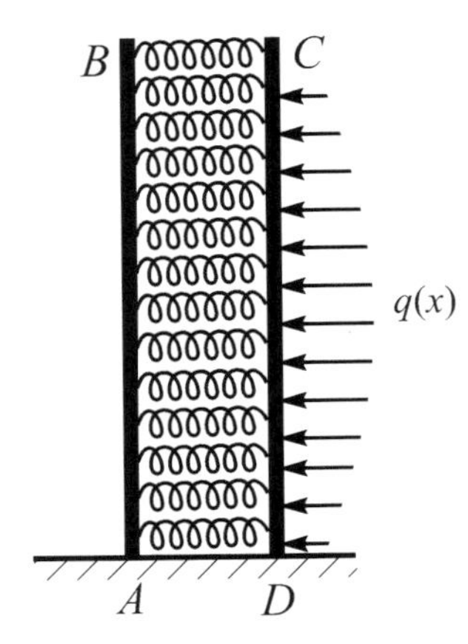

Fig. 2.17 Two beams related elastically.

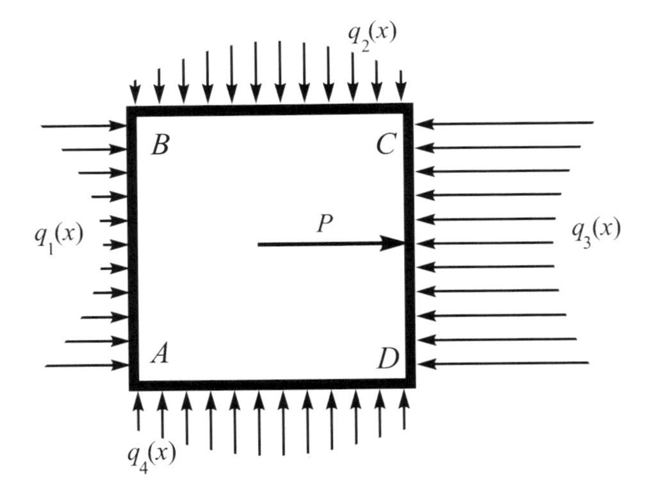

Fig. 2.18 A square constructed of beams under load.

2.9 Find the conservation laws for a functional of the form

$$F = \int_a^b f(y')\,dx.$$

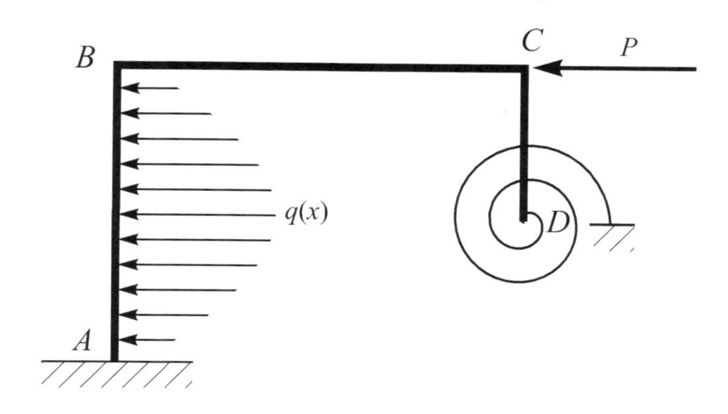

Fig. 2.19 Three beams with a supporting clock spring under load.

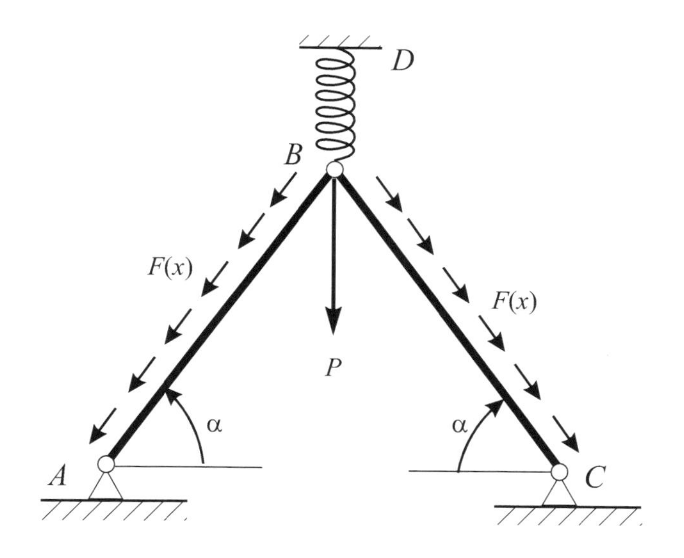

Fig. 2.20 Two beam system.

2.10 Find the conservation laws for the functional

$$F = \frac{1}{2}\int_a^b (y')^2\, dx.$$

2.11 Show that the functional

$$F = \frac{1}{2} \int_a^b (y')^2 \, dx,$$

has no variational symmetries other than those obtained in Exercises 2.9 and 2.10.

2.12 Find the variational symmetry condition for the functional

$$F = \frac{1}{2} \int_a^b f(x, y, y^{(k)}) \, dx,$$

where the constant $k > 2$.

2.13 Find divergence symmetries for the functional

$$F = \frac{1}{2} \int_a^b (y')^2 \, dx.$$

Chapter 3

Elements of Optimal Control Theory

3.1 A Variational Problem as a Problem of Optimal Control

Let us consider a special problem in the calculus of variations:

$$\int_a^b f(x, y(x), y'(x))\, dx \to \min_{\substack{y \in C^{(1)}(a,b) \\ y(a)=y_0}} \tag{3.1}$$

Let $y(x)$ be fixed for a moment. We introduce an equation for a new function $z = z(x)$:

$$z'(x) = f(x, y(x), y'(x)), \qquad z(a) = 0.$$

It is clear that

$$z(b) = z(b) - z(a) = \int_a^b z'(x)\, dx = \int_a^b f(x, y(x), y'(x))\, dx.$$

Now we introduce another function $u(x) = y'(x)$. Problem (3.1) can be formulated as follows:[1]

Problem of Terminal Control. Given ordinary differential equations

$$y'(x) = u(x), \qquad z'(x) = f(x, y(x), u(x)), \tag{3.2}$$

and initial conditions $y(a) = y_0$ and $z(a) = 0$ find $u = u(x) \in C(a, b)$ at which $z(b)$ attains the minimal value.

Since $z(b)$ is the value of the integral, this formulation is equivalent to the formulation of the problem of strong minimum of the functional (3.1).

[1]Thanks to Dr. K.V. Isaev of Rostov State University, who furnished the authors with a notebook of his lectures on control theory. The presentation of the terminal control problem follows, in large part, Dr. Isaev's lectures.

Note that the last formulation does not involve integration. The solution of the Cauchy problem for an ordinary differential equation (ODE) is less computationally intensive than the solution of the corresponding integral equation. This transformation of a variational problem to another form is numerically advantageous; moreover, it allows us to pose a new class of minimization problems along with new solution methods. Note that the new formulation should still give us the Euler equation for a minimizer and the natural boundary condition at $x = b$.

The formulation (3.1) is equivalent to the Problem of Terminal Control if f is sufficiently smooth. But the Problem of Terminal Control has brought us to a new class of problems that fall outside the calculus of variations. These problems also fall outside classical ODE theory, since for the Cauchy problem in the latter, the number of differential equations always equals the number of unknown functions. In our formulation we have two equations and three unknowns y, z, u. But if u is given we have a Cauchy problem in which y and z are uniquely determined. We solve a special minimum problem, seeking the minimum value of z at point b, changing u in the class of continuous functions. Continuity of u was stipulated by the tools of the calculus of variations. But for many problems having the form of the Problem of Terminal Control or something similar, this condition is too restrictive. We shall consider other tools for treating such problems — tools not equivalent to those of the calculus of variations.

The Problem of Terminal Control belongs to *optimal control theory*. The designation "terminal control" refers to the fact that something, namely z, is to be minimized at a final time instant $x = b$. A more general formulation is presented in § 3.2.

We have thus examined a variational problem as a problem of optimal control. Let us take a moment to compare the setups of these two problems. Each must provide a functional to be minimized. In the variational setup this functional is an integral that incorporates some information about the system structure. In the control problem these elements are separated: the system is governed by a set of ODEs relating internal parameters y, z to an external parameter u that can be changed at will (under some restrictions of course), while the "cost functional" is formulated separately. There are advantages in choosing to disentangle the elements of the problem setup in this way; in fact, many practical problems are so posed naturally and cannot be posed as variational problems. Consider, for example, a child on a playground swing. The amplitude of the oscillations is governed by the pendulum equation — an ODE — and the effective length $u = u(t)$ of the pendulum is under the child's control. There is no reason why this control parameter must be changed in a continuous fashion; every child knows that the best results can be obtained by sudden shifts in his or her center of gravity. Hence we should be able to accommodate discontinuities in u. Of course, it is easy to cite examples on a much larger scale of economic importance — examples ranging from space travel to the damping of a ship's oscillations in the ocean.

In short, we shall consider problems involving a "system" or "controlled object" having a control parameter u. In general we seek u that minimizes a cost functional G, which in turn depends on u through an initial or boundary value problem for a set of ODEs. We will not consider all aspects of standard mathematical optimal control theory, including existence theorems, etc. But we will present an introduction to certain practical aspects relating to the numerical solution of optimal control problems. The expression for the increment of the cost functional G which we will derive is analogous to the first variation in the calculus of variations, or to the differential in calculus. Its expression provides a basis for various numerical approaches to optimal control problems. It also brings us Pontryagin's maximum principle, which allows us to determine whether a governing function u is optimal.

3.2 General Problem of Optimal Control

First we generalize the Problem of Terminal Control. A controlled system is described by $n + m$ functions, which depend on a known variable. We shall call this latter variable t or x and regard it as the time variable. Given are n ordinary differential equations involving the first n parameters of the system $y_1, \ldots, y_n$ and their first derivatives. These equations are written in normal form. The vector $\mathbf{y} = (y_1, \ldots, y_n)$ is often called the *state vector*, and its component functions $y_1, \ldots, y_n$ the *state variables*. The remaining m parameters $u_1, \ldots, u_m$ are considered as free parameter-functions. We call $\mathbf{u} = (u_1, \ldots, u_m)$ the *control vector*, and its component functions the *control variables*. The differential equations are

$$\begin{aligned} y_1'(t) &= f_1(t, y_1(t), \ldots, y_n(t), u_1(t), \ldots, u_m(t)), \\ &\vdots \\ y_n'(t) &= f_n(t, y_1(t), \ldots, y_n(t), u_1(t), \ldots, u_m(t)), \end{aligned}$$

or

$$\mathbf{y}'(t) = \mathbf{f}(t, \mathbf{y}(t), \mathbf{u}(t)). \tag{3.3}$$

Equations (3.3) should be supplemented with conditions at the initial time $t = t_0$:

$$\mathbf{y}(t_0) = \mathbf{y}_0, \tag{3.4}$$

where $\mathbf{y}_0$ is a given *initial state.*

We now consider a problem of the form

$$G(\mathbf{y}(T)) \to \min$$

over the set of admissible $\mathbf{u}$, where T is a fixed (final) time instant. The quantity $G(\mathbf{y}(T))$ is a functional dependent on the values taken by $\mathbf{u}$ and $\mathbf{y}$ over $[t_0, T]$. The space in which these vector functions reside is an important issue to be discussed later. Whereas in variational problems we permit only smooth functions for comparison and consider non-smooth functions as exceptions, here we consider non-smooth control functions since these tend to be more useful in applications (and, often more importantly, allowed by the method of solution and investigation).

Many optimal control problems arise in classical mechanics. There a system, described by the equations of classical mechanics, can be acted upon by forces whose magnitudes and directions are subject to certain restrictions. We obtain a problem of terminal control if we attempt to minimize the value of a function, depending on the internal parameters of the system, at a certain (final) time instant. For example, we may wish to bring the system to a certain state with the best accuracy.

We can generalize the Problem of Terminal Control by supplanting the initial values (3.4) with n equations given at some fixed points $t_k \in [t_0, T]$:

$$B_k(\mathbf{y}(t_k)) = 0, \qquad k = 1, \dots, n.$$

The goal function can incorporate values of $\mathbf{y}$ at other points of $[t_0, T]$:

$$G(\mathbf{y}(\tau_1), \dots, \mathbf{y}(\tau_r)) \to \min.$$

Such a problem is solved practically by any system that has to meet some time schedule (e.g., by a flight team who must land at several airports at scheduled times during a flight).

Let us consider another type of optimal control problem:

Time-Optimal Control Problem. A system is described by (3.3). It is necessary to move the system from state $\mathbf{y}(t_0) = \mathbf{y}_0$ to state $\mathbf{y}(T) = \mathbf{y}_f$ in minimal time T.

Again, we leave the class of admissible $\mathbf{u}$ as an issue for the future. Note that for this problem an existence theorem is essential in many cases, since there are mechanical and other systems for which an initial-final pair of states $\mathbf{y}_0, \mathbf{y}_f$ is impossible to take on for any time.

We see that in Time-Optimal Control we have $2n$ given boundary conditions, but there is an additional unknown parameter T that must be determined as an outcome of the solution. We see a big difference in the number of boundary values for the state vector $\mathbf{y}$ in these problems. This is provided by the arbitrariness of the control vector $\mathbf{u}$, changes in which can lead to the requirement for new boundary conditions. The restriction on the number of boundary conditions r at each "boundary" (initial, final, or intermediate) point of time is that it cannot exceed n, the number of components of $\mathbf{y}$ and, in total, at any admissible fixed $\mathbf{u}$ we have to obtain a boundary value problem for our system of equations that is solvable (not necessarily uniquely). When the boundary value system has too few boundary conditions for uniqueness, then, in the same way there arise natural boundary conditions in the calculus of variations, there arise additional boundary conditions for $\mathbf{y}$ in the optimal control problems. In some versions of the numerical methods that are used for solving the corresponding problems, such natural conditions do not participate explicitly — as is also the case for natural conditions in the calculus of variations — however, an optimal solution obeys them.

These are not the only possible setups for optimal control problems. We can consider, for example, problems where the cost functional is given in an integral form which takes into account the values of $\mathbf{y}$ at all instants of time.

Above we mentioned restrictions on the control vector $\mathbf{u}$, but many problems require restrictions (frequently in the form of inequalities) on $\mathbf{y}$ as well. For example, the problem of manned spaceflight forces us to minimize expenses while restricting accelerations experienced during the flight.

Many real problems of optimal control require us to consider (nonlinear) systems of PDEs rather than ODEs. The interested reader can find this discussed elsewhere. Often, however, these problems can be reduced to the problems that appear in this chapter. Each practical problem for the same object can lead to a different mathematical setup, as well as to different theoretical and practical results. In this book we will consider only mathematical aspects of the problems of optimal control, leaving applications to many other sources. First we would like to slightly reduce the setup of the problems under consideration.

The system (3.3) is *autonomic* if f does not depend explicitly on t. Henceforth we shall consider only autonomic systems with $t_0 = 0$. We may do this without loss of generality. First, given $t_0 \neq 0$ we may shift the time origin by putting $x = t - t_0$. Let us consider the transformation to autonomic form. In principle there is nothing to limit the number of components that $\mathbf{y}$ may have. So we can always extend it by an additional component y_{n+1}, supplementing (3.3) with an additional equation $y'_{n+1}(x) = 1$ and initial condition $y_{n+1}(0) = 0$. Then (3.3) takes the form

$$\mathbf{y}'(x) = \mathbf{f}(y_{n+1}(x), \mathbf{y}(x), \mathbf{u}(x)).$$

Thus, redenoting $\mathbf{y} = (y_1, \ldots, y_{n+1})$ and the corresponding $\mathbf{f}$, we arrive again at (3.3) but in the form

$$\mathbf{y}'(t) = \mathbf{f}(\mathbf{y}(t), \mathbf{u}(t)). \tag{3.5}$$

This is the autonomic form we shall consider.

3.3 Simplest Problem of Optimal Control

So far we have said little about the restrictions to be placed on the behavior of $\mathbf{u}(t)$. We shall take the class of admissible controls to consist of those vector functions that are *piecewise* continuous on $[0, T]$. This is in contrast to what we saw in the calculus of variations. It is possible to relax this restriction on $\mathbf{u}(t)$, requiring it to be merely measurable in some sense, but we leave this and related questions of existence[2] for more advanced books.

What constitutes a "small" variation (increment) of a control function? In the calculus of variations we regarded a variation (increment) of a function as small if its norm in the space $C^{(1)}(0, T)$ was small. With such a small increment taken in its argument, the increment of a functional was also guaranteed to be small, and we were led to apply the tools of calculus. To obtain the Euler equation and the natural boundary conditions we linearized the functional with respect to the increment of the unknown function. Here we would like apply the same linearization idea and obtain necessary conditions for the objective functional to attain its minimal value, but at the same time introduce another notion of smallness of the increment of a control function.

[2]Such questions are more theoretical than we are able to treat here, but this does not mean they are unimportant. There are practical problems for which no optimal solution exists. In such cases, however, it is often possible to obtain a working approximation to an optimal solution.

When we linearize an expression we use the fact that a small increment in the independent variable brings a small increment in the value of the expression. We understand that if we change the control function in some small way then the increment of the output function will be small. But in Newtonian mechanics if a large force acts on a material point for a short time then the deviation of the point trajectory during a finite time is small — the shorter the time of action, the smaller the deviation. So "smallness" of the increment can be provided by a force of small magnitude *or* by a force of short duration. This situation is quite typical for disturbances to ODEs, and suggests a new class of "small" increments to control functions. From a more mathematical viewpoint, the norm of $C(0,T)$ is not the only norm with which we can define small increments while requiring that the change in a solution exhibit continuous dependence on changes in the control function. In particular, we may use the norm of $L(0,T)$.

Let us build a class of functions $\mathcal{U}$ in which we seek control functions $u = u(t)$. $\mathcal{U}$ is a set of functions piecewise continuous on $[0,T]$, and is restricted by some conditions: normally simultaneous linear inequalities given pointwise. An example of such a restriction is

$$0 \le u(t) \le 1.$$

The simplest problem of optimal control theory is the following problem of terminal control:

Simplest Problem of Optimal Control. Let a controlled object be described by the equation

$$y'(t) = f(y(t), u(t)) \tag{3.6}$$

subject to

$$y(0) = y_0. \tag{3.7}$$

Among all functions belonging to a class $\mathcal{U}$ described above, find a control function $u(t)$ that minimizes $g(y(t))$ at $t = T$:

$$g(y(T)) \to \min_{u(t)\in\mathcal{U}}.$$

Here $g(y)$ is a continuously differentiable function on the domain of all admissible values of $y = y(t)$.[3]

[3]Rather than formulating explicit restrictions on f and g, we simply assume they are sufficiently smooth. In particular we shall differentiate $g(y)$ and $f(y,u)$ with respect to y supposing that the corresponding derivatives are continuous, we shall assume a

First we define the main elementary increment of the control function, a so-called *needle-shaped increment*. This is where optimal control theory begins to depart from the calculus of variations. We choose some $u(t) \in \mathcal{U}$ and let $t = s$ be a point at which $u(t)$ is continuous. For definiteness we consider all the functions $u(t)$ to be continuous from the left on $[0, T]$. Consider another function $u^*(t)$ that differs from $u(t)$ only on the half-open segment $(s - \varepsilon, s]$ as shown in Fig. 3.1. Analytically this function is

$$u^*(t) = \begin{cases} u(t), & t \notin (s - \varepsilon, s], \\ v, & t \in (s - \varepsilon, s], \end{cases} \tag{3.8}$$

where $\varepsilon > 0$ is sufficiently small. The increment

$$\delta u(t) = u^*(t) - u(t),$$

which is zero everywhere except in the interval $(s - \varepsilon, s]$ of length ε, is what we term needle-shaped. Its smallness is characterized by

$$\|\delta u\|_{L(0,T)} = \int_0^T |\delta u| \, dt,$$

which is of order ε.

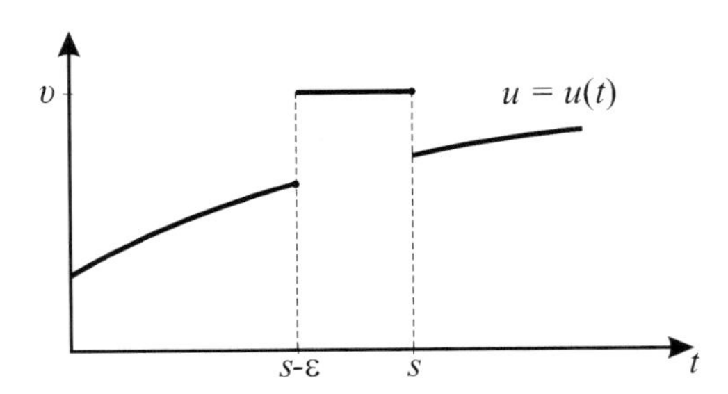

Fig. 3.1 A control function subject to a needle-shaped increment.

In what follows we suppose $u^*(t)$ belongs to $\mathcal{U}$. We also assume that together with some $u^*(t)$, defined by $\varepsilon_0 > 0$ and v_0, the class $\mathcal{U}$ contains all the $u^*(t)$ having the same final point s of the jump for which $\varepsilon < \varepsilon_0$. Since

continuous dependence of $f(y, u)$ on u, and we shall suppose that for any fixed admissible $u(t) \in \mathcal{U}$ the Cauchy problem (3.6)–(3.7) has a unique solution that depends continuously on the initial condition y_0. All this could be formulated purely in terms of the given functions f and g and it is possible that doing so would yield sharper results, but we choose clarity over rigor at this stage. In fact, the simple problem we have chosen to consider is not the most realistic one available. However, its investigation will open the way to general problems without obscuring the essential ideas.

the restrictions for $\mathcal{U}$ are usually given piecewise by simultaneous linear inequalities, this assumption does not place additional restrictions on such problems.

Many textbooks consider needle-shaped functions that are constant on the interval $(s-\varepsilon, s]$, but we consider them only for small ε so the norm in $L(0,T)$ of the difference between the above introduced and the traditional needle-shaped functions is of order higher than ε. We took our definition only for convenience. Note that we can approximate (in the uniform norm) any $u(t) \in \mathcal{U}$ with a finite linear combination of needle-shaped functions.

Since $g(y(T))$ is a number that depends on $u(t)$ through (3.6) and (3.7), we have a functional defined on $\mathcal{U}$. Experience suggests that we apply the ideas of calculus. We need to find the increment of the functional under that of the control function, introducing something like the first differential. Now $\delta u(t)$ is an elementary needle-shaped function whose smallness is determined by ε. From the corresponding increment of $g(y(T))$ we must select the part that is proportional to ε and neglect terms of higher order in ε.

As an intermediate step we will have to obtain the increment in $y(T)$ corresponding to $\delta u(t)$. Let us denote the solution of (3.6)–(3.7) corresponding to $u^*(t)$ by $y^*(t)$:

$$y^{*\prime}(t) = f(y^*(t), u^*(t)), \qquad y^*(0) = y_0.$$

We denote

$$\Delta y(t) = y^*(t) - y(t), \qquad J(u) = g(y(T)),$$

and seek the main (in ε) part of the increment

$$\Delta J_{\varepsilon,v}(u) = J(u^*) - J(u). \tag{3.9}$$

Again, this main part must be linear in ε; we neglect terms of higher order in ε. In this, we consider $u(t)$ as given and hence $y(t)$ is known uniquely as well.

Theorem 3.1. *Let $t = s$ be a point of continuity of a control function $u(t)$. We have*

$$\Delta J_{s,v}(u) = \varepsilon\, \delta_{s,v} J(u) + o(\varepsilon), \qquad \varepsilon > 0, \tag{3.10}$$

where

$$\delta_{s,v} J(u) = \psi(s)[f(y(s), u(s)) - f(y(s), v)] \tag{3.11}$$

and where $\psi(s)$ is a solution of the following Cauchy problem (in the reverse time):

$$\psi'(s) = -\frac{\partial f(y(s), u(s))}{\partial y}\psi(s), \qquad \psi(T) = -\frac{dg(y(T))}{dy}. \tag{3.12}$$

The quantity $\delta_{s,v}J(u)$ in (3.10) is called the ***variational derivative of the second kind****.*

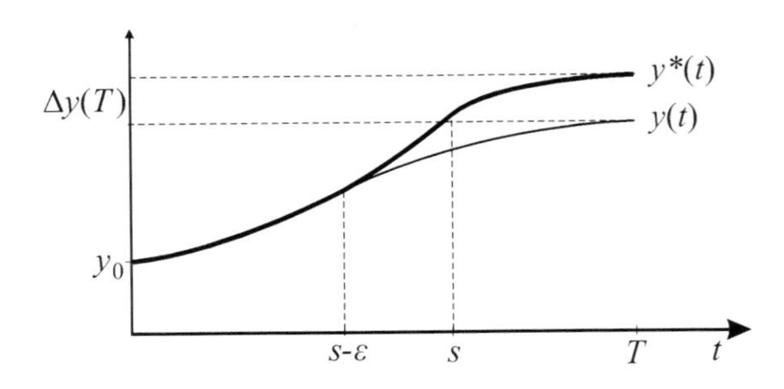

Fig. 3.2 The deviation of a trajectory $y(t)$ under a needle-shaped change of the control function on the time interval $[s - \varepsilon, s]$.

Proof. Take $\varepsilon > 0$ so small that all the points of $[s - \varepsilon, s]$ are points of continuity of $u(t)$. We require that $u^*(t)$, which differs from $u(t)$ by a needle-shaped increment, is admissible and has the form (3.8). We divide the proof into several steps.

Step 1. First let us find the main part in ε of the increment of $y(t)$, in particular at $t = T$. In Fig. 3.2 we show the behavior of $y(t)$ and $y^*(t)$. When $t < s - \varepsilon$ we have $u^*(t) = u(t)$ and thus $y^*(t) = y(t)$.

Let $t \in [s - \varepsilon, s]$. Subtracting the equations for y^* and y we get

$$y^{*\prime}(t) - y'(t) = f(y^*(t), v) - f(y(t), u(t))$$

or, since $\Delta y(t) = y^*(t) - y(t)$, the increment of $y(t)$ satisfies

$$\Delta y'(t) = f(y(t) + \Delta y(t), v) - f(y(t), u(t)). \tag{3.13}$$

Besides, we have the "initial" condition for this interval

$$\Delta y(s - \varepsilon) = 0 \tag{3.14}$$

since $y^*(s - \varepsilon) = y(s - \varepsilon)$. Integration of (3.13) gives us an equivalent integral equation on $[s - \varepsilon, s]$:

$$\Delta y(t) - \Delta y(s - \varepsilon) = \int_{s-\varepsilon}^{t} [f(y(\tau) + \Delta y(\tau), v) - f(y(\tau), u(\tau))]\, d\tau.$$

By (3.14) we have

$$\Delta y(t) = \int_{s-\varepsilon}^{t} [f(y(\tau) + \Delta y(\tau), v) - f(y(\tau), u(\tau))]\, d\tau \quad \text{on } [s-\varepsilon, s].$$

We assume $f(y, u)$ is continuous and bounded on the domain where the pair (y, u) takes its value, and thus when ε is small the modulus of the integral on the right is bounded by $M\varepsilon$ for $t \in [s-\varepsilon, s]$. So this integral has the first order of smallness in ε when $t \in [s-\varepsilon, s]$, and thus the same value bounds $|\Delta y(t)|$ on the same segment. Since ε is small and $y(t), u(t)$ are continuous on $[s-\varepsilon, s]$, the integrand is continuous as well, and we introduce in the values of this integral an error of order higher than the first in ε if we replace the integrand by the constant value $f(y(s), v) - f(y(s), u(s))$. So

$$\begin{aligned}\Delta y(t) &= \int_{s-\varepsilon}^{t} [f(y(s), v) - f(y(s), u(s))]\, d\tau + o(\varepsilon) \\ &= (t - s + \varepsilon)[f(y(s), v) - f(y(s), u(s))] + o(\varepsilon),\end{aligned}$$

and thus

$$\Delta y(s) = \varepsilon[f(y(s), v) - f(y(s), u(s))] + o(\varepsilon). \tag{3.15}$$

This gives us the "initial" value for the solution $y^*(t)$ on $[s, T]$. Note that on $[s-\varepsilon, s]$ the change of $\Delta y(t)$ in t is almost linear, which is expected since ε is small.

On $[s, T]$, subtracting the equations for $y(t)$ and $y^*(t)$ we get

$$\Delta y'(t) = f(y(t) + \Delta y(t), u(t)) - f(y(t), u(t)). \tag{3.16}$$

This is supplemented by the initial condition (3.15), which is small when ε is small. Since y and y^* obey the same equation on $[s, T]$ but their initial values differ by a small value $\Delta y(s)$ of order ε, we can expect that there is continuous dependence of the solution on the initial data and hence that the difference between y^* and y, which is Δy, remains of order ε when T is finite. So we linearize (3.16) using the first-order Taylor expansion

$$f(y(t) + \Delta y(t), u(t)) - f(y(t), u(t)) = \frac{\partial f(y(t), u(t))}{\partial y} \Delta y(t) + o(|\Delta y(t)|)$$

to get

$$\Delta y'(t) = \frac{\partial f(y(t), u(t))}{\partial y} \Delta y(t) + o(\varepsilon).$$

The main part of $\Delta y(t)$, denoted by $\delta y(t)$, satisfies

$$\delta y'(t) = \frac{\partial f(y(t), u(t))}{\partial y} \delta y(t). \tag{3.17}$$

This can be integrated explicitly since $y(t)$ and $u(t)$ and the initial condition for $\delta y(t)$ are defined by (3.15) as

$$\delta y(s) = \varepsilon[f(y(s), v) - f(y(s), u(s))].$$

However, we should allow for an extension to a system of ODEs. So we shall produce a mathematical trick of "finding" the solution in other terms. At this point we must interrupt the proof and cover some additional material.

3.4 Fundamental Solution of a Linear Ordinary Differential Equation

Consider a linear ODE

$$x'(t) = a(t)x(t). \tag{3.18}$$

This has a unique solution for any initial condition $x(s) = x_0$, $a(t)$ being a given continuous function (it can be continuous on an interval if we consider the equation on this interval or at any t). The *fundamental solution* is a function $\varphi(t,s)$ which at any fixed s satisfies

$$\frac{d\varphi(t,s)}{dt} = a(t)\varphi(t,s) \tag{3.19}$$

and the condition

$$\varphi(s,s) = 1. \tag{3.20}$$

This function in two variables has many useful properties, the first of which is trivial:

Proposition 3.2. *A solution of* (3.18) *satisfying the initial condition* $x(s) = x_0$ *is*

$$x(t) = x_0\varphi(t,s). \tag{3.21}$$

Proposition 3.3. *We have*

$$\varphi(t,s) = \varphi(t,\tau)\varphi(\tau,s) \tag{3.22}$$

for any t*,* s*, and* τ*.*

Proof. Indeed, for fixed τ, s the function $\varphi(t,\tau)\varphi(\tau,s)$ of the variable t is a solution to (3.18) when t is an independent variable, since $\varphi(\tau,s)$ does not depend on t. Thus we have two solutions to (3.18): the functions $\varphi(t,s)$ and $\varphi(t,\tau)\varphi(\tau,s)$. But for $t = \tau$ they correspondingly reduce to $\varphi(\tau,s)$ and

$\varphi(\tau, \tau)\varphi(\tau, s) = 1 \cdot \varphi(\tau, s)$, and thus at $t = \tau$ they coincide. By uniqueness of the solution to the Cauchy problem for (3.18) (the initial value is given at $t = \tau$) they coincide at any t. □

Since $\varphi(s, s) = 1$ we have $\varphi(s, t)\, \varphi(t, s) = 1$, hence

$$\varphi(t, s) = 1/\varphi(s, t). \tag{3.23}$$

In § 3.5 we shall need $\partial\varphi(t, s)/\partial s$. By (3.23) we have

Proposition 3.4. *The function $\varphi(t, s)$ considered*[4] *as a function in s satisfies*

$$\frac{d\varphi(t, s)}{ds} = -a(s)\, \varphi(t, s). \tag{3.24}$$

Proof. Using (3.23) we have

$$\begin{aligned} \frac{d\varphi(t, s)}{ds} &= \frac{d(\varphi^{-1}(s, t))}{ds} = -\varphi^{-2}(s, t)\frac{d(\varphi(s, t))}{ds} = -\varphi^{-2}(s, t)\, a(s)\, \varphi(s, t) \\ &= -a(s)\varphi^{-1}(s, t) = -a(s)\varphi(t, s). \end{aligned}$$

□

Now we can continue the proof of Theorem 3.1.

3.5 The Simplest Problem, Continued

Setting

$$a(t) = \frac{\partial f(y(t), u(t))}{\partial y}, \tag{3.25}$$

we apply the notion of fundamental solution to (3.17). So the solution[5] of (3.17) on $[s, T]$ satisfying (3.22) is

$$\delta y(t) = \varepsilon\, [f(y(s), v) - f(y(s), u(s))]\, \varphi(t, s).$$

Hence

$$\delta y(T) = \varepsilon\, [f(y(s), v) - f(y(s), u(s))]\, \varphi(T, s)$$

and we can write

$$\Delta y(T) = \varepsilon\, [f(y(s), v) - f(y(s), u(s))]\, \varphi(T, s) + o(\varepsilon). \tag{3.26}$$

[4] Here we consider t as a fixed parameter, which is why we use the notation for an ordinary derivative rather than a partial derivative.

[5] Of course, this is really just a useful representation rather than an explicit solution.

Step 2. The main part of the increment of $J(u) = g(y(T))$ can be found using the same idea of linearization and Taylor expansion:

$$\begin{aligned}\Delta J(u) &= J(u^*) - J(u)\\ &= g(y(T) + \Delta y(T)) - g(y(T))\\ &= \left.\frac{dg(y)}{dy}\right|_{y=y(T)} \Delta y(T) + o(|\Delta y(T)|).\end{aligned}$$

With regard for (3.26) this brings us to

$$\Delta J(u) = \varepsilon \left.\frac{dg(y)}{dy}\right|_{y=y(T)} [f(y(s), v) - f(y(s), u(s))]\, \varphi(T, s) + o(\varepsilon).$$

So we have found the main part of the increment of the objective functional; however, we must still represent it in the form shown in Theorem 3.1.

Step 3. Let

$$\psi(s) = -\left.\frac{dg(y)}{dy}\right|_{y=y(T)} \varphi(T, s). \tag{3.27}$$

With this notation $\Delta J(u)$ takes the form (3.10). It remains only to demonstrate that $\psi(s)$ satisfies (3.12). The second relation of (3.12) holds by definition of the fundamental solution:

$$\psi(T) = -\left.\frac{dg(y)}{dy}\right|_{y=y(T)} \varphi(T, T) = -\left.\frac{dg(y)}{dy}\right|_{y=y(T)}.$$

Let us show that $\psi(s)$ satisfies the first equation of (3.12):

$$\begin{aligned}\frac{d\psi(s)}{ds} &= \frac{d}{ds}\left[-\left.\frac{dg(y)}{dy}\right|_{y=y(T)} \varphi(T, s)\right]\\ &= -\left.\frac{dg(y)}{dy}\right|_{y=y(T)} \frac{d}{ds}\varphi(T, s)\\ &= -\left.\frac{dg(y)}{dy}\right|_{y=y(T)} [-a(s)\, \varphi(T, s)]\\ &= a(s)\left.\frac{dg(y)}{dy}\right|_{y=y(T)} \varphi(T, s).\end{aligned}$$

Here we used (3.24) to eliminate the derivative of $\varphi(T, s)$ with respect to the second argument. Finally, remembering (3.27) we obtain

$$\psi'(s) = -a(s)\psi(s).$$

This is the needed equation since $a(t)$ is given by (3.25).

3.6 Pontryagin's Maximum Principle for the Simplest Problem

What have we established in Theorem 3.1? To find the increment in the goal functional under a needle-shaped increment of the control function $u(t)$, we should do the following:

(1) Solve the Cauchy problem (3.6)–(3.7). In practice this is often done numerically (e.g., by the Runge–Kutta method).
(2) Having obtained $y(T)$, formulate equations (3.12) and solve this Cauchy problem with respect to $\psi(s)$ in the "reversed" time.
(3) Write out (3.10).

The second condition in (3.12) is analogous to the natural boundary condition in the calculus of variations. The first equation in (3.12) is called the *conjugate equation*; there is a weak analogy between this and the Euler equation. We also observe that in performing steps (1) and (2) we effectively solve a boundary value problem for the pair $y(t), \psi(s)$. A similar pair of equations arises for other types of optimal control problems, but in the terminal control problems they split.

Let us reformulate this problem, introducing a new function $H(y, \psi, u)$ in three variables:

$$H(y, \psi, u) = \psi f(y, u). \tag{3.28}$$

Because

$$\frac{\partial H(y, \psi, u)}{\partial \psi} = f(y, u), \qquad \frac{\partial H(y, \psi, u)}{\partial y} = \frac{\partial f(y, u)}{\partial y}\psi,$$

we can rewrite (3.6) and (3.12) as

$$y'(t) = \frac{\partial H(y(t), \psi(t), u(t))}{\partial \psi}, \qquad \psi'(t) = -\frac{\partial H(y(t), \psi(t), u(t))}{\partial y},$$

or

$$y'(t) = H_\psi(y(t), \psi(t), u(t)), \qquad \psi'(t) = -H_y(y(t), \psi(t), u(t)). \tag{3.29}$$

This is the so-called *Hamilton form* of a system of ODEs that is frequent in physics. L.S. Pontryagin called $H(y, \psi, u)$ the Hamilton function, but it was subsequently called the *Pontryagin function*. Again, we will obtain equations of the form (3.29) when we consider any sort of control problem for the system described by (3.6).

Let us rewrite the increment $\Delta J(u)$ under a needle-shaped increment with parameters ε, v given at $t = s$, which is presented by (3.10), in terms of $H(y, \psi, u)$:

$$\Delta J(u) = \varepsilon\, (H(y(s), \psi(s), u(s)) - H(y(s), \psi(s), v)) + o(\varepsilon).$$

Now we can formulate a necessary condition of minimum for $J(u)$, known as *Pontryagin's maximum principle*:

Theorem 3.5. *Let $u(t)$ be an optimal control function at which $J(u)$ attains its minimal value on $\mathcal{U}$, the set of all admissible control functions, and let $y(t)$ and $\psi(t)$ be solutions of the boundary value problem* (3.6), (3.7), (3.12). *At any point $t = s$ of continuity of $u(t)$, the function $H(y(s), \psi(s), v)$ considered as a function in the variable v takes its maximum value at $v = u(s)$.*

Proof. $J(u)$ attains its minimum at $u = u(t)$. For any admissible control function $u^*(t)$ given by (3.8) we have

$$J(u^*) - J(u) \geq 0.$$

For a point $t = s$ of continuity of $u = u(t)$, in terms of the Pontryagin function this is

$$\varepsilon\, (H(y(s), \psi(s), u(s)) - H(y(s), \psi(s), v)) + o(\varepsilon) \geq 0.$$

Note this is valid for any admissible v and small, nonnegative ε. It follows immediately that

$$H(y(s), \psi(s), u(s)) - H(y(s), \psi(s), v) \geq 0,$$

so for any admissible v we get $H(y(s), \psi(s), u(s)) \geq H(y(s), \psi(s), v)$. □

Let us consider the application of these results to a simple example.

Example 3.6. Find the form of the control function $u(t)$, $|u(t)| \leq 2$, that gives minimum deviation of $y(t)$ from 10 at $t = 1$ (described by the function $g(y(1)) = (10 - y(1))^2$) for a system governed by

$$y'(t) + y(t) = u(t), \qquad y(0) = 1.$$

Solution. We stay with our previous notation. Rewrite the equation as $y' = -y + u$ and construct Pontryagin's function

$$H(y, \psi, u) = \psi(-y + u).$$

We need to learn when this function takes its maximum value with respect to u along a solution. For this we need to know some properties of ψ. Let us establish how ψ changes. The conjugate equation for ψ is

$$\psi' = -\frac{\partial H}{\partial y} = \psi.$$

Its general solution is $\psi = ce^t$. For this example we need not find (y, ψ) for any control function, so we will not formulate the final value for ψ but merely note that its sign coincides with that of the constant c. This means that along any possible solution $y = y(t)$, at any point of continuity of y, the maximum is taken when $\psi(t)u(t)$ takes its maximum. Since this expression is linear in u, the maximum is taken when u takes one of its extreme values $u = \pm 2$ and, because of the constancy of sign of ψ, it cannot change from one extreme to another.[6]

So now we must consider the governing equation in two versions, with $u = 2$ and $u = -2$. These are

$$y' = -y + 2, \qquad y' = -y - 2.$$

The initial condition leads to the respective solutions

$$y_1(t) = -e^{-t} + 2, \qquad y_2(t) = 3e^{-t} - 2.$$

Comparing the values of the cost function $g(y)$ for y_1 and y_2 at $t = 1$, we see that $u = u(t) = 2$ is the optimal control. Correspondingly $y(t) = -e^{-t} + 2$, and the minimum value of g is $g(y(1)) = (8 + e^{-1})^2$. □

This example shows that not every optimal control problem has a solution. Indeed, if we pose the minimum time problem for the same equation with y beginning at $y = 1$ and ending at $y = 10$, under the restriction $|u| \le 2$, then there will be no solution; a solution starting from the point $y(0) = 1$ never takes the value 10.

Let us continue consideration of the same problem. We denote by $J(u)$ the value $g(y(1))$ so J is defined as a functional of the control function u.

Example 3.7. For the system of the previous example, find the main part of the increment of the goal functional under a needle-shaped disturbance of u if its value is $u(t) = 1$ for all t.

[6] A reader familiar with the elements of linear programming will note that the situation is the same as in that theory. Since many optimal control problems are described by relations containing a control vector in a linear manner, the reader sees that at this stage it is necessary to solve a linear programming problem in which we must maximize a linear function over a set in a finite vector space restricted by linear inequalities.

Solution. The governing equation of the system for $u = 1$ is $y' = -y + 1$. The solution that satisfies the initial condition is $y = 1$. Thus the final value for ψ is

$$\psi(1) = -\frac{\partial g(y(1))}{\partial y} = -2 \cdot 9(-1) = 18,$$

and the corresponding solution of the conjugate equation is

$$\psi(t) = de^t, \qquad d = 18e.$$

Thus the main part of the increment of the goal functional is

$$\begin{aligned} \varepsilon\delta_{s,v}J(u) &= \psi(s)[f(y(s), u(s) - f(y(s), v)] \\ &= 18\varepsilon e^{1+t}(0 + 1 - v) \\ &= 18\varepsilon e^{1+t}(1 - v) \end{aligned}$$

for any time s. It is clear that if we wish to decrease locally at any point s the value of the functional, then we should take the maximum admissible value of v, which is $v = 2$. □

This problem is important because it shows how we can improve an initial approximation to u. For sufficiently small ε, introducing a needle-shaped change of u at some s we reduce the value of $g(y(1))$. Choosing ε and s and decreasing correspondingly the value of $J(u)$ (of course, this happens only when $\varepsilon\delta_{s,v}J(u)$ has negative values on $[0, 1]$ — if there are no such values then a corresponding function u is optimal) we get a better approximation to the optimal control function. But the choice of ε, s is not uniquely defined even for this simple problem. If ε is small and fixed, it is clear that the maximal change in $J(u)$ happens (in this problem) when we take the maximum admissible value of v, that is $v = 2$. But what is the value of s? It is clear that we should introduce the needle-change into u at s where $\varepsilon\delta_{s,v}J(u)$ takes the lowest negative value. In this problem it is easy to see that it is the point $s = 1$. Changing u to 2 on $[1 - \varepsilon, 1]$ with some small ε we get a new control function u^* that is not optimal again. So we need to repeat the same steps: find $\varepsilon\delta_{s,v}J(u^*)$, choose ε and s, and introduce optimally a new needle-shaped perturbation into u to maximally decrease $J(u)$. This gives a second approximation to the optimal solution, and so on. In this simple case the approximation will be quite accurate. However, in practical problems, when we do not know the solution u in advance, it can be difficult to choose ε and s at each step.

Pontryagin's maximum principle allows us to test a given control function for optimality. In addition, we shall see later that for some relatively simple problems it can suggest an approach to finding solutions. Next we would like to note that formula (3.10) is the background for a class of numerical methods for finding an optimal solution. We shall discuss this for the general problem of terminal control, which should be further considered. In § 3.7 we present some essential mathematical tools.

3.7 Some Mathematical Preliminaries

When we considered the simplest problem of control theory we used the notions of fundamental solution and linearization. To extend these to vector functions one can use the tools of matrix theory, but the resulting formulas are much more compact and clear when presented in tensor notation. We therefore pause to present a small portion of tensor analysis. In doing so we shall confine ourselves to the simplest case involving only Cartesian frames having orthonormal basis vectors $\mathbf{e}_1, \ldots, \mathbf{e}_n$. In the general case the controlled functions $\mathbf{y}(t)$ take values in the n-dimensional vector space spanned by this basis, so we can represent $\mathbf{y}(t)$ as

$$\mathbf{y}(t) = \sum_{i=1}^{n} y_i(t)\, \mathbf{e}_i. \tag{3.30}$$

From now on we omit the summation symbol and write simply

$$\mathbf{y}(t) = y_i(t)\, \mathbf{e}_i. \tag{3.31}$$

This is the usual convention, due to Einstein, for dealing with Cartesian tensors: whenever we meet a repeated index (in this case i) we understand that summation is to be performed over this index from 1 to n. Now we shall demonstrate how this expansion can be used along with the dot product to produce representations of vectors, and to reproduce common operations involving vectors and matrices.

Matrices as the component representations of tensors and vectors

To perform operations with a vector $\mathbf{x}$ we must have a straightforward method of calculating its components $x_1, \ldots, x_n$ with respect to a basis $\mathbf{e}_1, \ldots, \mathbf{e}_n$. This can be done through simple dot multiplication. For additional clarity let us momentarily suspend use of the summation convention.

Dotting $\mathbf{x}$ with $\mathbf{e}_1$ we have

$$\begin{aligned}\mathbf{x}\cdot\mathbf{e}_1 &= (x_1\mathbf{e}_1+\cdots+x_n\mathbf{e}_n)\cdot\mathbf{e}_1\\ &= x_1(\mathbf{e}_1\cdot\mathbf{e}_1)+\cdots+x_n(\mathbf{e}_n\cdot\mathbf{e}_1).\end{aligned}$$

Because $\mathbf{e}_1\cdot\mathbf{e}_1 = 1$ and the remaining dot products vanish, we obtain

$$x_1 = \mathbf{x}\cdot\mathbf{e}_1.$$

Here the key observation is that

$$\mathbf{e}_i\cdot\mathbf{e}_j = \begin{cases}1, & j=i,\\ 0, & j\neq i,\end{cases} \tag{3.32}$$

and this same observation can be used in similar fashion to develop the formulas

$$x_2 = \mathbf{x}\cdot\mathbf{e}_2,\quad x_3 = \mathbf{x}\cdot\mathbf{e}_3,\quad \ldots,\quad x_n = \mathbf{x}\cdot\mathbf{e}_n.$$

In terms of the Kronecker delta symbol (page 39) we could have written

$$\begin{aligned}\mathbf{x}\cdot\mathbf{e}_1 &= (x_1\mathbf{e}_1+\cdots+x_n\mathbf{e}_n)\cdot\mathbf{e}_1\\ &= x_1\delta_{11}+\cdots+x_n\delta_{n1}\\ &= x_1\end{aligned}$$

to calculate x_1. We can now return to the summation convention and repeat these calculations in tensor notation. If $\mathbf{x}$ is given by

$$\mathbf{x} = x_k\mathbf{e}_k \tag{3.33}$$

then for $i = 1, 2, \ldots, n$ we have

$$x_i = \mathbf{x}\cdot\mathbf{e}_i \tag{3.34}$$

since $\mathbf{x}\cdot\mathbf{e}_i = x_k\mathbf{e}_k\cdot\mathbf{e}_i = x_k\delta_{ki} = x_i$ for each i. Thus in a given basis $\mathbf{e}_i$ the components x_i of the vector $\mathbf{x}$ are determined uniquely, and $\mathbf{x}$ is determined by these values x_i. It is convenient to display the components of $\mathbf{x}$ in a column matrix:

$$\begin{pmatrix}x_1\\ \vdots\\ x_n\end{pmatrix}.$$

Hence a matrix can act as the component representation of a vector. It is important to understand that a vector itself is an *objective* entity: it is independent of coordinate frame. Consequently if we expand the same vector $\mathbf{x}$ relative to a different Cartesian basis $\tilde{\mathbf{e}}_1, \tilde{\mathbf{e}}_2, \ldots, \tilde{\mathbf{e}}_n$ and repeat the above steps, we will in general arrive at a matrix representation

$$\begin{pmatrix} \tilde{x}_1 \\ \vdots \\ \tilde{x}_n \end{pmatrix}$$

whose entries $\tilde{x}_k$ differ from the x_k. We shall return to this issue later after examining tensors.

If $\mathbf{x}$ and $\mathbf{y}$ are two vectors, their dot product is a scalar:

$$c = \mathbf{x} \cdot \mathbf{y}. \tag{3.35}$$

When we represent each of $\mathbf{x}$ and $\mathbf{y}$ with respect to a basis $\mathbf{e}_i$ as

$$\mathbf{x} = x_i \mathbf{e}_i, \qquad \mathbf{y} = y_j \mathbf{e}_j,$$

we can easily calculate c as

$$\mathbf{x} \cdot \mathbf{y} = x_i \mathbf{e}_i \cdot y_j \mathbf{e}_j = x_i y_j (\mathbf{e}_i \cdot \mathbf{e}_j) = x_i y_j \delta_{ij} = x_i y_i.$$

Of course, this same result arises from the matrix multiplication

$$c = \begin{pmatrix} x_1 & \cdots & x_n \end{pmatrix} \begin{pmatrix} y_1 \\ \vdots \\ y_n \end{pmatrix}. \tag{3.36}$$

This familiar correspondence between dot multiplication of vectors and multiplication of the component matrices will be extended in what follows.

A vector is an example of a tensor of the first rank. The development of our subject will also require some simple work with tensors of the second rank. Just as a vector can be constructed as a linear combination of basis vectors $\mathbf{e}_i$, a tensor of the second rank can be constructed as a linear combination of *basis dyads*. These are in turn formed from pairs of vectors through the use of a *tensor product*. This operation, denoted $\otimes$, obeys laws analogous to those for ordinary multiplication: if $\mathbf{a}$, $\mathbf{b}$, and $\mathbf{c}$ are vectors then

$$\begin{aligned} (\lambda \mathbf{a}) \otimes \mathbf{b} &= \mathbf{a} \otimes (\lambda \mathbf{b}) = \lambda (\mathbf{a} \otimes \mathbf{b}), \\ (\mathbf{a} + \mathbf{b}) \otimes \mathbf{c} &= \mathbf{a} \otimes \mathbf{c} + \mathbf{b} \otimes \mathbf{c}, \\ \mathbf{a} \otimes (\mathbf{b} + \mathbf{c}) &= \mathbf{a} \otimes \mathbf{b} + \mathbf{a} \otimes \mathbf{c}, \end{aligned} \tag{3.37}$$

for any real number λ. We shall shorten the notation for the tensor product somewhat by omitting the $\otimes$ symbol: thus we write $\mathbf{ab}$ instead of $\mathbf{a} \otimes \mathbf{b}$.

The quantity $\mathbf{ab}$ is an example of a *dyad* of vectors. If we expand each of the vectors $\mathbf{a}$ and $\mathbf{b}$ in terms of a basis $\mathbf{e}_i$, the dyad $\mathbf{ab}$ becomes

$$\mathbf{ab} = a_i\mathbf{e}_i b_j\mathbf{e}_j = a_i b_j \, \mathbf{e}_i\mathbf{e}_j.$$

In this way the n^2 different basis dyads $\mathbf{e}_i\mathbf{e}_j$ make their appearance. The dyads $\mathbf{e}_i\mathbf{e}_j$ form the basis for a linear space called the space of tensors of the second rank. An element $\mathbf{A}$ of this space has a representation of the form

$$\mathbf{A} = a_{ij}\mathbf{e}_i\mathbf{e}_j \tag{3.38}$$

where the a_{ij} are called the components of $\mathbf{A}$ relative to the basis $\mathbf{e}_i\mathbf{e}_j$. Here we again use Einstein's summation rule. Note that we can write out the components of $\mathbf{A}$ as a square array

$$\begin{pmatrix} a_{11} & a_{12} & \cdots & a_{1n} \\ a_{21} & a_{22} & \cdots & a_{2n} \\ \vdots & \vdots & \ddots & \vdots \\ a_{n1} & a_{n2} & \cdots & a_{nn} \end{pmatrix},$$

and thus we get a correspondence between the tensor $\mathbf{A}$ and this matrix of its components.

One goal of the discussion is to demonstrate the usefulness of the dot product. The dot product of a dyad $\mathbf{ab}$ and a vector $\mathbf{c}$ is defined by the equation

$$(\mathbf{ab}) \cdot \mathbf{c} = \mathbf{a}(\mathbf{b} \cdot \mathbf{c}). \tag{3.39}$$

The result is a vector directed along $\mathbf{a}$. Analogously we can define the dot product from the left:

$$\mathbf{c} \cdot (\mathbf{ab}) = (\mathbf{c} \cdot \mathbf{a})\mathbf{b}. \tag{3.40}$$

These operations have matrix counterparts: (3.39) corresponds to multiplication of a matrix by a column vector and (3.40) corresponds to multiplication of a row vector by a matrix. For example let us write

$$\mathbf{v} = (\mathbf{ab}) \cdot \mathbf{c}, \tag{3.41}$$

expand $\mathbf{c}$ as $\mathbf{c} = c_k\mathbf{e}_k$, expand $\mathbf{ab}$ according to (3.37), and use (3.39) to write

$$\mathbf{v} = a_i b_j \mathbf{e}_i\mathbf{e}_j \cdot c_k\mathbf{e}_k = a_i b_j \delta_{jk} c_k \mathbf{e}_i = a_i b_j c_j \mathbf{e}_i.$$

Hence

$$v_i = a_i b_j c_j \tag{3.42}$$

for $i = 1, 2, \ldots, n$. Pausing to unpack the succinct tensor index notation, we see that (3.42) actually means the system of equalities

$$\begin{aligned}
v_1 &= a_1 b_1 c_1 + a_1 b_2 c_2 + \cdots + a_1 b_n c_n, \\
v_2 &= a_2 b_1 c_1 + a_2 b_2 c_2 + \cdots + a_2 b_n c_n, \\
&\vdots \\
v_n &= a_n b_1 c_1 + a_n b_2 c_2 + \cdots + a_n b_n c_n,
\end{aligned}$$

or, in matrix form,

$$\begin{pmatrix} v_1 \\ v_2 \\ \vdots \\ v_n \end{pmatrix} = \begin{pmatrix} a_1 b_1 & a_1 b_2 & \cdots & a_1 b_n \\ a_2 b_1 & a_2 b_2 & \cdots & a_2 b_n \\ \vdots & \vdots & \ddots & \vdots \\ a_n b_1 & a_n b_2 & \cdots & a_n b_n \end{pmatrix} \begin{pmatrix} c_1 \\ c_2 \\ \vdots \\ c_n \end{pmatrix}. \tag{3.43}$$

We now recall the analogy between (3.35) and (3.36), and examine (3.41) and (3.43) with similar thoughts in mind. Dot multiplication once again stands in correspondence with matrix multiplication; moreover, it is clear that the dyad **ab** is represented by the square matrix

$$\begin{pmatrix} a_1 b_1 & a_1 b_2 & \cdots & a_1 b_n \\ a_2 b_1 & a_2 b_2 & \cdots & a_2 b_n \\ \vdots & \vdots & \ddots & \vdots \\ a_n b_1 & a_n b_2 & \cdots & a_n b_n \end{pmatrix}.$$

We have seen that a dyad **ab** can map a vector **c** into another vector **v** through the dot product operation given in (3.41). This idea carries through to general tensors of the second rank, of which dyads are examples. If **A** is a tensor of second rank and **x** is a vector, then **A** can map **x** into an image vector **y** according to

$$\mathbf{y} = \mathbf{A} \cdot \mathbf{x}. \tag{3.44}$$

It is easy to check that the individual components of $\mathbf{A} = a_{ij}\mathbf{e}_i\mathbf{e}_j$ are given by

$$a_{ij} = \mathbf{e}_i \cdot \mathbf{A} \cdot \mathbf{e}_j, \tag{3.45}$$

and that (3.44) corresponds to

$$\begin{pmatrix} y_1 \\ y_2 \\ \vdots \\ y_n \end{pmatrix} = \begin{pmatrix} a_{11} & a_{12} & \cdots & a_{1n} \\ a_{21} & a_{22} & \cdots & a_{2n} \\ \vdots & \vdots & \ddots & \vdots \\ a_{n1} & a_{n2} & \cdots & a_{nn} \end{pmatrix} \begin{pmatrix} x_1 \\ x_2 \\ \vdots \\ x_n \end{pmatrix}.$$

A dot product operation known as pre-multiplication of a tensor by a vector is also considered: the quantity $\mathbf{y} \cdot \mathbf{A}$ is defined by the requirement that

$$(\mathbf{y} \cdot \mathbf{A}) \cdot \mathbf{x} = \mathbf{y} \cdot (\mathbf{A} \cdot \mathbf{x})$$

for all vectors $\mathbf{x}$. This can be also obtained as a consequence of the formal definition of left-dot-multiplication of a vector by a dyad:

$$\mathbf{a} \cdot \mathbf{bc} = (\mathbf{a} \cdot \mathbf{b})\mathbf{c}. \tag{3.46}$$

We see both dot product operations (pre-multiplication and post-multiplication) applied to the definition of the important *unit tensor* $\mathbf{E}$, which satisfies

$$\mathbf{E} \cdot \mathbf{x} = \mathbf{x} \cdot \mathbf{E} = \mathbf{x} \tag{3.47}$$

for any vector $\mathbf{x}$. It is easy to find the components of $\mathbf{E}$ from this definition. We start by writing $\mathbf{E} = e_{ij}\mathbf{e}_i\mathbf{e}_j$ and then apply (3.47) with $\mathbf{x} = \mathbf{e}_k$ to get

$$e_{ij}\mathbf{e}_i\mathbf{e}_j \cdot \mathbf{e}_k = \mathbf{e}_k.$$

Pre-multiplying by $\mathbf{e}_m$ we obtain

$$e_{ij}\delta_{mi}\delta_{jk} = \delta_{mk}$$

since $\mathbf{e}_m \cdot \mathbf{e}_i = \delta_{mi}$, $\mathbf{e}_j \cdot \mathbf{e}_k = \delta_{jk}$, and $\mathbf{e}_m \cdot \mathbf{e}_k = \delta_{mk}$. Hence $e_{mk} = \delta_{mk}$ and we have

$$\mathbf{E} = \delta_{ij}\mathbf{e}_i\mathbf{e}_j = \mathbf{e}_i\mathbf{e}_i.$$

Of course, the corresponding matrix representation is the $n \times n$ identity matrix

$$I = \begin{pmatrix} 1 & 0 & 0 & \cdots & 0 \\ 0 & 1 & 0 & \cdots & 0 \\ 0 & 0 & 1 & \cdots & 0 \\ \vdots & \vdots & \vdots & \ddots & \vdots \\ 0 & 0 & 0 & \cdots & 1 \end{pmatrix}.$$

Thus $\mathbf{E}$ is equivalent to the unit matrix.

The strong parallel that exists between tensors and matrices leads us to apply the notion of transposition to tensors of the second rank. Accordingly, if $\mathbf{A} = a_{ij}\mathbf{e}_i\mathbf{e}_j$ then we define

$$\mathbf{A}^T = a_{ji}\mathbf{e}_i\mathbf{e}_j = a_{ij}\mathbf{e}_j\mathbf{e}_i. \tag{3.48}$$

It is easy to see that

$$\mathbf{A} \cdot \mathbf{x} = \mathbf{x} \cdot \mathbf{A}^T \tag{3.49}$$

for any vector $\mathbf{x}$ and any tensor $\mathbf{A}$ of the second rank. It is even more obvious that $(\mathbf{A}^T)^T = \mathbf{A}$. If A is the matrix representation of $\mathbf{A}$, then A^T represents $\mathbf{A}^T$.

A dot product between two tensors is regarded as the composition of the two tensors viewed as operators. That is, $\mathbf{A} \cdot \mathbf{B}$ is defined by the equation

$$(\mathbf{A} \cdot \mathbf{B}) \cdot \mathbf{x} \equiv \mathbf{A} \cdot (\mathbf{B} \cdot \mathbf{x}). \tag{3.50}$$

A tensor $\mathbf{B}$ of the second rank is said to be the inverse of $\mathbf{A}$ if

$$\mathbf{A} \cdot \mathbf{B} = \mathbf{B} \cdot \mathbf{A} = \mathbf{E}. \tag{3.51}$$

In this case we write $\mathbf{B} = \mathbf{A}^{-1}$.

A central aspect of the study of tensors concerns how their components transform when the frame is changed. Although such frame transformations will not play a significant role in the discussion, the reader should understand that to express a tensor in another frame we would simply substitute the representation of the old basis vectors in terms of the new ones. As a simple example we may consider the case of a tensor of rank one: a vector. Let the components of $\mathbf{x}$ relative to the frame $\mathbf{e}_i$ be x_i so that $\mathbf{x} = x_i\mathbf{e}_i$. If a new frame $\tilde{\mathbf{e}}_i$ is introduced according to the set of linear relations $\mathbf{e}_i = A_{ij}\tilde{\mathbf{e}}_j$, then $\mathbf{x} = x_iA_{ij}\tilde{\mathbf{e}}_j$ and we see that $\mathbf{x} = \tilde{x}_j\tilde{\mathbf{e}}_j$ where $\tilde{x}_j = A_{ij}x_i$. The point is that we are not free to assign values to the $\tilde{x}_j$ in any way we wish: once the frame transformation is specified through the A_{ij}, the new components $\tilde{x}_i$ are completely determined by the old components x_i. The situation with tensors of higher order is the same.

Note that the correspondence between tensors and matrices is one-to-one only for a fixed basis. As soon as we change the basis, the matrix representation of a tensor changes by strictly defined rules. For example, if we take a non-Cartesian basis in space, the matrix representation of the tensor $\mathbf{E}$ is not the unit matrix, and thus $\mathbf{E}$ is not something we could call the unit tensor. Rather, it is known as the metric tensor.

Elements of calculus for vector and tensor fields

Now we consider how differentiation is performed on tensor and vector functions using tensor notation. Let us begin with a function $\mathbf{y}(t) = y_i(t)\,\mathbf{e}_i$. Since $\mathbf{e}_i$ does not depend on t, differentiation of $\mathbf{y}(t)$ with respect to t reduces to differentiation of the component scalar functions $y_i(t)$:

$$\mathbf{y}'(t) = y_i'(t)\,\mathbf{e}_i. \tag{3.52}$$

Similarly, the differential of a vector function $\mathbf{y}(t)$ is

$$d\mathbf{y}(t) = dy_i(t)\,\mathbf{e}_i. \tag{3.53}$$

Now suppose we wish to differentiate a composite function $f(\mathbf{y})(t)$ with respect to t. Writing this as $f(y_i(t)\,\mathbf{e}_i)$ or $f(y_1(t), \ldots, y_n(t))$, we have by the chain rule

$$\begin{aligned} \frac{d}{dt} f(\mathbf{y}(t)) &= \frac{d}{dt} f(y_1(t), \ldots, y_n(t)) \\ &= \sum_{i=1}^{n} \frac{\partial f(y_1(t), \ldots, y_n(t))}{\partial y_i} y_i'(t) \\ &= \frac{\partial f(\mathbf{y}(t))}{\partial y_i} y_i'(t). \end{aligned} \tag{3.54}$$

Let us write out the right side of (3.54) in vector form. For this we introduce ∇, a formal vector of differentiation (known as the gradient operator):

$$\nabla_{\mathbf{y}} = \sum_{i=1}^{n} \mathbf{e}_i \frac{\partial}{\partial y_i} = \mathbf{e}_i \frac{\partial}{\partial y_i}. \tag{3.55}$$

(We show the subscript $\mathbf{y}$ on ∇ to indicate the vector whose components participate in the differentiation. The subscript can be omitted if this is clear from the context.) When we apply $\nabla_{\mathbf{y}}$ to a function $f(\mathbf{y})$ we get a vector

$$\nabla_{\mathbf{y}} f(\mathbf{y}) = \sum_{i=1}^{n} \mathbf{e}_i \frac{\partial f(\mathbf{y})}{\partial y_i} = \frac{\partial f(\mathbf{y})}{\partial y_i} \mathbf{e}_i. \tag{3.56}$$

Let us dot multiply $\nabla_{\mathbf{y}} f(\mathbf{y}(t))$ by $\mathbf{y}'(t) = y_j'(t)\,\mathbf{e}_j$. Remembering that $\mathbf{e}_i \cdot \mathbf{e}_j = \delta_{ij}$, we get

$$\nabla_{\mathbf{y}} f(\mathbf{y}(t)) \cdot \mathbf{y}'(t) = \frac{\partial f(\mathbf{y}(t))}{\partial y_i} \mathbf{e}_i \cdot y_j'(t)\mathbf{e}_j = \frac{\partial f(\mathbf{y}(t))}{\partial y_i} y_j'(t)\delta_{ij} = \frac{\partial f(\mathbf{y}(t))}{\partial y_i} y_i'(t).$$

Since the right side of this coincides with that of (3.54), we have

$$\frac{d}{dt} f(\mathbf{y}(t)) = \nabla_{\mathbf{y}} f(\mathbf{y}(t)) \cdot \mathbf{y}'(t).$$

The differential of $f(\mathbf{y}(t))$ is given by

$$df(\mathbf{y}(t)) = \nabla_{\mathbf{y}} f(\mathbf{y}(t)) \cdot d\mathbf{y}(t). \tag{3.57}$$

Using this formula or, equivalently, the first-order Taylor approximation, we get

$$f(\mathbf{y}(t) + \Delta\mathbf{y}(t)) - f(\mathbf{y}(t)) = \nabla_{\mathbf{y}} f(\mathbf{y}(t)) \cdot \Delta\mathbf{y}(t) + o(\|\Delta\mathbf{y}(t)\|)$$

where $\Delta\mathbf{y}(t)$ is a small increment of $\mathbf{y}(t)$.

Now we would like to present the first-order Taylor approximation of the increment of a vector function $\mathbf{f}$ that depends on a vector function $\mathbf{y}(t)$. We assume that $\mathbf{f}$ takes values in the same space as $\mathbf{y}(t)$ and thus can be represented as $\mathbf{f} = f_i\,\mathbf{e}_i$ where $f_i = f_i(\mathbf{y}(t))$. For this we find the differential of $\mathbf{f}(\mathbf{y}(t))$ at $\mathbf{y}(t)$:

$$\begin{aligned} d\mathbf{f}(\mathbf{y}(t)) &= d\,(f_j(\mathbf{y}(t))\,\mathbf{e}_j) = df_j(\mathbf{y}(t))\,\mathbf{e}_j \\ &= \nabla_{\mathbf{y}} f_j(\mathbf{y}(t)) \cdot d\mathbf{y}(t)\,\mathbf{e}_j \\ &= \frac{\partial f_j(\mathbf{y}(t))}{\partial y_i}\,\mathbf{e}_i \cdot dy_k(t)\,\mathbf{e}_k\,\mathbf{e}_j \end{aligned}$$

The right side can be represented as

$$\left(\frac{\partial f_j(\mathbf{y}(t))}{\partial y_i}\mathbf{e}_j\mathbf{e}_i\right) \cdot dy_k(t)\,\mathbf{e}_k \quad \text{or} \quad dy_k(t)\,\mathbf{e}_k \cdot \left(\frac{\partial f_j(\mathbf{y}(t))}{\partial y_i}\mathbf{e}_i\mathbf{e}_j\right). \tag{3.58}$$

We see that in both brackets there is a sum of dyads so both of them are functions whose values are tensors of the second rank. A formal application of $\nabla_{\mathbf{y}}$ to $\mathbf{f}(\mathbf{y}(t))$ gives

$$\nabla_{\mathbf{y}} \mathbf{f}(\mathbf{y}(t)) = \mathbf{e}_i \frac{\partial}{\partial y_i} f_j(\mathbf{y}(t))\,\mathbf{e}_j = \frac{\partial f_j(\mathbf{y}(t))}{\partial y_i}\mathbf{e}_i\mathbf{e}_j.$$

Thus $\nabla_{\mathbf{y}} \mathbf{f}(\mathbf{y}(t))$ is the expression in brackets of the second equation (3.58) and the differential can be represented as

$$d\mathbf{f}(\mathbf{y}(t)) = d\mathbf{y}(t) \cdot \nabla_{\mathbf{y}} \mathbf{f}(\mathbf{y}(t)). \tag{3.59}$$

The term in the bracket of the first equation of (3.58) differs from the corresponding term of the second equation by a transposition of the vectors $\mathbf{e}_i$ and $\mathbf{e}_j$ so it can be written as $(\nabla_{\mathbf{y}} \mathbf{f}(\mathbf{y}(t)))^T$ and thus the differential can be presented in the other form

$$d\mathbf{f}(\mathbf{y}(t)) = (\nabla_{\mathbf{y}} \mathbf{f}(\mathbf{y}(t)))^T \cdot d\mathbf{y}(t). \tag{3.60}$$

The expression $\nabla_{\mathbf{y}} \mathbf{f}(\mathbf{y}(t))$ is called the *gradient* of $\mathbf{f}$. Let us see how it appears in more common matrix notation. We have said that a second rank tensor can be represented by a matrix of coefficients; in this representation the index i in the first position denotes the ith row of the matrix whereas the second index j denotes the jth column. Thus the matrix representation of the gradient of the vector function $\frac{\partial f_j(\mathbf{y}(t))}{\partial y_i}\mathbf{e}_i\mathbf{e}_j$ is

$$\begin{pmatrix} \dfrac{\partial f_1}{\partial y_1} & \dfrac{\partial f_2}{\partial y_1} & \cdots & \dfrac{\partial f_n}{\partial y_1} \\ \dfrac{\partial f_1}{\partial y_2} & \dfrac{\partial f_2}{\partial y_2} & \cdots & \dfrac{\partial f_n}{\partial y_2} \\ \vdots & \vdots & \ddots & \vdots \\ \dfrac{\partial f_1}{\partial y_n} & \dfrac{\partial f_2}{\partial y_n} & \cdots & \dfrac{\partial f_n}{\partial y_n} \end{pmatrix}.$$

Its determinant is the Jacobian of the transformation $\mathbf{z} = \mathbf{f}(\mathbf{y})$.

Now, using the formula for the differential (3.59) (or (3.60)) we are able to present an increment of a composite vector function $\mathbf{f}(\mathbf{y}(t))$ under the increment $\Delta\mathbf{y}(t)$ of the argument:

$$\mathbf{f}(\mathbf{y}(t) + \Delta\mathbf{y}(t)) - \mathbf{f}(\mathbf{y}(t)) = \Delta\mathbf{y}(t) \cdot \nabla_{\mathbf{y}}\mathbf{f}(\mathbf{y}(t)) + o(\|\Delta\mathbf{y}(t)\|).$$

Let the components of a tensor $\mathbf{A}(t) = a_{ij}(t)\mathbf{e}_i\mathbf{e}_j$ be continuously differentiable functions of t. Then by the rule for differentiating a matrix we have

$$\frac{d\mathbf{A}(t)}{dt} = \frac{da_{ij}(t)}{dt}\mathbf{e}_i\mathbf{e}_j. \tag{3.61}$$

The derivative of the dot product of two second-rank tensors obeys a formula similar to the ordinary product rule:

$$\frac{d}{dt}(\mathbf{A}(t) \cdot \mathbf{B}(t)) = \left(\frac{d}{dt}\mathbf{A}(t)\right) \cdot \mathbf{B}(t) + \mathbf{A}(t) \cdot \left(\frac{d}{dt}\mathbf{B}(t)\right). \tag{3.62}$$

A similar formula holds for the dot product of a tensor by a vector:

$$(\mathbf{A}(t) \cdot \mathbf{y}(t))' = \mathbf{A}'(t) \cdot \mathbf{y}(t) + \mathbf{A}(t) \cdot \mathbf{y}'(t). \tag{3.63}$$

If one factor does not depend on t then it can be removed from the symbol of differentiation:

$$(\mathbf{A} \cdot \mathbf{B}(t))' = \mathbf{A} \cdot \mathbf{B}'(t). \tag{3.64}$$

Fundamental solution of a linear system of ordinary differential equations

Consider a linear system of ODEs

$$\begin{aligned} y_1'(t) &= a_{11}(t)y_1(t) + a_{12}(t)y_2(t) + \cdots + a_{1n}(t)y_n(t), \\ y_2'(t) &= a_{21}(t)y_1(t) + a_{22}(t)y_2(t) + \cdots + a_{2n}(t)y_n(t), \\ &\vdots \\ y_n'(t) &= a_{n1}(t)y_1(t) + a_{n2}(t)y_2(t) + \cdots + a_{nn}(t)y_n(t). \end{aligned}$$

In terms of the tensor function $\mathbf{A}(t) = a_{ij}(t)\mathbf{e}_i\mathbf{e}_j$ and the vector $\mathbf{y}(t) = y_i(t)\mathbf{e}_i$ this system can be rewritten as

$$\mathbf{y}'(t) = \mathbf{A}(t) \cdot \mathbf{y}(t). \tag{3.65}$$

Definition 3.8. A tensor function $\mathbf{\Phi}(t, s)$ in two variables t, s is called the *fundamental solution*[7] of (3.65) if it satisfies two conditions:

(i) $\mathbf{\Phi}(t, s)$ is a solution of (3.65) in the first variable t:

$$\frac{d}{dt}\mathbf{\Phi}(t, s) = \mathbf{A}(t) \cdot \mathbf{\Phi}(t, s) \tag{3.66}$$

(here we use the symbol for the ordinary derivative, thinking of s as a fixed parameter).

(ii) For any s,

$$\mathbf{\Phi}(s, s) = \mathbf{E}. \tag{3.67}$$

This fundamental solution exists for any finite t, s if the tensor $\mathbf{A}(t)$ is continuous. The problem of finding it consists of n Cauchy problems for the same system of equations with n initial conditions given at $t = s$. Hence the fundamental solution is determined uniquely.

Now we would like to extend the results for the fundamental solution of a single linear ODE to the general case. We present them in a similar manner.

Proposition 3.9. *A solution of* (3.65) *satisfying the initial condition* $\mathbf{y}(s) = \mathbf{y}_0$ *is*

$$\mathbf{y}(t) = \mathbf{\Phi}(t, s) \cdot \mathbf{y}_0. \tag{3.68}$$

[7]The function $\mathbf{\Phi}(t, s)$ is also known as the *fundamental tensor* or *fundamental matrix*.

Indeed, dot-multiplying vector-equation (3.66) by $\mathbf{y}_0$ from the right we see that $\mathbf{\Phi}(t,s)\cdot\mathbf{y}_0$ satisfies (3.65). By (3.67) this solution satisfies the initial condition $\mathbf{y}(s)=\mathbf{y}_0$.

Proposition 3.10. *For any t,s and τ we have*

$$\mathbf{\Phi}(t,s)=\mathbf{\Phi}(t,\tau)\cdot\mathbf{\Phi}(\tau,s). \tag{3.69}$$

A consequence of this property and relation (3.67) *is the equation for the inverse*

$$\mathbf{\Phi}^{-1}(t,s)=\mathbf{\Phi}(s,t) \tag{3.70}$$

which follows when we write out a particular case of (3.69),

$$\mathbf{E}=\mathbf{\Phi}(t,t)=\mathbf{\Phi}(t,s)\cdot\mathbf{\Phi}(s,t).$$

Proof. Let us prove (3.69). Dot multiply (3.66) by $\mathbf{\Phi}(s,\tau)$ from the right. On the left we have

$$\left(\frac{d}{dt}\mathbf{\Phi}(t,s)\right)\cdot\mathbf{\Phi}(s,\tau)=\frac{d}{dt}\left(\mathbf{\Phi}(t,s)\cdot\mathbf{\Phi}(s,\tau)\right)$$

since $\mathbf{\Phi}(s,\tau)$ does not depend on t; on the right we have

$$\mathbf{A}(t)\cdot\mathbf{\Phi}(t,s)\cdot\mathbf{\Phi}(s,\tau)=\mathbf{A}(t)\cdot\left(\mathbf{\Phi}(t,s)\cdot\mathbf{\Phi}(s,\tau)\right).$$

So $\mathbf{\Phi}(t,s)\cdot\mathbf{\Phi}(s,\tau)$ satisfies $d\mathbf{\Psi}/dt=\mathbf{A}(t)\cdot\mathbf{\Psi}$ with parameters s,τ. Putting $t=s$ in this solution we get

$$\mathbf{\Phi}(t,s)\cdot\mathbf{\Phi}(s,\tau)|_{t=s}=\mathbf{\Phi}(s,s)\cdot\mathbf{\Phi}(s,\tau)=\mathbf{\Phi}(s,\tau).$$

So $\mathbf{\Phi}(t,s)\cdot\mathbf{\Phi}(s,\tau)$ coincides with $\mathbf{\Phi}(t,\tau)$ at $t=s$; by uniqueness of solution to the Cauchy problem, they coincide for all t. To complete the proof it remains to interchange s and τ. □

Proposition 3.11. *The equation*

$$\frac{\partial}{\partial s}\mathbf{\Phi}(t,s)=-\mathbf{\Phi}(t,s)\cdot\mathbf{A}(s)$$

holds.

Proof. It is easily verified that the derivative of the inverse to a differentiable tensor function $\mathbf{\Psi}(t)$ is given by

$$\left(\mathbf{\Psi}^{-1}(t)\right)'=-\mathbf{\Psi}^{-1}(t)\cdot\mathbf{\Psi}'(t)\cdot\mathbf{\Psi}^{-1}(t). \tag{3.71}$$

Hence by (3.70) we have

$$\begin{aligned}\frac{\partial \mathbf{\Phi}(t,s)}{\partial s} &= \frac{\partial(\mathbf{\Phi}^{-1}(s,t))}{\partial s}\\ &= -\mathbf{\Phi}^{-1}(s,t)\cdot\frac{\partial \mathbf{\Phi}(s,t)}{\partial s}\cdot\mathbf{\Phi}^{-1}(s,t)\\ &= -\mathbf{\Phi}(t,s)\cdot\frac{\partial \mathbf{\Phi}(s,t)}{\partial s}\cdot\mathbf{\Phi}(t,s).\end{aligned}$$

Finally, since s is the first argument in the derivative on the right we can change this derivative using (3.66):

$$\frac{\partial \mathbf{\Phi}(t,s)}{\partial s} = -\mathbf{\Phi}(t,s)\cdot\mathbf{A}(s)\cdot\mathbf{\Phi}(s,t)\cdot\mathbf{\Phi}(t,s) = -\mathbf{\Phi}(t,s)\cdot\mathbf{A}(s).$$

□

Proposition 3.12. *The solution of the Cauchy problem*

$$\mathbf{y}'(t) = \mathbf{A}(t)\cdot\mathbf{y}(t) + \mathbf{g}(t), \qquad \mathbf{y}(0) = 0,$$

with a given vector function $\mathbf{g}(t)$ *is*

$$\mathbf{y}(t) = \int_0^t \mathbf{\Phi}(t,s)\cdot\mathbf{g}(s)\,ds. \tag{3.72}$$

Proof. Let us find the derivative of $\mathbf{y}(t)$ given by (3.72):

$$\begin{aligned}\frac{d}{dt}\mathbf{y}(t) &= \frac{d}{dt}\int_0^t \mathbf{\Phi}(t,s)\cdot\mathbf{g}(s)\,ds\\ &= \mathbf{\Phi}(t,t)\cdot\mathbf{g}(t) + \int_0^t \frac{d}{dt}\mathbf{\Phi}(t,s)\cdot\mathbf{g}(s)\,ds\\ &= \mathbf{E}\cdot\mathbf{g}(t) + \int_0^t \mathbf{A}(t)\cdot\mathbf{\Phi}(t,s)\cdot\mathbf{g}(s)\,ds\\ &= \mathbf{A}(t)\cdot\int_0^t \mathbf{\Phi}(t,s)\cdot\mathbf{g}(s)\,ds + \mathbf{g}(t)\\ &= \mathbf{A}(t)\cdot\mathbf{y}(t) + \mathbf{g}(t).\end{aligned}$$

□

3.8 General Terminal Control Problem

We have stated the general problem of terminal control. Our understanding of the scope of the optimal control problem has changed, however, so it is appropriate to reexamine the setup of the terminal control problem.

The object of terminal optimal control is described by a vector function of time $\mathbf{y}(t)$ with values in Euclidean vector space E^n whose behavior is determined by a system of ODEs (or a vector ODE)

$$\mathbf{y}'(t) = \mathbf{f}(\mathbf{y}(t), \mathbf{u}(t)) \tag{3.73}$$

The vector function $\mathbf{f}(\mathbf{y}(t), \mathbf{u}(t))$ must be such that when the control function $\mathbf{u}(t)$ is given and admissible (i.e., belongs to the class $\mathcal{U}$), then the Cauchy problem for (3.73) supplemented with initial conditions has a unique continuous solution on a finite time interval $[0, T]$. Thus the history of the object determines uniquely its present state. Systems of this type are called *dynamical systems.*

The set $\mathcal{U}$ of admissible controls consists of vector functions $\mathbf{u}(t)$ taking values in the Euclidean space E^m that are piecewise continuous in t. In particular, $\mathcal{U}$ can consist of functions that take values in a finite set of vectorial values. The former is important when the control function describes several fixed positions that are taken by some governing device; it describes, say, the effect of some additional device that can exist only in "on–off" states.

Everything said so far in this section applies to all optimal control problems. The distinguishing feature of terminal control is the specification of the initial condition

$$\mathbf{y}(0) = \mathbf{y}_0 \tag{3.74}$$

and the form of the objective functional

$$J(\mathbf{u}) = G(\mathbf{y}(T)). \tag{3.75}$$

Thus we can consider terminal control as the problem of finding the minimal output value (3.75) when the input is determined by the initial vector $\mathbf{y}_0$ and the control function $\mathbf{u}(t)$ and the output is $G(\mathbf{y}(T))$. See Fig. 3.3. Our objective can be formulated as

$$G(\mathbf{y}(T)) \to \min_{\mathbf{u}(t) \in \mathcal{U}} . \tag{3.76}$$

This is known as the main setup of the problem (3.73)–(3.76). We can reduce various other other optimal control problems to this form.

Problem. For a system described by (3.73) whose initial state is given by (3.74), among all the admissible control vectors $\mathbf{u} \in \mathcal{U}$ find such for which an objective functional

$$\int_0^T g(\mathbf{y}(t), \mathbf{u}(t))\, dt$$

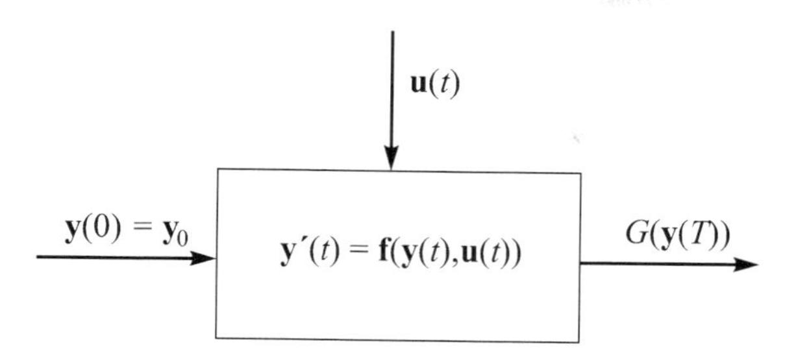

Fig. 3.3 A controlled object described by $\mathbf{y}' = f(\mathbf{y}, \mathbf{u})$: the input is $\mathbf{y}(0) = \mathbf{y}_0$, the control vector is $\mathbf{u}$, and the output is $G(\mathbf{y}(T))$.

takes its minimum value.

The reduction of this problem to the main form of the terminal control problem is done by introducing the additional component y_{n+1} for $\mathbf{y}$. Namely, we write down an additional scalar equation

$$y'_{n+1}(t) = g(\mathbf{y}(t), \mathbf{u}(t)), \qquad y'_{n+1}(0) = 0.$$

Now it is clear that

$$y_{n+1}(T) = \int_0^T g(\mathbf{y}(t), \mathbf{u}(t))\, dt \tag{3.77}$$

and thus the objective functional from (3.75) takes the form

$$J(\mathbf{u}) = y_{n+1}(T).$$

We can consider another version of the terminal control problem when it is necessary to minimize the objective functional

$$\int_0^T g(\mathbf{y}(t), \mathbf{u}(t))\, dt + G(\mathbf{y}(T))$$

for the same system described by (3.73)–(3.74). Then the same additional component for $\mathbf{y}$ given by (3.77) reduces the problem to the necessary form. The objective functional now is

$$J(\mathbf{u}) = y_{n+1}(T) + G(\mathbf{y}(T)).$$

Let us consider the main form of the terminal control problem (3.73)–(3.76) using an extension of the procedure for the simplest problem of optimal control. Much of the reasoning for the latter is simply reformulated to go from the scalar to the vector version. For the simplest problem, the main step involved finding the main part of the increment of $J(u)$ under a needle-shaped increment of a fixed control function. We shall do this here also. The next step involved establishing the condition under which a control function would be optimal for the problem. This led to Pontryagin's maximum principle. We shall extend this to the general problem. Finally we shall discuss how to use the formula for the increment of the functional, as well as the maximum principle, to find an optimal solution.

Let $t = s$ be a point of continuity of a control function $\mathbf{u}(t)$. Giving $\mathbf{u}(t)$ a needle-shaped increment (i.e., a vector whose components are all needle-shaped functions with perturbations in $(s - \varepsilon, s]$) we get a new control defined by

$$\mathbf{u}^*(t) = \begin{cases} \mathbf{u}(t), & t \notin (s - \varepsilon, s], \\ \mathbf{v}, & t \in (s - \varepsilon, s]. \end{cases} \tag{3.78}$$

We can continue to refer to Fig. 3.1. We can also refer to Fig. 3.3 for a representation of the function $\mathbf{y}^*(t)$ that satisfies the equation

$$(\mathbf{y}^*(t))' = \mathbf{f}(\mathbf{y}^*(t), \mathbf{u}^*(t)) \tag{3.79}$$

and the same initial condition $\mathbf{y}(0) = \mathbf{y}_0$. We suppose that at least for all positive ε less than some small fixed number ε_0, the incremented control function $\mathbf{u}^*(t)$ is admissible.

The main part of the increment $J(\mathbf{u}^*) - J(\mathbf{u})$, linear in small ε, is determined by

Theorem 3.13. *Let $t = s$ be a point of continuity of a control function $\mathbf{u}(t)$. The increment of $J(\mathbf{u})$ is*

$$J(\mathbf{u}^*) - J(\mathbf{u}) = \varepsilon\, \delta_{s,\mathbf{v}} J(\mathbf{u}) + o(\varepsilon) \tag{3.80}$$

where

$$\delta_{s,\mathbf{v}} J(\mathbf{u}) = \mathbf{\Psi}(s) \cdot [\mathbf{f}(\mathbf{y}(s), \mathbf{u}(s)) - \mathbf{f}(\mathbf{y}(s), \mathbf{v})] \tag{3.81}$$

and $\mathbf{\Psi}(s)$ is a solution of the following Cauchy problem (in the reverse time):

$$\mathbf{\Psi}'(s) = -\nabla_{\mathbf{y}} \mathbf{f}(\mathbf{y}(s), \mathbf{u}(s)) \cdot \mathbf{\Psi}(s), \qquad \mathbf{\Psi}(T) = -\nabla_{\mathbf{y}} G(\mathbf{y}(T)). \tag{3.82}$$

$\delta_{s,\mathbf{v}} J(\mathbf{u})$ is called the variational derivative of the second kind of the functional $J(\mathbf{u})$.

Proof. Take $\varepsilon > 0$ so small that all points of $[s - \varepsilon, s]$ are points of continuity of $\mathbf{u}(t)$ and the corresponding incremented control functions $\mathbf{u}^*(t)$ are admissible. We divide the proof into several steps.

Step 1, the main part of the increment of $\mathbf{y}(t)$. On $[0, s-\varepsilon]$ the control functions coincide. The initial conditions for $\mathbf{y}(t)$ and $\mathbf{y}^*(t)$ coincide as well, so on this segment we have $\mathbf{y}^*(t) = \mathbf{y}(t)$.

Let us find the increment of $\mathbf{y}(t)$ for $t \in [s-\varepsilon, s]$. Subtracting term by term (3.73) from (3.79) we have

$$(\mathbf{y}^*(t))' - \mathbf{y}'(t) = \mathbf{f}(\mathbf{y}^*(t), \mathbf{v}) - \mathbf{f}(\mathbf{y}(t), \mathbf{u}(t)).$$

Denoting $\Delta\mathbf{y}(t) = \mathbf{y}^*(t) - \mathbf{y}(t)$ we get

$$\Delta\mathbf{y}'(t) = \mathbf{f}(\mathbf{y}(t) + \Delta\mathbf{y}(t), \mathbf{v}) - \mathbf{f}(\mathbf{y}(t), \mathbf{u}(t)). \tag{3.83}$$

This equation, which holds on $(s-\varepsilon, s]$, is supplemented by the "initial" condition

$$\Delta\mathbf{y}(s-\varepsilon) = 0 \tag{3.84}$$

which follows from the above coincidence of $\mathbf{y}(t)$ and $\mathbf{y}^*(t)$. Let us reduce the Cauchy problem (3.83)–(3.84) for $\Delta\mathbf{y}(t)$, integrating (3.83) with respect to the time parameter:

$$\Delta\mathbf{y}(t) - \Delta\mathbf{y}(s-\varepsilon) = \int_{s-\varepsilon}^{t} [\mathbf{f}(\mathbf{y}(\tau) + \Delta\mathbf{y}(\tau), \mathbf{v}) - \mathbf{f}(\mathbf{y}(\tau), \mathbf{u}(\tau))]\, d\tau.$$

By (3.84) this reduces to

$$\Delta\mathbf{y}(t) = \int_{s-\varepsilon}^{t} [\mathbf{f}(\mathbf{y}(\tau) + \Delta\mathbf{y}(\tau), \mathbf{v}) - \mathbf{f}(\mathbf{y}(\tau), \mathbf{u}(\tau))]\, d\tau. \tag{3.85}$$

Since we assume $\mathbf{f}(\mathbf{y}, \mathbf{u})$ to be continuous and thus bounded, the integral on the right of (3.85) is of order ε and so is $\Delta\mathbf{y}(t)$. Thus replacing in the integrand the quantities $\mathbf{y}(\tau)$ and $\mathbf{u}(\tau)$ by $\mathbf{y}(s)$ and $\mathbf{u}(s)$ respectively, and placing $\Delta\mathbf{y}(\tau) = 0$, we introduce in the value of the integral an error of order $o(\varepsilon)$ for $t \in [s-\varepsilon, s]$. Hence (3.85) reduces to

$$\Delta\mathbf{y}(t) = \int_{s-\varepsilon}^{t} [\mathbf{f}(\mathbf{y}(s), \mathbf{v}) - \mathbf{f}(\mathbf{y}(s), \mathbf{u}(s))]\, d\tau + o(\varepsilon),$$

which can be rewritten as

$$\Delta\mathbf{y}(t) = (t - s + \varepsilon)[\mathbf{f}(\mathbf{y}(s), \mathbf{v}) - \mathbf{f}(\mathbf{y}(s), \mathbf{u}(s))] + o(\varepsilon),$$

and thus on this small segment $[s-\varepsilon, s]$ the difference $\Delta\mathbf{y}(t)$ changes almost linearly from zero, taking at $t = s$ the value

$$\Delta\mathbf{y}(s) = \varepsilon\, [\mathbf{f}(\mathbf{y}(s), \mathbf{v}) - \mathbf{f}(\mathbf{y}(s), \mathbf{u}(s))] + o(\varepsilon). \tag{3.86}$$

This is the initial value for the solution $\Delta\mathbf{y}(t)$ on $[s, T]$ of the equation

$$\Delta\mathbf{y}'(t) = \mathbf{f}(\mathbf{y}(t) + \Delta\mathbf{y}(t), \mathbf{u}(t)) - \mathbf{f}(\mathbf{y}(t), \mathbf{u}(t)) \tag{3.87}$$

(we recall that on this interval $\mathbf{u}^*(t) = \mathbf{u}(t)$ and it is considered to be known at this moment). Linearizing the right side of (3.87) with respect to $\Delta\mathbf{y}(t)$ (taking into account (3.60)) we have

$$\Delta\mathbf{y}'(t) = (\nabla_{\mathbf{y}}\mathbf{f}(\mathbf{y}(t), \mathbf{u}(t)))^T \cdot \Delta\mathbf{y}(t) + o(\|\Delta\mathbf{y}(t)\|). \tag{3.88}$$

Because of smallness of the initial condition of $\Delta\mathbf{y}(t)$ at $t = s$ and the form of (3.88) we expect the solution of the corresponding Cauchy problem on the finite interval $(s, T]$ to be of order ε and, up to terms of order higher than ε, equal to the solution of the following Cauchy problem:

$$\delta\mathbf{y}'(t) = (\nabla_{\mathbf{y}}\mathbf{f}(\mathbf{y}(t), \mathbf{u}(t)))^T \cdot \delta\mathbf{y}(t), \tag{3.89}$$

$$\delta\mathbf{y}(s) = \varepsilon\,[\mathbf{f}(\mathbf{y}(s), \mathbf{v}) - \mathbf{f}(\mathbf{y}(s), \mathbf{u}(s))], \tag{3.90}$$

which is the linearization of the complete initial problem (3.87), (3.86). By the linearity of this problem its solution is proportional to ε.

To find the main part of the increment $\Delta\mathbf{y}(T)$ it remains to solve the Cauchy problem (3.89)–(3.90). This can be integrated (often numerically) but we will use the notion of the fundamental solution from the previous section.

Let us denote $\mathbf{A}(t) = (\nabla_{\mathbf{y}}\mathbf{f}(\mathbf{y}(t), \mathbf{u}(t)))^T$ and leave the notation of § 3.7 for this fundamental solution, which satisfies

$$\frac{d}{dt}\mathbf{\Phi}(t, s) = \mathbf{A}(t) \cdot \mathbf{\Phi}(t, s)$$

and the "initial" condition $\mathbf{\Phi}(s, s) = \mathbf{E}$ for all s. By Property 3.9 of § 3.7 the solution to (3.89)–(3.90) is

$$\delta\mathbf{y}(t) = \varepsilon\,\mathbf{\Phi}(t, s) \cdot [\mathbf{f}(\mathbf{y}(s), \mathbf{v}) - \mathbf{f}(\mathbf{y}(s), \mathbf{u}(s))]$$

and thus, assuming "good" behavior of $\Delta\mathbf{y}(t)$ we have

$$\Delta\mathbf{y}(T) = \varepsilon\,\mathbf{\Phi}(T, s) \cdot [\mathbf{f}(\mathbf{y}(s), \mathbf{v}) - \mathbf{f}(\mathbf{y}(s), \mathbf{u}(s))] + o(\varepsilon). \tag{3.91}$$

Step 2, the main part of the increment of $J(\mathbf{u}) = G(\mathbf{y}(T))$. We again use the formula of the differential (3.57) for linearization of the increment of $J(\mathbf{u})$ with respect to $\Delta\mathbf{y}(t)$:

$$\begin{aligned}\Delta J(\mathbf{u}) &= J(\mathbf{u}^*) - J(\mathbf{u}) \\ &= G(\mathbf{y}(T) + \Delta\mathbf{y}(T)) - G(\mathbf{y}(T)) \\ &= \nabla_{\mathbf{y}} G(\mathbf{y})\big|_{\mathbf{y}=\mathbf{y}(T)} \cdot \Delta\mathbf{y}(T) + o(|\Delta\mathbf{y}(T)|).\end{aligned}$$

Using (3.91) we get

$$\Delta J(\mathbf{u}) = \varepsilon \nabla_{\mathbf{y}} G(\mathbf{y})\big|_{\mathbf{y}=\mathbf{y}(T)} \cdot \mathbf{\Phi}(T,s) \cdot [\mathbf{f}(\mathbf{y}(s),\mathbf{v}) - \mathbf{f}(\mathbf{y}(s),\mathbf{u}(s))] + o(\varepsilon).$$

This is the required formula. It remains to represent it in the form asserted by the theorem.

Step 3, the final step. Let us define a vector function $\mathbf{\Psi}(s)$ as

$$\mathbf{\Psi}(s) = -\nabla_{\mathbf{y}} G(\mathbf{y})\big|_{y=y(T)} \cdot \mathbf{\Phi}(T,s).$$

With this notation for $\Delta J(\mathbf{u})$ we do have the representation (3.80)–(3.81), so it remains to demonstrate that $\mathbf{\Psi}(s)$ satisfies (3.82). The second relation of (3.82) is a consequence of the equality $\mathbf{\Phi}(T,T) = \mathbf{E}$; indeed,

$$\mathbf{\Psi}(T) = -\nabla_{\mathbf{y}} G(\mathbf{y})\big|_{y=y(T)} \cdot \mathbf{\Phi}(T,T) = -\nabla_{\mathbf{y}} G(\mathbf{y})\big|_{y=y(T)}.$$

Let us show that it satisfies the first equation of (3.82) as well. The derivative of $\mathbf{\Psi}(s)$ is

$$\frac{d\mathbf{\Psi}(s)}{ds} = \frac{d}{ds}\left[-\nabla_{\mathbf{y}} G(\mathbf{y})\big|_{y=y(T)} \cdot \mathbf{\Phi}(T,s)\right] = -\nabla_{\mathbf{y}} G(\mathbf{y})\big|_{y=y(T)} \cdot \frac{d}{ds}\mathbf{\Phi}(T,s).$$

Let us now use the equation for the derivative with respect to the second argument of the fundamental solution, which is given by Property 3.11:

$$\begin{aligned}
\frac{d\mathbf{\Psi}(s)}{ds} &= -\nabla_{\mathbf{y}} G(\mathbf{y})\big|_{y=y(T)} \cdot (-\mathbf{\Phi}(T,s) \cdot \mathbf{A}(s)) \\
&= -\left(-\nabla_{\mathbf{y}} G(\mathbf{y})\big|_{y=y(T)} \cdot \mathbf{\Phi}(T,s)\right) \cdot \mathbf{A}(s) \\
&= -\mathbf{\Psi}(s) \cdot \mathbf{A}(s) = -(\mathbf{A}(s))^T \cdot \mathbf{\Psi}(s).
\end{aligned}$$

Remembering the above notation for $\mathbf{A}(s)$ we complete the proof. □

3.9 Pontryagin's Maximum Principle for the Terminal Optimal Problem

First we would like to discuss the statement of Theorem 3.13. When we seek a response of an object described by the problem

$$\mathbf{y}'(t) = \mathbf{f}(\mathbf{y}(t),\mathbf{u}(t)), \qquad \mathbf{y}(0) = \mathbf{y}_0, \tag{3.92}$$

to a needle-shaped disturbance of the control function $\mathbf{u}(t)$ we obtain a dual problem

$$\mathbf{\Psi}'(s) = -\nabla_{\mathbf{y}} \mathbf{f}(\mathbf{y}(s),\mathbf{u}(s)) \cdot \mathbf{\Psi}(s), \tag{3.93}$$

$$\mathbf{\Psi}(T) = -\nabla_{\mathbf{y}} G(\mathbf{y}(T)). \tag{3.94}$$

The dual equation (3.93) plays a role like that of the Euler equation of the calculus of variations, and the condition (3.94) is the condition of transversality. Together (3.92)–(3.94) compose a boundary value problem having a unique solution when $\mathbf{u}(t)$ is given. This splits into two "initial value problems" for $\mathbf{y}(t)$ and $\boldsymbol{\Psi}(s)$. For problems other than the problem of terminal control, other types of boundary conditions are given but the equations yielding a response to a needle-shaped disturbance are the same. Let us introduce an equivalent form of the equations for this boundary value problem. We define a scalar function in three variables $\mathbf{y}$, $\boldsymbol{\Psi}$, and $\mathbf{u}(t)$, called *Pontryagin's function*:

$$H(\mathbf{y}, \boldsymbol{\Psi}, \mathbf{u}) = \mathbf{f}(\mathbf{y}, \mathbf{u}) \cdot \boldsymbol{\Psi}. \tag{3.95}$$

Simple calculation demonstrates that

$$\nabla_{\mathbf{y}} H(\mathbf{y}, \boldsymbol{\Psi}, \mathbf{u}) = \nabla_{\mathbf{y}} \mathbf{f}(\mathbf{y}, \mathbf{u}) \cdot \boldsymbol{\Psi},$$
$$\nabla_{\boldsymbol{\Psi}} H(\mathbf{y}, \boldsymbol{\Psi}, \mathbf{u}) = \mathbf{f}(\mathbf{y}, \mathbf{u}),$$

where the second relation is a consequence of the equality

$$\nabla_{\mathbf{x}} \mathbf{x} = \mathbf{e}_i \frac{\partial}{\partial x_i}(x_j \mathbf{e}_j) = \mathbf{e}_i \mathbf{e}_i = \mathbf{E}.$$

It follows that (3.92) and (3.94) can be written as

$$\mathbf{y}'(t) = \nabla_{\boldsymbol{\Psi}} H(\mathbf{y}(t), \boldsymbol{\Psi}(t), \mathbf{u}(t)),$$
$$\boldsymbol{\Psi}'(t) = -\nabla_{\mathbf{y}} H(\mathbf{y}(t), \boldsymbol{\Psi}(t), \mathbf{u}(t)).$$

This is the Hamiltonian form.

In terms of Pontryagin's function the second kind derivative of $J(\mathbf{u})$ (3.81) can be written as

$$\delta_{s,\mathbf{v}} J(\mathbf{u}) = H(\mathbf{y}(s), \boldsymbol{\Psi}(s), \mathbf{u}(s)) - H(\mathbf{y}(s), \boldsymbol{\Psi}(s), \mathbf{v}). \tag{3.96}$$

Now we can formulate Pontryagin's maximum principle.

Theorem 3.14. *Let $\mathbf{u}(t)$ be an optimal control function at which $J(\mathbf{u})$ attains its minimal value on $\mathcal{U}$, the set of all admissible control functions, and let $\mathbf{y}(t)$ and $\boldsymbol{\Psi}(t)$ be a solution of the boundary value problem (3.92)–(3.94). At any point $t = s$ of continuity of $\mathbf{u}(t)$, the Pontryagin function $H(\mathbf{y}(t), \boldsymbol{\Psi}(t), \mathbf{v})$, considered as a function of the third argument $\mathbf{v}$, takes its maximum value at $\mathbf{v} = \mathbf{u}(s)$.*

Proof. Since $J(\mathbf{u})$ attains its minimum at $\mathbf{u}(t)$ then for any admissible control function $\mathbf{u}^*(t)$ we have

$$J(\mathbf{u}^*) - J(\mathbf{u}) \geq 0.$$

In particular it is valid for an admissible $\mathbf{u}^*(t)$ that is a disturbance of $\mathbf{u}(t)$ by a needle-shaped vector function

$$\mathbf{u}^*(t) = \begin{cases} \mathbf{u}(t), & t \notin (s-\varepsilon, s], \\ \mathbf{v}, & t \in (s-\varepsilon, s], \end{cases}$$

and thus, for sufficiently small ε because of (3.80) and (3.96) we have

$$J(\mathbf{u}^*) - J(\mathbf{u}) = \varepsilon\,(H(\mathbf{y}(s), \mathbf{\Psi}(s), \mathbf{u}(s)) - H(\mathbf{y}(s), \mathbf{\Psi}(s), \mathbf{v})) + o(\varepsilon) \geq 0.$$

From this it follows that $H(\mathbf{y}(s), \mathbf{\Psi}(s), \mathbf{u}(s)) - H(\mathbf{y}(s), \mathbf{\Psi}(s), \mathbf{v}) \geq 0.$ □

Pontryagin's principle of maximum gives us an effective tool to check whether $\mathbf{u}(t)$ is a needed control function at which $J(\mathbf{u})$ attains its minimum, but it does not show, except for quite simple problems, how to find this. However, (3.80) is the background of various numerical methods used to find this minimum. We shall discuss them in brief.

The formula (3.80) for the increment of $J(\mathbf{u})$, which can be rewritten as

$$J(\mathbf{u}) \approx J(\mathbf{u}^*) - \varepsilon\,\delta_{s,\mathbf{v}} J(\mathbf{u}), \tag{3.97}$$

generates an iterative procedure that begins with selection of a finite number of the time instants $(\tau_1, \ldots, \tau_r)$ at which one may introduce needle-shaped disturbances for finding a more effective control function. Next one must find an instant τ_i and a corresponding admissible value of $\mathbf{v}$, which we denote by $\mathbf{v}_i$, at which the maximum of the numerical set

$$\{\delta_{\tau_1,\mathbf{v}} J(\mathbf{u}), \ldots, \delta_{\tau_r,\mathbf{v}} J(\mathbf{u})\}$$

is attained. Denoting the control parameters of the previous step as $\mathbf{u}^{(i)}(t)$ and $\mathbf{u}^{(i)*}(t)$ where $\mathbf{u}^{(i)*}(t)$ is just determined, one must choose the value of ε, denoted by ε_i, at which (3.97) provides a sufficiently precise approximation. Then the next approximation of the value of $J(\mathbf{u})$ is given by the formula

$$J(\mathbf{u}^{(i+1)}) = J(\mathbf{u}^{(i)*}) - \varepsilon_i\,\delta_{\tau_i,\mathbf{v}_i} J(\mathbf{u}^{(i)}).$$

Versions of this procedure differ in their methods of determining each step, in particular the points τ_i. They are called the methods of coordinate-by-coordinate descent.

A modification is called the group descent procedure. We have found the main linear part of the increment of $J(\mathbf{u})$ under a needle-shaped disturbance of $\mathbf{u}(t)$ at $t = s$, which is characterized by the pair of parameters $\varepsilon, \mathbf{v}$. This means that if $\mathbf{u}(t)$ is disturbed by a finite set of N such needle-shaped variations, the ith of which is lumped at a point s_i of continuity of $\mathbf{u}(t)$ and is characterized by the pair $\varepsilon_i, \mathbf{v}_i$, then denoting by $\mathbf{u}^{**}(t)$ the corresponding control function we get the main part of the increment as the sum of increments of $J(\mathbf{u})$ due to each of the needle-shaped increments of $\mathbf{u}(t)$:

$$J(\mathbf{u}^{**}) - J(\mathbf{u}) = \sum_{i=1}^{N} \varepsilon_i [H(\mathbf{y}(s), \mathbf{\Psi}(s), \mathbf{u}(s)) - H(\mathbf{y}(s), \mathbf{\Psi}(s), \mathbf{v}_i)] + o\left(\max(\varepsilon_1, \ldots, \varepsilon_N)\right). \tag{3.98}$$

Then we can decrease the value of $J(\mathbf{u})$ on the next step of approximation using a group of needle-shaped increments and the formula (3.98).

3.10 Generalization of the Terminal Control Problem

Let us consider a generalized terminal control problem whose setup coincides with that of the usual problem except for the form of the objective function (functional). This set up is

Definition 3.15. From among the piecewise continuous control functions $\mathbf{u}(t) \in \mathcal{U}$ on $[0, T]$, find one that minimizes the functional $\mathcal{I}(\mathbf{u})$,

$$\mathcal{I}(\mathbf{u}) \to \min_{\mathbf{u} \in \mathcal{U}},$$

when $\mathcal{I}(\mathbf{u})$ is defined as

$$\mathcal{I}(\mathbf{u}) = G(\mathbf{y}(s_1), \mathbf{y}(s_2), \ldots, \mathbf{y}(s_N)),$$

$G(\mathbf{y}(s_1), \mathbf{y}(s_2), \ldots, \mathbf{y}(s_N))$ being a function continuously differentiable in all its variables, $0 < s_1 < s_2 < \cdots < s_N = T$ some fixed points of time, and $\mathbf{y}(t)$ satisfying the equations

$$\mathbf{y}'(t) = \mathbf{f}(\mathbf{y}(t), \mathbf{u}(t)), \qquad \mathbf{y}(0) = \mathbf{y}_0.$$

Such a form of the objective function can appear, for example, if the objective functional contains an integral depending on $\mathbf{y}(t)$ which is discretized according to some simple method such as Simpson's rule or the rectangular rule. To proceed further we need some additional material. We shall obtain a nonstandard Cauchy problem and then find a way to present it in a form that resembles the usual form for such a problem. For this we digress briefly to discuss the Dirac δ-function.

The δ-function concept was originated by physicists and used for many years before being given a rigorous footing (called the *theory of distributions*) by mathematicians. Although rigor has certain advantages, the heuristic viewpoint of the early physicists will suffice for our purposes. This viewpoint rests on the notion that $\delta(t)$ is a function of the argument t, taking the value zero for $t \neq 0$ and an infinite "value" at $t = 0$ such that

$$\int_{-\infty}^{+\infty} \delta(t)\, dt = 1. \tag{3.99}$$

Now from a mathematical viewpoint we are in trouble already because it can be shown that there is no such function. But we nonetheless proceed formally with the understanding that every step we take can be justified rigorously (with tremendous effort and with full chapters of extra explanation which, unfortunately, would not lend clarity to the topic).

The δ-function is a generalized derivative of the step function $h(t)$ given by

$$h(t) = \begin{cases} 1, & t \geq 0, \\ 0, & t < 0, \end{cases} \tag{3.100}$$

and we shall exploit this property. The introduction of the generalized derivative uses the main lemma of the calculus of variations and the formula for integration by parts. Let $\varphi(t)$ be a function infinitely differentiable on $(-\infty, +\infty)$ and with compact support (the support of $\phi(t)$ is the closure of the set of all t for which $\varphi(t) \neq 0$). Let us denote this class by $\mathcal{D}$. For any differentiable function $f(t)$ the formula for integration by parts holds:

$$\int_{-\infty}^{+\infty} f(t)\varphi'(t)\, dt = -\int_{-\infty}^{+\infty} f'(t)\varphi(t)\, dt. \tag{3.101}$$

The main lemma of the calculus of variations states that if the equality

$$\int_{-\infty}^{+\infty} f(t)\varphi'(t)\, dt = -\int_{-\infty}^{+\infty} g(t)\varphi(t)\, dt \tag{3.102}$$

holds for any $\varphi(t) \in \mathcal{D}$ then $g(t) = f'(t)$. This is valid for a differentiable function $f(t)$, but the same equation defines the generalized derivative of an integrable function $f(t)$: a function $g(t)$ is called the generalized derivative of $f(t)$ if (3.102) holds for any $\varphi(t) \in \mathcal{D}$. The generalized derivative is denoted by the usual differentiation symbols. The main lemma of the calculus of variations (more precisely, its variant) provides uniqueness of definition of the generalized derivative. Let us check that $h'(t) = \delta(t)$ in the generalized sense. Indeed,

$$\int_{-\infty}^{+\infty} h(t)\varphi'(t)\,dt = \int_0^{\infty} h(t)\varphi'(t)\,dt = \int_0^{\infty} \varphi'(t)\,dt = -\varphi(0)$$

and by the definition of δ-function

$$\int_{-\infty}^{+\infty} \delta(t)\varphi(t)\,dt = \varphi(0).$$

Thus for the pair $h(t), \delta(t)$ the definition of generalized derivative is valid and so $h'(t) = \delta(t)$. Using this property we can write out the Cauchy problem

$$y'(t) = g(t, y(t)), \qquad y(0) = y_0, \tag{3.103}$$

in an equivalent form

$$y'(t) = f(t, y(t)) + y_0\delta(t), \qquad y(t)\big|_{t\to -0} = 0. \tag{3.104}$$

Indeed, integration of (3.104) with respect to t (the starting point is $t = -0$) implies the equation

$$y(t) = \int_0^t f(s, y(s))\,ds + y_0 h(t),$$

which is equivalent to (3.103).

Now let us formulate the main theorem of this section, in which we keep the notation of § 3.8 for $\mathbf{u}^*(t)$ and $\mathbf{y}^*(t)$.

Theorem 3.16. *Let $t = s$ be a point of continuity of a control function $\mathbf{u}(t)$ that is different from $s_1, s_2, \ldots, s_N = T$. The increment of $\mathcal{I}(\mathbf{u})$ is*

$$\mathcal{I}(\mathbf{u}^*) - \mathcal{I}(\mathbf{u}) = \varepsilon\,\delta_{s,\mathbf{v}}\mathcal{I}(\mathbf{u}) + o(\varepsilon) \tag{3.105}$$

where

$$\delta_{s,\mathbf{v}}\mathcal{I}(\mathbf{u}) = \mathbf{\Psi}(s)\cdot[\mathbf{f}(\mathbf{y}(s), \mathbf{u}(s)) - \mathbf{f}(\mathbf{y}(s), \mathbf{v})] \tag{3.106}$$

and $\mathbf{\Psi}(s)$ is a solution of the following Cauchy problem (in the reverse time)

$$\begin{aligned}\mathbf{\Psi}'(s) &= -\nabla_{\mathbf{y}}\mathbf{f}(\mathbf{y}(s), \mathbf{u}(s))\cdot\mathbf{\Psi}(s)\\ &\quad + \sum_{i=1}^{N} \delta(s_i - s)\nabla_{\mathbf{y}(s_i)}G(\mathbf{y}(s_1), \mathbf{y}(s_2), \ldots, \mathbf{y}(s_N)),\\ \mathbf{y}(T+0) &= 0.\end{aligned} \tag{3.107}$$

Comparison with Theorem 3.13 shows that the current theorem differs only in the form of the problem for $\mathbf{\Psi}(s)$.

Proof. It is clear that $\mathbf{y}^*(t)$ for this problem coincides with that of § 3.8, so we can use the corresponding formulas of that section. In particular, for $t > s$ the main part of the increment $\Delta\mathbf{y}(t)$ of the corresponding solution $\mathbf{y}(t)$ on $(s, T]$, under the needle-shaped increment of the control vector $\mathbf{u}$, is

$$\delta\mathbf{y}(t) = \varepsilon\,\mathbf{\Phi}(t, s)\cdot[\mathbf{f}(\mathbf{y}(s), \mathbf{v}) - \mathbf{f}(\mathbf{y}(s), \mathbf{u}(s))]. \tag{3.108}$$

So we immediately go to the increment of the goal function. First we use the formula for the complete differential to get the main part of the increment of $\mathcal{I}(\mathbf{u}) = G(\mathbf{y}(s_1), \mathbf{y}(s_2), \ldots, \mathbf{y}(s_N))$, which is

$$\begin{aligned}\Delta\mathcal{I}(\mathbf{u}) &= \mathcal{I}(\mathbf{u}^*) - \mathcal{I}(\mathbf{u})\\ &= G(\mathbf{y}(s_1) + \Delta\mathbf{y}(s_1), \mathbf{y}(s_2) + \Delta\mathbf{y}(s_2), \ldots, \mathbf{y}(s_N) + \Delta\mathbf{y}(s_N))\\ &\quad - G(\mathbf{y}(s_1), \mathbf{y}(s_2), \ldots, \mathbf{y}(s_N))\\ &= \sum_{i=1}^{N} \nabla_{\mathbf{y}(s_i)} G(\mathbf{y}(s_1), \mathbf{y}(s_2), \ldots, \mathbf{y}(s_N))\cdot\Delta\mathbf{y}(s_i)\\ &\quad + o\left(\max_j \|\Delta\mathbf{y}(s_j)\|\right)\end{aligned} \tag{3.109}$$

To implement (3.108) we rewrite it in the form

$$\delta\mathbf{y}(t) = \varepsilon\,\mathbf{\Phi}(t, s)\cdot[\mathbf{f}(\mathbf{y}(s), \mathbf{v}) - \mathbf{f}(\mathbf{y}(s), \mathbf{u}(s))]\,h(t - s)$$

so it becomes valid for use in (3.109) for all $t \in [0, T]$ when the interval $[s - \varepsilon, s]$ does not contain any s_i (assumed). Then the increment of $\mathcal{I}(\mathbf{u})$ can be rewritten as

$$\begin{aligned}\Delta\mathcal{I}(\mathbf{u}) &= \mathcal{I}(\mathbf{u}^*) - \mathcal{I}(\mathbf{u})\\ &= \varepsilon\left\{\sum_{i=1}^{N} \nabla_{\mathbf{y}(s_i)} G(\mathbf{y}(s_1), \mathbf{y}(s_2), \ldots, \mathbf{y}(s_N))\cdot\mathbf{\Phi}(s_i, s)h(s_i - s)\right\}\cdot\\ &\quad \cdot[\mathbf{f}(\mathbf{y}(s), \mathbf{v}) - \mathbf{f}(\mathbf{y}(s), \mathbf{u}(s))] + o(\varepsilon).\end{aligned}$$

Denoting

$$\mathbf{\Psi}(s) = -\sum_{i=1}^{N} \nabla_{\mathbf{y}(s_i)} G(\mathbf{y}(s_1), \mathbf{y}(s_2), \ldots, \mathbf{y}(s_N))\cdot\mathbf{\Phi}(s_i, s)h(s_i - s) \tag{3.110}$$

we get, as in § 3.8,

$$\delta_{s,\mathbf{v}}\mathcal{I}(\mathbf{u}) = \mathbf{\Psi}(s)\cdot[\mathbf{f}(\mathbf{y}(s), \mathbf{u}(s)) - \mathbf{f}(\mathbf{y}(s), \mathbf{v})]$$

and for the increment of objective functional

$$\mathcal{I}(\mathbf{u}^*) - \mathcal{I}(\mathbf{u}) = \varepsilon\, \delta_{s,\mathbf{v}}\mathcal{I}(\mathbf{u}) + o(\varepsilon).$$

Note that the presence of $h(s_i - s)$ in the sum of the definition (3.110) means that at $s = s_i$ the value of $\mathbf{\Psi}(s)$ has some step change for an additional term in the sum.

It remains only to check the validity of (3.107). When $s > s_N = T$ we get $\mathbf{\Psi}(s) = 0$ so the second of (3.107) holds. To show that the first is valid let us find the derivative of $\mathbf{\Psi}(s)$. Taking into account Property 3.11 which in our terms is

$$\frac{d}{ds}\mathbf{\Phi}(s_i, s) = -\mathbf{\Phi}(s_i, s) \cdot (\nabla_{\mathbf{y}}\mathbf{f}(\mathbf{y}(s), \mathbf{u}(s)))^T$$

we get

$$\begin{aligned}
\frac{d\mathbf{\Psi}(s)}{ds} &= \sum_{i=1}^{N} \nabla_{\mathbf{y}(s_i)} G(\mathbf{y}(s_1), \mathbf{y}(s_2), \dots, \mathbf{y}(s_N)) \cdot \\
&\quad \cdot h(s_i - s)\mathbf{\Phi}(s_i, s) \cdot (\nabla_{\mathbf{y}}\mathbf{f}(\mathbf{y}(s), \mathbf{u}(s)))^T \\
&\quad + \sum_{i=1}^{N} \nabla_{\mathbf{y}(s_i)} G(\mathbf{y}(s_1), \mathbf{y}(s_2), \dots, \mathbf{y}(s_N)) \cdot \mathbf{\Phi}(s_i, s)\delta(s_i - s) \\
&= -\mathbf{\Psi}(s) \cdot (\nabla_{\mathbf{y}}\mathbf{f}(\mathbf{y}(s), \mathbf{u}(s)))^T \\
&\quad + \sum_{i=1}^{N} \nabla_{\mathbf{y}(s_i)} G(\mathbf{y}(s_1), \mathbf{y}(s_2), \dots, \mathbf{y}(s_N))\delta(s_i - s).
\end{aligned}$$

In the last transformation we used $\mathbf{\Phi}(s_i, s)\delta(s_i - s) = \mathbf{E}\delta(s_i - s)$. □

The form of Pontryagin's maximum principle for the generalized terminal control problem is the same as in the previous section. We leave its formulation to the reader.

This kind of generalized terminal control problem is used in practice and, as a rule, requires numerical solution of the problems when the formula for the increment (3.106) of the goal functional is used.

3.11 Small Variations of Control Function for Terminal Control Problem

The form of the increment of the objective functional for the generalized terminal control problem provides a hint that the conjugate equations and similar material should enter not only for needle-shaped variations of the control function, but for any small variations. We will see that this is really so, and for this case we will find the expression for the increment of the objective functional under the increment of control vector of other type. We reconsider the terminal control problem described by the dynamical system

$$\mathbf{y}'(t) = \mathbf{f}(\mathbf{y}(t), \mathbf{u}(t)), \qquad \mathbf{y}(0) = \mathbf{y}_0.$$

We wish to find the increment of the objective functional $J(\mathbf{u}) = G(\mathbf{y}(T))$ under a small increment $\Delta\mathbf{u}(t)$ of the control function $\mathbf{u}(t)$.

We demonstrated that one of the problems of the calculus of variations was covered by the setup of a problem of optimal control, but did not use the type of variations used in the calculus of variations until now. Here we will demonstrate how it can be done.

Let us define $\mathbf{v}(t) = \mathbf{u}(t) + \Delta\mathbf{u}(t)$ and require that $\mathbf{v}(t)$ is admissible. Smallness of $\Delta\mathbf{u}(t)$ means that $\sup_{[0,T]} \|\Delta\mathbf{u}(t)\|$ is sufficiently small. We suppose that the changed value $\mathbf{y}^*(t)$ satisfying the Cauchy problem

$$(\mathbf{y}^*(t))' = \mathbf{f}(\mathbf{y}^*(t), \mathbf{v}(t)), \qquad \mathbf{y}^*(0) = \mathbf{y}_0,$$

is such that $\Delta\mathbf{y}(t) = \mathbf{y}^*(t) - \mathbf{y}(t)$ is also small enough, that is $\max_{[0,T]} \|\Delta\mathbf{y}(t)\|$ is small.

Now we would like to find the increment of $J(\mathbf{u})$ under such a small admissible increment of $\mathbf{u}(t)$. The answer is given by

Theorem 3.17. *Suppose* $\sup_{[0,T]} \|\Delta\mathbf{u}(t)\| = \varepsilon$. *Then the increment of* $J(\mathbf{u})$ *is*

$$J(\mathbf{u}^*) - J(\mathbf{u}) = \delta J(\mathbf{u}) + o(\varepsilon)$$

where

$$\delta J(\mathbf{u}) = \int_0^T \mathbf{\Psi}(t) \cdot [\mathbf{f}(\mathbf{y}(t), \mathbf{u}(t)) - \mathbf{f}(\mathbf{y}(t), \mathbf{v}(t))]\, dt$$

and $\mathbf{\Psi}(s)$ *is a solution of the following Cauchy problem (in the reverse time):*

$$\mathbf{\Psi}'(s) = -\nabla_{\mathbf{y}}\mathbf{f}(\mathbf{y}(s), \mathbf{u}(s)) \cdot \mathbf{\Psi}(s), \qquad \mathbf{\Psi}(T) = -\nabla_{\mathbf{y}}G(\mathbf{y}(T)). \tag{3.111}$$

Proof. Let us note first that the conjugate equation (3.111) for $\mathbf{\Psi}(s)$ coincides with the conjugate equation we established for the terminal control problem in § 3.8. Much of that reasoning will apply here. Suppose for simplicity of notation that $\Delta\mathbf{y}(t)$ for all $t \in [0, T]$ is of order ε. The problem defining the increment $\Delta\mathbf{y}(t)$ is

$$\Delta\mathbf{y}'(t) = \mathbf{f}(\mathbf{y}(t) + \Delta\mathbf{y}(t), \mathbf{v}(t)) - \mathbf{f}(\mathbf{y}(t), \mathbf{u}(t)), \tag{3.112}$$

$$\Delta\mathbf{y}(0) = 0.$$

We need to find the main part of $\Delta\mathbf{y}(t)$ at $t = T$. Let us transform the right side of (3.112):

$$\begin{aligned}\mathbf{f}(\mathbf{y}+\Delta\mathbf{y},\mathbf{v}) - \mathbf{f}(\mathbf{y},\mathbf{u}) &= \mathbf{f}(\mathbf{y}+\Delta\mathbf{y},\mathbf{v}) - \mathbf{f}(\mathbf{y},\mathbf{v}) + [\mathbf{f}(\mathbf{y},\mathbf{v}) - \mathbf{f}(\mathbf{y},\mathbf{u})]\\ &= \nabla_{\mathbf{y}}\mathbf{f}(\mathbf{y},\mathbf{v})\cdot\Delta\mathbf{y} + [\mathbf{f}(\mathbf{y},\mathbf{v}) - \mathbf{f}(\mathbf{y},\mathbf{u})] + o(\|\Delta\mathbf{y}\|)\\ &= \nabla_{\mathbf{y}}\mathbf{f}(\mathbf{y},\mathbf{u})\cdot\Delta\mathbf{y} + [\mathbf{f}(\mathbf{y},\mathbf{v}) - \mathbf{f}(\mathbf{y},\mathbf{u})] + o(\|\Delta\mathbf{y}\|).\end{aligned}$$

Thus (3.112) becomes

$$\begin{aligned}(\Delta\mathbf{y}(t))' = \nabla_{\mathbf{y}}\mathbf{f}(\mathbf{y}(t),\mathbf{u}(t))\cdot\Delta\mathbf{y}(t) + [\mathbf{f}(\mathbf{y}(t),\mathbf{v}(t)) - \mathbf{f}(\mathbf{y}(t),\mathbf{u}(t))]\\ + o(\|\Delta\mathbf{y}(t)\|).\end{aligned}$$

The main linear part of $\Delta\mathbf{y}(t)$ is described by the following problem:

$$\begin{aligned}&(\delta\mathbf{y}(t))' = \nabla_{\mathbf{y}}\mathbf{f}(\mathbf{y}(t),\mathbf{u}(t))\cdot\delta\mathbf{y}(t) + [\mathbf{f}(\mathbf{y}(t),\mathbf{v}(t)) - \mathbf{f}(\mathbf{y}(t),\mathbf{u}(t))],\\ &\delta\mathbf{y}(0) = 0.\end{aligned}$$

Now we can use Property 3.12 and write out the form of the solution:

$$\delta\mathbf{y}(t) = \int_0^t \mathbf{\Phi}(t,s)\cdot[\mathbf{f}(\mathbf{y}(s),\mathbf{v}(s)) - \mathbf{f}(\mathbf{y}(s),\mathbf{u}(s))]\,ds.$$

So the main linear part of $\Delta\mathbf{y}(T)$ is

$$\delta\mathbf{y}(T) = \int_0^T \mathbf{\Phi}(T,s)\cdot[\mathbf{f}(\mathbf{y}(s),\mathbf{v}(s)) - \mathbf{f}(\mathbf{y}(s),\mathbf{u}(s))]\,ds.$$

Now we can find the main linear part of the increment of the objective functional $J(\mathbf{u})$:

$$\begin{aligned}\Delta J(\mathbf{u}) &= \nabla_{\mathbf{y}}G(\mathbf{y}(T))\cdot\Delta\mathbf{y}(T) + o(\|\Delta\mathbf{y}(T)\|)\\ &= \int_0^T \nabla_{\mathbf{y}}G(\mathbf{y}(T))\cdot\mathbf{\Phi}(T,s)\cdot[\mathbf{f}(\mathbf{y}(s),\mathbf{v}(s)) - \mathbf{f}(\mathbf{y}(s),\mathbf{u}(s))]\,ds + o(\varepsilon).\end{aligned}$$

Denote $\mathbf{\Psi}(s) = -\nabla_{\mathbf{y}}G(\mathbf{y}(T))\cdot\mathbf{\Phi}(T,s)$. Then the last relation takes the form

$$\Delta J(\mathbf{u}) = \int_0^T \mathbf{\Psi}(s)\cdot[\mathbf{f}(\mathbf{y}(s),\mathbf{u}(s)) - \mathbf{f}(\mathbf{y}(s),\mathbf{v}(s))]\,ds + o(\varepsilon)$$

as stated by the theorem. Since $\mathbf{\Psi}(s)$ is defined exactly as in § 3.8, we have completed the proof. □

3.12 A Discrete Version of Small Variations of Control Function for Generalized Terminal Control Problem

The formulas presented above for finding the change of the goal functional of a problem are used in practical calculations, but the problem itself should be discretized for this. Following the lecture of Dr. K.V. Isaev (Rostov State University) but in vector notation, let us consider one of the versions of possible discretization of the generalized terminal control problem. Let us recall the original problem. Given the governing equation

$$\mathbf{y}'(t) = \mathbf{f}(\mathbf{y}(t), \mathbf{u}(t)) \tag{3.113}$$

for $\mathbf{y} = \mathbf{y}(t)$ with the initial value $\mathbf{y}(0) = \mathbf{y}_0$, find an admissible control function $\mathbf{u} = \mathbf{u}(t)$ such that

$$\mathcal{I}(\mathbf{u}) \to \min_{\mathbf{u}\in U}$$

where

$$\mathcal{I}(\mathbf{u}) = G(\mathbf{y}(s_1), \mathbf{y}(s_2), \ldots, \mathbf{y}(s_N)). \tag{3.114}$$

We suppose that $\mathbf{u}(t)$ changes by a small variation $\delta\mathbf{u}(t)$ and would like to find the main part of the increment $\Delta\mathcal{I}(\mathbf{u}) = \mathcal{I}(\mathbf{u}+\delta\mathbf{u}) - \mathcal{I}(\mathbf{u})$ that is linear in $\delta\mathbf{u}$. We will not find the solution for this problem but will discretize the problem in whole and formulate the result for the latter.

Let us partition the interval $[0, s_N]$ by points $t_0 = 0 < t_1 < \ldots < t_R$, $t_R = s_N$, in such a way that the distance between two nearby points is small and the set $\{t_i\}$ contains all the points s_j from (3.114). On the segment $(t_{i-1}, t_i]$ we will approximate the control function $\mathbf{u}(t)$ by a constant value denoted $\mathbf{u}[i]$. Similarly, let us denote $\mathbf{y}[i] = \mathbf{y}(t_i)$. Considering $\mathbf{y}[i-1]$ as the initial value for equation (3.113) on $[t_{i-1}, t_i]$ with $\mathbf{u}(t) = \mathbf{u}[i]$, we can find the value $\mathbf{y}[i]$ that can be considered as a functional relation

$$\mathbf{y}[i] = \varphi_i(\mathbf{y}[i-1], \mathbf{u}[i]). \tag{3.115}$$

If all the $\mathbf{u}[i]$ are given, then starting with $\mathbf{y}[0] = \mathbf{y}_0$ we get, by (3.115), all the uniquely defined values $\mathbf{y}[i]$. In this way a discrete dynamical system is introduced. Note that it is not necessary to obtain (3.115) from (3.113); it can be formulated independently, and so the reasoning below is valid in a more general case that is not a consequence of the continuous dynamical system (3.113). The restriction for control function $\mathbf{u} \in U$ for discrete control functions is rewritten as $\mathbf{u} \in U^*$. Correspondingly the discrete generalized control problem can be reformulated as:

Problem. Given

$$\begin{aligned} \mathbf{y}[i] &= \varphi_i(\mathbf{y}[i-1], \mathbf{u}[i]), \quad \mathbf{y}[0] = \mathbf{y}_0, \\ \mathcal{I}(\mathbf{u}) &= G(\mathbf{y}[i_1], \mathbf{y}[i_2], \dots, \mathbf{y}[i_N]), \end{aligned} \tag{3.116}$$

find $\mathbf{u} \in U^*$ such that

$$\mathcal{I}(\mathbf{u}) \to \min_{\mathbf{u} \in U^*}.$$

The main part of the increment of $\mathcal{I}(\mathbf{u})$ that is linear in $\delta\mathbf{u}$ is given by the following

Theorem 3.18. *The main part of the increment of* $\mathcal{I}(\mathbf{u})$ *that is linear in* $\delta\mathbf{u} = \delta\mathbf{u}[i]$ *is*

$$\delta\mathcal{I}(\mathbf{u}) = \sum_{i=1}^{R} \nabla_{\mathbf{u}[i]}\mathcal{I}(\mathbf{u}) \cdot \delta\mathbf{u}[i] \tag{3.117}$$

where

$$\nabla_{\mathbf{u}[i]}\mathcal{I}(\mathbf{u}) = \left(\nabla_{\mathbf{u}[i]}\varphi_i(\mathbf{y}[i-1], \mathbf{u}[i])\right) \cdot \boldsymbol{\psi}[i] \tag{3.118}$$

and $\boldsymbol{\psi}[i]$ *satisfy the equations*

$$\begin{aligned} \boldsymbol{\psi}[i] &= \left(\nabla_{\mathbf{y}[i]}\varphi_{i+1}(\mathbf{y}[i], \mathbf{u}[i+1])\right) \cdot \boldsymbol{\psi}[i+1] \\ &\qquad + \nabla_{\mathbf{y}[i]}Q(\mathbf{y}[i_1], \mathbf{y}[i_2], \dots, \mathbf{y}[i_N]), \quad i = R-1, R-2, \dots, 1, \\ \boldsymbol{\psi}[R] &= \nabla_{\mathbf{y}[R]}Q(\mathbf{y}[i_1], \mathbf{y}[i_2], \dots, \mathbf{y}[i_N]). \end{aligned} \tag{3.119}$$

Proof. Before giving the proof we would like to point out the similarity between this and the result for the corresponding continuous control problem; in particular, there arises a system of equations for the complementary function $\boldsymbol{\psi}$ of the parameter i, whose solutions should be found in the reverse order, from $\boldsymbol{\psi}[R]$ to $\boldsymbol{\psi}[1]$. It is clear that it does not matter on which step and how we discretize the problem, the main features of solution should be the same. First let us mention that now $\mathcal{I}(\mathbf{u})$ is an ordinary function in many variables $\mathbf{u}[i]$ so all we need to find is the first differential of $\mathcal{I}(\mathbf{u})$ under constraints from (3.116). Thus the formula for the first differential gives us

$$\delta\mathcal{I}(\mathbf{u}) = \sum_{i=1}^{R} \nabla_{\mathbf{u}[i]}\mathcal{I}(\mathbf{u}) \cdot \delta\mathbf{u}[i]$$

which is (3.117). Next

$$\nabla_{\mathbf{u}[i]}\mathcal{I}(\mathbf{u}) = \nabla_{\mathbf{u}[i]}Q(\mathbf{y}[i_1], \mathbf{y}[i_2], \ldots, \mathbf{y}[i_N])$$
$$= \sum_{j=1}^{R} \nabla_{\mathbf{u}[i]}\mathbf{y}[j] \cdot \nabla_{\mathbf{y}[j]}Q(\mathbf{y}[i_1], \mathbf{y}[i_2], \ldots, \mathbf{y}[i_N]). \tag{3.120}$$

Here we used the chain rule for differentiation, formulated for the gradient. Let us find $\nabla_{\mathbf{u}[i]}\mathbf{y}[j]$. For this we introduce a new vector function $\mathbf{F}_{ji}$ induced by (3.115) that is defined for $j \geq i$:

$$\mathbf{y}[j] = \mathbf{F}_{ji}(\mathbf{y}[i]).$$

Let us formulate the properties of $\mathbf{F}_{ji}$. It is obvious that

$$\mathbf{F}_{ii}(\mathbf{y}[i]) = \mathbf{y}[i],$$
$$\mathbf{F}_{i+1\ i}(\mathbf{y}[i]) = \mathbf{y}[i+1] = \boldsymbol{\varphi}_{i+1}(\mathbf{y}[i], \mathbf{u}[i+1]).$$

Finally, it follows by the definition that

$$\mathbf{F}_{ji}(\mathbf{y}[i]) = \mathbf{F}_{j\ i+1}(\mathbf{y}[i+1]) = \mathbf{F}_{j\ i+1}\left(\varphi_{i+1}(\mathbf{y}[i], \mathbf{u}[i+1])\right). \tag{3.121}$$

It is evident that the components of $\mathbf{F}_{ji}$ depend only on the components $\mathbf{u}[i+1], \mathbf{u}[i+2], \ldots, \mathbf{u}[j]$ and do not depend on the rest of the components of $\mathbf{u}$. Let us return to finding $\nabla_{\mathbf{u}[i]}\mathbf{y}[j]$ using the chain rule again:

$$\nabla_{\mathbf{u}[i]}\mathbf{y}[j] = \nabla_{\mathbf{u}[i]}\mathbf{F}_{ji}(\mathbf{y}[i]) = \nabla_{\mathbf{u}[i]}\mathbf{y}[i] \cdot \nabla_{\mathbf{y}[i]}\mathbf{F}_{ji}(\mathbf{y}[i])$$
$$= \nabla_{\mathbf{u}[i]}\boldsymbol{\varphi}_i\left(\mathbf{y}[i-1], \mathbf{u}[i]\right) \cdot \nabla_{\mathbf{y}[i]}\mathbf{F}_{ji}(\mathbf{y}[i]).$$

Returning to (3.120) we get

$$\nabla_{\mathbf{u}[i]}\mathcal{I}(\mathbf{u}) = \sum_{j=i}^{R} \nabla_{\mathbf{u}[i]}\varphi_i\left(\mathbf{y}[i-1], \mathbf{u}[i]\right) \cdot \nabla_{\mathbf{y}[i]}\mathbf{F}_{ji}(\mathbf{y}[i])\cdot$$
$$\cdot \nabla_{\mathbf{y}[j]}Q(\mathbf{y}[i_1], \mathbf{y}[i_2], \ldots, \mathbf{y}[i_N]).$$

Denoting

$$\boldsymbol{\psi}[i] = \sum_{j=i}^{R} \nabla_{\mathbf{y}[i]}\mathbf{F}_{ji}(\mathbf{y}[i]) \cdot \nabla_{\mathbf{y}[j]}Q(\mathbf{y}[i_1], \mathbf{y}[i_2], \ldots, \mathbf{y}[i_N]) \tag{3.122}$$

we get

$$\nabla_{\mathbf{u}[i]}\mathcal{I}(\mathbf{u}) = \left(\nabla_{\mathbf{u}[i]}\varphi_i(\mathbf{y}[i-1], \mathbf{u}[i])\right) \cdot \boldsymbol{\psi}[i]$$

which is (3.118).

It remains to derive equations for $\psi[i]$. We begin with formula (3.121):

$$\mathbf{F}_{ji}(\mathbf{y}[i]) = \mathbf{F}_{j\ i+1}(\mathbf{y}[i+1]).$$

Applying the gradient by $\mathbf{y}[i]$ to both sides we get

$$\nabla_{\mathbf{y}[i]}\mathbf{F}_{ji}(\mathbf{y}[i]) = \nabla_{\mathbf{y}[i]}\varphi_{i+1}(\mathbf{y}[i], \mathbf{u}[i+1]) \cdot \nabla_{\mathbf{y}[i+1]}\mathbf{F}_{j\ i+1}\left(\mathbf{y}[i+1]\right).$$

Substituting this into (3.122) we get

$$\begin{aligned}
\psi[i] &= \sum_{j=i}^{R} \nabla_{\mathbf{y}[i]}\mathbf{F}_{ji}(\mathbf{y}[i]) \cdot \nabla_{\mathbf{y}[j]}Q(\mathbf{y}[i_1], \mathbf{y}[i_2], \ldots, \mathbf{y}[i_N]) \\
&= \sum_{j=i+1}^{R} \nabla_{\mathbf{y}[i]}\varphi_{i+1}(\mathbf{y}[i], \mathbf{u}[i+1]) \cdot \nabla_{\mathbf{y}[i+1]}\mathbf{F}_{j\ i+1}\left(\mathbf{y}[i+1]\right) \cdot \\
&\qquad \cdot \nabla_{\mathbf{y}[j]}Q(\mathbf{y}[i_1], \mathbf{y}[i_2], \ldots, \mathbf{y}[i_N]) \\
&\qquad + \nabla_{\mathbf{y}[i]}\mathbf{F}_{ii}(\mathbf{y}[i]) \cdot \nabla_{\mathbf{y}[i]}Q(\mathbf{y}[i_1], \mathbf{y}[i_2], \ldots, \mathbf{y}[i_N]) \\
&= \nabla_{\mathbf{y}[i]}\varphi_{i+1}(\mathbf{y}[i], \mathbf{u}[i+1]) \cdot \psi[i+1] + \nabla_{\mathbf{y}[i]}Q(\mathbf{y}[i_1], \mathbf{y}[i_2], \ldots, \mathbf{y}[i_N])
\end{aligned}$$

where we have used the fact that $\nabla_{\mathbf{y}[i]}\mathbf{F}_{ii}(\mathbf{y}[i]) = \mathbf{E}$. So we have obtained the first of (3.119). From the intermediate result of this equality chain the second of (3.119) follows. □

We now turn to another class of problems.

3.13 Optimal Time Control Problems

We recall that the problems of this type are as follows. The object is described by a dynamical system

$$\mathbf{y}'(t) = \mathbf{f}(\mathbf{y}(t), \mathbf{u}(t)) \tag{3.123}$$

for which we must find an admissible control function $\mathbf{u}(t)$ in such a way that the parameters of the system must be changed from the initial state

$$\mathbf{y}(0) = \mathbf{y}_0 \tag{3.124}$$

to the final state

$$\mathbf{y}(T) = \mathbf{y}_1 \tag{3.125}$$

in minimal time T. Unlike the terminal control problem, here the final state of the system is fixed but not the time interval.

Let us note that in this problem the set $\mathcal{U}$ of admissible control functions is limited not only by the external inequality restrictions, but also by the boundary conditions (3.124)–(3.125) because it may happen so that there are no admissible control vectors such that the system, starting with the initial state $\mathbf{y}_0$, can reach the final state $\mathbf{y}_1$ in finite time T.

Next we recall that for the terminal control problem we obtained a conjugate problem with an initial (i.e., "final") condition at T which was called the condition of transversality. The optimal time problem has both the boundary conditions for $\mathbf{y}$ of the same form as the condition at $t = 0$ of the terminal control problem. Thus we should expect that if Pontryagin's principle of maximum is valid in this or that form for the optimal control problem then any boundary conditions for $\mathbf{\Psi}(s)$ are absent. This means that the uniqueness for finding $\mathbf{\Psi}(s)$ needed for this problem is not provided by some explicit equations. The explicit formula for the increment of the objective functional for the optimal control problem is not obtained. So we formulate without proof the statement of Pontryagin's principle of maximum for the optimal control problem.

Theorem 3.19. *Let $\mathbf{u}(t)$ be a control function at which T, the length of the time interval, attains its minimal value among all the admissible control functions, for which* (3.123)–(3.125) *has a solution $\mathbf{y}(t)$. There is a nontrivial vector function $\mathbf{\Psi}(s)$ that is a solution of the conjugate equation*

$$\frac{d}{ds}\mathbf{\Psi}(s) = -\mathbf{\Psi}(s) \cdot \nabla_{\mathbf{y}}\mathbf{f}(\mathbf{y}(s), \mathbf{u}(s))$$

such that the Pontryagin function $H(\mathbf{y}, \mathbf{\Psi}, \mathbf{u}) = \mathbf{f}(\mathbf{y}, \mathbf{u}) \cdot \mathbf{\Psi}$, with respect to the third argument, takes its maximal value for all points of continuity of $\mathbf{u}(t)$:

$$H(\mathbf{y}(t), \mathbf{\Psi}(t), \mathbf{u}(t)) \geq H(\mathbf{y}(t), \mathbf{\Psi}(t), \mathbf{v}).$$

Let us note that in simple cases when $\mathbf{u}(t)$ comes into the equations linearly this theorem reduces the set of possible control functions to those which take values at boundaries of $\mathcal{U}$ at each time t. Indeed, then $\mathbf{u}(t)$ comes linearly into the presentation of $H(\mathbf{y}, \mathbf{\Psi}, \mathbf{u}) = \mathbf{\Psi} \cdot \mathbf{f}(\mathbf{y}, \mathbf{u})$ and thus its maximal value can be taken only at some extreme points of $\mathbf{u}(t)$.

Example 3.20. Consider the simplest optimal time problem. Let a material point of unit mass move along a straight line under the action of a force whose magnitude F cannot exceed unity. How should we vary F so that the point moves from one position to another in the shortest time?

Solution. If the velocity of the point at its initial and final states is zero then the solution is clear mechanically: first we need to accelerate the point with maximal force until it comes to the middle point between the initial and final state, and then to switch the force to the opposite direction leaving the maximal magnitude so the point is maximally decelerated. When the appointed initial and final velocities are not zero one must have good mechanical intuition to tell what the law for the force should be. Let us solve this problem using Theorem 3.18. The governing equation is

$$x''(t) = F(t), \quad x(0) = a_0, \quad x'(0) = a_1, \quad x(T) = b_0, \quad x'(T) = b_1, \tag{3.126}$$

and the restriction for $F(t)$ is

$$|F(t)| \leq 1. \tag{3.127}$$

Let us rewrite this using the notation we used above:

$$y_1(t) = x(t), \qquad y_2(t) = x_1'(t), \qquad u(t) = F(t).$$

Thus we introduce the phase coordinates of the point. Then equations (3.126)–(3.127) take the form

$$\begin{aligned} y_1'(t) &= y_2(t), \\ y_2'(t) &= u(t), \end{aligned}$$

the boundary conditions

$$\begin{aligned} y_1(0) &= a_0, \\ y_2(0) &= a_1, \end{aligned} \qquad \text{and} \qquad \begin{aligned} y_1(T) &= b_0, \\ y_2(T) &= b_1, \end{aligned}$$

and the restriction that defines the set $\mathcal{U}$ of piecewise continuous functions

$$-1 \leq u(t) \leq 1.$$

Let us first introduce the Pontryagin function $H = y_2\psi_1 + u\psi_2$. Let $\mathbf{y}(t)$ and $\mathbf{\Psi}(t)$ be the needed solutions of the main and conjugate systems of equations. The conjugate equations are

$$\begin{aligned} \psi_1' &= -\partial H/\partial y_1 = 0, \\ \psi_2' &= -\partial H/\partial y_2 = -\psi_1. \end{aligned}$$

The solution of this system results in $\psi_2 = d_1 t + d_2$ and thus may have no more than one point $t_0 \in [0, T]$ at which it changes sign. By Pontryagin's principle, it is the only point at which the control function u must switch sign as H can take its maximum when $\psi_2(t)u(t)$ takes its maximum. Thus t_0 splits $[0, T]$ into two parts having $u = \pm 1$. Thus the solution to our simplest optimal time control problem should be synthesized from trajectories of the two systems

$$\begin{aligned} y_1'(t) &= y_2(t), \\ y_2'(t) &= 1, \end{aligned} \qquad \text{and} \qquad \begin{aligned} y_1'(t) &= y_2(t), \\ y_2'(t) &= -1. \end{aligned}$$

The particle trajectories on the phase plane (y_1, y_2) are parabolas. For the first system $y_1 = t^2/2 + c_1 t + c_2$ and for the second $y_1 = -t^2/2 + c_3 t + c_4$. Geometrically it is evident that there are no more than two parabolas, one from each family, through the end points which intersect. That is the solution trajectory of the problem. Analytically we must compose five equations for unknown c_i and t_0. The first is that at t_0 the curves intersect, that is

$$t_0^2/2 + c_1 t_0 + c_2 = -t_0^2/2 + c_3 t_0 + c_4.$$

The other four equations (boundary conditions) depend on which of switched values of u goes first. If $u = 1$ on $[0, t_0]$ and thus $u = -1$ on the rest,

$$c_2 = a_0, \quad c_1 = a_1, \quad -T^2/2 + c_3 T + c_4 = b_0, \quad -T + c_3 = b_1.$$

If $u = -1$ on $[0, t_0]$ then

$$c_4 = a_0, \quad c_3 = a_1, \quad T^2/2 + c_1 T + c_2 = b_0, \quad T + c_1 = b_1.$$

Only one of these systems has a solution where real t_0 lies in $[0, T]$ and it is what we have sought. □

We would like to note that when the controlled object's equations are simple, the maximum principle of Pontryagin gives a good tool to find an optimal solution. For many industrial problems it is necessary to use other methods. In the same manner as Example 3.20, any optimal time problem for a system described by the equation $x'' + ax' + bx = u$ can be solved analytically. Textbooks are full of such problems from various areas of science, their analytical solutions as well as geometrical interpretation of some of their solutions.

Our next remark is the following. The terminal control problems and the optimal time problems are in a certain sense, the extremes of all control problems with respect to boundary conditions. For "intermediate" problems, with other types of boundary conditions at starting and ending moments, the conjugate system is supplemented with some conditions of transversality. The situation is similar to that for natural conditions in the calculus of variations.

3.14 Final Remarks on Control Problems

In this chapter we considered in large part the methods for finding optimal solutions. Of course it was an introductory chapter, and we limited ourselves to a small portion of the theory — that portion which is used in many industrial control processes and other applications. We did not touch on the problem of existence of solutions of control problems, which is extremely important since there are many practical problems that are formulated quite nicely from a common sense standpoint but that lack solutions.

We mention only another important part of control theory that is called dynamical programming. It was developed by R. Bellman and used quite successfully in many problems of optimal control. To give the reader some idea of what this theory is about and to lend vividness to the presentation we consider a very simple problem (in a form that might hold the attention of many undergraduate students):

Example 3.21. A racketeer has been drunk for three weeks and has failed to perform his job properly. One morning he receives a phone call from his boss, reminding him of a \$32,000 debt he owes the boss in one hour. Along with this reminder comes a suitable threat about one lost tooth for each \$1000 he fails to bring in. The racketeer lives quite far from his boss, and wishes to collect as much additional money as possible on the way. He has a street map showing how much money he can collect on each possible route. He is constrained to move ahead only, and cannot turn back.

Solution. We draw the map as a graph (Fig. 3.4) that should begin at point O and end at B. To get a more convenient presentation at the final point B we split all routes to B and draw them along the final line B_0-B_1 as shown on the picture. On the lines connecting the nodes we put the amounts of money that the racketeer expects to be able to collect from the peaceful citizenry. □

Let us discuss this problem. Of course, for this small map the racketeer could test all the possibilities and find the optimal way quite quickly. There are six levels at each of the way can branch so there are few possibilities. Let us imagine that this map has 1000 such levels; then the number of possible ways grows to 2^{1000} and simple experimentation would not bring a quick result. So it becomes necessary to propose a procedure for which the number of operations could be sufficiently small, say several million.

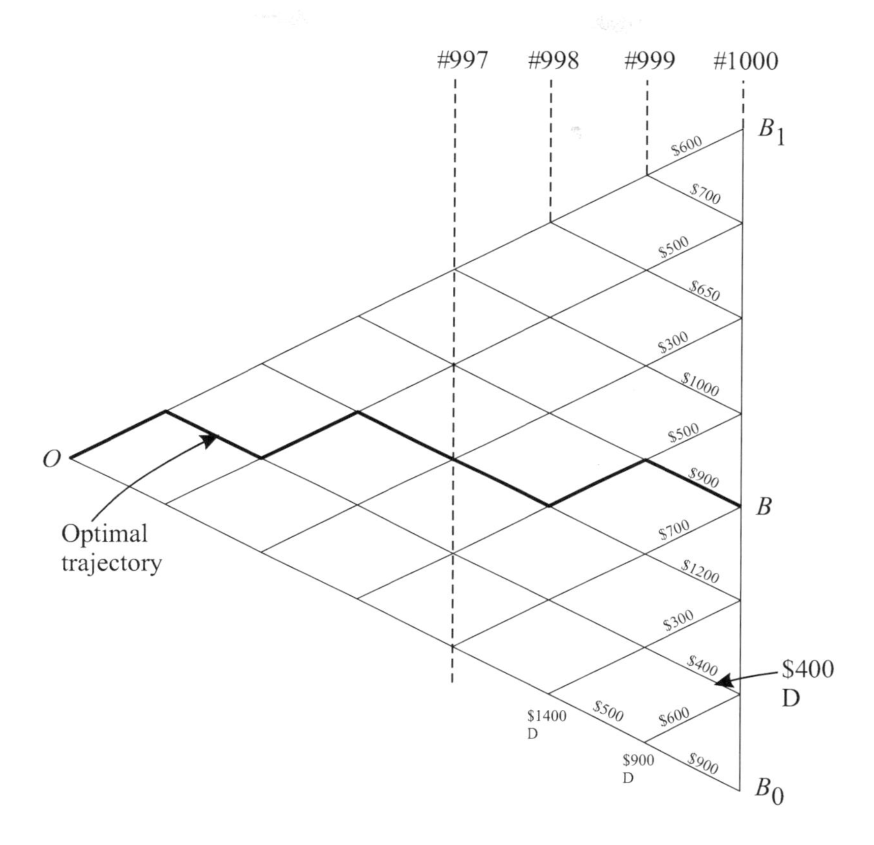

Fig. 3.4 A racketeer's possible routes; optimal trajectory shown as the thick line.

Any cross-section of the map would not bring the needed optimal result since the optimal trajectory can be quite strange. The crucial step to the solution is to choose the first step as follows. Suppose that we are at the 999th level of nodes. From each node of this level we exactly know where to move since it is a choice between two possibilities. Near each node of this level we write down where we should move (Down or Up) and the amount. On the 998th level we again should fulfill few operations at each node: moving along the upper street we then add the figure of this street with the price of corresponding 999th node after which we should decide between the two possibilities and to write near the node Up or Down (showing where to go next) and the optimal cost. On the 997th level everything will be repeated: the finding of two sums of two numbers, the choosing of the bigger one, and the placement of the necessary information near the node.

This is must be done at each level. In this way we come to the initial point, getting the optimal sum of money as the resulting figure at it, and the optimal trajectory moving along signs Up and Down.

At first glance this seems to be a nice problem for a high school math competition, since it is solved using only "common sense". However, its solution is based on a hard mathematical idea: when we come to some point of the optimal trajectory, the remainder of the optimal trajectory is optimal for the "reduced" problem whose initial point is this one at which we just stopped.

We shall not discuss the many fruitful applications of this principle of Bellman. As the central principle of dynamical programming it has brought many results, both theoretical and practical, in discrete and continuous problems.

We leave it to the reader to explore other books, and thereby to discover other ways to view problems in optimal control and the calculus of variations. These are indeed part of the more general branch of mathematics known as Mathematical Programming.

3.15 Exercises

3.1 Show that the coefficients of the squared gradient

$$\nabla^2_{\mathbf{y}} = \nabla_{\mathbf{y}} \left(\nabla_{\mathbf{y}}\right)$$

applied to a scalar valued function $f(\mathbf{y}(t))$ constitute the Hessian matrix of f.

3.2 Establish the formula (3.71).

3.3 Formulate the form of the main linear part of the increment of $J(\mathbf{u})$ under the sum of the increments of the control function by the needle-shaped vector function and a small increment as discussed in § 3.3.

3.4 (A harder problem.) Let the objective functional for the terminal control problem be changed to

$$J^*(\mathbf{u}) = \int_0^T G(\mathbf{y}(t))\, dt.$$

What is the form of the main part of its increment in this case?

3.5 A mechanical oscillator (a mass on a spring) oscillates under force $|F(t)|$ such that $|F(t)| \leq 1$. The governing equation is $mx'' + kx = F$, $m = 1$, $k = 1$. Find the law of the change of the force when the mass goes from state $x(0) = a$, $x'(0) = b$ to the state of equilibrium, $x(T) = 0$, $x'(T) = 0$ in the shortest time T.

Chapter 4

Functional Analysis

Over the long history of engineering, numerous analytical and theoretical tools were developed for the approximate solution of complex mathematical problems. These techniques are still meaningful and tend to dominate engineering textbooks. However, modern engineering incorporates models so complex that their theoretical solution is practically impossible. A reliance on computers has pushed aside analytic methods even when the latter can be applied. Approaches such as the finite element method enable solutions of extremely complex problems that are impossible to solve analytically and, moreover, produce numerical results that may be presented in an attractive graphical manner. An engineer may have the impression that these programs can solve nearly any problem — at least in his or her range of interests. But this is not the case. Computers are finite automata; they act with finite sets of numbers. Computers reduce differential equations and other continuous models to equations in finite dimensional spaces.

Engineers working with computer programs may become accustomed to the idea that finite dimensional results are good approximations to experimental results. In mathematical terms, they begin to view the ideology of linear algebra in $\mathbb{R}^n$ as infallible. This viewpoint is not entirely invalid. We may take, for example, the equations of the finite element method, obtained from boundary value problems for differential equations. If deduced while taking proper account of the principal physical laws, these equations are indeed finite dimensional models of certain objects or processes. Hence they can sometimes approximate real objects no worse than differential (infinite dimensional) models that are also merely approximate. But the phrase "deduced while taking proper account of the principal physical laws" is essential, and finite dimensional models may lack certain important properties. Furthermore, restrictions on calculation time will necessitate truncation errors, the effects of which can be gauged only through comparison with experiment (which is often impractical) or with results from infinite dimensional models.

We can expect manual solution methods to continue their slow decline in engineering practice. Within a few generations, students may be unable to integrate simple functions or solve ordinary differential equations (consider what happened with the logarithmic slide rule). But a grasp of the basic properties of finite dimensional and infinite dimensional operators and their relations will be needed for understanding what computers do with numerical models and what can be expected from them.

As finite dimensional and infinite dimensional problems appear to be related, it is attractive to use the methods and ideology elaborated in finite dimensions for the infinite dimensional case. Many aspects of standard matrix analysis remain valid under the transition to infinite dimensional problems. But this is not uniformly the case. For example, closed and bounded sets are guaranteed to be compact only in finite dimensional spaces. Another example concerns infinite dimensional vectors, often written in the harmless looking form $(x_1, x_2, \ldots)$, which do not conform to all the rules that apply to ordinary vectors. Many such examples could be given.

Although these sorts of issues are studied in the portion of mathematics that still lies outside the typical engineering curriculum, the situation is bound to change. Engineers must understand the background of the tools they employ.

A principal tool in the modern analysis of partial differential equations, *functional analysis* allows us to shift our perspective on functions from the viewpoint of ordinary calculus to a viewpoint in which we deal with a function (such as a differential or integral operator) as a whole entity. We accomplish this conceptual shift by extending the notion of an ordinary three-dimensional vector so that a function can be viewed as an element of a linear vector space. Because this extension involves some subtle points regarding the dimension of a vector space, the present chapter is devoted to a suitable introduction.

As a branch of mathematics, functional analysis is in large part delineated by the tools it offers to the practitioner. Important applications arise in a variety of areas: differential and integral equations, the theory of integration, probability theory, etc. It has been said that functional analysis is not a special branch of mathematics at all, but rather a united point of view on mathematical objects of differing natures. A full presentation of functional analysis would require many volumes. The goal of the present chapter is to offer the reader a relatively brief but still self-contained treatment, and therefore to provide all the tools necessary for the study of boundary value problems.

It is worth noting that the pioneers of functional analysis were not all pure mathematicians. Stefan Banach received a polytechnical diploma and for many years taught courses in theoretical mechanics. He also published an interesting textbook on mechanics. John von Neumann, who pioneered the application of computers to engineering practice, wrote a fundamental textbook on functional analysis that had important influences on quantum mechanics and other areas of physics. Functional analysis spawned many important applications of mathematics to physics and engineering. While some of its subtopics did arise in pure mathematics, they are often powerfully applicable. The approaches taken in books on functional analysis depend strongly on the interests of the authors. Many are deeply theoretical. In this short chapter, we consider a portion of functional analysis used to study mathematical problems in mechanics.

Before we begin, recall two standard theorems from ordinary calculus.

Theorem 4.1. *Suppose a sequence* $\{f_n(\mathbf{x})\}$ *of functions continuous on a compact set* $\Omega \subset \mathbb{R}^k$ *converges uniformly; that is, for any* $\varepsilon > 0$ *there is an integer* $N = N(\varepsilon)$ *such that* $|f_n(\mathbf{x}) - f_m(\mathbf{x})| < \varepsilon$ *whenever* $n, m > N$ *and* $\mathbf{x} \in \Omega$. *Then the limit function*

$$f(\mathbf{x}) = \lim_{n\to\infty} f_n(\mathbf{x})$$

is continuous on Ω.

This is called Weierstrass' theorem. The next one shows the properties of a continuous function on a compact set.

Theorem 4.2. *Suppose* $f(\mathbf{x})$ *is continuous on a compact set* $\Omega \subset \mathbb{R}^k$. *Then* $f(\mathbf{x})$ *is uniformly continuous on* Ω; *that is, for any* $\varepsilon > 0$ *there is a* $\delta > 0$ *(dependent only on* ε*) such that* $|f(\mathbf{x}) - f(\mathbf{y})| < \varepsilon$ *whenever* $\|\mathbf{x} - \mathbf{y}\| < \delta$ *and* $\mathbf{x}, \mathbf{y} \in \Omega$.

4.1 A Normed Space as a Metric Space

Regarding a function as a single object (a viewpoint which functional analysis inherited from the calculus of variations), we must provide a way to quantify the difference between two functions. The simplest and most convenient way to do this is to use the tools of normed spaces. First of all a normed space, consisting of elements of any nature (of functions in particular), must be a *linear space.* This means we can add or subtract any two elements of the space, or multiply an element of the space by a number, and the result will always be an element of the same space. If complex numbers are used as multipliers then the linear space is called a *complex linear space*; if purely real numbers are used then the space is a *real linear space.* The definition of a linear space can be stated rigorously in terms of axioms and the reader has undoubtedly seen these in a linear algebra course. The main distinction between a general linear space and a normed space is the existence of a norm on the latter. A *norm* is a real-valued function $\|x\|$ that is determined (which means it carries a unit and takes a finite value) at each element x of the space and satisfies the following axioms:

(1) $\|x\| \geq 0$ for all x; $\|x\| = 0$ if and only if $x = 0$;
(2) $\|\lambda x\| = |\lambda| \, \|x\|$ for any x and any real number λ;
(3) $\|x + y\| \leq \|x\| + \|y\|$ for all x, y.

The first of these is called the axiom of positiveness, the second is the axiom of homogeneity, and the third is the triangle inequality.

Definition 4.3. A *normed linear space* is a linear space X on which a norm $\|\cdot\|$ is defined.

More specifically, $\|\cdot\|$ is "defined" on X if the number $\|x\|$ exists and is finite for every element $x \in X$.

In classical functional analysis one deals with dimensionless quantities. In applications this restriction is not necessary: one can use numbers with dimensional units and get norms having dimensional units. Although this causes no theoretical complications and is sometimes useful, we follow the classical procedure and consider all elements to be dimensionless.

Example 4.4. Show that if $\|x\|$ is any norm on X and $x, y \in X$, then

$$\big| \, \|x\| - \|y\| \, \big| \leq \|x - y\| \, . \tag{4.1}$$

We shall find this inequality useful later. In particular, by the definition of continuity it means that the norm is continuous with respect to the norm itself.

Solution. Let us begin by replacing x with $x - y$ in norm axiom 3:

$$\|x\| - \|y\| \leq \|x - y\| \, .$$

Interchanging the roles of x and y in this inequality, we get

$$\|y\| - \|x\| \leq \|y - x\| .$$

But the right sides of these two inequalities are the same; indeed, $\|y - x\| = \|(-1)(x - y)\| = \|x - y\|$ by norm axiom 2. So the quantity $\|x - y\|$ is greater than or equal to both $\|x\| - \|y\|$ and $\|y\| - \|x\|$. This means it is greater than or equal to $|\, \|x\| - \|y\| \,|$. □

We have introduced the normed space $C^{(k)}(\Omega)$ of functions that are k times continuously differentiable on a compact set Ω with the norm

$$\|f(\mathbf{x})\|_{C^{(k)}(\Omega)} = \max_{\mathbf{x}\in\Omega} |f(\mathbf{x})| + \sum_{|\alpha|\leq k} \max_{\mathbf{x}\in\Omega} |D^\alpha f(\mathbf{x})|, \tag{4.2}$$

where

$$D^\alpha f = \frac{\partial^{|\alpha|} f}{\partial x_1^{\alpha_1} \cdots \partial x_n^{\alpha_n}}, \qquad |\alpha| = \alpha_1 + \cdots + \alpha_n. \tag{4.3}$$

As with any other proposed norm, the reader should verify satisfaction of the axioms.[1] A particular case is the space of all functions continuous on Ω with the norm

$$\|f(\mathbf{x})\|_{C(\Omega)} = \max_{\mathbf{x}\in\Omega} |f(\mathbf{x})|. \tag{4.4}$$

In the space of functions continuous on a compact Ω we can impose another norm:

$$\|f(\mathbf{x})\| = \left(\int_\Omega |f(\mathbf{x})|^p \, d\Omega \right)^{1/p} \qquad (p \geq 1). \tag{4.5}$$

The norm axioms can be verified here also (the triangle inequality being known as Minkowski's inequality).[2] Hence on the same set (linear space) of elements we can impose one of several norms. On the same compact Ω we can consider the set of all bounded functions and introduce the norm

$$\|f(\mathbf{x})\| = \sup_{\mathbf{x}\in\Omega} |f(\mathbf{x})| . \tag{4.6}$$

The resulting space will be called $M(\Omega)$. The space $C(\Omega)$ is a subspace of $M(\Omega)$ (note that for a continuous function the norm (4.6) reduces to (4.4)). Note that a normed space is defined by the set of elements and the form of

[1] For example one could take the set of functions continuous on $[0, 1]$ and try to impose a "norm" using the formula $\|f(x)\| = |f(0.5)|$. Which norm axiom would fail?

[2] We assume Ω is Jordan measurable. This is a safe assumption for our purposes, because we consider only domains occupied by physical bodies having comparatively simple shape.

the norm imposed on it. So to refer properly to a space, we must display a pair $(X, \|\cdot\|)$ consisting of the set of elements X and the norm. For the most frequently used spaces it is common to use shorthand notation such as $C(\Omega)$ where the norm is understood. This is especially appropriate when there is a unique norm imposed on a set, and we shall adopt the practice. When it is necessary to distinguish different norms, we indicate the space by a subscript on the norm symbol as in (4.2) and (4.4).

The functional

$$d(x, y) = \|x - y\| \,, \tag{4.7}$$

defined for each pair of elements of a normed space, satisfies the axioms of a *metric*:

(1) $d(x, y) \geq 0$ for all x, y, and $d(x, y) = 0$ if and only if $x = y$;
(2) $d(x, y) = d(y, x)$ for all x, y;
(3) $d(x, y) \leq d(x, z) + d(z, y)$ for all x, y, z.

If such a functional (metric) d is defined for any pair of elements of a set X, then we have a *metric space*.

Definition 4.5. A *metric space* is a set X on which a metric $d(x, y)$ is defined.

Hence every normed space is a metric space (the metric (4.7) is called the *natural metric* and is said to be *induced* by the norm). The notion of metric space is more general than that of normed space. Not all metric spaces can be normed: first of all a metric space need not be a linear space (a fact which is sometimes important, as in applications of the contraction mapping principle). Note that the use of elements with dimensional units would give a metric having dimensions as well; although the metric is a generalization of the notion of distance, this distance can be expressed in units of force, power, etc.

The axioms of a metric replicate the essential properties of distance from ordinary geometry: (1) distance is nonnegative, the distance from a point to itself is zero, and the distance between two distinct points is nonzero; (2) the distance between two points does not depend on the order in which the points are considered; and (3) the triangle inequality holds, meaning that for a triangle the length of any side does not exceed the sum of the lengths of the other two sides. In this way, the more general notion of metric preserves many terms and concepts from ordinary geometry.

Definition 4.6. An *open ball* with center x_0 and radius R is the set of points $x \in X$ such that $d(x_0, x) < R$. The corresponding *closed ball* is the set of all $x \in X$ such that $d(x_0, x) \leq R$, and the corresponding *sphere* of radius R is the set of all $x \in X$ such that $d(x_0, x) = R$.

Note that the term "ball" can denote various objects depending on the metric chosen: if we impose the metric

$$d(x, y) = \max_{1 \leq i \leq 3} |x_i - y_i|$$

in ordinary three-dimensional space where $x = (x_1, x_2, x_3)$ and $y = (y_1, y_2, y_3)$, then a ball is really shaped like a cube. The other abstract space structures also provide notions that correspond to those of ordinary geometry. In a linear space of vectors we can determine a straight line through the points x_1 and x_2 by

$$tx_1 + (1 - t)x_2, \qquad t \in (-\infty, \infty),$$

and can obtain the segment having x_1 and x_2 as endpoints by restricting t to the interval $[0, 1]$. It is especially important that we can use the notion of metric to introduce the tools of calculus in such a way that functions can be dealt with as whole objects. (Metric spaces are not linear in general, so they include spaces that cannot be normed. However, even some linear metric spaces cannot be normed.)

Armed with a notion of distance in a normed space, we can introduce any of the notions from calculus that are connected with the notion of distance. The first is convergence.

Definition 4.7. A sequence $\{x_n\}$ is *convergent* to an element x if to each positive number ε there corresponds a number $N = N(\varepsilon)$ such that $d(x_k, x) < \varepsilon$ whenever $k > N$.

The reader can easily phrase this definition in terms of the norm, using (4.7). As in calculus, we call x the *limit* of $\{x_k\}$ and write

$$\lim_{k \to \infty} x_k = x$$

or $x_k \to x$ as $k \to \infty$.

Example 4.8. (a) Show that every convergent sequence in a metric space has a unique limit. (b) Show that if $x_n \to x$ and $y_n \to y$, then $d(x_n, y_n) \to d(x, y)$ as $n \to \infty$.

Solution. (a) We suppose that $x_n \to x$ and $x_n \to x'$, then show that $x' = x$ follows. Let ε be an arbitrarily small positive number. By assumption we can choose N so large that the inequalities $d(x_N, x) < \varepsilon/2$ and $d(x_N, x') < \varepsilon/2$ both hold. Hence, by the triangle inequality,

$$d(x, x') \leq d(x, x_N) + d(x_N, x') < \varepsilon.$$

Since the distance $d(x, x')$ is both nonnegative and smaller than any preassigned positive number, it must equal zero. According to metric axiom 1, we conclude that $x = x'$. (b) The generalized triangle inequality

$$d(x_1, x_n) \leq d(x_1, x_2) + d(x_2, x_3) + \cdots + d(x_{n-1}, x_n)$$

is easily established through the use of mathematical induction. We can use this fact as follows. We write

$$d(x, y) \leq d(x_n, x) + d(x_n, y_n) + d(y_n, y)$$

and

$$d(x_n, y_n) \leq d(x_n, x) + d(x, y) + d(y_n, y),$$

and then combine these two inequalities into the form

$$|d(x_n, y_n) - d(x, y)| \leq d(x_n, x) + d(y_n, y).$$

Now for any $\varepsilon > 0$ there exists N so large that $n > N$ implies both $d(x_n, x) < \varepsilon/2$ and $d(y_n, y) < \varepsilon/2$. This means that $|d(x_n, y_n) - d(x, y)| < \varepsilon$, as desired. □

Clearly, a sequence of functions continuous on $[0, 1]$ and convergent in the norm (4.4) is also convergent in the norm

$$\|f(x)\| = 2 \max_{x \in [0,1]} |f(x)| .$$

However there are other norms, of $L^p(0, 1)$ say, under which the meaning of convergence is different. If two norms $\|\cdot\|_1$ and $\|\cdot\|_2$ satisfy the inequalities

$$m \|x\|_1 \leq \|x\|_2 \leq M \|x\|_1 \tag{4.8}$$

for some positive constants m and M that do not depend on x, then the two resulting notions of convergence on the set of elements are the same.

Definition 4.9. Two norms $\|\cdot\|_1$ and $\|\cdot\|_2$ that satisfy (4.8) for all $x \in X$ are *equivalent* on X.

We shall not distinguish between normed spaces consisting of the same elements and having equivalent norms.

4.2 Dimension of a Linear Space and Separability

The *dimension* of a linear space is the maximal number of linearly independent elements of the space. Recall that the elements x_k, $k = 1, 2, \ldots, n$, are linearly independent if the equation

$$c_1 x_1 + c_2 x_2 + \cdots + c_n x_n = 0$$

with respect to the unknowns c_k implies that $c_k = 0$ for all $k = 1, 2, \ldots, n$. We shall deal for the most part with infinite dimensional spaces. An important example is the space $C(0, 1)$ of functions $f(x)$ continuous on $[0, 1]$. Indeed, any set of monomials $f_k(x) = x^k$ is linearly independent in this space, since for any integer n the equation

$$c_1 x + c_2 x^2 + \cdots + c_n x^n = 0$$

cannot hold for any x unless $c_k = 0$ for all $k = 1, 2, \ldots, n$. Therefore the dimension of $C(0, 1)$ cannot be finite.

Let us discuss the problem of the number of elements in an abstract set. We say that two sets have *equal power* if we can place their elements in one-to-one correspondence. The simplest known infinite sets are those whose elements can be placed in one-to-one correspondence with the set of natural numbers. Such sets are said to be *countable.* An example is the set of all integers. It is clear that a finite union of countable sets is countable, since we can successively count first the elements standing at the first position of each of the sets, then the elements at standing at the second position, etc. There is a sharper result:

Theorem 4.10. *A countable union of countable sets is countable.*

Proof. Let X_n be the nth countable set and denote its kth element by x_{nk}, $k = 1, 2, \ldots$. The union of the X_n is the set of all elements x_{nk}. We need only to show how to recount them; this can be done as follows. The first element is x_{11}. The second and third elements are x_{12} and x_{21}, i.e., the elements whose indices sum to 3. The next three elements are the elements whose indices sum to 4: x_{13}, x_{22} x_{31}. We proceed to the elements whose indices sum to 5, 6, etc. In this way we can associate any element of the union with an integer. □

It follows that the set $\mathbb{Q}$ of all rational numbers is countable. Recall that a rational number can be represented as i/j where i and j are integers; denoting $x_{ij} = i/j$, we obtain the proof. Thus a countable set can have a great many elements. However, it can be shown that

Theorem 4.11. *The points of the interval* $[0, 1]$ *are not countable.*

The reader is referred to books on real analysis for a proof. We say that the points of $[0, 1]$ form a *continuum*. One might wonder whether there exist any sets intermediate in power between the countable sets and continuum sets. It turns out that the existence or non-existence of such a set is an independent axiom of arithmetic, a fact which points to the interesting (and sometimes mysterious) nature of the real numbers.

Example 4.12. Show that the set P_r of all polynomials with rational coefficients is countable.

Solution. For each fixed nonnegative integer n, denote by P_r^n the set of all polynomials of degree n having rational coefficients. The set P_r^n can be put into one-to-one correspondence with the countable set

$$\underbrace{\mathbb{Q} \times \mathbb{Q} \times \cdots \times \mathbb{Q}}_{n+1 \text{ times}}$$

where $\mathbb{Q}$ is the set of all rational numbers. Finally, the set P_r is given by

$$P_r = \bigcup_{n=0}^{\infty} P_r^n,$$

and this is a countable union of countable sets. □

Another example of a countable set is the collection of all finite trigonometric polynomials of the form

$$a_0 + \sum_{k=1}^{n} (a_k \cos kx + b_k \sin kx)$$

with rational coefficients a_0, a_k, b_k.

Let us discuss the real numbers further, keeping in mind that many of our remarks also apply to the complex numbers. Any real number can be obtained as a limit point of some sequence of rational numbers. This fundamental fact is, of course, the reason why a computer can approximate a real number by a rational number. The ability to approximate the elements of a given set by elements from a certain subset is important in general.

Definition 4.13. Let S be a set in a metric space X. A set $Y \subset S$ is *dense in* S if for each point $s \in S$ and $\varepsilon > 0$, there is a point $y \in Y$ such that $d(s, y) < \varepsilon$.

Alternatively, Y is dense in S if for any $s \in S$ there is a sequence $\{y_n\} \subset Y$ that converges to s. The set of rational numbers is dense in the set of real numbers.

Example 4.14. Let A, B, C be sets in a metric space. Show that if A is dense in B, and B is dense in C, then A is dense in C.

Solution. Suppose A is dense in B and B is dense in C. Let c be a given point of the set C, and let $\varepsilon > 0$ be given. There is a point $b \in B$ such that $d(c, b) < \varepsilon/2$. Similarly, there is a point $a \in A$ such that $d(b, a) < \varepsilon/2$. Since

$$d(c, a) \leq d(c, b) + d(b, a) < \varepsilon/2 + \varepsilon/2 = \varepsilon,$$

there is a point $a \in A$ that lies within distance ε of $c \in C$. □

Definition 4.15. If a metric space X contains a countable subset that is dense in X, then X is *separable.*

Example 4.16. Demonstrate that the set of all complex numbers with the natural metric (induced by the absolute value of a number) is a separable metric space.

Solution. Consider the subset of complex numbers having rational real and imaginary parts. This set is clearly countable (it can be placed into one-to-one correspondence with the countable set $\mathbb{Q} \times \mathbb{Q}$). We must still show that it is dense in $\mathbb{C}$. Let $z = u + iv$ be a given point of $\mathbb{C}$, $i = \sqrt{-1}$, and let $\varepsilon > 0$ be given. Since u and v are real numbers, and the rationals are dense in the reals, there are rational numbers $\bar{u}$ and $\bar{v}$ such that

$$|u - \bar{u}| < \varepsilon/\sqrt{2}, \qquad |v - \bar{v}| < \varepsilon/\sqrt{2}.$$

The number $\bar{z} = \bar{u} + i\bar{v}$ is a complex number with rational real and imaginary parts. Noting that

$$d(z, \bar{z}) = \sqrt{(u - \bar{u})^2 + (v - \bar{v})^2} < \sqrt{(\varepsilon/\sqrt{2})^2 + (\varepsilon/\sqrt{2})^2} = \varepsilon,$$

we are finished. □

Theorem 4.17. *Every finite dimensional normed space is separable.*

Proof. Every finite dimensional linear space has a finite basis, and the set of all finite linear combinations of the basis elements with rational coefficients is countable and dense in the space. □

The following result is important in practice.

Theorem 4.18. *Every subspace of a separable space is separable.*

Proof. Let E be a subspace of a separable space X. Consider a countable set consisting of $(x_1, x_2, \ldots)$ which is dense in X. Let B_{ki} be a ball of radius $1/k$ about x_i. By Theorem 4.10, the set of all B_{ki} is countable.

For any fixed k the union $\cup_i B_{ki}$ covers X and thus E. For every B_{ki}, take an element of E which lies in B_{ki} (if it exists). Denote this element by e_{ki}. For any $e \in B_{ki} \cap E$, the distance $d(e, e_{ki})$ is less than $2/k$. It follows that the set of all e_{ki} is, on the one hand, countable, and, on the other hand, dense in E. □

Recall that a subspace of a linear space X is a subset of X whose elements satisfy the linear space axioms. Normally the separability of function spaces is proved via the approximation of functions by polynomials with rational coefficients. These polynomials constitute a countable set. When we establish separability of spaces used for the setup of boundary value problems, certain conditions on the boundary are involved in defining the useful subspaces. The polynomials usually do not satisfy these conditions and therefore are not included in the needed subspace. However, we can prove separability of the principal space without boundary restrictions. Separability of the subspace with the restrictions follows from Theorem 4.18.

An important result from analysis is the *Weierstrass approximation theorem*: if f is continuous on a compact domain in $\mathbb{R}^n$, then there is a sequence of polynomials that can "uniformly approximate" f on that domain. Upon this result rests

Theorem 4.19. *If Ω is a compact domain in $\mathbb{R}^n$, then the space $C(\Omega)$ is separable.*

Proof. The set of all polynomials with rational coefficients is dense in the set of all polynomials. Then the Weierstrass theorem implies that the set P_r of all polynomials with rational coefficients is dense in $C(\Omega)$. Since P_r is countable, $C(\Omega)$ is separable. □

We also have

Theorem 4.20. *The space $C^{(k)}(\Omega)$ is separable for any integer k.*

4.3 Cauchy Sequences and Banach Spaces

If $x_n \to x$, then the triangle inequality

$$d(x_{n+m}, x_n) \le d(x_{n+m}, x) + d(x, x_n)$$

shows that for any $\varepsilon > 0$ there is a number $N = N(\varepsilon)$ such that for any $n > N$ and any positive integer m,

$$d(x_{n+m}, x_n) \le \varepsilon.$$

In calculus such a sequence is given a special name:

Definition 4.21. A sequence $\{x_n\}$ is a *Cauchy sequence* if to each $\varepsilon > 0$ there corresponds $N = N(\varepsilon)$ such that for every pair of numbers m, n the inequalities $m > N$ and $n > N$ together imply that $d(x_m, x_n) < \varepsilon$.

Every convergent sequence is a Cauchy sequence. According to a famous theorem of calculus, any Cauchy sequence of real numbers is necessarily convergent to some real number, so in $\mathbb{R}$ the notions of Cauchy sequence and convergent sequence are equivalent. In a general metric space this is not so, as is demonstrated next.

Example 4.22. Show that the sequence of functions

$$f_n(x) = \begin{cases} 0, & 0 \le x \le \frac{1}{2}, \\ nx - \frac{n}{2}, & \frac{1}{2} \le x \le \frac{1}{2} + \frac{1}{n}, \\ 1 & \frac{1}{2} + \frac{1}{n} \le x \le 1, \end{cases} \qquad (n = 2, 3, 4, \ldots)$$

continuous on $[0, 1]$ is a Cauchy sequence in $L(0, 1)$ but has no continuous limit. Note: the norm in the space $L(0, 1)$ is given by $\|f(x)\| = \int_0^1 |f(x)|\,dx$. Is this a Cauchy sequence in the norm of $C(0, 1)$?

Solution. Each $f_n(x)$ is continuous on $[0, 1]$. To see that $\{f_n\}$ is a Cauchy sequence, assume $m > n$ and calculate

$$\begin{aligned} d(f_n(x), f_m(x)) &= \int_{\frac{1}{2}}^{\frac{1}{2}+\frac{1}{m}} \left|\left(mx - \frac{m}{2}\right) - \left(nx - \frac{n}{2}\right)\right| dx \\ &\quad + \int_{\frac{1}{2}+\frac{1}{m}}^{\frac{1}{2}+\frac{1}{n}} \left|1 - \left(nx - \frac{n}{2}\right)\right| dx \\ &= \frac{1}{2}\left(\frac{1}{n} - \frac{1}{m}\right) \to 0 \quad \text{as } m, n \to \infty. \end{aligned}$$

However, $f_n \to f$ where

$$f(x) = \begin{cases} 0, & 0 \le x \le \frac{1}{2}, \\ 1, & \frac{1}{2} < x \le 1, \end{cases}$$

because

$$d(f_n(x), f(x)) = \int_{\frac{1}{2}}^{\frac{1}{2}+\frac{1}{n}} \left|1 - \left(nx - \frac{n}{2}\right)\right| dx = \frac{1}{2n} \to 0 \quad \text{as } n \to \infty.$$

The function $f(x)$ is clearly not continuous. □

Example 4.23. Show that if a sequence converges, then any of its subsequences also converges and has the same limit.

Solution. Let $\{x_{n_k}\}$ be a subsequence of $\{x_n\}$ where $x_n \to x$. Given $\varepsilon > 0$, we can find N such that $n \ge N$ implies $d(x_n, x) < \varepsilon$. Since $n_k \ge k$ for all k, we have $d(x_{n_k}, x) < \varepsilon$ whenever $k \ge N$. □

Example 4.24. Show that if some subsequence of a Cauchy sequence has a limit, then the entire sequence must converge to the same limit.

Solution. Suppose $\{x_{n_k}\}$ is a convergent subsequence of a Cauchy sequence $\{x_n\}$. We show that if $x_{n_k} \to x$, then $x_n \to x$. Let $\varepsilon > 0$ be given and choose N such that $d(x_n, x_m) < \varepsilon/2$ for $n, m > N$. Since $x_{n_k} \to x$, there exists $n_k > N$ such that $d(x_{n_k}, x) < \varepsilon/2$. So for $n > N$ we have $d(x_n, x) \le d(x_n, x_{n_k}) + d(x_{n_k}, x) < \varepsilon/2 + \varepsilon/2 = \varepsilon$. □

Example 4.25. A set S in a normed space X is *bounded* if there exists $R > 0$ such that $\|x\| \le R$ whenever $x \in S$. Show that every Cauchy sequence is bounded.

Solution. Let $\{x_n\}$ be a Cauchy sequence. There exists N such that

$$\|x_n - x_{N+1}\| < 1$$

whenever $n > N$. For all $n > N$ we have

$$\|x_n\| \le \|x_n - x_{N+1}\| + \|x_{N+1}\| < \|x_{N+1}\| + 1.$$

Hence an upper bound for $\|x_n\|$ for any n is given by

$$B = \max\{\|x_1\|, \ldots, \|x_N\|, \|x_{N+1}\| + 1\}.$$

Therefore $\{x_n\}$ is a bounded sequence. □

The property that any Cauchy sequence of a metric space has a limit element belonging to the space is so important that a metric space having this property is called *complete*. If a normed space is complete, it is called a *Banach space* in honor of the Polish mathematician Stefan Banach who discovered many important properties of normed spaces.

Definition 4.26. A metric space X is *complete* if every Cauchy sequence in X converges to a point in X. A *Banach space* is a complete normed space.

Example 4.27. Show that $\mathbb{R}^n$ is complete.

Solution. Let $\{\mathbf{x}^{(k)}\}$ be a Cauchy sequence in $\mathbb{R}^n$. The kth term of this sequence is an n-tuple

$$\mathbf{x}^{(k)} = (x_1^{(k)}, \dots, x_n^{(k)}).$$

Since $\{\mathbf{x}^{(k)}\}$ is a Cauchy sequence, for each $\varepsilon > 0$ there exists N such that $m > N$ and $p > 0$ imply

$$d(\mathbf{x}^{(m+p)}, \mathbf{x}^{(m)}) = \left[\sum_{i=1}^{n} \left|x_i^{(m+p)} - x_i^{(m)}\right|^2\right]^{1/2} \leq \varepsilon.$$

Since all terms in the sum are nonnegative, we have

$$\left|x_i^{(m+p)} - x_i^{(m)}\right| < \varepsilon \quad \text{for each } i = 1, \dots, n \tag{4.9}$$

whenever $m > N$ and $p > 0$. Hence $x_i^{(j)}$ is a Cauchy sequence of reals for any $i = 1, \dots, n$. By the completeness of $\mathbb{R}$ we know that $x_i^{(j)}$ converges (as $j \to \infty$) to a limit, say x_i^*, in $\mathbb{R}$. Now let

$$\mathbf{x}^* = (x_1^*, \dots, x_n^*).$$

We will show that

$$\mathbf{x}^{(k)} \to \mathbf{x}^* \tag{4.10}$$

where convergence is understood in the sense of the Euclidean metric on $\mathbb{R}^n$. Fix $m > N$; by (4.9) we get

$$\lim_{p\to\infty} \left|x_i^{(m+p)} - x_i^{(m)}\right| \leq \varepsilon,$$

hence

$$|x_i^* - x_i^{(m)}| \leq \varepsilon \quad \text{for each } i = 1, \dots, n.$$

So

$$\left(\sum_{i=1}^{n} \left| x_i^* - x_i^{(m)} \right|^2 \right)^{1/2} = d(\mathbf{x}^*, \mathbf{x}^{(m)}) \le \sqrt{n}\varepsilon$$

for $m > N$, and (4.10) is proved. Since every Cauchy sequence in $\mathbb{R}^n$ converges to a point of $\mathbb{R}^n$, the space $\mathbb{R}^n$ is complete. □

In applications we encounter solutions to many problems expressed in the form of functional series. To deal with them as with series of elements in the usual calculus, let us introduce series in a Banach space. By definition, a series of the form

$$\sum_{k=1}^{\infty} x_k \qquad (x_k \in X)$$

converges to an element $s \in X$ if the sequence $\{s_n\}$ of partial sums

$$s_n = \sum_{k=1}^{n} x_k$$

converges to $s \in X$ in the norm of X. The notion of absolute convergence may also be adapted to series in Banach spaces.

Definition 4.28. The series $\sum_{k=1}^{\infty} x_k$ *converges absolutely* if the numerical series $\sum_{k=1}^{\infty} \|x_k\|$ converges.

In a Banach space, as in ordinary calculus, absolute convergence implies convergence:

Theorem 4.29. *Let $\{x_k\}$ be a sequence of elements in a Banach space X. If the series $\sum_{k=1}^{\infty} x_k$ converges absolutely, then it converges.*

Proof. By the triangle inequality we have, for any n and $p \ge 1$,

$$\left\| \sum_{k=1}^{n+p} x_k - \sum_{k=1}^{n} x_k \right\| \le \left| \sum_{k=1}^{n+p} \|x_k\| - \sum_{k=1}^{n} \|x_k\| \right|.$$

By hypothesis the sequence $\sum_{k=1}^{n} \|x_k\|$ converges and is therefore a Cauchy sequence. By the inequality above, $\sum_{k=1}^{n} x_k$ is a Cauchy sequence and will converge to an element of X by completeness. □

Example 4.30. Show that under the conditions of the previous theorem,

$$\left\| \sum_{k=1}^{\infty} x_k \right\| \le \sum_{k=1}^{\infty} \|x_k\| .$$

Solution. We have

$$\left\| \sum_{k=1}^{\infty} x_k \right\| = \left\| \lim_{n\to\infty} \sum_{k=1}^{n} x_k \right\| = \lim_{n\to\infty} \left\| \sum_{k=1}^{n} x_k \right\| \le \lim_{n\to\infty} \sum_{k=1}^{n} \|x_k\| = \sum_{k=1}^{\infty} \|x_k\| .$$

We used the continuity of the norm, and then the triangle inequality for finite sums. □

Many of the other results from ordinary calculus also carry over to series in Banach spaces. We can add convergent series termwise:

$$\sum_{k=1}^{\infty} x_k + \sum_{k=1}^{\infty} y_k = \sum_{k=1}^{\infty} (x_k + y_k). \tag{4.11}$$

We can also multiply a series by a scalar constant λ in the usual way:

$$\lambda \sum_{k=1}^{\infty} x_k = \sum_{k=1}^{\infty} \lambda x_k. \tag{4.12}$$

Definition 4.31. An element x of a metric space X is a *limit point* of a set S if any ball centered at x contains a point of S different from x. The set S is *closed in* X if it contains all its limit points.

Limit points are sometimes called *points of accumulation.* The following result provides a useful alternative characterization for a closed subset of a complete metric space.

Theorem 4.32. *A subset S of a complete metric space X supplied with the metric of X is a complete metric space if and only if S is closed in X.*

Proof. Assume S is complete. If x is a limit point of S, then there is a sequence $\{x_n\} \subset S$ such that $x_n \to x$. But every convergent sequence is a Cauchy sequence, hence by completeness $\{x_n\}$ converges to a point of S. From this and uniqueness of the limit we conclude that $x \in S$. Hence S contains all its limit points and is therefore a closed set.

Now assume S is closed. If $\{x_n\}$ is any Cauchy sequence in S, then $\{x_n\}$ is also a Cauchy sequence in X and converges to a point $x \in X$. This point x is also a limit point of S however, hence $x \in S$. So every Cauchy sequence in S converges to a point of S, and S is complete. □

We turn to some examples of Banach and normed spaces. The simplest kind of Banach space is formed by imposing a norm on the linear space

$\mathbb{R}^n$ of n-dimensional vectors $\mathbf{x} = (x_1, \ldots, x_n)$. A standard norm defined on this space is the Euclidean norm

$$\|\mathbf{x}\|_e = \left(\sum_{i=1}^{n} x_i^2 \right)^{1/2} . \tag{4.13}$$

The resulting Banach space $(\mathbb{R}^n, \|\cdot\|_e)$ is finite dimensional. The following result allows us to ignore the distinction between different normed spaces that are formed from the same underlying finite dimensional vector space by imposing different norms:

Theorem 4.33. *On a finite dimensional space all norms are equivalent.*

Proof. It is enough to prove that any norm is equivalent to the Euclidean norm $\|\cdot\|_e$. Take any basis i_k that is orthonormal in the Euclidean inner product. We can express any x as $x = \sum_{k=1}^{n} c_k i_k$. Then

$$\|x\|_e = \left(\sum_{k=1}^{n} c_k^2 \right)^{1/2} .$$

For an arbitrary norm $\|\cdot\|$,

$$\|x\| = \left\| \sum_{k=1}^{n} c_k i_k \right\| \le \sum_{k=1}^{n} |c_k| \, \|i_k\| \le \sum_{k=1}^{n} \left(\sum_{j=1}^{n} |c_j|^2 \right)^{1/2} \|i_k\| = m \, \|x\|_e$$

where $m = \sum_{k=1}^{n} \|i_k\|$ is finite. So one side is proved. For the other side, consider $\|x\|$ as a function of the n variables c_k. Because of the above inequality it is a continuous function in the usual sense. Indeed

$$| \, \|x_1\| - \|x_2\| \, | \le \|x_1 - x_2\| \le m \, \|x_1 - x_2\|_e ,$$

which for $x_1 = \sum_{k=1}^{n} c_k i_k$ and $x = \sum_{k=1}^{n} (c_k + \Delta_k) i_k$ can be rewritten as

$$\left| \left\| \sum_{k=1}^{n} c_k i_k \right\| - \left\| \sum_{k=1}^{n} (c_k + \Delta_k) i_k \right\| \right| \le m \left\| \sum_{k=1}^{n} \Delta_k i_k \right\| = m \sum_{k=1}^{n} {\Delta_k}^2$$

and from which we get ordinary ε-δ definition of continuity of the function at any point $(c_1, \ldots, c_n)$. Now it is enough to show that on the sphere $\|x\|_e = 1$ we have $\inf \|x\| = a > 0$ (because of homogeneity of norms). Being a continuous function, $\|x\|$ achieves its minimum on the compact set $\|x\|_e = 1$ at a point x_0. So $\|x_0\| = a$. If $a = 0$ then $x_0 = 0$ and thus x_0 does not belong to the unit sphere (in the Euclidean norm). Thus $a > 0$ and for any x we have $\|x\| / \|x\|_e \ge a$. □

The notion of an infinite dimensional vector $\mathbf{x} = (x_1, x_2, \ldots)$ with a countable number of components is, of course, a straightforward generalization of the notion of a finite dimensional vector $\mathbf{x} = (x_1, \ldots, x_n)$. Such a vector can be encountered by considering a numerical sequence $\{x_i\}$ as a whole entity; the individual terms x_i of the sequence become the components of a vector $\mathbf{x}$. We shall use the terms *infinite dimensional vector* and *sequence* interchangeably. Another way to introduce vectors with infinitely many components is to consider expansions of functions, such as Fourier or Taylor expansions. The expansion coefficients can be collected into something like a vector with infinitely many components.

Simple infinite dimensional Banach spaces can be formed by imposing suitable norms on spaces of infinite dimensional vectors. The results are *sequence spaces.* For example, we may take the set c of all convergent numerical sequences and impose the norm

$$\|\mathbf{x}\| = \sup_i |x_i|.$$

Note that an infinite dimensional vector $\mathbf{x}$ does not belong to c if a subsequence $\{x_{i_k}\}$ of its components satisfies $x_{i_k} \to \infty$ as $i_k \to \infty$. So c contains only a subset of all infinite dimensional vectors.

An interesting family of sequence spaces can be defined, one for each integer $p \geq 1$. The space ℓ^p is the set of all vectors $\mathbf{x}$ such that $\sum_{i=1}^{\infty} |x_i|^p < \infty$, and its norm is taken to be

$$\|\mathbf{x}\| = \left(\sum_{i=1}^{\infty} |x_i|^p \right)^{1/p}. \tag{4.14}$$

The fact that (4.14) is a norm is a consequence of the *Minkowski inequality*

$$\left(\sum_{i=1}^{\infty} |x_i + y_i|^p \right)^{1/p} \leq \left(\sum_{i=1}^{\infty} |x_i|^p \right)^{1/p} + \left(\sum_{i=1}^{\infty} |y_i|^p \right)^{1/p} \tag{4.15}$$

since satisfaction of the other norm axioms for (4.14) is evident. An important special case is the space ℓ^2 of *square summable* sequences $\mathbf{x}$ with $\sum_{i=1}^{\infty} |x_i|^2 < \infty$ and norm

$$\|\mathbf{x}\| = \left(\sum_{i=1}^{\infty} |x_i|^2 \right)^{1/2}. \tag{4.16}$$

Looking ahead, we mention that any element in a separable Hilbert space H (it is a complete space with an inner product that is similar to the dot product in a Euclidean space) can be represented as a Fourier expansion with respect to an orthonormal basis of H, and there is a one-to-one correspondence between the elements of H and ℓ^2. So all the general properties we could establish for the elements of ℓ^2 can be reformulated for a separable Hilbert space H and vice versa. We can add that ℓ^2 was the first space introduced by David Hilbert and initiated functional analysis as a branch of mathematics.

We emphasize that the normed spaces c and ℓ^p are not defined on the same underlying set of vectors. For example, the vector $\mathbf{x} = (1, 1, 1, \ldots)$ obviously belongs to c but not to ℓ^p for any $p \geq 1$. Moreover, there is no analog to Theorem 4.33 for infinite dimensional spaces.

There is a subspace of c denoted by c_0 that consists of vectors (sequences) having zero limit. Note that a set of sequences converging to some fixed nonzero limit could not be a linear space. If we wish to consider the set of all convergent sequences with some nonzero limit, we call it a *cone*. We can restrict a cone to some of its subsets by placing additional conditions on the components of vectors.

It is also possible to study weighted spaces of sequences with norms of the form

$$\|\mathbf{x}\| = \left(\sum_{i=1}^{\infty} k_i |x_i|^2 \right)^{1/2} \tag{4.17}$$

where the $k_i \geq 0$ are constants used to weight the terms of the sequence.

We can show that all of the spaces mentioned above are Banach spaces.

Example 4.34. Show that c is a Banach space.

Solution. We use the fact that the normed space consisting of the set $\mathbb{R}$ of real numbers under the usual norm $|x|$ is a Banach space. Let $\{\mathbf{x}^{(k)}\}$ be a Cauchy sequence in c. The kth term of this sequence is a numerical sequence:

$$\mathbf{x}^{(k)} = (x_1^{(k)}, x_2^{(k)}, x_3^{(k)}, \ldots).$$

To each $\varepsilon > 0$ there corresponds $N = N(\varepsilon)$ such that

$$\|\mathbf{x}^{(n+m)} - \mathbf{x}^{(n)}\|_c = \sup_i |x_i^{(n+m)} - x_i^{(n)}| \leq \varepsilon$$

whenever $n > N$ and $m > 0$. This implies that

$$|x_i^{(n+m)} - x_i^{(n)}| \leq \varepsilon \quad \text{for each } i \tag{4.18}$$

whenever $n > N$ and $m > 0$. Hence $\{x_i^{(j)}\}$ is a Cauchy sequence of real numbers for any fixed i. By the completeness of the normed space $(\mathbb{R}, \|\cdot\|)$

we know that $\{x_i^{(j)}\}$ converges (as $j \to \infty$) to a limit, say x_i^*, in $\mathbb{R}$. Now let

$$\mathbf{x}^* = (x_1^*, x_2^*, x_3^*, \ldots).$$

We will show that

$$\mathbf{x}^{(k)} \to \mathbf{x}^*. \tag{4.19}$$

Fix $n > N$; by (4.18) and continuity

$$\lim_{m\to\infty} |x_i^{(n+m)} - x_i^{(n)}| \le \varepsilon$$

which gives

$$|x_i^* - x_i^{(n)}| \le \varepsilon \quad \text{for each } i.$$

Hence

$$\sup_i |x_i^* - x_i^{(n)}| = \|\mathbf{x}^* - \mathbf{x}^{(n)}\|_c \le \varepsilon$$

for $n > N$, so (4.19) is established. Finally we must show that $\mathbf{x}^* \in c$ by showing that $\{x_i^*\}$ converges. Since every Cauchy sequence of real numbers converges, it suffices to show that $\{x_i^*\}$ is a Cauchy sequence. Let us consider the difference

$$|x_n^* - x_m^*| \le |x_n^* - x_n^{(k)}| + |x_n^{(k)} - x_m^{(k)}| + |x_m^{(k)} - x_m^*|$$

and use an $\varepsilon/3$ argument. Let $\varepsilon > 0$ be given. We can make the first and third terms on the right side less than $\varepsilon/3$ for any n, m by fixing k sufficiently large. For this k, $\{x_j^{(k)}\}$ is a Cauchy sequence; therefore we can make the second term on the right side less than $\varepsilon/3$ by taking n and m sufficiently large. □

Note the general pattern of these completeness proofs. We take an arbitrary Cauchy sequence $\{x_n\}$ in (X, d), construct an element x that appears to be the limit of $\{x_n\}$, prove that $x \in X$, and prove that $x_n \to x$ with respect to d.

Example 4.35. Show that c_0 is a Banach space.

Solution. Let $\{\mathbf{x}^{(k)}\}$ be a Cauchy sequence in c_0. The kth term of this sequence is a numerical sequence

$$\mathbf{x}^{(k)} = (x_1^{(k)}, x_2^{(k)}, x_3^{(k)}, \ldots)$$

that converges to 0. As with a Cauchy sequence in the space c, we can show that

$$\mathbf{x}^{(k)} \to \mathbf{x}^* = (x_1^*, x_2^*, x_3^*, \ldots) \qquad \text{where} \qquad x_i^* = \lim_{j\to\infty} x_i^{(j)}.$$

(As before, in the process we find that by fixing n sufficiently large we can get the inequality $|x_i^* - x_i^{(n)}| \le \varepsilon$ to hold for all i.) To complete the proof we must show that $\mathbf{x}^* \in c_0$, i.e., that $x_i^* \to 0$ as $i \to \infty$. Let $\varepsilon > 0$ be given. We have

$$|x_i^*| \le |x_i^* - x_i^{(k)}| + |x_i^{(k)}|.$$

We can fix k large enough that the first term on the right is less than $\varepsilon/2$ for all i. For this k, we can choose i large enough that the second term on the right is less than $\varepsilon/2$. □

Now let us turn to function spaces. We have introduced the space $C(\Omega)$. If Ω is a compact set in $\mathbb{R}^n$, then $C(\Omega)$ is a Banach space. Indeed, the Weierstrass theorem states that a uniformly convergent sequence of functions defined on a compact set has as a limit a continuous function. A sequence of functions $\{f_k(\mathbf{x})\}$ is a Cauchy sequence in $C(\Omega)$ if to each $\varepsilon > 0$ there corresponds $N = N(\varepsilon)$ such that

$$\max_{\mathbf{x}\in\Omega} |f_{n+m}(\mathbf{x}) - f_n(\mathbf{x})| \le \varepsilon$$

for any $n > N$ and any positive integer m. This definition means that $\{f_n(\mathbf{x})\}$ converges uniformly on Ω and thus its limit point exists and belongs to $C(\Omega)$. (Note that the uniform convergence of a sequence of functions in calculus and convergence with respect to the norm of $C(\Omega)$, Ω being compact, are the same.) That is, by definition, $C(\Omega)$ is a Banach space. Similarly, $C^{(k)}(\Omega)$ is a Banach space.

We mentioned earlier that on the set of functions continuous on a compact set Ω we can impose

$$\|f(\mathbf{x})\|_{L^p(\Omega)} = \left(\int_\Omega |f(\mathbf{x})|^p \, d\Omega\right)^{1/p} \tag{4.20}$$

for $p \ge 1$. Writing out the corresponding Riemann sums for the integral and then using the limit passage, we may show that the triangle inequality holds (this is Minkowski's inequality for integrals). Fulfillment of the remaining norm axioms is evident. Example 4.22 shows that the set of continuous functions under this norm, for the case $p = 1$, is not a Banach space. The situation is the same for any $p > 1$ and for any dimension of Ω.

On the set of differentiable functions we can impose an important class of norms called *Sobolev's norms.* A simple but useful example is the norm of $W^{1,2}(0,1)$:

$$\|f(x)\|_{W^{1,2}(0,1)} = \left(\int_0^1 \left(|f(x)|^2 + |f'(x)|^2 \right) dx \right)^{1/2}$$

This was first studied by Banach. The general form of a Sobolev norm is

$$\|f(\mathbf{x})\|_{W^{l,p}(\Omega)} = \left(\int_\Omega \sum_{|\alpha| \le l} |D^\alpha f(\mathbf{x})|^p \, d\Omega \right)^{1/p}, \qquad p \ge 1. \tag{4.21}$$

The set of l-times continuously differentiable functions on Ω is not complete in the norm (4.21). Under this norm, as with the L^p norm, the difference between "close" functions can be very large on subdomains of small area. Later we shall study Banach spaces having these norms.

Example 4.36. The Cartesian product $X \times Y$ of two linear spaces X and Y can form a linear space under suitable definitions of vector addition and scalar multiplication. If X and Y are also normed spaces with norms $\|\cdot\|_X$, $\|\cdot\|_Y$, respectively, then $X \times Y$ is a normed space under the norm

$$\|(x, y)\| = \|x\|_X + \|y\|_Y \, . \tag{4.22}$$

Show that if X and Y are Banach spaces, then so is $X \times Y$.

Solution. Choose any Cauchy sequence $\{(x_k, y_k)\} \subset X \times Y$. Then

$$\begin{aligned}
\|(x_m, y_m) - (x_n, y_n)\|_{X \times Y} &= \|(x_m - x_n, y_m - y_n)\|_{X \times Y} \\
&= \|x_m - x_n\|_X + \|y_m - y_n\|_Y \to 0
\end{aligned}$$

as $m, n \to \infty$, hence

$$\|x_m - x_n\|_X \to 0 \quad \text{and} \quad \|y_m - y_n\|_Y \to 0 \quad \text{as } m, n \to \infty.$$

So $\{x_k\}$ and $\{y_k\}$ are each Cauchy sequences in their respective spaces X, Y; since these are Banach spaces there exist $x \in X$ and $y \in Y$ such that $x_k \to x$ and $y_k \to y$. Finally, $(x_k, y_k) \to (x, y)$ in the norm of $X \times Y$:

$$\begin{aligned}
\|(x_k, y_k) - (x, y)\|_{X \times Y} &= \|(x_k - x, y_k - y)\|_{X \times Y} \\
&= \|x_k - x\|_X + \|y_k - y\|_Y \to 0 \quad \text{as } k \to \infty.
\end{aligned}$$

We see that $X \times Y$ is complete. □

4.4 The Completion Theorem

Incomplete spaces can be inconvenient. For example, using only rational numbers we leave out such numbers as $\sqrt{2}$ and π, and so cannot obtain exact solutions for many quadratic equations or geometry problems. Various approaches can be used to introduce irrational numbers. To define an irrational number π, we can define a sequence of approximations such as 3, 3.1, 3.14, 3.141, and so on. The limit of this sequence is what we call π. But the approximating sequence $4, 3.2, 3.142, \ldots$ also consists of rational numbers and can be used to define the same number π. There are infinitely many sequences having this same limit, and we can collect this set of Cauchy sequences together as an entity that defines π. We call such sequences *equivalent.* The same can be done with any irrational number. If we then regard a real number as something defined by a set of all equivalent sequences, a rational number can be represented as a set of all equivalent sequences one of which is a *stationary* sequence having all terms equal to the rational number. We shall use this idea to "extend" an incomplete space to one that is complete. In advance we shall introduce several notions.

Definition 4.37. Two sequences $\{x_n\}, \{y_n\}$ in a metric space (M, d) are *equivalent* if $d(x_n, y_n) \to 0$ as $n \to \infty$. If $\{x_n\}$ is a Cauchy sequence in M, we can collect into an equivalence class X all Cauchy sequences in M that are equivalent to $\{x_n\}$. Any Cauchy sequence from X is a *representative* of X. To any $x \in M$ there corresponds a *stationary* equivalence class containing the Cauchy sequence $x, x, x, \ldots$.

Definition 4.38. A mapping $F\colon M_1 \to M_2$ is an *isometry* between (M_1, d_1) and (M_2, d_2) if $d_1(x, y) = d_2(F(x), F(y))$ for all $x, y \in M_1$. Distances are obviously preserved under such a mapping. If F is also a one-to-one correspondence between M_1 and M_2, then it is a *one-to-one isometry* and the two metric spaces are said to be *isometric.* Isometric spaces are essentially the same, the isometry amounting to a mere relabeling of the points in each space.

Now we can state the *completion theorem:*

Theorem 4.39. *For a metric space M, there is a one-to-one isometry between M and a set $\tilde{M}$ which is dense in a complete metric space M^*. We call M^* the* **completion** *of M.*

Proof. As we said, we shall use the same idea as above for introducing the needed space. The proof consists of four steps: (1) introduction of the elements of the space M^*; (2) introduction of a metric on this space and verification of the axioms; (3) demonstration that the new space is complete; (4) verification of the remaining statements of the theorem.

1. As indicated in Definition 4.37, we collect into an equivalence class X all Cauchy sequences in M that are equivalent to a given Cauchy sequence $\{x_n\}$. We denote the set of all the equivalence classes by M^*, and the set of all stationary equivalence classes by $\tilde{M}$.

2. We impose a metric on M^*. Given $X, Y \in M^*$, we choose any representatives $\{x_n\} \in X$ and $\{y_n\} \in Y$ and define

$$d(X, Y) = \lim_{n\to\infty} d(x_n, y_n). \tag{4.23}$$

This same metric is applied to the subspace $\tilde{M}$ of M^*. To see that $d(X, Y)$ is actually a metric, we must first check that the limit in (4.23) exists and is independent of the choice of representatives. Metric axiom D4 implies

$$d(x_n, y_n) \le d(x_n, x_m) + d(x_m, y_m) + d(y_m, y_n)$$

so that

$$d(x_n, y_n) - d(x_m, y_m) \le d(x_n, x_m) + d(y_m, y_n).$$

Interchanging m and n we obtain a similar inequality; combining the two, we obtain

$$|d(x_n, y_n) - d(x_m, y_m)| \le d(x_n, x_m) + d(y_n, y_m).$$

But $d(x_n, x_m) \to 0$ and $d(y_n, y_m) \to 0$ as $m, n \to \infty$ because $\{x_n\}$ and $\{y_n\}$ are Cauchy sequences. Thus

$$|d(x_n, y_n) - d(x_m, y_m)| \to 0 \quad \text{as } m, n \to \infty$$

and $\{d(x_n, y_n)\}$ is a Cauchy sequence in $\mathbb{R}$. By completeness of $\mathbb{R}$, the limit in (4.23) exists. To show that it does not depend on the choice of representatives, we take any $\{x'_n\} \in X$ and $\{y'_n\} \in Y$ and show that

$$\lim_{n\to\infty} d(x'_n, y'_n) = \lim_{n\to\infty} d(x_n, y_n). \tag{4.24}$$

Because $\lim_{n\to\infty} d(x_n, x'_n) = 0 = \lim_{n\to\infty} d(y_n, y'_n)$, the inequality

$$|d(x_n, y_n) - d(x'_n, y'_n)| \le d(x_n, x'_n) + d(y_n, y'_n)$$

gives

$$\lim_{n\to\infty} |d(x_n, y_n) - d(x'_n, y'_n)| = 0$$

which implies (4.24). We check that the metric axioms are satisfied by $d(X, Y)$:

D1: Since $d(x_n, y_n) \geq 0$ for all n, it follows that

$$d(X, Y) = \lim_{n\to\infty} d(x_n, y_n) \geq 0.$$

D2: If $X = Y$ then $d(X, Y) = 0$ (we can choose the same Cauchy sequence $\{x_n\}$ from both X and Y, and since the limit is unique we get the needed conclusion). Conversely, if $d(X, Y) = 0$ then any two Cauchy sequences $\{x_n\} \in X$ and $\{y_n\} \in Y$ satisfy $\lim_{n\to\infty} d(x_n, y_n) = 0$. By definition they are equivalent, hence $X = Y$.

D3: We have

$$d(X, Y) = \lim_{n\to\infty} d(x_n, y_n) = \lim_{n\to\infty} d(y_n, x_n) = d(Y, X).$$

D4: For $x_n, y_n, z_n \in M$ the triangle inequality gives

$$d(x_n, y_n) \leq d(x_n, z_n) + d(z_n, y_n);$$

as $n \to \infty$ we have

$$d(X, Y) \leq d(X, Z) + d(Z, Y)$$

for the equivalence classes X, Y, Z containing $\{x_n\}$, $\{y_n\}$, $\{z_n\}$, respectively.

3. To see that M^* is complete, we must show that for any Cauchy sequence $\{X^i\} \subset M^*$, there exists

$$X = \lim_{i\to\infty} X^i \in M^*. \tag{4.25}$$

Indeed, from each X^i we choose a Cauchy sequence $\{x_j^{(i)}\}$ and from this an element denoted x_i such that $d(x_i, x_j^{(i)}) < 1/i$ whenever $j > i$. To see that $\{x_i\}$ is a Cauchy sequence, denote by X_i the equivalence class containing the stationary sequence $(x_i, x_i, \ldots)$ and write

$$\begin{aligned} d(x_i, x_j) &= d(X_i, X_j) \\ &\leq d(X_i, X^i) + d(X^i, X^j) + d(X^j, X_j) \\ &\leq \frac{1}{i} + d(X^i, X^j) + \frac{1}{j}. \end{aligned}$$

As $i, j \to \infty$, $d(x_i, x_j) \to 0$ as required. Finally, denote by X the equivalence class containing $\{x_i\}$. Because $\{x_i\}$ is a Cauchy sequence,

$$\begin{aligned} d(X^i, X) &\le d(X^i, X_i) + d(X_i, X) \\ &\le \frac{1}{i} + d(X_i, X) \\ &= \frac{1}{i} + \lim_{j\to\infty} d(x_i, x_j) \to 0 \quad \text{as } i \to \infty. \end{aligned}$$

This proves (4.25).

4. $\tilde{M}$ is dense in M^*. To see this, choose $X \in M^*$. Selecting a representative $\{x_n\}$ from X, we denote by X_n the stationary equivalence class containing the stationary sequence $(x_n, x_n, \ldots)$. Then

$$d(X_n, X) = \lim_{m\to\infty} d(x_n, x_m) \to 0 \quad \text{as } n \to \infty$$

since $\{x_n\}$ is a Cauchy sequence.

The equality

$$d(X, Y) = d(x, y)$$

if X and Y are stationary classes corresponding to x and y, respectively, demonstrates the one-to-one isometry between M and $\tilde{M}$. □

Corollary 4.40. *If M is a linear space, the isometry preserves algebraic operations.*

Since a normed space is a linear metric space we immediately have

Theorem 4.41. *Any normed space X can be completed in its natural metric $d(x, y) = \|x - y\|$, resulting in a Banach space X^*.*

We will also make use of the following result:

Theorem 4.42. *The completion of a separable metric space is separable.*

Proof. Suppose X is a separable metric space, containing a countable, dense subset S. The completion theorem places X into one-to-one correspondence with a set $\tilde{X}$ that is dense in the completion X^*. Let $\tilde{S}$ be the image of S under this correspondence. Since the correspondence is also an isometry, $\tilde{S}$ is dense in $\tilde{X}$. So we have $\tilde{S} \subseteq \tilde{X} \subseteq X^*$, where each set is dense in the next; therefore $\tilde{S}$ is dense in X^*. Since $\tilde{S}$ is evidently countable, the proof is complete. □

We have lingered over the completion theorem because it is the background for many important notions, including the Lebesgue integral and the Sobolev and energy spaces.

4.5 L^p Spaces and the Lebesgue Integral

To introduce the Lebesgue integral and the $L^p(\Omega)$ spaces, we will apply the completion theorem to the set of functions continuous on a closed and bounded (i.e., compact) subset Ω of $\mathbb{R}^n$. Fix $p \geq 1$. The set S of functions $f(\mathbf{x})$ continuous on Ω becomes a normed space under the norm

$$\|f(\mathbf{x})\|_p = \|f(\mathbf{x})\|_{L^p(\Omega)} = \left(\int_\Omega |f(\mathbf{x})|^p \, d\Omega\right)^{1/p} \tag{4.26}$$

(recall Convention 1.4 on page 16). It is therefore also a metric space under the natural metric

$$d_p(f(\mathbf{x}), g(\mathbf{x})) = \|f(\mathbf{x}) - g(\mathbf{x})\|_p \, . \tag{4.27}$$

In these equations the integral is an ordinary Riemann integral. We saw in Example 4.22 that a sequence of continuous functions on $[0, 1]$ can be a Cauchy sequence with respect to the metric

$$\|f(x) - g(x)\| = \int_0^1 |f(x) - g(x)| \, dx$$

and yet lack a continuous limit. More generally, the metric space formed using S and the metric $d_p(f, g)$ for $p \geq 1$ is incomplete. The completion of this space is called $L^p(\Omega)$. The elements of $L^p(\Omega)$ can be integrated in a certain sense; although we have used Riemann integration in the definition, on the resulting space we shall end up introducing a more general type of integration. Our approach to the *Lebesgue integral* will be different from, but equivalent to, the classical one due to Lebesgue. The Lebesgue integral extends the notion of the Riemann integral in the sense that for an element corresponding to a usual continuous function the Lebesgue integral equals the Riemann integral.

In this section we shall denote an element of $L^p(\Omega)$ using uppercase notation such as $F(\mathbf{x})$. An element $F(\mathbf{x}) \in L^p(\Omega)$ is, of course, an equivalence class of Cauchy sequences of continuous functions. In this case "Cauchy" means Cauchy in the norm $\|\cdot\|_p$, and two sequences $\{f_n(\mathbf{x})\}$ and $\{g_n(\mathbf{x})\}$ are equivalent if

$$\|f_n(\mathbf{x}) - g_n(\mathbf{x})\|_p \to 0 \quad \text{as } n \to \infty.$$

Linear space operations may be carried out in the space $L^p(\Omega)$. If $F(\mathbf{x}) \in L^p(\Omega)$ and λ is a scalar, we take $\lambda F(\mathbf{x}) \in L^p(\Omega)$ to be the element for which $\{\lambda f_n(\mathbf{x})\}$ is a representative whenever $\{f_n(\mathbf{x})\}$ is a representative of $F(\mathbf{x})$. A sum such as $F(\mathbf{x}) + G(\mathbf{x})$ is interpreted similarly, in terms of representative Cauchy sequences.

The main goal of this section is to define the Lebesgue integral

$$\int_\Omega F(\mathbf{x})\, d\Omega \quad \text{for } F(\mathbf{x}) \in L^p(\Omega).$$

We will do this in such a way that if $F(\mathbf{x})$ belongs to the dense set in $L^p(\Omega)$ that corresponds to the initial set of continuous functions, then the value of this new integral is equal to the Riemann integral of the continuous preimage. In the process we shall make use of *Hölder's inequality*

$$\int_\Omega |f(\mathbf{x})g(\mathbf{x})|\, d\Omega \le \left(\int_\Omega |f(\mathbf{x})|^p\, d\Omega\right)^{1/p} \left(\int_\Omega |g(\mathbf{x})|^q\, d\Omega\right)^{1/q} \tag{4.28}$$

which holds under the conditions

$$\frac{1}{p} + \frac{1}{q} = 1, \qquad p > 1.$$

This is an integral analogue of Hölder's inequality for series

$$\sum_{n=1}^{\infty} |f_n g_n| \le \left(\sum_{n=1}^{\infty} |f_n|^p\right)^{1/p} \left(\sum_{n=1}^{\infty} |g_n|^q\right)^{1/q}. \tag{4.29}$$

See [10] for further details. Let us mention that for nontrivial $f(\mathbf{x})$ and $g(\mathbf{x})$ the sign of equality in (4.28) holds if and only if there is a positive constant λ such that $|f(\mathbf{x})| = \lambda\, |g(\mathbf{x})|$ almost everywhere. A consequence of (4.28) is *Minkowski's inequality*

$$\|f(\mathbf{x}) + g(\mathbf{x})\|_p \le \|f(\mathbf{x})\|_p + \|g(\mathbf{x})\|_p\,, \tag{4.30}$$

from which the useful result

$$\left| \|f(\mathbf{x})\|_p - \|g(\mathbf{x})\|_p \right| \le \|f(\mathbf{x}) - g(\mathbf{x})\|_p \tag{4.31}$$

is easily obtained.

We begin by defining the integral

$$\int_\Omega |F(\mathbf{x})|^p\, d\Omega, \qquad F(\mathbf{x}) \in L^p(\Omega).$$

We take a representative Cauchy sequence $\{f_n(\mathbf{x})\}$ from $F(\mathbf{x})$ and consider the sequence $\{K_n\}$ given by

$$K_n = \|f_n(\mathbf{x})\|_p\,.$$

This is a Cauchy sequence of numbers; indeed

$$|K_m - K_n| = \left| \|f_m(\mathbf{x})\|_p - \|f_n(\mathbf{x})\|_p \right|$$
$$\le \|f_m(\mathbf{x}) - f_n(\mathbf{x})\|_p \to 0 \quad \text{as } m, n \to \infty.$$

Because $\{K_n\}$ is a Cauchy sequence in $\mathbb{R}$ or $\mathbb{C}$, by completeness there exists a number

$$K = \lim_{n\to\infty} K_n = \lim_{n\to\infty} \left(\int_\Omega |f_n(\mathbf{x})|^p \, d\Omega \right)^{1/p}.$$

It can also be shown that K is independent of the choice of representative sequence. If $\{\tilde{f}_n(\mathbf{x})\}$ is another representative of $F(\mathbf{x})$, i.e., if

$$\|f_n(\mathbf{x}) - \tilde{f}_n(\mathbf{x})\|_p \to 0,$$

then we can set

$$\tilde{K} = \lim_{n\to\infty} \tilde{K}_n = \lim_{n\to\infty} \|\tilde{f}_n(\mathbf{x})\|_p$$

but subsequently find that

$$|K - \tilde{K}| = \left| \lim_{n\to\infty} \|f_n(\mathbf{x})\|_p - \lim_{n\to\infty} \|\tilde{f}_n(\mathbf{x})\|_p \right|$$
$$= \lim_{n\to\infty} \left| \|f_n(\mathbf{x})\|_p - \|\tilde{f}_n(\mathbf{x})\|_p \right|$$
$$\le \lim_{n\to\infty} \|f_n(\mathbf{x}) - \tilde{f}_n(\mathbf{x})\|_p = 0.$$

The uniquely determined number K^p,

$$K^p = \left[\lim_{n\to\infty} \left(\int_\Omega |f_n(\mathbf{x})|^p \, d\Omega \right)^{1/p} \right]^p = \lim_{n\to\infty} \int_\Omega |f_n(\mathbf{x})|^p \, d\Omega,$$

is defined as the Lebesgue integral of $|F(\mathbf{x})|^p$. That is, we have

$$\int_\Omega |F(\mathbf{x})|^p \, d\Omega = \lim_{n\to\infty} \int_\Omega |f_n(\mathbf{x})|^p \, d\Omega$$

where $\{f_n(\mathbf{x})\}$ is any representative of $F(\mathbf{x})$.

We show that when Ω is compact the L^p spaces are nested in the sense that

$$L^p(\Omega) \subseteq L^r(\Omega) \quad \text{whenever} \quad 1 \le r \le p. \tag{4.32}$$

Let q be such that $1/q + r/p = 1$ and apply Hölder's inequality:

$$\left| \int_\Omega 1 \cdot |f(\mathbf{x})|^r \, d\Omega \right| \le \left(\int_\Omega 1^q \, d\Omega \right)^{1/q} \left(\int_\Omega |f(\mathbf{x})|^p \, d\Omega \right)^{r/p}$$

$$= (\operatorname{mes}\Omega)^{1-\frac{r}{p}} \left(\int_\Omega |f(\mathbf{x})|^p \, d\Omega \right)^{r/p},$$

or

$$\|f(\mathbf{x})\|_r \le (\operatorname{mes}\Omega)^{\frac{1}{r}-\frac{1}{p}} \|f(\mathbf{x})\|_p \tag{4.33}$$

where $\operatorname{mes}\Omega = \int_\Omega 1 \, d\Omega$ is the measure[3] of Ω. Putting $f(\mathbf{x}) = f_n(\mathbf{x}) - f_m(\mathbf{x})$ in (4.33), we see that $\{f_n(\mathbf{x})\}$ is a Cauchy sequence in the norm $\|\cdot\|_r$ if it is a Cauchy sequence in the norm $\|\cdot\|_p$. Putting $f(\mathbf{x}) = f_n(\mathbf{x}) - g_n(\mathbf{x})$, we see that any two Cauchy sequences equivalent in the norm $\|\cdot\|_p$ are equivalent in the norm $\|\cdot\|_r$. Hence

$$F(\mathbf{x}) \in L^p(\Omega) \implies F(\mathbf{x}) \in L^r(\Omega)$$

for $1 \le r \le p$, and we have established (4.32). We thus observe that if $F(\mathbf{x}) \in L^p(\Omega)$ then $\int_\Omega |F(\mathbf{x})|^r \, d\Omega$ is defined for any r such that $1 \le r \le p$. Moreover, putting $f(\mathbf{x}) = f_n(\mathbf{x})$ in (4.33) we see that passage to the limit as $n \to \infty$ gives

$$\|F(\mathbf{x})\|_r \le (\operatorname{mes}\Omega)^{\frac{1}{r}-\frac{1}{p}} \|F(\mathbf{x})\|_p, \qquad 1 \le r \le p. \tag{4.34}$$

Subsequently will interpret this by saying that $L^p(\Omega)$ imbeds continuously into $L^r(\Omega)$. That is, the elements of $L^p(\Omega)$ belong to $L^r(\Omega)$ as well, and the inequality means continuity of the correspondence (imbedding operator) between the elements of $L^p(\Omega)$ and the same elements considered as elements of $L^r(\Omega)$. In a similar way we can show that many inequalities satisfied by the Riemann integral are also satisfied by the Lebesgue integral.

It is now time to introduce the Lebesgue integral

$$\int_\Omega F(\mathbf{x}) \, d\Omega, \qquad F(\mathbf{x}) \in L^p(\Omega).$$

---

[3]Because we use the Riemann integral to construct the Lebesgue integral, we must exclude some "exotic" domains Ω that are actually permitted in Lebesgue integration. Physical problems involve relatively simple domains for which Riemann integration generally suffices. In particular we assume the Riemann integral $\int_\Omega 1 \, d\Omega$ exists for all of our purposes, giving the quantity we call the "measure" of Ω. The full notion of Lebesgue measure is far too involved to consider here; fortunately, our domains are all simple enough that we can use the notation "$\operatorname{mes}\Omega$" without a full chapter of explanation.

Taking a representative $\{f_n(\mathbf{x})\}$ from $F(\mathbf{x})$, we use the modulus inequality

$$\left| \int_\Omega f(\mathbf{x})\, d\Omega \right| \le \int_\Omega |f(\mathbf{x})|\, d\Omega, \tag{4.35}$$

to show that the numerical sequence $\{\int_\Omega f_n(\mathbf{x})\, d\Omega\}$ is a Cauchy sequence:

$$\begin{aligned} \left| \int_\Omega f_n(\mathbf{x})\, d\Omega - \int_\Omega f_m(\mathbf{x})\, d\Omega \right| &= \left| \int_\Omega [f_n(\mathbf{x}) - f_m(\mathbf{x})]\, d\Omega \right| \\ &\le \int_\Omega |f_n(\mathbf{x}) - f_m(\mathbf{x})|\, d\Omega \\ &\le (\mathrm{mes}\,\Omega)^{1-\frac{1}{p}} \left\| f_n(\mathbf{x}) - f_m(\mathbf{x}) \right\|_p \\ &\to 0 \quad \text{as } m, n \to \infty. \end{aligned}$$

The quantity

$$\int_\Omega F(\mathbf{x})\, d\Omega = \lim_{n\to\infty} \int_\Omega f_n(\mathbf{x})\, d\Omega \tag{4.36}$$

is uniquely determined by $F(\mathbf{x})$ and is called the Lebesgue integral of $F(\mathbf{x})$ over Ω. If the element $F(\mathbf{x})$ happens to correspond to a continuous function, then the Lebesgue integral equals the corresponding Riemann integral. Of course, it is important to understand that $F(\mathbf{x})$ is not a function in the ordinary sense: it is an equivalence class of Cauchy sequence of continuous functions. Nevertheless, for manipulative purposes it often does no harm to treat an element like $F(\mathbf{x})$ as if it were an ordinary function; we may justify this by our ability to choose and work with a representative function that is defined uniquely by some limit passage. With proper understanding we can also relax our notational requirements and employ lowercase notation such as $f(\mathbf{x})$ for an element of $L^p(\Omega)$. We shall do this whenever convenient.

The Lebesgue integral satisfies the inequality

$$\left| \int_\Omega F(\mathbf{x})\, d\Omega \right| \le (\mathrm{mes}\,\Omega)^{1/q} \left\| F(\mathbf{x}) \right\|_p, \qquad \frac{1}{p} + \frac{1}{q} = 1. \tag{4.37}$$

This results directly from passage to the limit $n \to \infty$ in

$$\left| \int_\Omega f_n(\mathbf{x})\, d\Omega \right| \le (\mathrm{mes}\,\Omega)^{1-\frac{1}{p}} \left\| f_n(\mathbf{x}) \right\|_p.$$

It can also be shown that a sufficient condition for existence of the integral

$$\int_\Omega F(\mathbf{x}) G(\mathbf{x})\, d\Omega$$

is that $F(\mathbf{x}) \in L^p(\Omega)$ and $G(\mathbf{x}) \in L^q(\Omega)$ for some p and q such that $1/p + 1/q = 1$. In this case Hölder's inequality

$$\left| \int_\Omega F(\mathbf{x})G(\mathbf{x})\,d\Omega \right| \leq \left(\int_\Omega |F(\mathbf{x})|^p\,d\Omega \right)^{1/p} \left(\int_\Omega |G(\mathbf{x})|^q\,d\Omega \right)^{1/q} \tag{4.38}$$

holds, with equality if and only if $F(\mathbf{x}) = \lambda G(\mathbf{x})$ for some λ.

If $p \geq 1$, then $L^p(\Omega)$ is a Banach space under the norm

$$\|F(\mathbf{x})\|_p = \left(\int_\Omega |F(\mathbf{x})|^p\,d\Omega \right)^{1/p}. \tag{4.39}$$

Verification of the norm axioms for $\|F(\mathbf{x})\|_p$ is mostly straightforward, depending on familiar limiting operations. To verify the triangle inequality

$$\|F(\mathbf{x}) + G(\mathbf{x})\|_p \leq \|F(\mathbf{x})\|_p + \|G(\mathbf{x})\|_p,$$

for instance, we write

$$\|f_n(\mathbf{x}) + g_n(\mathbf{x})\|_p \leq \|f_n(\mathbf{x})\|_p + \|g_n(\mathbf{x})\|_p$$

for representatives $\{f_n(\mathbf{x})\}$ and $\{g_n(\mathbf{x})\}$ of $F(\mathbf{x})$ and $G(\mathbf{x})$, and then let $n \to \infty$. In fact the validity of this is a consequence of the completion theorem, but we wished to prove it independently. The only norm axiom that warrants further mention is

$$\|F(\mathbf{x})\|_p = 0 \iff F(\mathbf{x}) = 0.$$

The statement "$F(\mathbf{x}) = 0$" on the right means that the stationary sequence $(0, 0, 0, \ldots)$, where 0 is the zero function on Ω, belongs to the equivalence class $F(\mathbf{x})$. So $L^p(\Omega)$ is indeed a normed linear space. That it is a Banach space follows from its construction via the metric space completion theorem. According to Theorem 4.39, $L^p(\Omega)$ is complete in the metric

$$\begin{aligned} d(F(\mathbf{x}), G(\mathbf{x})) &= \lim_{n\to\infty} \left(\int_\Omega |f_n(\mathbf{x}) - g_n(\mathbf{x})|^p\,d\Omega \right)^{1/p} \\ &= \left(\int_\Omega |F(\mathbf{x}) - G(\mathbf{x})|^p\,d\Omega \right)^{1/p}, \end{aligned}$$

which of course coincides with the metric induced by the norm (4.39).

We began our development with the base set S of continuous functions on Ω, and introduced $L^p(\Omega)$ as the completion of S in the norm (4.26). We defined the Lebesgue integral so that for any element of $L^p(\Omega)$ it is the unique number that coincides with Riemann integral of f if F corresponds to a continuous function f in the base set. In addition to the fact that the

Lebesgue integral is defined for a wider set of functions than the Riemann integral, the Lebesgue integral is more convenient for performing limit passages. These operations include taking the limit of an integral with respect to a parameter in the integrand (Lebesgue's theorem) and interchanging the order of integration in a repeated integral (Fubini's theorem). The theory of Riemann integration is based on the notion of Jordan measurability of a set in $\mathbb{R}^n$. The classical theory of Lebesgue integration starts with a wider notion of measurability of a set in $\mathbb{R}^n$. Under this definition the set of all rational points on the segment $[0, 1]$ is measurable and its Lebesgue measure is zero. These considerations fall outside our scope, and the interested reader should consult standard textbooks on real analysis for details. Lebesgue integration is not only useful in itself; it finds applications to Sobolev spaces and to the generalized setup of boundary value problems.

Example 4.43. Show that $L^p(\Omega)$ is separable for compact Ω.

Solution. First we show that the space of continuous functions with the L^p metric is separable. We know that the set $P_r(\Omega)$ of polynomials defined on Ω and having rational coefficients is dense in $C(\Omega)$, where $C(\Omega)$ is the space of continuous functions under the metric

$$\|f(\mathbf{x}) - g(\mathbf{x})\|_{C(\Omega)} = \max_{\mathbf{x}\in\Omega} |f(\mathbf{x}) - g(\mathbf{x})|. \tag{4.40}$$

This follows from the classical Weierstrass theorem. Hence for any $f(\mathbf{x})$ continuous on Ω we can find $p_\varepsilon(\mathbf{x}) \in P_r(\Omega)$ such that

$$\max_{\mathbf{x}\in\Omega} |f(\mathbf{x}) - p_\varepsilon(\mathbf{x})| \le \frac{\varepsilon}{(\operatorname{mes}\Omega)^{1/p}}.$$

(This is why the domain Ω was required to be compact.) Therefore we have

$$\|f(\mathbf{x}) - p_\varepsilon(\mathbf{x})\| = \left(\int_\Omega |f(\mathbf{x}) - p_\varepsilon(\mathbf{x})|^p \, d\Omega\right)^{1/p} \le \left(\frac{\varepsilon^p}{\operatorname{mes}\Omega}\int_\Omega d\Omega\right)^{1/p} = \varepsilon.$$

So imposing the L^p metric on the space of functions continuous on Ω, we get a separable metric space. Furthermore, $L^p(\Omega)$ is the completion of this space. Since the completion of a separable metric space is separable, the conclusion follows. □

4.6 Sobolev Spaces

We proceed to some normed spaces that play an important role in the modern treatment of partial differential equations. On the set of l times continuously differentiable functions $f(\mathbf{x})$ given on a compact set Ω, we have defined the family of norms

$$\|f(\mathbf{x})\|_{l,p} = \|f(\mathbf{x})\|_{W^{l,p}(\Omega)} = \left(\int_\Omega \sum_{|\alpha|\le l} |D^\alpha f(\mathbf{x})|^p \, d\Omega\right)^{1/p}, \qquad p \ge 1 \tag{4.41}$$

(again, recall Convention 1.4 on page 16). The resulting normed spaces are incomplete in their natural metrics. Applying the completion theorem to this case (in the same way we produced the Lebesgue spaces $L^p(\Omega)$), we obtain a family of Banach spaces known as the *Sobolev spaces* $W^{l,p}(\Omega)$. The form of the norm (4.41) suggests that the elements of a Sobolev space possess something like derivatives. We shall discuss these *generalized derivatives* momentarily, but at this point (4.41) seems to indicate that they belong to the space $L^p(\Omega)$. Because $W^{l,p}(\Omega)$ is a completion of the separable space $C^{(l)}(\Omega)$, Theorem 4.42 gives us

Theorem 4.44. *$W^{l,p}(\Omega)$, $p \ge 1$, is a separable normed space.*

We can use the following definition for a generalized derivative. For $u \in L^p(\Omega)$, K.O. Friedrichs called $v \in L^p(\Omega)$ a *strong derivative* $D^\alpha(u)$ if there exists a sequence $\{\varphi_n\}$, $\varphi_n \in C^{(\infty)}(\Omega)$, such that

$$\int_\Omega |u(\mathbf{x}) - \varphi_n(\mathbf{x})|^p \, d\Omega \to 0 \quad \text{and}$$
$$\int_\Omega |v(\mathbf{x}) - D^\alpha \varphi_n(\mathbf{x})|^p \, d\Omega \to 0 \quad \text{as } n \to \infty.$$

Since $C^{(\infty)}(\Omega)$ is dense in any $C^{(k)}(\Omega)$, an element of $W^{m,p}(\Omega)$ has all strong derivatives up to the order m lying in $L^p(\Omega)$. Note that in this definition we need not define intermediate derivatives as is done for standard derivatives. But this definition does not seem too classical or familiar. In [27], the notion of generalized derivative was introduced using variational ideas. Sobolev introduced this for elements of $L^p(\Omega)$ (not for just any element of course, but for those elements for which it can be done). He called $v \in L^p(\Omega)$ a *weak derivative* $D^\alpha u$ of $u \in L^p(\Omega)$ if for every function $\varphi(\mathbf{x}) \in \mathcal{D}$ the relation

$$\int_\Omega u(\mathbf{x}) D^\alpha \varphi(\mathbf{x}) \, d\Omega = (-1)^{|\alpha|} \int_\Omega v(\mathbf{x}) \varphi(\mathbf{x}) \, d\Omega \tag{4.42}$$

holds. Here $\mathcal{D}$ is the set of functions that are infinitely differentiable on Ω and that vanish in some neighborhood of the boundary of Ω (the neighborhood can vary from function to function). This definition of derivative inherits some ideas from the calculus of variations: in particular, the fundamental lemma insures that we are defining the derivative in a unique way. For elements of $W^{l,p}(\Omega)$ it can be demonstrated that the two notions of generalized derivative are equivalent. Of course, the name "generalized derivative" is warranted because classical derivatives (say, of functions continuous on Ω) are also generalized derivatives, but not vice versa.

Later we will discuss the Sobolev imbedding theorem.

4.7 Compactness

Definition 4.45. Let S be a subset of a metric space. We say that S is *precompact* if every sequence taken from S contains a Cauchy subsequence.

In many textbooks, the term "relatively compact" is used instead of "precompact."

Any bounded set in $\mathbb{R}^n$ is precompact. We know this from calculus, where the classical Bolzano–Weierstrass theorem asserts that any bounded sequence from $\mathbb{R}^n$ contains a Cauchy subsequence. This is not necessarily the case in other spaces, however (see Theorem 4.52). In § 4.3 we introduced c, the space of convergent numerical sequences with norm

$$\|\mathbf{x}\| = \sup_i |x_i|. \tag{4.43}$$

The sequence of elements

$$\begin{aligned}
\mathbf{x}_1 &= (1, 0, 0, 0, \ldots), \\
\mathbf{x}_2 &= (0, 1, 0, 0, \ldots), \\
\mathbf{x}_3 &= (0, 0, 1, 0, \ldots), \\
&\vdots
\end{aligned}$$

taken from c has no Cauchy subsequence, since for any pair of distinct elements $\mathbf{x}_i, \mathbf{x}_j$ we have $\|\mathbf{x}_i - \mathbf{x}_j\| = 1$. Nonetheless, this sequence is bounded: we have $\|\mathbf{x}_i\| = 1$ for each i. So the Bolzano–Weierstrass theorem for $\mathbb{R}^n$ does not automatically extend to all other normed spaces.

What is the principal difference between a bounded set in c and a bounded set in $\mathbb{R}^n$? In $\mathbb{R}^n$, using, say, three decimal places, we can approximate all the coordinates of any point of the unit ball up to an accuracy of 0.001. There are a finite number of points lying within the unit ball whose coordinates are the approximated coordinates of the actual points (the reader could calculate the actual number of such points for a space of n dimensions). Increasing accuracy through the use of m decimal places, $m > 3$, we again have a finite number of points with which we can better approximate any point of the unit ball. In c, as is shown by the above example, such an approximation of all the points of the unit ball by a finite number of elements within a prescribed precision is impossible.

Let us consider the abstract variant of an approximating finite set for some given set of points:

Definition 4.46. Let S and E be subsets of a metric space. We call E a *finite ε-net* for S if E is finite and for every $x \in S$ there exists $e \in E$ such that $d(x, e) < \varepsilon$. We say that S is *totally bounded* if there is a finite ε-net for S for every $\varepsilon > 0$.

Note that a set is totally bounded if when we draw a ball of radius ε about each point of an ε-net of the set, then the set is covered by the union of these balls (i.e., any point of the set is a point of one of the balls).

In particular, if a set is totally bounded, it is bounded. Indeed taking a 1-net we get a finite collection of balls that covers the set. It is clear that there exists some ball of finite radius that contains all these balls inside itself, and so all the points of the initial set, and this implies that the initial set is bounded.

Total boundedness of a set is exactly the same property we described for a ball of $\mathbb{R}^n$, on the existence of finite sets of points with which we can approximate the coordinates of any point of the ball within any prescribed accuracy. We said this was a crucial property in determining whether a set is compact. This is confirmed by the following *Hausdorff criterion.*

Theorem 4.47. *A subset of a metric space is precompact if and only if it is totally bounded.*

Proof. Let S be a precompact subset of a metric space X. To show that S is totally bounded, we prove the contrapositive statement. Suppose S has no finite ε_0-net for some particular $\varepsilon_0 > 0$. This means that no finite union of balls of radius ε_0 can contain S. Taking $x_1 \in S$ and a ball B_1 of radius ε_0 about x_1, we know that there exists $x_2 \in S$ such that $x_2 \notin B_1$ (otherwise x_1 by itself would generate a finite ε_0-net for S). Constructing the ball B_2 of radius ε_0 about x_2, we know that there exists $x_3 \in S$ such that $x_3 \notin B_1 \cup B_2$. Continuing in this way, we construct a sequence $\{x_n\}$ such that $d(x_n, x_m) \geq \varepsilon_0$ whenever $n \neq m$. Because $\{x_n\}$ cannot contain a Cauchy subsequence, S is not precompact.

Conversely, suppose S is totally bounded and take any sequence $\{x_n\}$ from S. We begin to select a Cauchy subsequence from $\{x_n\}$ by taking $\varepsilon_1 = 1/2$ and constructing a finite ε_1-net for S. One of the balls, say B_1, must contain infinitely many elements of $\{x_n\}$. Choose one of these elements and call it x_{i_1}. Then construct a finite ε_2-net for S with $\varepsilon_2 = 1/2^2$. One of the balls, say B_2, must contain infinitely many of those elements of $\{x_n\}$ which belong to B_1. Choose one of these elements and call it x_{i_2}. Note that $d(x_{i_1}, x_{i_2}) \le (2)(1/2) = 1$ since both x_{i_2} and x_{i_1} belong to B_1. Continuing in this way we obtain a subsequence $\{x_{i_k}\} \subset \{x_n\}$ where, by construction, x_{i_k} and $x_{i_{k+1}}$ reside in a ball B_k of radius $\varepsilon_k = 1/2^k$ so that

$$d(x_{i_k}, x_{i_{k+1}}) \le 2\left(\frac{1}{2^k}\right) = \frac{1}{2^{k-1}}.$$

Thus

$$\begin{aligned} d(x_{i_k}, x_{i_{k+m}}) &\le d(x_{i_k}, x_{i_{k+1}}) + d(x_{i_{k+1}}, x_{i_{k+2}}) + \cdots + d(x_{i_{k+m-1}}, x_{i_{k+m}}) \\ &\le \frac{1}{2^{k-1}} + \frac{1}{2^k} + \cdots + \frac{1}{2^{k+m-2}} < \frac{1}{2^{k-2}} \end{aligned}$$

for any $m \ge 1$, and $\{x_{i_k}\}$ is a Cauchy sequence. □

Definition 4.48. Let S be a subset of a metric space. We say that S is *compact* if every sequence taken from S contains a Cauchy subsequence that converges to a point of S.

Note that a compact subset of a metric space is closed. But a closed set is not, in general, compact. (In $\mathbb{R}^n$ a closed *and* bounded set is compact according to the present definition.) Let us reformulate the Hausdorff criterion for compactness:

Theorem 4.49. *A subset of a complete metric space is compact if and only if it is closed and totally bounded.*

The proof is left as an exercise.

Example 4.50. Show that the *Hilbert cube*

$$S = \{\mathbf{x} = (\xi_1, \xi_2, \ldots) \in \ell^2 \colon |\xi_n| \le \tfrac{1}{n} \text{ for } n = 1, 2, \ldots\}$$

is a compact subset of ℓ^2.

Solution. We show that S is closed and totally bounded in the complete space ℓ^2. Let $\mathbf{y} = (\eta_1, \eta_2, \ldots)$ be a limit point of S. There is a sequence

$\{\mathbf{x}^{(j)}\} \subset S$ such that

$$\|\mathbf{y} - \mathbf{x}^{(j)}\|_{\ell^2}^2 = \sum_{k=1}^{\infty} |\eta_k - \xi_k^{(j)}|^2 \to 0 \quad \text{as } j \to \infty.$$

Hence for each k we have $|\eta_k - \xi_k^{(j)}| \to 0$ as $j \to \infty$. By the triangle inequality

$$|\eta_k| \leq |\eta_k - \xi_k^{(j)}| + |\xi_k^{(j)}| \leq |\eta_k - \xi_k^{(j)}| + \frac{1}{k},$$

and passage to the limit as $j \to \infty$ gives $|\eta_k| \leq \frac{1}{k}$ for each k. This shows that $\mathbf{y} \in S$, hence S is closed. Next we show that S is totally bounded. Let $\varepsilon > 0$ be given. We begin to construct a finite ε-net by noting that the nth component of any element $\mathbf{z} = (\zeta_1, \zeta_2, \ldots) \in S$ differs from zero by no more than $1/n$. Since the series $\sum 1/n^2$ is convergent we can choose N such that

$$\sum_{n=N+1}^{\infty} |\zeta_n|^2 < \varepsilon^2/2.$$

Now take the first N components and consider the corresponding bounded closed hypercube in $\mathbb{R}^N$. For this there certainly exists a finite $\varepsilon^2/2$-net of N-tuples, and we can select $(\xi_1, \ldots, \xi_N)$ such that

$$\sum_{n=1}^{N} |\zeta_n - \xi_n|^2 < \varepsilon^2/2.$$

We construct a corresponding element $\mathbf{x}_\varepsilon \in \ell^2$ by appending zeros:

$$\mathbf{x}_\varepsilon = (\xi_1, \ldots, \xi_N, 0, 0, \ldots).$$

For this element

$$\|\mathbf{z} - \mathbf{x}_\varepsilon\|_{\ell^2}^2 = \sum_{n=1}^{N} |\zeta_n - \xi_n|^2 + \sum_{n=N+1}^{\infty} |\zeta_n|^2 < \varepsilon^2/2 + \varepsilon^2/2 = \varepsilon^2$$

as desired. □

Theorem 4.51. *Every precompact metric space is separable.*

Proof. Let X be a precompact metric space. For each $k = 1, 2, 3, \ldots$, let $\varepsilon_k = 1/k$ and construct a finite ε_k-net $(x_{k1}, x_{k2}, \ldots, x_{kN})$ for X. (Here N depends on k.) The union of these nets is countable and dense in X. □

Theorem 4.52. *Every closed and bounded subset of a Banach space is compact if and only if the Banach space has finite dimension.*

The proof of Theorem 4.52 requires *Riesz's lemma*:

Lemma 4.53. *Let M be a proper closed subspace of a normed space X. If $0 < \varepsilon < 1$, then there is an element $x_\varepsilon \notin M$ having unit norm such that*

$$\inf_{y \in M} \|y - x_\varepsilon\| > 1 - \varepsilon.$$

(Here we use the term "proper" to exclude the case $M = X$.)

Proof. Take an element $x_0 \in X$ that does not belong to M and let

$$d = \inf_{y \in M} \|x_0 - y\|.$$

We have $d > 0$; indeed, the assumption $d = 0$ leads to a contradiction because it implies the existence of a sequence $\{y_k\} \subset M$ such that $\|x_0 - y_k\| \to 0$, hence $y_k \to x_0$, hence $x_0 \in M$ because M is closed. By definition of infimum, for any $\varepsilon > 0$ there exists $y_\varepsilon \in M$ such that

$$d \le \|x_0 - y_\varepsilon\| < \frac{d}{1 - \varepsilon/2}.$$

The normalized element

$$x_\varepsilon = \frac{x_0 - y_\varepsilon}{\|x_0 - y_\varepsilon\|}$$

has the properties specified in the lemma. It clearly has unit norm and does not belong to M. Moreover, for any $y \in M$ we have

$$\begin{aligned} \|x_\varepsilon - y\| &= \left\| \frac{x_0 - y_\varepsilon}{\|x_0 - y_\varepsilon\|} - y \right\| = \frac{\|x_0 - (y_\varepsilon + \|x_0 - y_\varepsilon\| \, y)\|}{\|x_0 - y_\varepsilon\|} \\ &> d \Big/ \frac{d}{1 - \varepsilon/2} = 1 - \frac{\varepsilon}{2} \end{aligned}$$

where the intermediate inequality holds because $y_\varepsilon + \|x_0 - y_\varepsilon\| \, y$ belongs to M. □

As an application of Riesz's lemma, let us show that the unit ball

$$B = \{x \in X \colon \|x\| \le 1\} \tag{4.44}$$

is not compact if X is infinite dimensional. (This is the "only if" part of Theorem 4.52.) Take $y_1 \in B$. This element generates a proper closed subspace E_1 of X given by $E_1 = \{\alpha y_1 \colon \alpha \in \mathbb{C}\}$. By Riesz's lemma (with $\varepsilon = 1/2$) there exists y_2 such that $y_2 \in B$, $y_2 \notin E_1$, and $\|y_1 - y_2\| > 1/2$. The elements y_1, y_2 generate a proper closed subspace E_2 of X, and by Riesz's lemma there exists y_3 such that $y_3 \in B$, $y_3 \notin E_2$, and $\|y_i - y_3\| > 1/2$

for $i = 1, 2$. Since X is infinite dimensional we can continue this process indefinitely, producing a sequence $\{y_n\} \subset B$ any two distinct points of which are separated by a distance exceeding $1/2$. Since no subsequence of $\{y_n\}$ is a Cauchy sequence, B is not compact.

Definition 4.54. Let M be a set of functions continuous on a compact set $\Omega \subset \mathbb{R}^n$. We say that M is

(1) *uniformly bounded* if there is a constant c such that for every $f(\mathbf{x}) \in M$, $|f(\mathbf{x})| \leq c$ for all $\mathbf{x} \in \Omega$.
(2) *equicontinuous* if for any $\varepsilon > 0$ there exists $\delta > 0$, dependent on ε, such that whenever $|\mathbf{x} - \mathbf{y}| < \delta$, $\mathbf{x}, \mathbf{y} \in \Omega$, then $|f(\mathbf{x}) - f(\mathbf{y})| < \varepsilon$ holds for every $f(\mathbf{x}) \in M$.

Uniform boundedness simply means that the set of functions lies in a ball of radius c in $C(\Omega)$ (in Arzelà's time the normed space terminology was not yet in full use). Since the space $C(\Omega)$ is infinite dimensional, this cannot be the sole condition for compactness. We also note that any finite set of continuous functions is equicontinuous by Weierstrass's theorem from calculus. Given $\varepsilon > 0$, we can find the required $\delta > 0$ for each individual function. We then take the minimum of these values, which is not zero, and use it as δ for the whole set. An infinite set of continuous functions need not be equicontinuous.

The space of continuous functions is one of the main objects of calculus, differential equations, and many other branches of mathematics. It is important to have a set of practical criteria under which a subset of this space must be precompact. This is provided by *Arzelà's theorem.*

Theorem 4.55. *Let Ω be a compact set in $\mathbb{R}^n$, and let M be a set of functions continuous on Ω. Then M is precompact in $C(\Omega)$ if and only if it is uniformly bounded and equicontinuous.*

Proof. Suppose M is precompact in $C(\Omega)$. By Theorem 4.47 there is a finite ε-net for M with $\varepsilon = 1$; i.e., there is a finite set of continuous functions $\{g_i(\mathbf{x})\}_{i=1}^k$ such that to any $f(\mathbf{x})$ there corresponds $g_i(\mathbf{x})$ for which

$$\|f(\mathbf{x}) - g_i(\mathbf{x})\| = \max_{\mathbf{x} \in \Omega} |f(\mathbf{x}) - g_i(\mathbf{x})| \leq 1.$$

Since the $g_j(\mathbf{x})$ are continuous there is a constant c_1 such that $\|g_j(\mathbf{x})\| < c_1$ for each j. Using the inequality $\|f(\mathbf{x})\| \leq \|g_i(\mathbf{x})\| + \|f(\mathbf{x}) - g_i(\mathbf{x})\|$, we have

$$\max_{\mathbf{x} \in \Omega} |f(\mathbf{x})| \leq c_1 + 1.$$

It follows that M is uniformly bounded with $c = c_1 + 1$. We proceed to verify equicontinuity. Let $\varepsilon > 0$ be given, and choose a finite $\varepsilon/3$-net for M, say $\{g_i(\mathbf{x})\}_{i=1}^m$. Since the number of $g_i(\mathbf{x})$ is finite and, by a calculus theorem, each of them is equicontinuous on Ω, there exists $\delta > 0$ such that $|\mathbf{x} - \mathbf{y}| < \delta$ implies

$$|g_i(\mathbf{x}) - g_i(\mathbf{y})| < \varepsilon/3, \qquad i = 1, \ldots, m.$$

For each $f(\mathbf{x}) \in M$, there exists $g_r(\mathbf{x})$ such that

$$|f(\mathbf{x}) - g_r(\mathbf{x})| < \varepsilon/3 \text{ for all } \mathbf{x} \in \Omega.$$

Whenever $\mathbf{x}, \mathbf{y} \in \Omega$ are such that $|\mathbf{x} - \mathbf{y}| < \delta$ then, we have

$$\begin{aligned} |f(\mathbf{x}) - f(\mathbf{y})| &\le |f(\mathbf{x}) - g_r(\mathbf{x})| + |g_r(\mathbf{x}) - g_r(\mathbf{y})| + |g_r(\mathbf{y}) - f(\mathbf{y})| \\ &< \varepsilon/3 + \varepsilon/3 + \varepsilon/3 = \varepsilon \end{aligned}$$

as desired.

Conversely suppose M is uniformly bounded and equicontinuous. We must show that from any sequence of functions $\{f_k(\mathbf{x})\} \subset M$ we can choose a Cauchy subsequence. Let $\{\mathbf{x}_k\}$ be the set of all rational points of Ω (enumerated somehow); this set is countable and dense in Ω. Consider the sequence $\{f_k(\mathbf{x}_1)\}$. Because this numerical sequence is bounded, we can choose a Cauchy subsequence $\{f_{k_1}(\mathbf{x}_1)\}$. We have thus chosen a subsequence $\{f_{k_1}(\mathbf{x})\} \subset \{f_k(\mathbf{x})\}$ that is a Cauchy sequence at $\mathbf{x} = \mathbf{x}_1$. From the bounded numerical sequence $\{f_{k_1}(\mathbf{x}_2)\}$ we can choose a Cauchy subsequence $\{f_{k_2}(\mathbf{x}_2)\}$. The subsequence $\{f_{k_2}(\mathbf{x})\}$ is thus a Cauchy sequence at both $\mathbf{x} = \mathbf{x}_1$ and $\mathbf{x} = \mathbf{x}_2$. We continue in this way, taking subsequences of previously constructed subsequences, so that on the nth step the subsequence $\{f_{k_n}(\mathbf{x}_n)\}$ is a Cauchy sequence and, since it is a subsequence of any previous subsequence, the sequences obtained by evaluating $\{f_{k_n}(\mathbf{x})\}$ at $\mathbf{x}_1, \ldots, \mathbf{x}_{n-1}$ are Cauchy sequences as well.

The diagonal sequence $\{f_{n_n}(\mathbf{x})\}$ is a Cauchy sequence at $\mathbf{x} = \mathbf{x}_i$ for all i. We show that it is a Cauchy sequence in the norm of $C(\Omega)$. Let $\varepsilon > 0$ be given. According to equicontinuity we can find $\delta > 0$ such that $|\mathbf{x} - \mathbf{y}| < \delta$ gives for every n

$$|f_{n_n}(\mathbf{x}) - f_{n_n}(\mathbf{y})| < \varepsilon/3.$$

Take $\delta' < \delta$ and construct a finite δ'-net for Ω with nodes $\{\mathbf{z}_i\}_{i=1}^r \subset \{\mathbf{x}_i\}$. Since r is finite we can find N such that whenever $n, m > N$ we have

$$|f_{n_n}(\mathbf{z}_i) - f_{m_m}(\mathbf{z}_i)| < \varepsilon/3, \quad i = 1, \ldots, r.$$

Choose any $\mathbf{x} \in \Omega$ and let $\mathbf{z}_k$ be the point of the δ'-net nearest $\mathbf{x}$ so that $|\mathbf{x} - \mathbf{z}_k| < \delta'$. Then $n, m > N$ implies

$$|f_{n_n}(\mathbf{x}) - f_{m_m}(\mathbf{x})| \le |f_{n_n}(\mathbf{x}) - f_{n_n}(\mathbf{z}_k)| + |f_{n_n}(\mathbf{z}_k) - f_{m_m}(\mathbf{z}_k)| \\ + |f_{m_m}(\mathbf{z}_k) - f_{m_m}(\mathbf{x})| < \varepsilon/3 + \varepsilon/3 + \varepsilon/3 = \varepsilon,$$

hence

$$\max_{\mathbf{x} \in \Omega} |f_{n_n}(\mathbf{x}) - f_{m_m}(\mathbf{x})| = \|f_{n_n}(\mathbf{x}) - f_{m_m}(\mathbf{x})\| < \varepsilon$$

for all $n, m > N$. □

Remark 4.56. In the proof we made use of the *diagonal sequence* idea. Since this is a standard technique in analysis and will be used again in this chapter, we take a moment to clarify the ideas involved.

Suppose we start with a sequence $\{x_n\}$ and want to extract a subsequence that satisfies some set of convergence-related criteria p_k ($k = 1, 2, 3, \ldots$). Let us agree to write $\{x_{nk}\}$ for the subsequence we select at the kth step of the process ($k = 1, 2, 3, \ldots$), and x_{nk} for the nth element of that subsequence ($n = 1, 2, 3, \ldots$).

The process begins with the selection of successive subsequences:

1. From $\{x_n\}$ we select $\{x_{n1}\}$ that satisfies p_1. It is clear that the whole sequence $\{x_{n1}\}$ as well as each of its subsequences satisfies p_1.
2. Then from $\{x_{n1}\}$ we take $\{x_{n2}\}$ that satisfies p_2. The whole sequence as well as each of its subsequences satisfies p_2. Being a subsequence of $\{x_{n1}\}$, it and all of its subsequences satisfy p_1 as well.
3. The same is done with $\{x_{n2}\}$: choose $\{x_{n3}\}$ that satisfies p_3, so all of its subsequences satisfy p_3 and, simultaneously, p_1 and p_2.

⋮

k. Choose $\{x_{nk}\}$ that satisfies p_k and $p_1, \ldots, p_{k-1}$.

⋮

We now form the sequence

$$\{x_{nn}\}_{n=1}^{\infty} = x_{11},\ x_{22},\ x_{33},\ \ldots. \tag{4.45}$$

This is the desired diagonal sequence.

The sequence (4.45) is automatically contained in $\{x_{n1}\}$. Except possibly for the first term, it is also contained in $\{x_{n2}\}$; the first term is a non-issue because the behavior of a finite number of terms has no impact on the satisfaction of p_2. Except possibly for the first two terms, (4.45) is also contained in $\{x_{n3}\}$, and so on. So the diagonal sequence, except for finite numbers of terms, is contained in $\{x_{nk}\}$ for each k. It therefore satisfies p_k for $k = 1, 2, 3, \ldots$. □

Example 4.57. Let Ω be a compact subset of $\mathbb{R}^n$, and suppose S is a collection of functions $\{f_k(\mathbf{x})\}$ continuous on Ω. Further, suppose S is bounded in $C(\Omega)$ and that $K(\mathbf{x}, \mathbf{y})$ is a function continuous on $\Omega \times \Omega$. Show that the set

$$A = \left\{ \int_\Omega K(\mathbf{x}, \mathbf{y}) f_k(\mathbf{y})\, d\Omega_\mathbf{y} \right\}$$

is precompact in $C(\Omega)$.

Solution. The members of A clearly belong to $C(\Omega)$. Uniform boundedness of A is shown by the inequality

$$\max_{\mathbf{x}\in\Omega} \left| \int_\Omega K(\mathbf{x}, \mathbf{y}) f_k(\mathbf{y})\, d\Omega_\mathbf{y} \right| \le \max_{\mathbf{x}\in\Omega} |f_k(\mathbf{x})| \cdot \max_{(\mathbf{x},\mathbf{y})\in\Omega\times\Omega} |K(\mathbf{x}, \mathbf{y})| \cdot \operatorname{mes}\Omega,$$

since the set $\{f_k(\mathbf{x})\}$ is itself uniformly bounded so that $\max_{\mathbf{x}\in\Omega} |f_k(\mathbf{x})| \le c$ where c is some constant that does not depend on k. Equicontinuity of A follows from the inequality

$$\left| \int_\Omega K(\mathbf{x}, \mathbf{y}) f_k(\mathbf{y})\, d\Omega_\mathbf{y} - \int_\Omega K(\mathbf{x}', \mathbf{y}) f_k(\mathbf{y})\, d\Omega_\mathbf{y} \right|$$
$$\le c \cdot \int_\Omega |K(\mathbf{x}, \mathbf{y}) - K(\mathbf{x}', \mathbf{y})|\, d\Omega_\mathbf{y}.$$

Indeed, for any $\varepsilon > 0$ there exists $\delta = \delta(\varepsilon)$ such that

$$|K(\mathbf{x}, \mathbf{y}) - K(\mathbf{x}', \mathbf{y})| \le \frac{\varepsilon}{c \operatorname{mes}\Omega}$$

whenever $|\mathbf{x} - \mathbf{x}'| < \delta$ (independent of $\mathbf{y} \in \Omega$). Because A is a uniformly bounded and equicontinuous subset of $C(\Omega)$, it is precompact in $C(\Omega)$ by Arzelà's theorem. □

People working in application areas often prefer to have crude but convenient sufficient conditions for the fulfillment of some properties. In the case of $C(a, b)$, the space of functions continuous on $[a, b]$, a sufficient condition is given by

Theorem 4.58. *A set of continuously differentiable functions bounded in the space $C^{(1)}(a, b)$ is precompact in the space $C(a, b)$.*

Proof. The proof follows from the classical Lagrange theorem which for any continuously differentiable function $f(x)$ and arbitrary x, y guarantees the existence of $z \in [x, y]$ such that $f(x) - f(y) = f'(z)(x - y)$. Equicontinuity of a bounded subset of $C^{(1)}(a, b)$ is a consequence of this. Uniform boundedness of the set is evident. □

The reader can formulate and prove the similar statement for the more general space $C^{(1)}(\Omega)$. Indeed there is an analogue of the mean value theorem for multivariable functions belonging to $C^{(1)}(\Omega)$ where Ω is compact and convex. A region Ω is said to be convex if for any two points $\mathbf{x}, \mathbf{y} \in \Omega$ the connecting segment $A = \{t\mathbf{y} + (1 - t)\mathbf{x}\}$, $t \in [0, 1]$, lies in Ω. Consider a function $f(\mathbf{x}) \in C^{(1)}(\Omega)$. For fixed $\mathbf{x}, \mathbf{y}$, the function

$$F(t) = f(t\mathbf{y} + (1 - t)\mathbf{x})$$

of the real argument t belongs to $C^{(1)}(0, 1)$, hence the one-dimensional form of Lagrange's formula yields

$$F(1) - F(0) = F_t(t)|_{t=\xi}(1 - 0) \quad \text{for some } \xi \in [0, 1].$$

Rewriting this in terms of f we get

$$f(\mathbf{y}) - f(\mathbf{x}) = \nabla f(\mathbf{z})\big|_{\mathbf{z}=\xi\mathbf{y}+(1-\xi)\mathbf{x}} \cdot (\mathbf{y} - \mathbf{x}),$$

which is also called Lagrange's formula. The estimate

$$|f(\mathbf{y}) - f(\mathbf{x})| \le \max_{\mathbf{z}\in A} |\nabla f(\mathbf{z})||\mathbf{y} - \mathbf{x})|$$

follows immediately. In the same way, beginning with the Newton–Leibniz formula

$$F(1) - F(0) = \int_0^1 F_t(t)\, dt$$

it is easy to prove the integral formula

$$f(\mathbf{y}) - f(\mathbf{x}) = \int_0^1 \nabla f(\mathbf{z})\big|_{\mathbf{z}=t\mathbf{y}+(1-t)\mathbf{x}} \cdot (\mathbf{y} - \mathbf{x})\, dt.$$

From this we can derive the above estimate as well.

Note that now we consider the same continuously differentiable functions as elements of different spaces, $C^{(1)}(\Omega)$ and $C(\Omega)$. When we consider the correspondence between an element in $C^{(1)}(\Omega)$ and the same element in $C(\Omega)$, it is not an identity mapping since the spaces are different and the properties of the operator are defined not only by the elements but also by the properties of the spaces. This a typical example of an operator of imbedding (we imbed a set of $C^{(1)}(\Omega)$ into $C(\Omega)$). Using this term and the notion of compact operator given in § 4.16, we can reformulate the last theorem as follows:

Theorem 4.59. *Let Ω be a compact set in $\mathbb{R}^n$. The imbedding operator from $C^{(1)}(\Omega)$ into $C(\Omega)$ is compact.*

The concept of an imbedding operator between normed spaces will be covered formally in § 4.18.

4.8 Inner Product Spaces, Hilbert Spaces

The existence of the dot product in Euclidean space offers many advantages with respect to the operations that may be performed in the space. The dot product also generates the norm in Euclidean space. In order that there might exist a functional defined on each pair of elements of a normed space and possessing the properties of the dot product, a linear space X should have quite special properties. Let us define what we call an *inner product.* This is a functional (x, y) defined (i.e., always giving a uniquely defined finite result) for any pair of elements x, y of the space X, and having the following properties:

(1) $(x, x) \geq 0$ for all $x \in X$, with $(x, x) = 0$ if and only if $x = 0$.
(2) $(y, x) = \overline{(x, y)}$ for all $x, y \in X$.
(3) $(\lambda x + \mu y, z) = \lambda(x, z) + \mu(y, z)$ for all $x, y, z \in X$ and any complex scalars λ, μ.

We have defined this for a complex space. If X is a real space instead, then property 2 must be changed to

2. $(y, x) = (x, y)$ for all $x, y \in X$

and in property 3 we must use only real scalars λ, μ. Note that the inner product is linear in the first argument and *conjugate linear* in the second argument:

$$(\alpha_1 x_1 + \alpha_2 x_2, y) = \alpha_1(x_1, y) + \alpha_2(x_2, y), \tag{4.46}$$

$$(x, \alpha_1 y_1 + \alpha_2 y_2) = \overline{\alpha}_1(x, y_1) + \overline{\alpha}_2(x, y_2). \tag{4.47}$$

Example 4.60. Let X be an inner product space under the inner product $(\cdot, \cdot)$. Show that $(x, z) = (y, z)$ holds for arbitrary $z \in X$ if and only if

$x = y$.

Solution. The "if" part of the proposition is trivial. To prove the "only if" part, assume $(x, z) = (y, z)$ for all $z \in X$. Rearranging this as

$$(x, z) - (y, z) = 0,$$

we can use property 3 to get $(x - y, z) = 0$. Since this holds for all $z \in X$, it holds in particular for $z = x - y$:

$$(x - y, x - y) = 0.$$

By property 1 we conclude that $x - y = 0$ or $x = y$. □

Since this functional, the inner product, is defined by copying the main properties of the dot product, we preserve the terminology connected with the dot product in Euclidean space. In particular there is the notion of orthogonality. We say that two elements x, y are mutually orthogonal if $(x, y) = 0$. We say that x is orthogonal to Y, a subspace of X, if x is orthogonal to each element of Y.

Definition 4.61. A linear space with an inner product possessing the properties listed above is an *inner product space* or a *pre-Hilbert space.*

First we demonstrate

Theorem 4.62. *A pre-Hilbert space is a normed space.*

Proof. By similarity to Euclidean space let us introduce a functional denoted as a norm

$$\|x\| = (x, x)^{1/2}. \tag{4.48}$$

This functional is defined for any element of X. Let us demonstrate that it satisfies all the axioms of the norm. Norm axiom 1 is fulfilled by virtue of inner product axiom 1. We verify norm axiom 2 by noting that

$$\begin{aligned}
\|\lambda x\| &= [(\lambda x, \lambda x)]^{1/2} = [\lambda(x, \lambda x)]^{1/2} = [\lambda\overline{(\lambda x, x)}]^{1/2} \\
&= [(\lambda\overline{\lambda})\overline{(x, x)}]^{1/2} = [|\lambda|^2(x, x)]^{1/2} \\
&= |\lambda|(x, x)^{1/2}.
\end{aligned}$$

Verification of norm axiom 3 requires us to use the *Schwarz inequality*

$$|(x, y)| \le \|x\| \, \|y\|, \tag{4.49}$$

in which for nonzero x and y the equality holds if and only if there is a number λ such that $x = \lambda y$. Using it we have

$$\begin{aligned} \|x+y\|^2 &= (x+y, x+y) \\ &= (x,x) + (x,y) + (y,x) + (y,y) \\ &\le \|x\|^2 + \|x\| \, \|y\| + \|x\| \, \|y\| + \|y\|^2 \\ &= (\|x\| + \|y\|)^2 \end{aligned}$$

as required. □

It remains to establish (4.49). We start by noting that if $x = 0$ or $y = 0$ then (4.49) is evidently valid. So let $y \neq 0$. If λ is any scalar, then $(x + \lambda y, x + \lambda y) \geq 0$ and expansion gives

$$(x + \lambda y, x + \lambda y) = (x,x) + \lambda(y,x) + \overline{\lambda}(x,y) + \lambda\overline{\lambda}(y,y).$$

The particular choice $\lambda = -(x,y)/(y,y)$ reduces this to

$$\|x\|^2 - 2\frac{|(x,y)|^2}{\|y\|^2} + \frac{|(x,y)|^2 \|y\|^2}{\|y\|^4} \geq 0,$$

and (4.49) follows directly.

Example 4.63. Show that

$$\|x+y\|^2 + \|x-y\|^2 = 2\|x\|^2 + 2\|y\|^2 . \tag{4.50}$$

This is known as the *parallelogram equality.*

Solution. We write

$$\begin{aligned} \|x+y\|^2 + \|x-y\|^2 &= (x+y, x+y) + (x-y, x-y) \\ &= (x, x+y) + (y, x+y) + (x, x-y) - (y, x-y) \\ &= \overline{(x+y, x)} + \overline{(x+y, y)} + \overline{(x-y, x)} - \overline{(x-y, y)} \\ &= \overline{(x,x)} + \overline{(y,x)} + \overline{(x,y)} + \overline{(y,y)} + \\ &\quad + \overline{(x,x)} - \overline{(y,x)} - \overline{(x,y)} + \overline{(y,y)} \\ &= 2(x,x) + 2(y,y) \\ &= 2\|x\|^2 + 2\|y\|^2 \end{aligned}$$

and have the desired result. □

Example 4.64. Show that if x and y are orthogonal vectors in an inner product space, then

$$\|x+y\|^2 = \|x\|^2 + \|y\|^2 . \tag{4.51}$$

This is known as the *Pythagorean theorem.*

Solution. We write

$$\begin{aligned}\|x+y\|^2 &= (x+y, x+y)\\ &= (x, x+y) + (y, x+y)\\ &= (x,x) + (x,y) + (y,x) + (y,y)\end{aligned}$$

and simply note that $(x, y) = (y, x) = 0$ for orthogonal vectors. □

Example 4.65. Assume the norm is induced by the inner product, and suppose that $x_n \to x$ and $y_n \to y$. Show that $(x_n, y_n) \to (x, y)$. That is, any inner product is a continuous functional in each of its arguments.

Solution. Let us write

$$\begin{aligned}|(x_n, y_n) - (x, y)| &= |(x_n, y_n) - (x_n, y) + (x_n, y) - (x, y)|\\ &= |(x_n, y_n - y) + (x_n - x, y)|\\ &\le |(x_n, y_n - y)| + |(x_n - x, y)|\\ &\le \|x_n\| \, \|y_n - y\| + \|x_n - x\| \, \|y\| .\end{aligned}$$

Since $\{x_n\}$ is convergent it is bounded. The other n-dependent quantities can be made as small as desired by choosing n sufficiently large. □

Example 4.66. Let M be a dense subset of an inner product space X, and let $v \in X$. Show that if $(v, m) = 0$ for all $m \in M$, then $v = 0$.

Solution. Use continuity of the inner product. Let $v \in X$ be fixed. Since M is dense in X there is a sequence of elements $m_k \in M$ such that $m_k \to v$ as $k \to \infty$. Since $0 = (v, m_k)$ for all k, we can take the limit as $k \to \infty$ on both sides and use continuity of the inner product to obtain

$$0 = \lim_{k\to\infty} (v, m_k) = \left(v, \lim_{k\to\infty} m_k\right) = (v, v).$$

Hence $v = 0$. □

Definition 4.67. A complete pre-Hilbert space is a *Hilbert space.*

Let us consider some Hilbert spaces. The space ℓ^2 is the space of infinite sequences having inner product

$$(\mathbf{x}, \mathbf{y}) = \sum_{i=1}^{\infty} x_i \overline{y_i} \tag{4.52}$$

in the complex case and

$$(\mathbf{x}, \mathbf{y}) = \sum_{i=1}^{\infty} x_i y_i \tag{4.53}$$

in the real case. The corresponding generated (induced) norm is

$$\|\mathbf{x}\| = (x, x)^{1/2} = \left(\sum_{i=1}^{\infty} |x_i|^2 \right)^{1/2}. \tag{4.54}$$

As we noted earlier, the theory of the space ℓ^2 was the predecessor of functional analysis. It plays an extremely important role in the functional analysis of Hilbert spaces because, as we shall see later, with any separable Hilbert space we have a one-to-one isometric correspondence with ℓ^2 that preserves algebraic operations in the spaces. This is done by Fourier expansion of elements of the Hilbert space.

In the space $L^2(\Omega)$ an inner product can be introduced as

$$(f(\mathbf{x}), g(\mathbf{x})) = \int_{\Omega} f(\mathbf{x})\overline{g(\mathbf{x})}\, d\Omega \tag{4.55}$$

in the complex case and

$$(f(\mathbf{x}), g(\mathbf{x})) = \int_{\Omega} f(\mathbf{x})g(\mathbf{x})\, d\Omega \tag{4.56}$$

in the real case. We have defined the inner product in such a way that the induced norm coincides with the norm imposed earlier on $L^2(\Omega)$. This raises the question of how to introduce an inner product in any Sobolev space $W^{l,2}(\Omega)$. We use

$$(f(\mathbf{x}), g(\mathbf{x})) = \int_{\Omega} \sum_{|\alpha| \le l} D^{\alpha} f(\mathbf{x}) \overline{D^{\alpha} g(\mathbf{x})}\, d\Omega.$$

The induced norm is the norm we introduced earlier in $W^{l,2}(\Omega)$.

4.9 Operators and Functionals

Definition 4.68. A correspondence between two sets (metric spaces) X and Y, under which to any element of X there corresponds no more than one element of Y, is an *operator*. Frequent synonyms for "operator" are *map, mapping, function, correspondence*, and *transformation*.

Let A be an operator. The set of elements $x \in X$ for which there is a corresponding element $y \in Y$ is the *domain* of A and is denoted $D(A)$. We write $y = A(x)$ and call y the *image* of x under A. It is not necessarily true that each element $y \in Y$ is the image of some element $x \in X$ under A; the set of all elements of Y that *are* images of elements of X is known as the *range* of A and is denoted $R(A)$. We say that A *acts from* X *to* Y. If $Y = X$, we say that A *acts in* the set X.

Definition 4.69. If Y is $\mathbb{C}$ (or $\mathbb{R}$), then an operator acting from X to Y is a complex (or real) *functional* defined on X.

An important role in functional analysis is played by linear operators. To explore this notion we need X and Y to be linear spaces.

Definition 4.70. An operator A from a linear space X to a linear space Y is a *linear operator* if for any elements x_1 and x_2 of X and any scalars λ and μ we have

$$A(\lambda x_1 + \mu x_2) = \lambda A(x_1) + \mu A(x_2). \tag{4.57}$$

For a linear operator A, we often write Ax instead of $A(x)$. Linear operators are not as elementary as they may seem. Many physical problems are linear. We now extend the definition of function continuity to operators:

Definition 4.71. Let A be an operator from a normed space X to a normed space Y. We say that A is continuous at $x_0 \in X$ if to each $\varepsilon > 0$ there corresponds $\delta = \delta(\varepsilon) > 0$ such that $\|Ax - Ax_0\|_Y < \varepsilon$ whenever $\|x - x_0\|_X < \delta$.

Example 4.72. Show that any norm is a continuous mapping from X to $\mathbb{R}$. Note, however, that it is not a linear functional.

Solution. Using the inequality of Example 4.4 we can write

$$\big| \|x\| - \|x_0\| \big| \le \|x - x_0\| \,.$$

Given $\varepsilon > 0$ then, we can choose $\delta = \varepsilon$ in the definition of continuity. □

For linear operators there is a convenient theorem:

Theorem 4.73. *A linear operator defined on a normed space X is continuous if and only if it is continuous at $x = 0$.*

Proof. Immediate from the relation $Ax - Ax_0 = A(x - x_0)$. □

There is a central theorem that shows how to check whether a linear operator is continuous:

Theorem 4.74. *A linear operator A from a normed space X to a normed space Y is continuous if and only if there is a constant c such that for all $x \in D(A)$,*

$$\|Ax\| \leq c\,\|x\| \tag{4.58}$$

Proof. Assume (4.58) holds. Then with $\delta = \varepsilon/c$ in the definition of continuity we see that A is continuous at $x = 0$. Conversely, suppose A is continuous at $x = 0$. Take $\varepsilon = 1$; by definition there exists $\delta > 0$ such that $\|Ax\| \leq 1$ whenever $\|x\| < \delta$. For every nonzero $x \in X$, the norm of $x^* = \delta x/(2\,\|x\|)$ is

$$\|x^*\| = \|\delta x/(2\,\|x\|)\| = \delta/2 < \delta,$$

so $\|Ax^*\| \leq 1$. By linearity of A this gives us

$$\|Ax\| \leq \frac{2}{\delta}\,\|x\|\,,$$

which is (4.58) with $c = 2/\delta$. □

So continuous linear operators are often called *bounded* linear operators.

Definition 4.75. The least constant c satisfying (4.58) is the *norm* of A, denoted $\|A\|$.

Note that $\|A\|$ satisfies all the norm axioms:

(1) $\|A\|$ is clearly nonnegative. If $\|A\| = 0$ then $\|Ax\| = 0$ for all $x \in X$, i.e., $A = 0$. Conversely, if $A = 0$ then $\|A\| = 0$.
(2) It is obvious that $\|\lambda A\| = |\lambda|\,\|A\|$.
(3) From

$$\|(A+B)x\| = \|Ax + Bx\| \leq \|Ax\| + \|Bx\| \leq \|A\|\,\|x\| + \|B\|\,\|x\|$$

we get $\|A + B\| \leq \|A\| + \|B\|$.

Let $L(X, Y)$ denote the normed linear space consisting of the set of all continuous linear operators from X to Y under this operator norm.

There is also a notion of *sequential continuity*; as in ordinary calculus, it is equivalent to the notion of continuity according to Definition 4.71:

Theorem 4.76. *An operator A from X to Y is continuous at $x_0 \in X$ if and only if $A(x_n) \to A(x_0)$ whenever $x_n \to x_0$.*

The proof is easily adapted from the corresponding proof that appears in any calculus book. This result justifies manipulations of the form

$$A\left(\lim_{n\to\infty} x_n\right) = \lim_{n\to\infty} Ax_n \tag{4.59}$$

for continuous operators A.

Suppose A is a continuous operator acting in a Banach space X. The convergent series $s = \sum_{k=1}^{\infty} x_k$ may be defined by the limiting operation

$$s = \lim_{n\to\infty} \sum_{k=1}^{n} x_k.$$

But (4.59) allows us to write

$$A\left(\sum_{k=1}^{\infty} x_k\right) = \lim_{n\to\infty} A\left(\sum_{k=1}^{n} x_k\right).$$

If A is also linear, then

$$A\left(\sum_{k=1}^{\infty} x_k\right) = \lim_{n\to\infty} \sum_{k=1}^{n} Ax_k = \sum_{k=1}^{\infty} Ax_k.$$

Hence interchanges of the form

$$A\sum_{k=1}^{\infty} x_k = \sum_{k=1}^{\infty} Ax_k$$

are permissible with convergent series and continuous linear operators.

The most frequent operation in mathematical physics is that of finding a solution x to the equation

$$Ax = y \tag{4.60}$$

when y is given. Let us introduce the notion of the inverse to A. If for any $y \in Y$ there is no more than one solution $x \in X$ of (4.60), then the correspondence from Y to X defined by (4.60) is an operator; this operator is the *inverse* to A and is denoted A^{-1}.

Lemma 4.77. *If A and B are each invertible, then the composition BA is invertible with $(BA)^{-1} = A^{-1}B^{-1}$.*

The proof is left to the reader.

Theorem 4.78. *Let X, Y be normed spaces. A linear operator A on $D(A) \subseteq X$ admits a continuous inverse on $R(A) \subseteq Y$ if and only if there is a positive constant c such that*

$$\|Ax\| \geq c\,\|x\| \qquad \text{for all } x \in D(A). \tag{4.61}$$

Proof. Assuming (4.61) holds, $Ax = 0$ implies $x = 0$ so the inverse A^{-1} exists. Then the same inequality means that the inverse is bounded (hence continuous) on $R(A)$. The converse is immediate. □

An operator A that satisfies (4.61) is said to be *bounded below.*

Example 4.79. Show that a bounded linear operator maps Cauchy sequences into Cauchy sequences.

Solution. Let $\{x_n\}$ be a Cauchy sequence in X. Let $\varepsilon > 0$ be given and choose N so that $n, m > N$ implies $\|x_n - x_m\| < \varepsilon / \|A\|$. For $n, m > N$ we have

$$\|Ax_n - Ax_m\| = \|A(x_n - x_m)\| \leq \|A\| \, \|x_n - x_m\| < \varepsilon,$$

so $\{Ax_n\}$ is a Cauchy sequence in Y. □

Example 4.80. Show that every bounded linear operator has a closed null space.

Solution. Let A be a bounded linear operator. The *null space* of A, often denoted by $N(A)$, is the set of elements x such that $Ax = 0$. Let $\{x_n\}$ be a sequence of points in $N(A)$ with $x_n \to x_0$ as $n \to \infty$. It is easy to see that x_0 belongs to $N(A)$:

$$Ax_0 = A\left(\lim_{n\to\infty} x_n\right) = \lim_{n\to\infty} Ax_n = \lim_{n\to\infty} 0 = 0.$$

Hence $N(A)$ is a closed set. □

Example 4.81. Show that if $k(x, \xi)$ is a continuous, real-valued function of the real variables x, ξ on $[a, b] \times [a, b]$, then the operator A given by

$$Af = \int_a^b k(x, \xi) f(\xi) \, d\xi$$

is a bounded linear operator from $C(a, b)$ to itself.

Solution. The linearity of A is obvious. To see that A is bounded, observe that

$$\begin{aligned}
\|Af\| &= \max_{x\in[a,b]} \left| \int_a^b k(x,\xi)f(\xi)\,d\xi \right| \\
&\le \max_{x\in[a,b]} \left[\int_a^b |k(x,\xi)|\,|f(\xi)|\,d\xi \right] \\
&\le \max_{x\in[a,b]} \left[\max_{x\in[a,b]} |f(x)| \int_a^b |k(x,\xi)|\,d\xi \right] \\
&= \|f(x)\| \max_{x\in[a,b]} \int_a^b |k(x,\xi)|\,d\xi.
\end{aligned}$$

So $\|Af\| \le \alpha \|f\|$, where

$$\alpha = \max_{x\in[a,b]} \int_a^b |k(x,\xi)|\,d\xi.$$

□

Example 4.82. Show that if a linear operator is invertible, then its inverse is a linear operator.

Solution. Suppose A is linear and A^{-1} exists. Let $y_1, y_2 \in R(A)$ where $y_i = Ax_i$ $(i = 1, 2)$ and let a_1, a_2 be scalars. We have

$$a_1y_1 + a_2y_2 = a_1Ax_1 + a_2Ax_2 = A(a_1x_1 + a_2x_2)$$

so that

$$A^{-1}(a_1y_1 + a_2y_2) = a_1x_1 + a_2x_2 = a_1A^{-1}y_1 + a_2A^{-1}y_2$$

as required.

□

4.10 Contraction Mapping Principle

We know that the iterative Newton method of tangents for finding zeros of a differentiable function $g(x)$ demonstrates fast convergence and is widely used in practice. In this method we reduce a given problem to a problem of the form

$$x = f(x) \tag{4.62}$$

and the procedure for finding zeros of $g(x)$ is

$$x_{n+1} = f(x_n). \tag{4.63}$$

A solution x^* of (4.62) is such that the value of $f(x)$ at x^* is x^*, so a solution is a *fixed point* of the mapping f. There are different ways in which an equation $g(x) = 0$ may be reduced to the form (4.62), the simplest but not the best of which is to represent the equation as $x = x - g(x)$. Such a transformation is good only when the iterative procedure of solution converges fast enough. It turns out that we can reduce various equations of different natures, from systems of equations to boundary value problems and integral equations, to forms of the type (4.62) so that the iterative procedure gives us a good approximation to a solution with few iterations required. The methods of reduction of a general equation $G(x) = 0$ extend those known for the simple equation $g(x) = 0$. In this section we discuss a class of problems of the general form

$$x = F(x) \tag{4.64}$$

where $F(x)$ is a mapping on a metric space M, i.e.,

$$F\colon M \to M,$$

and $x \in M$ is the desired unknown. If x is to satisfy (4.64) then the image of x under F must be x itself, so we continue to use the term "fixed point" in this more general case.

We would like to use an iterative process to solve equation (4.64). The iteration begins with an initial value $x_0 \in M$ (sometimes called the *seed* element) and proceeds via use of the recursion

$$x_{k+1} = F(x_k) \qquad k = 0, 1, 2, \ldots. \tag{4.65}$$

Under suitable conditions the resulting values $x_0, x_1, x_2, \ldots$ will form a sequence of *successive approximations* to the desired solution. That is, if the approach works we will have

$$\lim_{k\to\infty} x_k = x^* \tag{4.66}$$

where x^* is a fixed point of F. With this background, let us formulate conditions providing the applicability of the method.

Definition 4.83. Let $F(x)$ be a mapping on M. We say that $F(x)$ is a *contraction mapping* if there exists a number $\alpha \in [0, 1)$ such that

$$d(F(x), F(y)) \le \alpha\, d(x, y) \tag{4.67}$$

for every pair of elements $x, y \in M$.

Repeated application of (4.67) yields

$$d(x_{k+1}, x_k) \le \alpha^k d(x_1, x_0), \qquad k = 0, 1, 2, \ldots,$$

and with $0 \le \alpha < 1$ the successive iterates will land closer and closer together in M. We might expect these iterates to converge to a solution; rigorous confirmation that they do is provided by the following celebrated result due to Banach. It is known as the *contraction mapping theorem.*

Theorem 4.84. *A contraction mapping F with constant α, $0 \le \alpha < 1$, on a complete metric space M has a unique fixed point. Convergence of successive approximations to the fixed point is independent of the choice of seed element.*

Proof. We choose an arbitrary seed element $x_0 \in M$ for the recursion (4.65). Using the triangle inequality for several elements, for $m > n$ we have

$$\begin{aligned} d(x_m, x_n) \le d(x_m, x_{m-1}) &+ d(x_{m-1}, x_{m-2}) + \cdots \\ &+ d(x_{n+2}, x_{n+1}) + d(x_{n+1}, x_n) \end{aligned}$$

hence

$$\begin{aligned} d(x_m, x_n) &\le (\alpha^{m-1} + \alpha^{m-2} + \cdots + \alpha^{n+1} + \alpha^n)\, d(x_1, x_0) \\ &= \alpha^n (1 + \alpha + \cdots + \alpha^{m-n-2} + \alpha^{m-n-1})\, d(x_1, x_0) \\ &\le \alpha^n (1 - \alpha)^{-1} d(x_1, x_0) \\ &\to 0 \quad \text{as } n \to \infty. \end{aligned}$$

In this, we summed up the geometrical progression. So $\{x_k\}$ is a Cauchy sequence, and by completeness of M there is a point $x^* \in M$ such that $x_k \to x^*$ as $k \to \infty$. From the contraction condition for F it follows that $F(x)$ is continuous on M, hence

$$x^* = \lim_{k \to \infty} F(x_k) = F\left(\lim_{k \to \infty} x_k\right) = F(x^*).$$

We have therefore established the existence of a fixed point of $F(x)$. Uniqueness is proved by assuming the existence of another such point y^*. Then

$$d(x^*, y^*) = d(F(x^*), F(y^*)) \le \alpha\, d(x^*, y^*)$$

so that

$$(1 - \alpha) d(x^*, y^*) = 0.$$

But $\alpha < 1$, so we must have $d(x^*, y^*) = 0$ and hence $x^* = y^*$. □

The proof of Theorem 4.84 provides information concerning the rate of convergence of the iterates x_k to x^*. Specifically, we have

Corollary 4.85. *Let $F(x)$ be a contraction mapping on a complete metric space M. Then the estimates*

$$d(x_n, x^*) \leq \frac{\alpha^n}{1-\alpha} d(x_1, x_0) \tag{4.68}$$

and

$$d(x_n, x^*) \leq \frac{\alpha}{1-\alpha} d(x_n, x_{n-1}) \tag{4.69}$$

both hold for $n = 0, 1, 2, \ldots$, where α is the contraction constant for $F(x)$ and x^* is the fixed point of F.

Proof. In the inequality

$$d(x_m, x_n) \leq \frac{\alpha^n}{1-\alpha} d(x_1, x_0)$$

we can let $m \to \infty$ and obtain (4.68). If on the right side of (4.68) we take x_0 to be x_{n-1}, then x_1 becomes x_n and we obtain (4.69). □

Inequality (4.68) is called an *a priori* error estimate, since it provides an upper bound on $d(x_n, x^*)$ in terms of quantities known at the start of the iteration procedure. Inequality (4.69) is called an *a posteriori* error estimate, and can be used to monitor convergence as the iteration proceeds.

The contraction mapping principle can be applied to a variety of problems.

Example 4.86. Consider a (possibly finite dimensional) system of linear equations

$$x_i = \sum_{j=1}^{\infty} a_{ij} x_j + c_i \qquad (i = 1, 2, 3, \ldots).$$

Solution. To solve this problem by iteration we can write

$$\mathbf{x}^{(k+1)} = F(\mathbf{x}^{(k)}) = A(\mathbf{x}^{(k)}) + \mathbf{c}$$

where $\mathbf{c} = (c_1, c_2, c_3, \ldots)$ is a given vector, $\{\mathbf{x}^{(k)}\}$ is a sequence of vector iterates

$$\begin{aligned}
\mathbf{x}^{(0)} &= (x_1^{(0)}, x_2^{(0)}, x_3^{(0)}, \ldots),\\
\mathbf{x}^{(1)} &= (x_1^{(1)}, x_2^{(1)}, x_3^{(1)}, \ldots),\\
\mathbf{x}^{(2)} &= (x_1^{(2)}, x_2^{(2)}, x_3^{(2)}, \ldots),\\
&\vdots
\end{aligned}$$

and A is the mapping given by

$$A(\mathbf{x}^{(k)}) = \left(\sum_{j=1}^{\infty} a_{1j}x_j^{(k)}, \sum_{j=1}^{\infty} a_{2j}x_j^{(k)}, \sum_{j=1}^{\infty} a_{3j}x_j^{(k)}, \ldots \right).$$

We should note that the possibility to employ iteration (and even simply to solve the system) depends on the space in which we seek a solution. Here we will study the iteration procedure in ℓ^∞, and therefore suppose that $\mathbf{c} \in \ell^\infty$. We recall that ℓ^∞ is the space of all bounded sequences under the norm

$$\|\mathbf{x}\|_\infty = \sup_{i\geq 1} |x_i|.$$

For the operator A to act in ℓ^∞ it is sufficient that the quantity

$$K = \sup_{i\geq 1} \sum_{j=1}^{\infty} |a_{ij}|$$

is finite. This follows from the fact that $\mathbf{c} \in \ell^\infty$ and the next chain of inequalities, with which we will determine when F is a contraction on ℓ^∞. We have

$$\begin{aligned}
\|F(\mathbf{x}) - F(\mathbf{x}')\|_\infty &= \sup_{i\geq 1} \left| \left(\sum_{j=1}^{\infty} a_{ij}x_j + c_i \right) - \left(\sum_{j=1}^{\infty} a_{ij}x_j' + c_i \right) \right|\\
&= \sup_{i\geq 1} \left| \sum_{j=1}^{\infty} a_{ij}(x_j - x_j') \right| \leq \sup_{i\geq 1} \sum_{j=1}^{\infty} |a_{ij}||x_j - x_j'|\\
&\leq \sup_{i\geq 1} \left[\left(\sup_{j\geq 1} |x_j - x_j'| \right) \left(\sum_{j=1}^{\infty} |a_{ij}| \right) \right]\\
&= \left(\sup_{i\geq 1} \sum_{j=1}^{\infty} |a_{ij}| \right) \left(\sup_{j\geq 1} |x_j - x_j'| \right)
\end{aligned}$$

hence

$$\|F(\mathbf{x}) - F(\mathbf{x}')\|_\infty \leq K \|\mathbf{x} - \mathbf{x}'\|_\infty .$$

With $K = \sup_{i\geq 1} \sum_{j=1}^{\infty} |a_{ij}| < 1$ we have a contraction and Banach's theorem applies. □

In other sequence spaces the appropriate conditions for a_{ij} are different. The reader can treat the problem for iterations and a solution in ℓ^2.

Let us state another corollary to the contraction mapping theorem. By F^k we denote the k-fold composition of the mapping F: that is, we have

$$F^{n+1}(x) = F(F^n(x)), \qquad n = 1, 2, 3, \ldots,$$

where it is understood that $F^1 = F$.

Corollary 4.87. *If F^k is a contraction mapping on a complete metric space for some integer $k \geq 1$, then F has a unique fixed point. Convergence of successive approximations is independent of the choice of seed element.*

Proof. F^k has a unique fixed point x^* by Theorem 4.84; moreover,

$$\lim_{n\to\infty} (F^k)^n(x) = x^*$$

for any $x \in M$. Putting $x = F(x^*)$ we obtain

$$x^* = \lim_{n\to\infty} (F^k)^n F(x^*) = \lim_{n\to\infty} F(F^k)^n(x^*) = \lim_{n\to\infty} F(x^*) = F(x^*),$$

hence x^* is also a fixed point of F. (Here we have used the assumption that x^* is a fixed point of F^k, hence it is a fixed point of $(F^k)^n$, hence $(F^k)^n(x^*) = x^*$.) If y^* is another fixed point of F, then y^* is also fixed point of F^k, hence $y^* = x^*$. □

Let us proceed to a second example.

Example 4.88. An integral equation of the form

$$\psi(x) = g(x) + \lambda \int_a^x K(x,t)\psi(t)\,dt, \qquad x \in [a,b], \tag{4.70}$$

where $\psi(x)$ is unknown, is said to be a *Volterra integral equation.* Suppose $g(x)$ is continuous on $[a,b]$, and that the kernel $K(x,t)$ is continuous on the closed, triangular region $a \leq t \leq x$, $a \leq x \leq b$. Show that the mapping F given by

$$F[\psi(x)] = g(x) + \lambda \int_a^x K(x,t)\psi(t)\,dt$$

will generate convergent iterates in $C(a,b)$.

Solution. The approach will be to prove that F^n is a contraction mapping for some integer $n \geq 1$. First, let $u(x)$ and $v(x)$ be any two elements of $C(a,b)$ and observe that

$$|F[v(x)] - F[u(x)]| \leq |\lambda| \int_a^x |K(x,t)||v(t) - u(t)|\, dt.$$

Now $K(x,t)$, being continuous on a compact set, is bounded by some number M. So

$$\begin{aligned} |F[v(x)] - F[u(x)]| &\leq |\lambda| M \int_a^x |v(t) - u(t)|\, dt \\ &\leq |\lambda| M \max_{t\in[a,b]} |v(t) - u(t)| \int_a^x dt \\ &= |\lambda| M (x-a)\, d(v,u). \end{aligned} \tag{4.71}$$

We show by induction that

$$|F^n[v(x)] - F^n[u(x)]| \leq |\lambda|^n M^n \frac{(x-a)^n}{n!} d(v,u), \quad n = 1, 2, 3, \ldots. \tag{4.72}$$

The case $n = 1$ was established in (4.71). Assuming (4.72) holds for $n = k$, we have

$$\begin{aligned} \left|F^{k+1}[v(x)] - F^{k+1}[u(x)]\right| &\leq |\lambda| \int_a^x |K(x,t)| \left|F^k[v(t)] - F^k[u(t)]\right| dt \\ &\leq |\lambda| M \int_a^x |\lambda|^k M^k \frac{(t-a)^k}{k!} d(v,u)\, dt \\ &= |\lambda|^{k+1} M^{k+1} \frac{(x-a)^{k+1}}{(k+1)!} d(v,u), \end{aligned}$$

which is the corresponding statement for $n = k+1$. Taking the maximum of (4.72) over $x \in [a,b]$ we get

$$d(F^n[v], F^n[u]) \leq |\lambda|^n M^n \frac{(b-a)^n}{n!} d(v,u).$$

For any λ we can choose n so large that

$$|\lambda|^n M^n \frac{(b-a)^n}{n!} < 1,$$

so F^n is a contraction mapping for sufficiently large n. By Corollary 4.87 then, (4.70) has a unique solution that can be found by successive approximations starting with any seed element. The usual choice for seed element is $\psi(x) = g(x)$. □

Example 4.89. An integral equation of the form

$$\psi(x) = g(x) + \lambda \int_a^b K(x,t)\psi(t)\,dt \qquad (a \le x \le b),$$

is called a *Fredholm equation of the second kind.* Suppose that $g(x)$ is continuous on $[a,b]$, and that $K(x,t)$ is continuous on the square $[a,b] \times [a,b]$. Find a condition on λ for the equation to be uniquely solvable by iteration in the space $C(a,b)$.

Solution. We need the integral operator

$$F(\psi(x)) = g(x) + \lambda \int_a^b K(x,t)\psi(t)\,dt$$

to be a contraction mapping on $C(a,b)$. Now $K(x,t)$, being continuous on a compact set, is bounded: $|K(x,t)| \le B$ on $[a,b] \times [a,b]$ where B is some constant. Hence if $u(x)$ and $v(x)$ be arbitrary elements of $C(a,b)$, we have

$$\begin{aligned} d(F(u), F(v)) &= \max_{x\in[a,b]} \left| \lambda \int_a^b K(x,t)[u(t) - v(t)]\,dt \right| \\ &\le \max_{x\in[a,b]} |\lambda| \int_a^b |K(x,t)||u(t) - v(t)|\,dt \\ &\le B|\lambda| \max_{x\in[a,b]} \int_a^b |u(t) - v(t)|\,dt \\ &\le B|\lambda|(b-a) \max_{x\in[a,b]} |u(x) - v(x)| \\ &= B|\lambda|(b-a)\,d(u(x), v(x)). \end{aligned}$$

So F will be a contraction on $C(a,b)$ if $|\lambda| < 1/B(b-a)$. □

Note that for application of the Banach principle we do not need the space to be linear. This fact is used in the solution of nonlinear problems which can have several solutions. The principle applies when it is possible to find a domain M_1 in the original space M such that M_1 is a complete metric space, the operator A acts in M_1, and is a contraction on it.

4.11 Some Approximation Theory

Let X be a normed space. Given $x \in X$ and a set of elements $g_1, \ldots, g_n \in X$, it is reasonable to seek scalars $\lambda_1, \ldots, \lambda_n$ that will minimize the distance between x and the linear combinations $\sum_{i=1}^n \lambda_i g_i$. So we would like to find the best approximation of x from among all the linear combinations $\sum_{i=1}^{n} \lambda_i g_i$. This *general problem of approximation* can be rephrased as

$$\phi(\lambda_1, \ldots, \lambda_n) \to \min_{\lambda_1, \ldots, \lambda_n} \tag{4.73}$$

where ϕ is the functional given by

$$\phi(\lambda_1, \ldots, \lambda_n) = \left\| x - \sum_{i=1}^{n} \lambda_i g_i \right\|. \tag{4.74}$$

We take the g_i to be linearly independent. If they are not linearly independent, the solution of the approximation problem will not be unique. Note that $\phi(\lambda_1, \ldots, \lambda_n)$ is a usual function in the n variables λ_i, so we can employ the usual tools of calculus.

Theorem 4.90. *For any $x \in X$ there exists $x^* = \sum_{i=1}^{n} \lambda_i^* g_i$ such that*

$$\|x - x^*\| = \inf_{\lambda_1, \ldots, \lambda_n} \phi(\lambda_1, \ldots, \lambda_n). \tag{4.75}$$

Proof. An application of the inequality

$$\|x - y\| \geq | \, \|x\| - \|y\| \, | \tag{4.76}$$

permits us to show that $\phi(\lambda_1, \ldots, \lambda_n)$ is continuous in the n scalar variables $\lambda_1, \ldots, \lambda_n$:

$$\begin{aligned} &|\phi(\lambda_1 + h_1, \ldots, \lambda_n + h_n) - \phi(\lambda_1, \ldots, \lambda_n)| \\ &\quad = \left| \left\| x - \sum_{i=1}^{n} (\lambda_i + h_i) g_i \right\| - \left\| x - \sum_{i=1}^{n} \lambda_i g_i \right\| \right| \\ &\quad \leq \left\| \left[x - \sum_{i=1}^{n} (\lambda_i + h_i) g_i \right] - \left[x - \sum_{i=1}^{n} \lambda_i g_i \right] \right\| \\ &\quad = \left\| \sum_{i=1}^{n} h_i g_i \right\| \leq \sum_{i=1}^{n} |h_i| \, \|g_i\| \, . \end{aligned}$$

Continuity of the function

$$\psi(\lambda_1, \ldots, \lambda_n) = \left\| \sum_{i=1}^{n} \lambda_i g_i \right\|$$

is also apparent since it is a particular case of $\phi(\lambda_1, \ldots, \lambda_n)$ at $x = 0$, and $\psi(\lambda_1, \ldots, \lambda_n)$ must therefore reach a minimum on the sphere $\sum_{i=1}^{n} |\lambda_i|^2 = 1$ at some point $(\lambda_{10}, \ldots, \lambda_{n0})$. By linear independence of the g_i we have $\psi(\lambda_{10}, \ldots, \lambda_{n0}) = d > 0$. Also note that ψ is a homogeneous function,

$$\psi(k\lambda_1, \cdots, k\lambda_n) = |k| \, \psi(\lambda_1, \cdots, \lambda_n),$$

which means that

$$\psi(\lambda_1, \cdots, \lambda_n) \geq Rd \quad \text{when} \quad \left(\sum_{i=1}^{n} |\lambda_i|^2 \right)^{1/2} = R,$$

and that $\psi(\lambda_1, \ldots, \lambda_n) > Rd$ for $(\lambda_1, \ldots, \lambda_n)$ outside a sphere of radius R. We wish to show that $\phi(\lambda_1, \ldots, \lambda_n)$ actually attains its minimum value at some finite point.

Since

$$\phi(\lambda_1, \ldots, \lambda_n) \geq \psi(\lambda_1, \ldots, \lambda_n) - \|x\|$$

by (4.76), we see that for $(\lambda_1, \ldots, \lambda_n)$ *outside* a ball of radius R we have

$$\phi(\lambda_1, \ldots, \lambda_n) > Rd - \|x\| .$$

Outside of the sphere of radius $R = R_0 = 3\|x\|/d$ we have

$$\phi(\lambda_1, \ldots, \lambda_n) > 2\|x\|$$

whereas inside this sphere $\phi(0, \ldots, 0) = \|x\|$. Hence when $x \neq 0$ (to the reader: what happens when $x = 0$?) the minimum of ϕ *is* inside the sphere of radius R_0 with center at the origin. Thus the corresponding closed ball of radius R_0 contains the minimum point. □

The preceding proof holds in a complex space X as well.

Uniqueness can be addressed with the help of the following concepts.

Definition 4.91. A normed space X is *strictly normed* if from the equality

$$\|x + y\| = \|x\| + \|y\| , \quad x \neq 0, \tag{4.77}$$

it follows that $y = \lambda x$ for some nonnegative λ.

Not all normed spaces are strictly normed. For example, the space $C(\Omega)$ is not strictly normed. But some important classes of spaces are strictly normed, including $L^p(\Omega)$ and $W^{l,p}(\Omega)$. Later we shall show that every inner product space is strictly normed.

Definition 4.92. A subset S of a linear space is *convex* if for any pair $x, y \in S$ it contains the whole segment

$$\lambda x + (1 - \lambda)y, \qquad 0 \leq \lambda \leq 1.$$

Theorem 4.93. *Let X be a strictly normed space, and let M be a closed convex subset of X. For any $x \in X$, there is at most one $y \in M$ that minimizes the distance $\|x - y\|$.*

Proof. Suppose that y_1 and y_2 are each minimizers:

$$\|x - y_1\| = \|x - y_2\| = \inf_{y \in M} \|x - y\| \equiv d. \tag{4.78}$$

If $x \in M$, we obtain that $y_1 = y_2 = x$. Suppose $x \notin M$. Then $d > 0$. By convexity $(y_1 + y_2)/2 \in M$, hence

$$\left\| x - \frac{y_1 + y_2}{2} \right\| \geq d.$$

But

$$\left\| x - \frac{y_1 + y_2}{2} \right\| = \left\| \frac{x - y_1}{2} + \frac{x - y_2}{2} \right\| \leq \frac{1}{2} \|x - y_1\| + \frac{1}{2} \|x - y_2\| = d,$$

so

$$\left\| \frac{x - y_1}{2} + \frac{x - y_2}{2} \right\| = \left\| \frac{x - y_1}{2} \right\| + \left\| \frac{x - y_2}{2} \right\|.$$

Because X is strictly normed we have $x - y_1 = \lambda(x - y_2)$ for some $\lambda \geq 0$, hence $\|x - y_1\| = \lambda \|x - y_2\|$. From (4.78) we deduce that $\lambda = 1$, thus $y_1 = y_2$. □

By this theorem we see that, for a strictly normed space, a solution to the general problem of approximation is unique. A set of spaces important in applications are included here, as shown next.

Lemma 4.94. *Every inner product space is strictly normed.*

Proof. Let X be an inner product space and suppose $x, y \in X$ satisfy (4.77). We have $\|x + y\|^2 = (\|x\| + \|y\|)^2$; rewriting this as

$$\|x\|^2 + 2\operatorname{Re}(x, y) + \|y\|^2 = \|x\|^2 + 2\|x\|\|y\| + \|y\|^2,$$

we obtain

$$\operatorname{Re}(x, y) = \|x\| \|y\|.$$

This and the Schwarz inequality show that $\operatorname{Im}(x, y) = 0$ so that

$$(x, y) = \|x\| \|y\|.$$

But this last equation represents the case of equality holding in the Schwarz inequality, which can happen only if $y = \lambda x$ for some λ. Making this replacement for y we obtain $(x, \lambda x) = \|x\| \|\lambda x\|$, hence $\overline{\lambda} \|x\|^2 = |\lambda| \|x\|^2$. Since $x \neq 0$ we have $\overline{\lambda} = |\lambda|$, and therefore $\lambda \geq 0$. □

The subspace H_n of an inner product space H that is spanned by g_i, $i = 1, \ldots, n$, is finite dimensional. We know that for any $x \in H$ there is a unique element that minimizes the distance $\|x - y\|$ over $y \in H_n$. In a Euclidean space this element is a projection of the element onto the subspace H_n. Let us show that this result on the unique existence of the projection extends to a Hilbert space. This extension is the basis for an important part of the theory of Hilbert spaces connected with Fourier expansions and many other questions.

Theorem 4.95. *Let H be a Hilbert space and let M be closed convex subset of H. For every $x \in H$, there is a unique $y \in M$ that minimizes $\|x - y\|$.*

Proof. Fix $x \in H$. By definition of infimum there is a sequence $\{y_k\} \subset M$ such that

$$\lim_{k \to \infty} \|x - y_k\| = \inf_{y \in M} \|x - y\| .$$

By the parallelogram law

$$\|2x - y_i - y_j\|^2 + \|y_i - y_j\|^2 = 2\left(\|x - y_i\|^2 + \|x - y_j\|^2\right),$$

hence

$$\|y_i - y_j\|^2 = 2\left(\|x - y_i\|^2 + \|x - y_j\|^2\right) - 4\left\|x - \frac{y_i + y_j}{2}\right\|^2 .$$

Since $\|x - y_j\|^2 = d^2 + \varepsilon_j$ where $\varepsilon_j \to 0$ as $j \to \infty$, it follows that

$$\|y_i - y_j\|^2 \le 2(d^2 + \varepsilon_i + d^2 + \varepsilon_j) - 4d^2 = 2(\varepsilon_i + \varepsilon_j) \to 0 \quad \text{as } i, j \to \infty.$$

Therefore $\{y_k\}$ is a Cauchy sequence, and converges to an element $y \in M$ since M is closed. This minimizer y is unique by Theorem 4.93. □

4.12 Orthogonal Decomposition of a Hilbert Space and the Riesz Representation Theorem

Definition 4.96. Let M be a subspace of a Hilbert space H. An element $n \in H$ is *orthogonal to* M if n is orthogonal to every element of M.

In $\mathbb{R}^3$ we may imagine a straight line segment inclined with respect to a plane and with one end touching the plane. We may then define the projections of the segment onto the plane and onto the normal, respectively. The length of the normal projection is the shortest distance from the other end of the segment to the surface. The next result extends this fact to inner product spaces.

Lemma 4.97. *Let H be a Hilbert space and M a closed linear subspace of H. Given $x \in H$, the unique minimizer $m \in M$ guaranteed by Theorem 4.95 is such that $(x - m)$ is orthogonal to M.*

Proof. Let $v \in M$. The function

$$f(\alpha) = \|x - m - \alpha v\|^2$$

of the real variable α takes its minimum value at $\alpha = 0$, hence

$$\left.\frac{df}{d\alpha}\right|_{\alpha=0} = 0.$$

This gives

$$\left.\frac{d}{d\alpha}(x - m - \alpha v, x - m - \alpha v)\right|_{\alpha=0} = -2\operatorname{Re}(x - m, v) = 0.$$

Replacing v by iv we get $\operatorname{Im}(x - m, v) = 0$, hence $(x - m, v) = 0$. □

Definition 4.98. Two subspaces M and N of H are *mutually orthogonal* if every $n \in N$ is orthogonal to M and every $m \in M$ is orthogonal to N. In this case we write $M \perp N$. If, furthermore, any $x \in H$ can be uniquely represented in the form

$$x = m + n, \qquad m \in M,\ n \in N, \tag{4.79}$$

then we write $H = M \dot{+} N$ and speak of an *orthogonal decomposition* of H into M and N.

Note that mutually orthogonal subspaces have zero as their only point of intersection.

Theorem 4.99. *Let M be a closed subspace of a Hilbert space H. There is a closed subspace N of H such that $M \dot{+} N$ is an orthogonal decomposition of H.*

Proof. Let N be the set of all elements of H that are orthogonal to M. We assume $M \neq H$, hence N has nonzero elements. If $n_1, n_2 \in N$ so that $(n_1, m) = (n_2, m) = 0$ for every $m \in M$, then $(\lambda_1 n_1 + \lambda_2 n_2, m) = 0$ for any scalars λ_1, λ_2. Hence N is a subspace of H. To see that N is closed, let $\{n_k\}$ be a Cauchy sequence in N. The limit element $n^* = \lim_{k\to\infty} n_k$ exists; it belongs to N because

$$(n^*, m) = \lim_{k\to\infty}(n_k, m) = 0 \quad \text{for all } m \in M$$

by continuity of the inner product.

For any element $x \in H$ the representation (4.79) exists because we can project x onto M to obtain the element m, then obtain n from $n = x - m$. To show uniqueness, assume that for some x there are two such representations:

$$x = m_1 + n_1, \qquad x = m_2 + n_2.$$

Equating these, we obtain

$$m_1 - m_2 = n_1 - n_2.$$

Taking the inner product of both sides of this equality with $m_1 - m_2$ and then with $n_1 - n_2$, we get $\|m_1 - m_2\|^2 = 0$ and $\|n_1 - n_2\|^2 = 0$. □

Let us turn to a principal fact we shall need from the theory of Hilbert spaces. We consider a simple case first. Let $\{\mathbf{e}_1, \ldots, \mathbf{e}_n\}$ be an orthonormal basis of $\mathbb{R}^n$ so that any vector $\mathbf{x} \in \mathbb{R}^n$ can be expressed as

$$\mathbf{x} = \sum_{i=1}^{n} x_i \mathbf{e}_i.$$

Now suppose $F(\mathbf{x})$ is a linear functional defined on $\mathbb{R}^n$. It is easy to see that $F(\mathbf{x})$ has a representation of the form

$$F(\mathbf{x}) = \sum_{i=1}^{n} x_i c_i \tag{4.80}$$

where the c_i are scalars independent of $\mathbf{x}$; indeed, with $c_i \equiv F(\mathbf{e}_i)$ we have

$$F(\mathbf{x}) = F\left(\sum_{i=1}^{n} x_i \mathbf{e}_i\right) = \sum_{i=1}^{n} x_i F(\mathbf{e}_i) = \sum_{i=1}^{n} x_i c_i$$

by linearity of F. We can write (4.80) as

$$F(\mathbf{x}) = (\mathbf{x}, \mathbf{c})$$

where $\mathbf{c}$ is a vector in $\mathbb{R}^n$, independent of $\mathbf{x}$, whose value is uniquely determined by F; in this sense we can say that F has been "represented by an inner product." More generally, we have the following important result known as the *Riesz representation theorem*:

Theorem 4.100. *Let $F(x)$ be a continuous linear functional given on a Hilbert space H. There is a unique element $f \in H$ such that*

$$F(x) = (x, f) \quad \text{for every } x \in H. \tag{4.81}$$

Moreover, $\|F\| = \|f\|$.

Hence any bounded linear functional defined on a Hilbert space can be represented by an inner product. The element f is sometimes called the *representer* of $F(x)$.

Proof. Let M be the set of all x for which

$$F(x) = 0. \tag{4.82}$$

By linearity of $F(x)$ any finite linear combination of elements of M also belongs to M, hence M is a subspace of H. M is also closed; indeed, a Cauchy sequence $\{m_k\} \subset M$ is convergent in H to some $m^* = \lim_{k\to\infty} m_k$, and by continuity of $F(x)$ we see that m^* satisfies (4.82). By Theorem 4.99, there is a closed subspace N of H such that $N \perp M$ and such that any $x \in H$ can be uniquely represented as $x = m + n$ for some $m \in M$ and $n \in N$. We can deduce the dimension of N. If n_1 and n_2 are any two elements of N, then so is $n_3 = F(n_1)n_2 - F(n_2)n_1$. Since $F(n_3) = F(n_1)F(n_2) - F(n_2)F(n_1) = 0$ we have $n_3 \in M$. But the only element that belongs to both N and M is the zero vector. This means that n_2 is a scalar multiple of n_1, hence N is one-dimensional.

Now choose $n \in N$ and define $n_0 = n/\|n\|$. Any $x \in H$ can be represented as

$$x = m + \alpha n_0, \qquad m \in M,$$

where $\alpha = (x, n_0)$, and therefore

$$F(x) = F(m) + \alpha F(n_0) = \alpha F(n_0) = F(n_0)(x, n_0) = (x, \overline{F(n_0)}n_0).$$

Denoting $\overline{F(n_0)}n_0$ by f we obtain the representation (4.81). To establish its uniqueness, let f_1 and f_2 be two representers:

$$F(x) = (x, f_1) = (x, f_2).$$

So $(x, f_1 - f_2) = 0$ for all x. Setting $x = f_1 - f_2$ we have $\|f_1 - f_2\|^2 = 0$, hence $f_1 = f_2$.

Finally, we must establish $\|F\| = \|f\|$. Since this certainly holds for $F = 0$ we assume $F \neq 0$. Then $f \neq 0$, and

$$\|f\|^2 = (f, f) = F(f) \le \|F\| \, \|f\|$$

gives $\|f\| \le \|F\|$. On the other hand

$$\|F\| = \sup_{\|x\|\neq 0} \frac{|F(x)|}{\|x\|} = \sup_{\|x\|\neq 0} \frac{|(x, f)|}{\|x\|} \le \sup_{\|x\|\neq 0} \frac{\|x\| \, \|f\|}{\|x\|} = \|f\|$$

by the Schwarz inequality. $\square$

The Riesz representation theorem states that a continuous linear functional on a Hilbert space H is identified with an element of H; this correspondence is one-to-one, isometric, and preserves algebraic operations with respect to the elements and functionals. The set of all continuous linear functionals on a normed space X is called the *dual space* to X and is denoted by X'. In these terms, the Riesz theorem states that X' is isometrically isomorphic to X.

Example 4.101. (a) Let a functional in $L^2(0,2)$ be given by

$$F(f) = \int_0^1 f(x)g(x)\,dx$$

where $g(x) \in L^2(0,1)$ is given. What is the representer of this functional given by Theorem 4.100 in $L^2(0,2)$? (b) Define on $L^2(0,1)$ a linear functional by the formula

$$G(f) = f(0.5).$$

What is the Riesz representer of this functional?

Solution. (a) We can use

$$G(x) = \begin{cases} g(x), & x \in [0,1], \\ 0, & x \in (1,2], \end{cases}$$

as a representer. (b) The functional G is linear but not continuous in $L^2(0,1)$, so Theorem 4.100 does not apply. The functional by its form relates to the δ-function, which lies outside $L^2(0,1)$. □

The Riesz representation theorem will play a key role when we consider the generalized setup of some problems in mechanics.

4.13 Basis, Gram–Schmidt Procedure, and Fourier Series in Hilbert Space

If Y is an n-dimensional linear space, then there are n linearly independent elements $g_1, \ldots, g_n \in Y$ such that every $y \in Y$ can be uniquely represented in the form

$$y = \sum_{k=1}^{n} \alpha_k g_k \tag{4.83}$$

for scalars $\alpha_1, \ldots, \alpha_n$. The scalars are called the components of x. We refer to the finite set $\{g_i\}_{i=1}^n$ as a *basis* of Y. A basis is not unique. The concept of basis can be extended to infinite dimensional normed spaces.

Definition 4.102. Let X be a normed linear space. A system of elements $\{e_i\}$ is a *basis* (or *Schauder basis*) of X if any $x \in X$ can be represented uniquely as

$$x = \sum_{k=1}^{\infty} \alpha_k e_k \tag{4.84}$$

for scalars $\{\alpha_k\}$.

The elements e_i of a basis play the role of coordinate vectors of the space. Every such basis is linearly independent. Indeed, with $x = 0$ equation (4.84) holds with $\alpha_k \equiv 0$, and the α_k are unique by assumption.

A normed space X having a basis is separable. To see this, we note that the set of all linear combinations $\sum_{k=1}^{\infty} q_k e_k$ with rational coefficients q_k is countable and dense in X. Countability is evident. To show denseness let $x \in X$ and $\varepsilon > 0$ be given. Write $x = \sum_{k=1}^{\infty} \alpha_k e_k$. Let $e = \sum_{k=1}^{\infty} r_k e_k$ where r_k is a rational number such that

$$|\alpha_k - r_k| < \frac{\varepsilon}{2^k \|e_k\|}.$$

Then

$$\|x - e\| = \left\| \sum_{k=1}^{\infty} (\alpha_k - r_k) e_k \right\| \le \sum_{k=1}^{\infty} |\alpha_k - r_k| \, \|e_k\| < \varepsilon \sum_{k=1}^{\infty} \frac{1}{2^k} = \varepsilon$$

as required.

In practice we often use finite approximations of quantities. Finite linear combinations of basis elements are appropriate.

Definition 4.103. Let X be a normed space. A countable system $\{g_i\} \subset X$ is *complete in* X if for every $x \in X$ and $\varepsilon > 0$ there is a finite linear combination $\sum_{i=1}^{n(\varepsilon)} \alpha_i(\varepsilon) g_i$ such that $\left\| x - \sum_{i=1}^{n(\varepsilon)} \alpha_i(\varepsilon) g_i \right\| < \varepsilon$.

Note that the coefficients α_i need not be continuous in ε.

The space X is separable if it has a countable complete system: the set of finite linear combinations with rational coefficients is dense in the set of all linear combinations, and thus in the space.

Among all the bases of $\mathbb{R}^n$ an orthonormal basis has some advantages for calculation. The same can be said of an infinite dimensional Hilbert space. A system of elements $\{g_k\} \subset H$ is said to be *orthonormal* if

$$(g_m, g_n) = \begin{cases} 1, & m = n, \\ 0, & m \neq n. \end{cases} \tag{4.85}$$

If we have an arbitrary basis $\{f_i\}_{i=1}^{\infty}$ of a Hilbert space, we sometimes need to construct an orthonormal basis of the space. An orthonormal basis of a Hilbert space is not unique. One way to produce such a basis is the *Gram–Schmidt procedure*. The process is straightforward. A linearly independent set of elements cannot contain the zero vector, so we may obtain g_1 by normalizing f_1:

$$g_1 = f_1 / \|f_1\| .$$

To obtain g_2, we first generate a vector e_2 by subtracting from f_2 the "component" of f_2 that is the projection of f_2 on the direction of g_1:

$$e_2 = f_2 - (f_2, g_1)g_1$$

(recall that g_1 is a unit vector). We then normalize e_2 to obtain g_2:

$$g_2 = e_2 / \|e_2\| .$$

(Note that $e_2 \neq 0$, otherwise f_1 and f_2 are linearly dependent. The same applies to the rest of the e_i).

We obtain g_3 from f_3 by subtracting the components of f_3 that are the projections of f_3 on both g_1 and g_2:

$$e_3 = f_3 - (f_3, g_1)g_1 - (f_3, g_2)g_2, \qquad g_3 = e_3 / \|e_3\| .$$

In general we set

$$g_i = \frac{e_i}{\|e_i\|} \quad \text{where} \quad e_i = f_i - \sum_{k=1}^{i-1} (f_i, g_k)g_k, \qquad i = 2, 3, 4, \ldots .$$

The reader should verify directly that the Gram–Schmidt procedure yields an orthogonal set of elements.

In linear algebra it is shown that a system $\{f_i\}_{i=1}^{n}$ is linearly independent in $\mathbb{R}^n$ if and only if

$$\begin{vmatrix} (f_1, f_1) & (f_1, f_2) & \cdots & (f_1, f_n) \\ (f_2, f_1) & (f_2, f_2) & \cdots & (f_2, f_n) \\ \vdots & & & \\ (f_n, f_1) & (f_n, f_2) & \cdots & (f_n, f_n) \end{vmatrix} \neq 0.$$

The determinant on the left is the *Gram determinant*. A finite dimensional inner product space stands in a one-to-one correspondence with $\mathbb{R}^n$, a correspondence in which inner products are preserved. Thus the same Gram criterion is valid for an inner product space as well. It is easy to see that every finite orthonormal system is linearly independent, since the Gram determinant would reduce to $+1$ in that case.

In the space $\mathbb{R}^n$ we find the components of a vector $\mathbf{x}$ with respect to the orthonormal frame vectors $\mathbf{i}_k$ by direct projection of $\mathbf{x}$ onto $\mathbf{i}_k$: $x_k = \mathbf{x} \cdot \mathbf{i}_k$. Similarly, the components of an element in a Hilbert space are given by

Definition 4.104. Let $\{g_i\}$ be an orthonormal system in a complex Hilbert space H. Given $f \in H$, the numbers α_k defined by

$$\alpha_k = (f, g_k), \qquad k = 1, 2, 3, \ldots, \tag{4.86}$$

are the *Fourier coefficients* of f with respect to the system $\{g_i\}$.

We use the same terms as in the classical Fourier theory of expansion of functions, because all the results and even their proofs parallel the results for Fourier expansions established in the space $L^2(a, b)$.

Theorem 4.105. *Let H be a Hilbert space. A complete orthonormal system $\{g_i\} \subset H$ is a basis of H; with respect to $\{g_i\}$, any $f \in H$ has the unique representation*

$$f = \sum_{k=1}^{\infty} \alpha_k g_k \tag{4.87}$$

where $\alpha_k = (f, g_k)$ is the kth Fourier coefficient of f. The series (4.87) is called the **Fourier series** *of f with respect to $\{g_i\}$.*

Proof. Let $f \in H$ be given, and consider approximating f by a finite linear combination $\sum_{k=1}^{n} c_k g_k$ of the elements $\{g_i\}_{i=1}^{n}$. The approximation error is given by

$$\left\| f - \sum_{k=1}^{n} c_k g_k \right\|^2 = \left(f - \sum_{k=1}^{n} c_k g_k, f - \sum_{k=1}^{n} c_k g_k \right),$$

and manipulation of the right side allows us to put this in the form

$$\left\| f - \sum_{k=1}^{n} c_k g_k \right\|^2 = \|f\|^2 - \sum_{k=1}^{n} |\alpha_k|^2 + \sum_{k=1}^{n} |c_k - \alpha_k|^2.$$

Clearly the error is minimized when $c_k = \alpha_k$ for each k, so the best approximation is the element given by

$$f_n = \sum_{k=1}^{n} (f, g_k) g_k.$$

We call f_n the nth partial sum of the Fourier series for f. Since the error is nonnegative we also have

$$\sum_{k=1}^{n} |(f, g_k)|^2 \le \|f\|^2,$$

known as *Bessel's inequality*. This shows that

$$\|f_{n+m} - f_n\|^2 = \left\| \sum_{k=n+1}^{n+m} (f, g_k) g_k \right\|^2 = \sum_{k=n+1}^{n+m} |(f, g_k)|^2 \to 0 \quad \text{as } n \to \infty,$$

hence $\{f_n\}$ is a Cauchy sequence in H. Since H is a Hilbert space the sequence has a limit. We need to show that it coincides with f. Indeed, by completeness of $\{g_i\}$, for any $\varepsilon > 0$ there exists $N = N(\varepsilon)$ and coefficients $c_k(\varepsilon)$ such that

$$\left\| f - \sum_{k=1}^{N} c_k(\varepsilon) g_k \right\|^2 < \varepsilon.$$

But f_N is at least as good an approximation to f, so

$$\|f - f_N\|^2 = \left\| f - \sum_{k=1}^{N} \alpha_k g_k \right\|^2 \le \left\| f - \sum_{k=1}^{N} c_k(\varepsilon) g_k \right\|^2 < \varepsilon$$

and we conclude that $f_N \to f$. From this we obtain

$$f = \lim_{n \to \infty} f_n,$$

and the proof is complete. □

Corollary 4.106. ***Parseval's equality***

$$\sum_{k=1}^{\infty} |(f, g_k)|^2 = \|f\|^2 \tag{4.88}$$

holds for any $f \in H$ and any complete orthonormal system $\{g_i\}$.

Proof. We established above that

$$\left\| f - \sum_{k=1}^{n} (f, g_k) g_k \right\|^2 = \|f\|^2 - \sum_{k=1}^{n} |(f, g_k)|^2. \tag{4.89}$$

Passage to the limit as $n \to \infty$ yields (4.88). □

The proof shows that the sequence of partial Fourier sums is a Cauchy sequence regardless of whether $\{g_k\}$ is a complete system. We formulate

Corollary 4.107. *Let $\{g_k\}$ be an arbitrary orthonormal system in H (not necessarily complete). The sequence of partial Fourier sums f_n of $f \in H$ converges to an element f^* such that $\|f^*\| \le \|f\|$. If the system is complete, then $f^* = f$.*

Definition 4.108. We say that $\{g_i\} \subset H$ is *closed in H* if the system of equations

$$(f, g_k) = 0 \text{ for all } k = 1, 2, 3, \ldots \tag{4.90}$$

implies that $f = 0$.

Theorem 4.109. *An orthonormal system $\{g_i\}$ in a Hilbert space H is complete in H if and only if it is closed in H.*

Proof. If $\{g_i\}$ is a complete orthonormal system in H, then any $f \in H$ can be written as

$$f = \sum_{k=1}^{\infty} (f, g_k) g_k$$

by Theorem 4.105. Enforcement of the condition (4.90) obviously does yield $f = 0$, hence $\{g_i\}$ is closed. Conversely, assume $\{g_i\}$ is a closed orthonormal system in H. We established previously (Corollary 4.107) that for any $f \in H$ the sequence of partial Fourier sums $f_n = \sum_{k=1}^{n} \alpha_k g_k$ is a Cauchy sequence converging to some $f^* \in H$ since H is a Hilbert space. We have

$$(f - f^*, g_m) = \lim_{n \to \infty} \left(f - \sum_{k=1}^{n} \alpha_k g_k, g_m \right) = \alpha_m - \alpha_m = 0$$

hence

$$(f - f^*, g_m) = 0 \text{ for all } m = 1, 2, 3, \ldots.$$

It follows that $f^* = f$ since $\{g_i\}$ is closed. Because $f_n = \sum_{k=1}^{n} \alpha_k g_k$ converges to f, the system $\{g_i\}$ is complete by Definition 4.103. □

The existence of the Gram–Schmidt process implies

Theorem 4.110. *A system of elements $\{g_i\}$ (not necessarily orthonormal) in a Hilbert space H is complete in H if and only if it is closed in H.*

Theorem 4.111. *A Hilbert space H has a countable orthonormal basis if and only if H is separable.*

Proof. We saw earlier that the existence of a countable basis in a Hilbert space provides for separability. Conversely, assume H is separable and select a countable set that is dense in H. To this set the Gram–Schmidt procedure can be applied (removing any linearly dependent elements) to produce an orthonormal system. Since the initial set was dense it was complete, hence the Gram–Schmidt procedure yields an orthonormal basis of H. □

One advantage afforded by the tools of functional analysis is that we can discuss many common procedures of numerical analysis in terms to which we are accustomed in finite dimensional spaces. A knowledge of this theory gives us an understanding, without long deliberation, of when we can do so and when we cannot — some nice finite dimensional pictures become invalid or doubtful in spaces of infinite dimension.

The following result will be used in § 4.21.

Theorem 4.112. *Any bounded subset of a Hilbert space H is precompact if and only if H is finite dimensional.*

Proof. If H is finite dimensional then we can place it in one-to-one correspondence with $\mathbb{R}^n$ for some n. Then precompactness of any bounded set follows from calculus.

Next let us suppose that any bounded set of H is precompact but, to the contrary, that H is infinite dimensional. We can construct an infinite Fourier basis $\{e_k\}$. Since $\|e_k - e_n\|^2 = 2$ for $k \neq n$, the sequence $\{e_k\}$ cannot contain a Cauchy subsequence, hence the unit ball of H cannot be precompact. □

Example 4.113. Show that every separable, infinite dimensional, complex Hilbert space is isometrically isomorphic to ℓ^2.

Solution. Let X be a Hilbert space as described. By separability X has a countable, complete orthonormal set $E = \{e_k\}_{k=1}^{\infty}$. For any $x \in X$, denote the nth Fourier coefficient with respect to E by α_n. Since E is complete we have $\|x\|^2 = \sum_{n=1}^{\infty} |\alpha_n|^2 < \infty$, hence $\alpha = (\alpha_1, \alpha_2, \ldots) \in \ell^2$. Define a transformation A from X to ℓ^2 by $Ax = \alpha$. Because A is clearly linear we can show that it is injective by showing that $N(A) = \{0\}$. But $Ax = 0$ implies $\alpha = 0$, hence each $\alpha_k = 0$, hence $(x, e_k) = 0$ for each k, hence $x = 0$ since the orthonormal set E is closed. Next we show that A is surjective. Choose any $y = (\eta_1, \eta_2, \ldots) \in \ell^2$; since $\sum_{n=1}^{\infty} |\eta_n|^2 < \infty$, the series $\sum_{n=1}^{\infty} \eta_n e_n = x$ for some $x \in X$. Moreover we have $\eta_n = (x, e_n)$ for all n, and hence $\|Ax\|^2 = \|y\|^2 = \sum_{n=1}^{\infty} |\eta_n|^2 = \sum_{n=1}^{\infty} |(x, e_n)|^2 = \|x\|^2$. That is, A is also an isometry. □

4.14 Weak Convergence

It is easy to show that $\{\mathbf{x}_k\}$ is a Cauchy sequence in $\mathbb{R}^n$ if and only if each of its component sequences $\{(\mathbf{x}_k, \mathbf{i}_j)\}$, $j = 1, \ldots, n$, is a numerical Cauchy sequence. So in $\mathbb{R}^n$, norm convergence is equivalent to component-wise convergence. Remember that, besides, all the norms in $\mathbb{R}^n$ are equivalent. Unlike $\mathbb{R}^n$, in an infinite dimensional Hilbert space, where the role of components is played by the Fourier coefficients of an element, the component-wise convergence of a sequence does not guarantee strong convergence of the same sequence. Indeed, consider the sequence composed of the elements of an orthonormal basis $\{g_k\}$. The sequence of the jth Fourier component $(g_k, g_j) \to 0$ as $k \to \infty$ because of the mutual orthogonality of the elements of the basis; hence, by similarity to the case of $\mathbb{R}^n$, we could conclude that the zero element is a limit. But $\{g_k\}$ does not have a strong limit, because $\|g_k - g_m\| = \sqrt{2}$ whenever $k \neq m$. However, component-wise convergence in a Hilbert space is still important, and we need to introduce a suitable notion. A component in Hilbert space is given by the Fourier coefficient, which is found through the use of an inner product. This coefficient is a continuous linear functional on H. So a natural extension of the definition of component-wise convergence is

Definition 4.114. Let $\{x_k\} \subset H$ where H is a Hilbert space. We say that $\{x_k\}$ is a *weak Cauchy sequence* if $\{F(x_k)\}$ is a (numerical) Cauchy sequence for every continuous linear functional $F(x)$ defined on H.

In contrast, we know that $\{x_k\}$ is a Cauchy sequence in H if

$$\|x_n - x_m\| \to 0 \quad \text{as } m, n \to \infty.$$

In this latter case we shall refer to $\{x_k\}$ as a *strong Cauchy sequence* whenever there is danger of ambiguity. It is apparent that every strong Cauchy sequence is a weak Cauchy sequence. We also observe that, by Theorem 4.100, $\{x_k\}$ is a weak Cauchy sequence if the numerical sequence $\{(x_n, f)\}$ is a Cauchy sequence for every element $f \in H$. But above we showed the existence of a sequence that is a weak Cauchy sequence but not a strong Cauchy sequence. Thus we have defined a new kind of convergence in a Hilbert space. We shall rephrase all the notions of strong continuity for the weak version.

Definition 4.115. Let $x_0 \in H$. If $F(x_n) \to F(x_0)$ for every continuous linear functional $F(x)$ defined on H, we write

$$x_n \rightharpoonup x_0$$

and say that $\{x_n\}$ is *weakly convergent* to x_0. Alternatively, by the Riesz representation theorem we have $x_n \rightharpoonup x_0$ if and only if $(x_n, f) \to (x_0, f)$ for every element $f \in H$.

Recalling that the strong limit of a sequence is unique, we might wonder whether weak limits also share this property. The answer is affirmative:

Theorem 4.116. *If a sequence in a Hilbert space has a weak limit, the limit is unique.*

Proof. Suppose there are two weak limits x^* and x^{**} of a sequence $\{x_k\}$. An arbitrary continuous linear functional, by Theorem 4.100, is $F(x) = (x, f)$. When k tends to infinity the numerical sequence (x_k, f) can have only one limit (by calculus), so $(x^{**}, f) = (x^*, f)$. This holds for any $f \in H$, and thus for $f = x^{**} - x^*$. But then it follows that $\|x^{**} - x^*\|^2 = 0$. □

There is a simple and convenient sufficient condition for a weakly convergent sequence to be strongly convergent:

Theorem 4.117. *Suppose $x_k \rightharpoonup x_0$ in a Hilbert space H. Then $\|x_k\| \to \|x_0\|$ implies that $x_k \to x_0$ as $k \to \infty$.*

Proof. For each k we have

$$\|x_k - x_0\|^2 = (x_k - x_0, x_k - x_0) = \|x_k\|^2 - (x_0, x_k) - (x_k, x_0) + \|x_0\|^2 .$$

But as $k \to \infty$ both (x_0, x_k) and (x_k, x_0) approach $\|x_0\|^2$ by definition of weak convergence, and we have $\|x_k\| \to \|x_0\|$ by assumption. So $\|x_k - x_0\| \to 0$ as $k \to \infty$. □

A strong Cauchy sequence is bounded, but it is not immediately apparent that a weak Cauchy sequence has this property.

Theorem 4.118. *In a Hilbert space, every weak Cauchy sequence is bounded.*

Proof. Suppose $\{x_n\}$ is a weak Cauchy sequence in H with $\|x_n\| \to \infty$ as $n \to \infty$. Before seeking a contradiction we establish an auxiliary fact: if $B(y_0, \varepsilon)$ is a closed ball of radius $\varepsilon > 0$ and arbitrary center $y_0 \in H$, then there is a sequence $\{y_n\} \subset B(y_0, \varepsilon)$ such that the numerical sequence

$$(x_n, y_n) \to \infty \quad \text{as } n \to \infty. \tag{4.91}$$

The sequence $\{y_n\}$ given by

$$y_n = y_0 + \varepsilon \frac{x_n}{2\|x_n\|}$$

is suitable. Indeed

$$\|y_n - y_0\| = \left\| \frac{\varepsilon x_n}{2\|x_n\|} \right\| = \frac{\varepsilon}{2} < \varepsilon$$

shows that $y_n \in B(y_0, \varepsilon)$ for each n. Furthermore,

$$(x_n, y_n) = (x_n, y_0) + \frac{\varepsilon}{2\|x_n\|}(x_n, x_n) = (x_n, y_0) + \frac{\varepsilon}{2}\|x_n\|$$

establishes (4.91) since the numerical sequence $\{(x_n, y_0)\}$ is a Cauchy sequence by definition of weak convergence of $\{x_n\}$, and every Cauchy sequence is bounded.

We are now ready to obtain a contradiction. Starting with $\varepsilon_1 = 1$ and $y_0 = 0$, we can find x_{n_1} and $y_1 \in B(y_0, \varepsilon_1)$ such that

$$(x_{n_1}, y_1) > 1. \tag{4.92}$$

By continuity of the inner product in the second argument, there is a ball $B(y_1, \varepsilon_2) \subset B(y_0, \varepsilon_1)$ such that (4.92) holds not only for y_1 but for all $y \in B(y_1, \varepsilon_2)$:

$$(x_{n_1}, y) > 1 \text{ for all } y \in B(y_1, \varepsilon_2).$$

Similarly, we can find x_{n_2} (with $n_2 > n_1$) and $y_2 \in B(y_1, \varepsilon_2)$ such that

$$(x_{n_2}, y_2) > 2,$$

and, by continuity, a ball $B(y_2, \varepsilon_3) \subset B(y_1, \varepsilon_2)$ such that

$$(x_{n_2}, y) > 2 \text{ for all } y \in B(y_2, \varepsilon_3).$$

Continuing this process we generate a nested sequence of balls $B(y_k, \varepsilon_{k+1})$ and a corresponding subsequence $\{x_{n_k}\}$ of $\{x_n\}$ such that

$$(x_{n_k}, y) > k \text{ for all } y \in B(y_k, \varepsilon_{k+1}).$$

Since H is a Hilbert space the intersection $\bigcap_k B(y_k, \varepsilon_{k+1})$ is nonempty, hence there exists y^* such that $(x_{n_k}, y^*) > k$ for each k. For the continuous linear functional $F^*(x) = (x, y^*)$ then, the numerical sequence $\{F^*(x_{n_k})\}$ is not a Cauchy sequence. Because $\{x_{n_k}\}$ is not a weak Cauchy sequence, neither is $\{x_n\}$. This is the contradiction sought. $\square$

As a byproduct of this proof we have

Lemma 4.119. *If $\{x_k\}$ is an unbounded sequence in H, i.e., $\|x_k\| \to \infty$ as $k \to \infty$, then there exists $y^* \in H$ and a subsequence $\{x_{n_k}\}$ such that $(x_{n_k}, y^*) \to \infty$ as $k \to \infty$.*

We now present another important theorem with which we can show boundedness of some sets in a Hilbert space. Set boundedness plays an important role in the applications of functional analysis to mathematical physics. The present result is the *principle of uniform boundedness*:

Theorem 4.120. *Let $\{F_k(x)\}_{k=1}^{\infty}$ be a family of continuous linear functionals defined on a Hilbert space H. If $\sup_k |F_k(x)| < \infty$ for each $x \in H$, then $\sup_k \|F_k\| < \infty$.*

Proof. Each $F_k(x)$ has Riesz representation $F_k(x) = (x, f_k)$ for a unique $f_k \in H$ such that $\|f_k\| = \|F_k\|$. So it suffices to show that if $\sup_k |(x, f_k)| < \infty$ for each $x \in H$, then $\sup_k \|f_k\| < \infty$. We prove the contrapositive of this. If $\sup_k \|f_k\| = \infty$, then Lemma 4.119 guarantees the existence of $x_0 \in H$ and a subsequence $\{f_{k_n}\}$ such that $|(x_0, f_{k_n})| \to \infty$ as $k \to \infty$. This completes the proof. $\square$

Corollary 4.121. *Let $\{F_k(x)\}$ be a sequence of continuous linear functionals given on H. If for every $x \in H$ the numerical sequence $\{F_k(x)\}$ is a Cauchy sequence, then there is a continuous linear functional $F(x)$ on H such that*

$$F(x) = \lim_{k\to\infty} F_k(x) \quad \textit{for all } x \in H \tag{4.93}$$

and

$$\|F\| \le \liminf_{k\to\infty} \|F_k\| < \infty. \tag{4.94}$$

Proof. The limit in (4.93) exists by hypothesis and clearly defines a linear functional $F(x)$. By Theorem 4.120 we have $\sup_k \|F_k\| < \infty$; from

$$|F(x)| = \lim_{k\to\infty} |F_k(x)| \le \sup_k \|F_k\| \, \|x\|$$

it follows that $F(x)$ is continuous. Writing

$$|F(x)| = \lim_{k\to\infty} |F_k(x)| \le \liminf_{k\to\infty} \|F_k\| \, \|x\| \, ,$$

we establish (4.94). □

Because of the Riesz representation theorem we can rephrase this as

Theorem 4.122. *A weak Cauchy sequence in a Hilbert space has a weak limit belonging to the space. Hence any Hilbert space is* ***weakly complete****.*

It is therefore unnecessary for us to define weak completeness for a Hilbert space separately.

Theorem 4.123. *A sequence $\{x_n\} \subset H$ is a weak Cauchy sequence if and only if the following two conditions hold:*

(i) $\{x_n\}$ is bounded in H;

(ii) for any element from a complete system $\{f_\alpha\}$ in H, the sequence of numbers $\{(x_n, f_\alpha)\}$ is a Cauchy sequence.

Proof. Since necessity of the two conditions follows from Theorem 4.118 and Definition 4.115, we proceed to prove sufficiency. Suppose conditions (i) and (ii) hold, and let $\varepsilon > 0$ be given. Condition (i) means that $\|x_n\| \le M$ for all n. Take an arbitrary continuous linear functional defined by its Riesz representer $f \in H$ as (x, f). By (ii) there is a linear combination $f_\varepsilon = \sum_{k=1}^{N} c_k f_k$ such that

$$\|f - f_\varepsilon\| < \varepsilon/3M.$$

We have

$$\begin{aligned} |(x_n - x_m, f)| &= |(x_n - x_m, f_\varepsilon + f - f_\varepsilon)| \\ &\le |(x_n - x_m, f_\varepsilon)| + |(x_n - x_m, f - f_\varepsilon)| \\ &\le \sum_{k=1}^{N} |c_k||(x_n - x_m, f_k)| + (\|x_n\| + \|x_m\|)\,\|f - f_\varepsilon\|\,. \end{aligned}$$

By (ii), $\{(x_n, f_k)\}$ is a Cauchy sequence for each k. Therefore for sufficiently large m, n we have

$$\sum_{k=1}^{N} |c_k||(x_n - x_m, f_k)| < \varepsilon/3.$$

So

$$|(x_n - x_m, f)| \le \varepsilon/3 + 2M\varepsilon/(3M) = \varepsilon$$

for sufficiently large m, n, as required. □

Definition 4.124. A set S in an inner product space X is *weakly closed* if $x_n \rightharpoonup x_0 \in X$ implies that $x_0 \in S$.

Lemma 4.125. *In a Hilbert space, any closed ball with center at the origin is weakly closed.*

Proof. From the ball $\|x\| \leq M$, choose a sequence $\{x_n\}$ that converges weakly to $x_0 \in H$. We shall show that $\|x_0\| \leq M$. The formula

$$F(y) = \lim_{n\to\infty} (y, x_n)$$

defines a linear functional on H. It is bounded (continuous) because

$$|F(y)| = \lim_{n\to\infty} |(y, x_n)| \leq M \|y\| ,$$

and we have $\|F\| \leq M$. Applying Theorem 4.100 we obtain $F(y) = (y, f)$ for a unique $f \in H$ such that $\|f\| \leq M$. So

$$\lim_{n\to\infty} (y, x_n) = (y, f)$$

for any $y \in H$, and conclude that $x_n \rightharpoonup f$. □

A result known as *Mazur's theorem* (see, for example, [32]) states that every closed convex set in a Hilbert space is weakly closed. This would apply to the previous case, as well as to any closed subspace of a Hilbert space.

Definition 4.126. Let S be a subset of an inner product space. We say that S is *weakly precompact* if every sequence taken from S contains a weak Cauchy subsequence. We say that S is *weakly compact* if every sequence taken from S contains a weak Cauchy subsequence that converges weakly to a point of S.

Next, we see that a bounded set in a separable Hilbert space is weakly precompact.

Theorem 4.127. *Every bounded sequence in a separable Hilbert space contains a weak Cauchy subsequence.*

Proof. Let $\{x_n\}$ be a bounded sequence in a separable Hilbert space H, and let $\{g_n\}$ be an orthonormal basis of H. By Theorem 4.123 it suffices to show that there is a subsequence $\{x_{n_k}\}$ such that, for any fixed g_m, the numerical sequence $\{(x_{n_k}, g_m)\}$ is a Cauchy sequence. Let us demonstrate its existence. From the bounded numerical sequence $\{(x_n, g_1)\}$ we can choose a Cauchy subsequence $\{(x_{n_1}, g_1)\}$. Then, from the bounded numerical sequence $\{(x_{n_1}, g_2)\}$ we can choose a Cauchy subsequence $\{(x_{n_2}, g_2)\}$. We can continue this process, on the kth step obtaining a Cauchy subsequence $\{(x_{n_k}, g_k)\}$. The diagonal sequence $\{x_{n_n}\}$ has the property that for any fixed g_m the numerical sequence $\{(x_{n_n}, g_m)\}$ is a Cauchy sequence. Hence $\{x_{n_n}\}$ is a weak Cauchy sequence. □

A simple but important corollary of this and Lemma 4.125 is

Theorem 4.128. *In a Hilbert space, any closed ball with center at the origin is weakly compact.*

That is, a bounded sequence $\{x_n\}$ with $\|x_n\| \le M$ has a subsequence that converges weakly to some x^* with $\|x^*\| \le M$. We shall use this fact in the next chapter.

Example 4.129. Prove the following assertions. (a) If $\{x_n\}$ is a (strong) Cauchy sequence, then it is a weak Cauchy sequence. (b) Let $\{x_n\}$ be a weak Cauchy sequence, and suppose that one of its subsequences converges (strongly) to x_0. Then $\{x_n\}$ converges weakly to x_0. (c) If $\{x_n\}$ converges weakly to x_0, so do each of its subsequences. (d) Suppose $x_k \rightharpoonup x$ and $y_k \rightharpoonup y$. Then $x_k + y_k \rightharpoonup x + y$, and $\alpha x_k \rightharpoonup \alpha x$ for any scalar α. (e) Let $x_n \rightharpoonup x_0$ and $y_n \to y_0$. Then $(x_n, y_n) \to (x_0, y_0)$ as $n \to \infty$.

Solution. Let F be an arbitrary continuous linear functional. (a) Let $\varepsilon > 0$ be given, and choose N so large that $n, m > N$ imply $\|x_n - x_m\| < \varepsilon / \|F\|$. Then for $n, m > N$ we have

$$|F(x_n) - F(x_m)| = |F(x_n - x_m)| \le \|F\| \, \|x_n - x_m\| < \varepsilon.$$

(b) Since $\{x_n\}$ is weakly Cauchy, the sequence $\{F(x_n)\}$ is Cauchy. Also, $x_{n_k} \to x_0$ implies that $F(x_{n_k}) \to F(x_0)$. Because the Cauchy sequence $\{F(x_n)\}$ has a subsequence $\{F(x_{n_k})\}$ that converges to $F(x_0)$, the whole sequence converges to $F(x_0)$. This shows that x_n converges to x_0 weakly. (c) If $x_n \rightharpoonup x_0$, then $F(x_n) \to F(x_0)$. But then $F(x_{n_k}) \to F(x_0)$ for every subsequence $\{F(x_{n_k})\}$ of $\{F(x_n)\}$. (d) We have $F(x_k) \to F(x)$ and $F(y_k) \to F(y)$. Hence

$$F(x_k + y_k) = F(x_k) + F(y_k) \to F(x) + F(y) = F(x + y)$$

and

$$F(\alpha x_k) = \alpha F(x_k) \to \alpha F(x) = F(\alpha x).$$

(e) We have

$$\begin{aligned}|(x_n, y_n) - (x_0, y_0)| &\le |(x_n, y_n) - (x_n, y_0)| + |(x_n, y_0) - (x_0, y_0)| \\ &= |(x_n, y_n - y_0)| + |(x_n, y_0) - (x_0, y_0)| \\ &\le \|x_n\| \, \|y_n - y_0\| + |(x_n, y_0) - (x_0, y_0)|.\end{aligned}$$

The first term tends to zero as $n \to \infty$ because the weakly convergent sequence $\{x_n\}$ is bounded and $\|y_n - y_0\| \to 0$. The second term tends to zero by weak convergence of $\{x_n\}$ to x_0. □

4.15 Adjoint and Self-Adjoint Operators

In the theory of matrices, for a matrix A the equality

$$(A\mathbf{x}, \mathbf{y}) = (\mathbf{x}, A^T\mathbf{y})$$

which is valid for any $\mathbf{x}, \mathbf{y}$, defines a dual (conjugate) matrix A^T. The formula for integration by parts (when $g(0) = 0 = g(1)$),

$$\int_0^1 f'(x)g(x)\,dx = -\int_0^1 f(x)g'(x)\,dx,$$

introduces a correspondence between the operator of differentiation (of the first argument f) and a dual operator, $-d/dx$, for the second argument. For a linear differential operator with constant coefficients, integration by parts can be used to find a corresponding dual operator that plays an important role in the theory of differential equations. An extension of these ideas to the general case brings us to the notion of adjoint operator.

Let H be a Hilbert space and A a continuous linear operator from H to H. For any fixed $y \in H$, we can view the inner product (Ax, y) as a functional with respect to the variable $x \in H$. This functional is linear:

$$(A(\lambda x_1 + \mu x_2), y) = (\lambda A x_1 + \mu A x_2, y) = \lambda\,(Ax_1, y) + \mu\,(Ax_2, y)\,.$$

It is also bounded (i.e., continuous) since

$$|(Ax, y)| \le \|Ax\| \, \|y\| \le \|A\| \, \|y\| \, \|x\|$$

by the Schwarz inequality and the fact that A is bounded. By Theorem 4.100 we can write

$$(Ax, y) = (x, z)$$

where $z \in H$ is uniquely determined by y and A. The correspondence $y \mapsto z$ defines an operator that we shall denote by A^*.

Definition 4.130. Let A be a continuous linear operator acting in H. The operator A^* from H to H given by

$$(Ax, y) = (x, A^*y) \qquad \text{for all } x \in H$$

is the *adjoint* of A.

Let us verify that A^* is a linear operator. For any $y_1, y_2 \in H$ we have

$$(Ax, y_1) = (x, A^*y_1), \qquad (Ax, y_2) = (x, A^*y_2),$$

and, if λ and μ are any scalars, $(Ax, \lambda y_1 + \mu y_2) = (x, A^*(\lambda y_1 + \mu y_2))$. Hence

$$\begin{aligned}(x, A^*(\lambda y_1 + \mu y_2)) &= \overline{\lambda}(Ax, y_1) + \overline{\mu}(Ax, y_2)\\ &= \overline{\lambda}(x, A^*y_1) + \overline{\mu}(x, A^*y_2)\\ &= (x, \lambda A^*y_1) + (x, \mu A^*y_2).\end{aligned}$$

Therefore, since $x \in H$ is arbitrary,

$$A^*(\lambda y_1 + \mu y_2) = \lambda A^*y_1 + \mu A^*y_2$$

Let us proceed to some other properties of A^*.

Lemma 4.131. *We have*

$$(A + B)^* = A^* + B^*, \qquad (AB)^* = B^*A^*,$$

for any continuous linear operators A, B acting in H.

Proof. The first property is evident. We write

$$\begin{aligned}(x, (AB)^*y) &= ((AB)x, y) = (A(Bx), y) = (Bx, A^*y) = (x, B^*(A^*y))\\ &= (x, (B^*A^*)y)\end{aligned}$$

to establish the second property. □

Lemma 4.132. *If A is a continuous linear operator, then so is A^*; moreover, we have $\|A^*\| = \|A\|$.*

Proof. Define[4]

$$M = \sup_{x,y \in H} \frac{|(Ax, y)|}{\|x\| \, \|y\|}.$$

By the Schwarz inequality

$$M \le \sup_{x,y \in H} \frac{\|A\| \, \|x\| \, \|y\|}{\|x\| \, \|y\|} = \|A\| \, .$$

[4]Here it is evident that the sup should be taken only over $x, y \neq 0$, so we suppress this condition to simplify the notation.

But we also have

$$M = \sup_{x,y \in H} \frac{|(x, A^*y)|}{\|x\| \, \|y\|}$$

and can put $x = A^*y$ to obtain a new value

$$M_1 = \sup_{y \in H} \frac{|(A^*y, A^*y)|}{\|A^*y\| \, \|y\|} = \sup_{y \in H} \frac{\|A^*y\|}{\|y\|}.$$

Since $M_1 \leq M$ we see that A^* is bounded and

$$M_1 = \|A^*\| \leq M \leq \|A\| .$$

So A^* is continuous with $\|A^*\| \leq \|A\|$. The reverse inequality, obtained as

$$\|A\| = \|(A^*)^*\| \leq \|A^*\| ,$$

rests on the next lemma. □

Lemma 4.133. $(A^*)^* = A$.

Proof. Since A^* is continuous we have

$$(x, (A^*)^*y) = (A^*x, y) = \overline{(y, A^*x)} = \overline{(Ay, x)} = (x, Ay)$$

for any $x, y \in H$. □

We are now ready to consider some specific examples. In preparation for this it will be helpful to have

Definition 4.134. An operator A is *self-adjoint* if $A^* = A$.

Let us note that for boundary value problems the equality $A^* = A$ means not only coincidence of the form of the operators, but coincidence of their domains as well. This remark becomes important when in mathematical physics one introduces the notion of the adjoint to an operator having a domain that is only dense in the space. Then one may introduce symmetrical operators (these are such that the form of the adjoint operator remains the same) and self-adjoint operators for which there is complete coincidence with the original operator.

On the space ℓ^2 having elements $\mathbf{x} = (x_1, x_2, \ldots)$, we can define a matrix operator A by

$$(A\mathbf{x})_i = \sum_{j=1}^{\infty} a_{ij} x_j .$$

It follows from

$$\|A\mathbf{x}\|_{\ell^2} = \left[\sum_{i=1}^{\infty}\left(\sum_{j=1}^{\infty} a_{ij}x_j\right)^2\right]^{1/2} \le \left[\sum_{i=1}^{\infty}\sum_{j=1}^{\infty}|a_{ij}|^2\sum_{k=1}^{\infty}|x_k|^2\right]^{1/2}$$

that

$$\|A\| \le \left(\sum_{i=1}^{\infty}\sum_{j=1}^{\infty}|a_{ij}|^2\right)^{1/2}.$$

Suppose

$$\left(\sum_{i=1}^{\infty}\sum_{j=1}^{\infty}|a_{ij}|^2\right)^{1/2} \le M$$

so A becomes continuous. From

$$(A\mathbf{x},\mathbf{y}) = \sum_{i=1}^{\infty}\sum_{j=1}^{\infty} a_{ij}x_j\overline{y_i} = \sum_{j=1}^{\infty} x_j \overline{\left(\sum_{i=1}^{\infty}\overline{a}_{ij}y_i\right)} = (\mathbf{x}, A^*\mathbf{y})$$

we see that A^* is defined by

$$(A^*\mathbf{y})_j = \sum_{i=1}^{\infty}\overline{a}_{ij}y_i.$$

It is evident that A is self-adjoint if $a_{ij} = \overline{a}_{ji}$ for all indices i, j. A continuous analogue is the integral operator B acting in $L^2(0,1)$ defined by

$$(Bf)(x) = \int_0^1 k(x,s)f(s)\,ds$$

where $k(x,s)$ is known as the *kernel* of the operator. The inequality

$$\begin{aligned}
\|Bf\|_{L^2(0,1)} &= \left(\int_0^1\left|\int_0^1 k(x,s)f(s)\,ds\right|^2 dx\right)^{1/2} \\
&\le \left(\int_0^1\left(\int_0^1 |k(x,s)|^2\,ds\int_0^1|f(s)|^2\,ds\right)dx\right)^{1/2} \\
&= \left(\int_0^1\int_0^1|k(x,s)|^2\,ds\,dx\right)^{1/2}\|f\|_{L^2(0,1)}
\end{aligned}$$

shows that B is bounded if $k(x,s) \in L^2([0,1]\times[0,1])$ and that

$$\|B\| \le \left(\int_0^1\int_0^1|k(x,s)|^2\,ds\,dx\right)^{1/2}.$$

Manipulations analogous to those done for the matrix example above show that B^* is given by

$$(B^*g)(s) = \int_0^1 \overline{k(x,s)}g(x)\,dx.$$

Clearly B is self-adjoint if $k(x,s) = \overline{k(s,x)}$ and $k(x,s) \in L^2([0,1] \times [0,1])$.

Definition 4.135. An operator acting in a Hilbert space is *weakly continuous* if it maps every weakly convergent sequence into a weakly convergent sequence.

Lemma 4.136. *A continuous linear operator acting in a Hilbert space is also weakly continuous.*

Proof. Let A be continuous on H and choose $\{x_n\}$ such that $x_n \rightharpoonup x_0$ in H. An arbitrary continuous linear functional $F(x)$ takes the form $F(x) = (x, f)$ for some $f \in H$, hence we must show that $(Ax_n - Ax_0, f) \to 0$ as $n \to \infty$. But

$$(Ax_n - Ax_0, f) = (x_n - x_0, A^*f) \to 0 \quad \text{as } n \to \infty$$

since $A^*f \in H$ and $\{x_n\}$ converges weakly to x_0. □

The proof shows that

$$x_n \rightharpoonup x_0 \implies Ax_n \rightharpoonup Ax_0,$$

analogous to the case with ordinary (strong) continuity.

The following lemma plays an important role in justifying many numerical methods for the solution of boundary value problems.

Lemma 4.137. *Assume that A is a continuous linear operator acting in a Hilbert space H. If $x_n \rightharpoonup x_0$ and $y_n \to y_0$ in H, then $(Ax_n, y_n) \to (Ax_0, y_0)$.*

Proof. We will show that $(Ax_n, y_n) - (Ax_0, y_0) \to 0$. We have

$$\begin{aligned}(Ax_n, y_n) - (Ax_0, y_0) &= (x_n, A^*y_n) - (x_0, A^*y_0) \\ &= (x_n, A^*y_n) - (x_n, A^*y_0) + (x_n, A^*y_0) - (x_0, A^*y_0) \\ &= (x_n, A^*(y_n - y_0)) + (x_n - x_0, A^*y_0).\end{aligned}$$

The first term on the right tends to zero because

$$|(x_n, A^*(y_n - y_0))| \le \|x_n\|\,\|A^*\|\,\|y_n - y_0\|$$

and $y_n \to y_0$ (here $\|x_n\|$ is bounded since $\{x_n\}$ is weakly convergent); the second term tends to zero because $x_n \rightharpoonup x_0$. □

Sometimes it is important to obtain an exact value or accurate bound for the norm of an operator. For a self-adjoint operator this can be done via the following theorem.

Theorem 4.138. *If A is a self-adjoint continuous linear operator given on a Hilbert space H, then*

$$\|A\| = \sup_{\|x\|\le 1} |(Ax, x)|. \tag{4.95}$$

Proof. We denote the right side of (4.95) by γ. By the Schwarz inequality

$$\gamma \le \sup_{\|x\|\le 1} \{\|Ax\| \, \|x\|\} \le \sup_{\|x\|\le 1} \{\|A\| \, \|x\|^2\} = \|A\| .$$

The reverse inequality, which completes the proof, takes a bit more effort to establish. First, by definition of γ we have $|(Ax, x)| \le \gamma$ whenever $\|x\| \le 1$. Hence, replacing x by $x/\|x\|$, we can write

$$|(Ax, x)| \le \gamma \|x\|^2$$

for any $x \in H$. Setting $x_1 = y + \lambda z$ and $x_2 = y - \lambda z$ where $\lambda \in \mathbb{R}$ and $y, z \in H$, we have

$$\begin{aligned} C &\equiv |(Ax_1, x_1) - (Ax_2, x_2)| \\ &= |2\lambda| \, |(Ay, z) + (Az, y)| \\ &= |2\lambda| \, |(Ay, z) + (z, Ay)|. \end{aligned}$$

On the other hand

$$\begin{aligned} C &\le |(Ax_1, x_1)| + |(Ax_2, x_2)| \\ &\le \gamma(\|x_1\|^2 + \|x_2\|^2) \\ &= 2\gamma(\|y\|^2 + \lambda^2 \|z\|^2) \end{aligned}$$

by the parallelogram equality, so

$$|2\lambda| \, |(Ay, z) + (z, Ay)| \le 2\gamma(\|y\|^2 + \lambda^2 \|z\|^2).$$

Since this holds for all $y, z \in H$ we may set $z = Ay$ to obtain

$$|4\lambda| \, \|Ay\|^2 \le 2\gamma(\|y\|^2 + \lambda^2 \|Ay\|^2).$$

With $\lambda = \|y\| / \|Ay\|$ this reduces to $\|Ay\| \le \gamma \|y\|$ and so $\|A\| \le \gamma$. □

The theorem implies that a self-adjoint continuous operator A in a Hilbert space is zero if and only if $(Ax, x) = 0$ for all $x \in H$.

4.16 Compact Operators

Using computers we can successfully solve finite systems of linear algebraic equations. A computer performs a finite number of operations, so if we need to solve a problem with some accuracy it should have a structure close to that of finite algebraic equations. An important class of operators with which problems of this kind arise is the class of compact operators. In this section we take X to be a normed space and Y a Banach space.

Definition 4.139. A linear operator A from X to Y is *compact* if it maps bounded subsets of X into precompact subsets of Y.

It suffices to show that A maps the *unit ball* of X into a precompact subset of Y. (By "the unit ball" of a space, if nothing is said about its center, we mean a ball of unit radius centered at the origin of the space.) This follows from the linearity of A. It is also evident that A is compact if and only if every bounded sequence $\{x_n\}$ in X has a subsequence whose image under A is a Cauchy sequence in Y.

In the space $\mathbb{R}^n$ with a fixed basis, a matrix A defines a continuous linear operator that is denoted by A as well. Such an operator A maps a closed and bounded subset of $\mathbb{R}^n$ into a closed and bounded subset of $\mathbb{R}^n$; so the image is compact, and A is a compact operator. In an infinite dimensional space a continuous linear operator is not in general compact. For example, the identity operator I on $C(0,1)$ performs the simple mapping $f(x) \mapsto f(x)$. Therefore I maps the unit ball of $C(0,1)$ into itself, but the unit ball of $C(0,1)$ is not precompact.

Theorem 4.140. *Every compact linear operator is bounded (continuous).*

Proof. Suppose A is not bounded. Then we can find a bounded sequence $\{x_n\}$ in X such that $\|Ax_n\| \to \infty$. As $\{Ax_n\}$ contains no convergent subsequence, A is not compact. □

It is clear that the zero operator is compact. Let us present a nontrivial example of a compact linear operator. Consider the operator A from $C(0,1)$ to $C(0,1)$ given by

$$(Af)(t) = \int_0^1 h(t,\tau) f(\tau)\, d\tau,$$

where the kernel function $h(t,\tau)$ is continuous on the square $[0,1] \times [0,1]$. Let B_1 be the unit ball of $C(0,1)$, and let $S = A(B_1)$. Because h is continuous there exists $\alpha > 0$ such that $|h(t,\tau)| \le \alpha$, and thus

$$\max_{t\in[0,1]} |(Af)(t)| \le \alpha \max_{t\in[0,1]} |f(t)| \le \alpha$$

whenever $f(t) \in B_1$ (i.e., whenever $|f(t)| \le 1$ on $[0,1]$). We conclude that S is uniformly bounded. S is also equicontinuous: we have

$$\begin{aligned} |(Af)(t_2) - (Af)(t_1)| &\le \int_0^1 |h(t_2,\tau) - h(t_1,\tau)|\,|f(\tau)|\,d\tau \\ &\le \max_{\tau\in[0,1]} |h(t_2,\tau) - h(t_1,\tau)| \end{aligned}$$

for $f(t) \in B_1$, and, given $\varepsilon > 0$, the uniform continuity of $h(t,\tau)$ guarantees that we can find δ such that $|h(t_1,\tau) - h(t_1,\tau)| < \varepsilon$ whenever $|t_2 - t_1| < \delta$ and $\tau \in [0,1]$. So by Arzelà's theorem S is precompact, and we conclude that A is a compact operator.

Let us consider a practically important class of compact linear operators. An operator is called *one dimensional* if its image is a one dimensional subspace. The general form of a continuous one dimensional linear operator T is evidently

$$Tx = (F(x))y_0$$

where F is a continuous linear functional and y_0 is some fixed element of the image. A one dimensional linear operator is compact. Indeed, the functional F maps the unit ball B with center at the origin into a bounded numerical set $F(B)$, so it is precompact. Thus the set $F(B)y_0$ is precompact in the space Y as well. A linear operator T_n is called *finite dimensional* if

$$T_n x = \sum_{k=1}^{n} (F_k(x))y_k$$

where the F_k are linear functionals in X and the y_k are some elements of Y. If the F_k are continuous then so is T_n. Because each component of T_n is a compact linear operator, so is T_n; this is a consequence of the following general theorem.

Theorem 4.141. *If A_1 and A_2 are compact linear operators from X to Y, then so is each operator of the form $\lambda A_1 + \mu A_2$ where λ, μ are scalars.*

Proof. If $\{x_n\}$ is a bounded sequence in X, it has a subsequence $\{x_{n_1}\}$ for which $\{A_1 x_{n_1}\}$ is a Cauchy sequence in Y. Because this subsequence is itself bounded, it has a subsequence $\{x_{n_2}\}$ for which $\{A_2 x_{n_2}\}$ is a Cauchy sequence. The image subsequences $\{A_1 x_{n_2}\}$ and $\{A_2 x_{n_2}\}$ are both Cauchy sequences then. Weighting by the scalars λ and μ does not affect whether a sequence is a Cauchy sequence, and the sum of two Cauchy sequences is a Cauchy sequence. Hence the operator $\lambda A_1 + \mu A_2$ is compact. □

This theorem means that the set of compact linear operators from X to Y is a linear subspace of $L(X, Y)$.

Lemma 4.142. *Let A and B be linear operators in X. If A is compact and B is continuous, then the composition operators AB and BA are compact.*

Proof. First consider the operator AB. If M is any bounded subset of X, then $B(M)$ is bounded. But the compact operator A maps bounded sets to precompact sets, so $AB(M)$ is precompact as required. Now consider the operator BA. Let $\{x_n\}$ be a bounded sequence in X. Then $\{Ax_n\}$ has a Cauchy subsequence $\{Ax_{n_k}\}$. But a bounded linear operator maps Cauchy sequences into Cauchy sequences. So $\{BAx_{n_k}\}$ is a Cauchy sequence, as required. □

Theorem 4.143. *If $A \in L(X, Y)$ is compact, then A maps weak Cauchy sequences from X into strong Cauchy sequences in Y.*

Proof. Let $\{x_n\}$ be a weak Cauchy sequence in X. Then $\{x_n\}$ is bounded and, since A is a compact operator, the sequence $\{Ax_n\}$ contains a strong Cauchy subsequence $\{Ax_{n_1}\}$. This subsequence converges to some $y \in Y$ since Y is a Banach space. It is easy to show that $\{Ax_n\}$ is a weak Cauchy sequence in Y; furthermore, because one of its subsequences converges strongly to y, the whole sequence $\{Ax_n\}$ converges weakly to $y \in Y$.

We now show that $\{Ax_n\}$ converges strongly to y. Suppose to the contrary that it does not. Then there is a subsequence $\{Ax_{n_2}\}$ and $\varepsilon > 0$ such that

$$\|Ax_{n_2} - y\| > \varepsilon \tag{4.96}$$

for each n_2. But from $\{Ax_{n_2}\}$ we can select a subsequence $\{Ax_{n_3}\}$ that is a strongly Cauchy sequence in Y and thus has a limit $y_1 \in Y$. This subsequence converges weakly to the same element y_1. By the paragraph above it also converges weakly to y. But we must have $y_1 = y$ by uniqueness of the weak limit; hence $Ax_{n_3} \to y$, and this contradicts (4.96). □

In a separable Hilbert space this result can be strengthened:

Theorem 4.144. *A linear operator A acting in a separable Hilbert space H is compact if and only if it takes every weak Cauchy sequence $\{x_n\}$ into the strong Cauchy sequence $\{Ax_n\}$ in H.*

Proof. Suppose A maps each weak Cauchy sequence $\{x_n\} \subset H$ into a strong Cauchy sequence $\{Ax_n\} \subset H$. To show that A is compact, we take a bounded set $M \subset H$ and show that $A(M)$ is precompact. Take a sequence $\{y_n\} \subset A(M)$ and consider its preimage $\{x_n\} \subset M$ (i.e., the sequence for which $Ax_n = y_n$). Since $\{x_n\}$ is bounded it has a weak Cauchy subsequence $\{x_{n_k}\}$. By hypothesis $\{Ax_{n_k}\}$ is a strong Cauchy sequence in H, hence $A(M)$ is precompact.

The converse was proved in Theorem 4.143. □

Example 4.145. Show that if $x_n \rightharpoonup x_0$, and A from X to Y is compact, then $Ax_n \to Ax_0$ as $n \to \infty$.

Solution. If $\{x_n\}$ is weakly convergent then it is weakly Cauchy and by Theorem 4.143 we have $Ax_n \to y$ for some $y \in Y$. Since strong convergence implies weak convergence we have $Ax_n \rightharpoonup y$ for some $y \in Y$. On the other hand A is compact, hence continuous, hence weakly continuous, so $x_n \rightharpoonup x_0$ implies $Ax_n \rightharpoonup Ax_0$. Finally, $y = Ax_0$ by uniqueness of the weak limit. □

Recall that $L(X, Y)$ is a normed linear space under the operator norm $\|\cdot\|$. If $\{A_n\}$ is a sequence of linear operators such that

$$\lim_{n\to\infty} \|A_n - A\| = 0,$$

then $\{A_n\}$ is said to be *uniformly convergent* and the operator A is known as the *uniform operator limit* of the sequence $\{A_n\}$.

Theorem 4.146. *A uniform operator limit of a sequence of compact linear operators is a compact linear operator.*

Proof. Let $\{A_n\} \subset L(X, Y)$ be a sequence of compact linear operators and suppose $\|A_n - A\| \to 0$ as $n \to \infty$. The approach is to take any bounded sequence $\{x_n\} \subset X$ and show that we can select a subsequence whose image under A is a Cauchy sequence in Y. By compactness of A_1 we can select from $\{x_n\}$ a subsequence $\{x_{n_1}\}$ such that $\{A_1 x_{n_1}\}$ is a Cauchy sequence. Similarly, by compactness of A_2 we can select from $\{x_{n_1}\}$ a subsequence $\{x_{n_2}\}$ such that $\{A_2 x_{n_2}\}$ is a Cauchy sequence. Continuing in this way, after the kth step we have a subsequence $\{x_{n_k}\}$ for which $\{A_k x_{n_k}\}$ is a Cauchy sequence. The diagonal sequence $\xi_n \equiv x_{n_n}$ has the property that $\{A_k \xi_n\}$ is a Cauchy sequence for each fixed k. Then for any $m \geq 1$ we have

$$\begin{aligned}
&\|A\xi_{n+m} - A\xi_n\| \\
&\quad = \|(A\xi_{n+m} - A_k\xi_{n+m}) + (A_k\xi_{n+m} - A_k\xi_n) + (A_k\xi_n - A\xi_n)\| \\
&\quad \le \|A - A_k\| \, \|\xi_{n+m}\| + \|A_k\xi_{n+m} - A_k\xi_n\| + \|A_k - A\| \, \|\xi_n\| \\
&\quad \le 2b \, \|A - A_k\| + \|A_k\xi_{n+m} - A_k\xi_n\|
\end{aligned}$$

where $\|\xi_n\| \le b$ for all n. Given $\varepsilon > 0$ we can choose and fix p so that $\|A - A_p\| < \varepsilon/4b$; then

$$\|A\xi_{n+m} - A\xi_n\| \le \varepsilon/2 + \|A_p\xi_{n+m} - A_p\xi_n\| ,$$

and we can finish the proof by choosing N so large that the second term on the right is less than $\varepsilon/2$ for $n > N$ and any $m \ge 1$. □

Thus the set of all compact linear operators from X to Y is a closed linear subspace of $L(X, Y)$.

Above we introduced the set of finite dimensional linear operators; these, being continuous, are compact. The importance of this class is given by the following theorem, which states that this class is dense in the set of compact linear operators in a Hilbert space.

Theorem 4.147. *If A is a compact operator acting in a separable Hilbert space, then there is a sequence of finite dimensional continuous linear operators $\{A_n\}$ having uniform operator limit A.*

Proof. A Hilbert space H has an orthonormal basis $\{g_n\}$, in terms of which any $f \in H$ can be represented as

$$f = \sum_{k=1}^{\infty} (f, g_k) g_k .$$

Since A is a continuous operator we have

$$Af = \sum_{k=1}^{\infty} (f, g_k) A g_k .$$

We define A_n by

$$A_n f = \sum_{k=1}^{n} (f, g_k) A g_k$$

and show that

$$\lim_{n\to\infty} \|A - A_n\| = 0. \tag{4.97}$$

By definition

$$\|A - A_n\| = \sup_{\|f\| \le 1} \|(A - A_n)f\|.$$

First we show that there exists f_n^* such that

$$\|f_n^*\| \le 1 \quad \text{and} \quad \|A - A_n\| = \|(A - A_n)f_n^*\|. \tag{4.98}$$

By definition of the supremum there is a sequence $\{f_k\}$ such that

$$\|f_k\| \le 1 \quad \text{and} \quad \lim_{k\to\infty} \|(A - A_n)f_k\| = \|A - A_n\|.$$

This bounded sequence in a separable Hilbert space has a weak Cauchy subsequence $\{f_{k_1}\}$, and this subsequence converges weakly to an element f_n^*; moreover, by the proof of Lemma 4.125 we have $\|f_n^*\| \le 1$. Because $A - A_n$ is compact the sequence $\{(A - A_n)f_{k_1}\}$ converges strongly to $(A - A_n)f_n^*$, i.e., a subsequence of the convergent sequence $\{\|(A - A_n)f_k\|\}$ converges to the number $\|(A - A_n)f_n^*\|$ as $k \to \infty$. So the second relation in (4.98) also holds. But

$$(A - A_n)f_n^* = A\left(\sum_{k=1}^{\infty}(f_n^*, g_k)g_k\right) - \sum_{k=1}^{n}(f_n^*, g_k)Ag_k = A\left(\sum_{k=n+1}^{\infty}(f_n^*, g_k)g_k\right)$$

so taking the norm of both sides we have, by (4.98),

$$\|A - A_n\| = \|A\varphi_n\| \quad \text{where} \quad \varphi_n = \sum_{k=n+1}^{\infty}(f_n^*, g_k)g_k. \tag{4.99}$$

The sequence $\{\varphi_n\} \subset H$ converges weakly to zero. Indeed for any $f \in H$ we can write

$$\begin{aligned}
(\varphi_n, f) &= \left(\sum_{k=n+1}^{\infty}(f_n^*, g_k)g_k, \sum_{m=1}^{\infty}(f, g_m)g_m\right) \\
&= \left(\sum_{k=n+1}^{\infty}(f_n^*, g_k)g_k, \sum_{m=n+1}^{\infty}(f, g_m)g_m\right) \\
&= \sum_{k=n+1}^{\infty}(f_n^*, g_k)\overline{(f, g_k)},
\end{aligned}$$

hence

$$|(\varphi_n, f)| \leq \left(\sum_{k=n+1}^{\infty} |(f_n^*, g_k)|^2 \right)^{1/2} \left(\sum_{k=n+1}^{\infty} |(f, g_k)|^2 \right)^{1/2}$$
$$\leq \left(\sum_{k=n+1}^{\infty} |(f, g_k)|^2 \right)^{1/2} \|f_n^*\| \to 0 \quad \text{as } n \to \infty$$

since $\|f_n^*\| \leq 1$ and $\sum_{k=1}^{\infty} |(f, g_k)|^2 = \|f\|^2 < \infty$ by Parseval's equality (i.e., the parenthetical quantity represents the tail of a convergent series). Since $\varphi_n \rightharpoonup 0$ and A is compact we have

$$\lim_{n\to\infty} \|A\varphi_n\| = 0.$$

By (4.99) this proves (4.97). □

We will need the following simple theorem.

Theorem 4.148. *If A is a compact linear operator acting in a Hilbert space, then A^* is compact.*

Proof. We take a sequence $\{f_n\}$ such that $f_n \rightharpoonup f_0$ and show that $A^* f_n \to A^* f_0$. We have

$$\begin{aligned}
\|A^* f_n - A^* f_0\|^2 &= (A^* f_n - A^* f_0, A^* f_n - A^* f_0) \\
&= (f_n - f_0, AA^*(f_n - f_0)) \\
&\leq \|f_n - f_0\| \, \|AA^*(f_n - f_0)\| \\
&\leq (\|f_n\| + \|f_0\|) \, \|AA^*(f_n - f_0)\| .
\end{aligned}$$

But $\|f_n\| \leq M$ for some constant M, and the product AA^* is compact since A^* is continuous. Hence $AA^*(f_n - f_0) \to 0$ as $n \to \infty$, and so

$$\|A^* f_n - A^* f_0\|^2 \to 0 \quad \text{as } n \to \infty.$$

This completes the proof. □

Sobolev's imbedding theorem states that some imbedding operators from a Sobolev space are compact. A simple illustration can serve to clarify this idea. Let us consider the mapping under which a continuously differentiable function $f(x)$ (we show this as $f(x) \in C^{(1)}(0,1)$) is regarded as an element of the space $C(0,1)$, the space of functions continuous on $[0,1]$. Although this mapping is an operator, we cannot call it an identity operator since its domain and range are different spaces. Instead, we refer to it as the imbedding operator from $C^{(1)}(0,1)$ to $C(0,1)$.

Theorem 4.149. *The imbedding operator from $C^{(1)}(0,1)$ to $C(0,1)$ is compact.*

Proof. We need to check that the image S of the unit ball of the domain is a precompact set in $C(0,1)$. By Arzelà's theorem we need to show that the set of functions S is uniformly bounded and equicontinuous. It is uniformly bounded since a function of the unit ball of $C^{(1)}(0,1)$ satisfies $|f(x)| \leq 1$ and thus is inside the unit ball of $C(0,1)$. The Lagrange mean value theorem then states that for any $x_1 < x_2$ from $[0,1]$ where the function is continuously differentiable there exists $\xi \in [x_1, x_2]$ such that

$$f(x_2) - f(x_1) = f'(\xi)(x_2 - x_1).$$

Since $|f'(\xi)| \leq 1$ for any $f \in S$, we have

$$|f(x_2) - f(x_1)| \leq |x_2 - x_1|.$$

This implies the equicontinuity of S. □

4.17 Closed Operators

We have considered the case of a continuous linear operator whose domain is the whole space. However, the differentiation operator d/dx acting on the space of functions continuous on $[0,1]$ does not have the entire space $C(0,1)$ as its domain, since there are continuous functions that are nowhere differentiable on $[0,1]$. But this operator, as we shall see below, has some properties that are "better" than the properties of a general operator with an arbitrary domain. We shall show that it resides in a class of operators that is wider than the class of continuous operators, but such that there remains the possibility for us to perform some limit passages with it. The class is given by the following definition.

Definition 4.150. Let A be a linear operator mapping elements of a Banach space X into elements of a Banach space Y. We say that A is *closed* if for any sequence $\{x_n\} \subset D(A)$ such that $x_n \to x$ and $Ax_n \to y$ as $n \to \infty$, it follows that $x \in D(A)$ and $y = Ax$.

It is evident that A is closed if A is continuous and $D(A) = X$. There are, however, closed operators that are not continuous. An example is the derivative operator $A = d/dt$ acting from $C(0,1)$ to $C(0,1)$. The domain of A is the subset of $C(0,1)$ consisting of those functions having continuous first derivatives on $[0, 1]$. To see that A is closed, we first assume that

$$x_n(t) \to x(t) \quad \text{as } n \to \infty$$

in the norm of $C(0, 1)$, where each $x'_n(t)$ is continuous, and that

$$Ax_n(t) = x'_n(t) \to y(t) \quad \text{as } n \to \infty,$$

also in the norm of $C(0, 1)$. Realizing that convergence in the max norm is uniform convergence, we recall a theorem from ordinary calculus:

Theorem 4.151. *If $f_n(t)$ is continuous for each n and $f_n(t) \to f(t)$ uniformly on $[0, 1]$, then*

(1) $f(t)$ is continuous on $[0, 1]$, and
(2) uniform convergence of the sequence $\{f'_n(t)\}$ of derivatives that are continuous on $[0, 1]$ implies that $f'(t)$ exists, is continuous on $[0, 1]$, and that $f'_n(t) \to f'(t)$.

By this theorem $A = d/dx$ on $C(0, 1)$ meets the definition of a closed operator. To see that A is not continuous, consider its action on the set of functions $\{t^n\}$. This set is bounded with

$$\|t^n\| = 1 \quad \text{for each } n,$$

but its image under A is unbounded with

$$\left\| \frac{d}{dt} x_n(t) \right\| = \|nt^{n-1}\| = n.$$

So A does not map every bounded set into a bounded set.

If $\Omega \subset \mathbb{R}^n$ is compact, then the more general differential operator A given by

$$Af(\mathbf{x}) = \sum_{|\alpha| \leq n} c_\alpha(\mathbf{x}) D^\alpha f(\mathbf{x}), \tag{4.100}$$

with continuous coefficients $c_\alpha(\mathbf{x})$ and acting from $C^{(n)}(\Omega)$ to $C(\Omega)$, is a closed operator.

Definition 4.152. Let A be an operator from X to Y. Suppose that an operator B, also from X to Y, satisfies the following two conditions:

(1) $D(A) \subseteq D(B)$, and
(2) $B(x) = A(x)$ for all $x \in D(A)$.

Then B is an *extension of* A.

Lemma 4.153. *A linear operator A acting from a Banach space X to a Banach space Y has a closed extension if and only if from the condition*

(*) $\{x_n\} \subset D(A)$ is an arbitrary sequence such that $x_n \to 0$ and $Ax_n \to y$

it follows that $y = 0$.

Proof. Necessity follows from Definition 4.150. To prove sufficiency let us explicitly construct a closed extension B of A.

We define B, then verify its properties. Let $D(B)$ consist of those elements x for which there exists $\{x_n\} \subset D(A)$ such that $x_n \to x$ and $Ax_n \to y$ as $n \to \infty$; for each such x, define $Bx = y$. Condition (*) ensures that y is uniquely defined by x. Indeed, suppose two sequences $\{x_n\}$ and $\{z_n\}$ in $D(A)$ both converge to x, and $Ax_n \to y$ while $Az_n \to y'$. Then

$$x_n - z_n \to 0, \qquad A(x_n - z_n) = Ax_n - Az_n \to y - y',$$

and from (*) it follows that $y - y' = 0$.

To see that B is linear, we take two elements $x, \tilde{x}$ in $D(B)$ and any two scalars λ, μ. By definition of $D(B)$ there are sequences $\{x_n\}$ and $\{\tilde{x}_n\}$ in $D(A)$ such that

$$x_n \to x, \quad Ax_n \to y, \qquad \tilde{x}_n \to \tilde{x}, \quad A\tilde{x}_n \to \tilde{y},$$

and we define $Bx = y$, $B\tilde{x} = \tilde{y}$. But $\lambda x + \mu\tilde{x} \in D(B)$ because

$$\lambda x_n + \mu\tilde{x}_n \to \lambda x + \mu\tilde{x}, \quad A(\lambda x_n + \mu\tilde{x}_n) = \lambda Ax_n + \mu A\tilde{x}_n \to \lambda y + \mu\tilde{y},$$

and we therefore *define* $B(\lambda x + \mu\tilde{x}) = \lambda y + \mu\tilde{y} = \lambda Bx + \mu B\tilde{x}$.

Finally, let $\{u_n\} \subset D(B)$ be such that $u_n \to u$ and $Bu_n \to v$. According to Definition 4.150 we must prove that $u \in D(B)$ and $Bu = v$. Let us construct a sequence $\{x_n\} \subset D(A)$ that is equivalent to $\{u_n\}$, and then verify the desired properties for $\{x_n\}$. Fix u_n. By definition of B there exists $\{w_{nk}\} \subset D(A)$ such that $w_{nk} \to u_n$ and $Aw_{nk} \to Bu_n$ as $k \to \infty$. Hence there exists N such that for all $k > N$ we have both $\|w_{nk} - u_n\| < 1/n$ and $\|Aw_{nk} - Bu_n\| < 1/n$. Choose one of the points w_{nk_0} where $k_0 > N$, and denote this point x_n. Now consider the sequence of points $\{x_n\} \subset D(A)$. The inequalities $\|x_n - u_n\| < 1/n$ and $\|Ax_n - Bu_n\| < 1/n$ show that $x_n \to u$ and $Ax_n \to v$ as $n \to \infty$. By definition of B we have $u \in D(B)$ and $Bu = v$. □

It sometimes happens that we can establish boundedness of an operator directly on a subspace that is everywhere dense in the space. To establish that it is continuous on the whole space, we may employ

Theorem 4.154. *Let A be a closed linear operator whose domain is a Banach space X and whose range lies in a Banach space Y. Assume there is a set M which is dense in X and a positive constant c such that*

$$\|Ax\| \leq c\,\|x\| \quad \text{for all } x \in M.$$

Then A is continuous on the whole space X.

Proof. For any $x_0 \in X$, we can find $\{x_n\} \subset M$ such that $\|x_n - x_0\| < 1/n$ for each n. The inequality

$$\|Ax_{k+m} - Ax_k\| \leq c\,\|x_{k+m} - x_k\| \leq c(\|x_{k+m} - x_0\| + \|x_k - x_0\|) \leq 2c/k$$

shows that $\{Ax_k\}$ is a Cauchy sequence in Y. We have $Ax_k \to y$ for some $y \in Y$ since Y is a Banach space; since A is closed, $Ax_0 = y$. Now we can write

$$\|Ax_0\| = \lim_{k\to\infty} \|Ax_k\| \leq \lim_{k\to\infty} c\,\|x_k\| = c\,\|x_0\|\,.$$

Since x_0 is an arbitrary element of X and c does not depend on x_0, the proof is complete. □

Closed operators can be considered from another viewpoint. If X and Y are Banach spaces over the same scalar field, then the Cartesian product space $X \times Y$ with algebraic operations defined by

$$(x_1, y_1) + (x_2, y_2) = (x_1 + x_2, y_1 + y_2), \qquad \alpha(x, y) = (\alpha x, \alpha y),$$

and norm defined by

$$\|(x, y)\| = (\|x\|_X^2 + \|y\|_Y^2)^{1/2},$$

is also a Banach space.

Definition 4.155. Let A be an operator acting from $D(A) \subset X$ to Y. Then the set

$$G(A) = \{(x, Ax) \in X \times Y : x \in D(A)\} \tag{4.101}$$

is the *graph* of A.

Theorem 4.156. *A linear operator A acting from $D(A) \subset X$ to Y is closed if and only if $G(A)$ is a closed linear subspace of $X \times Y$.*

Proof. Suppose A is a closed operator. Let (x, y) be a limit point of $G(A)$. Then there is a sequence $\{(x_n, Ax_n)\} \subset G(A)$ that converges to (x, y) in the norm of $X \times Y$. Evidently this implies that as $n \to \infty$ we have $x_n \to x$ in X and $Ax_n \to y$ in Y. Because A is closed, $x \in D(A)$ and $y = Ax$. Hence $(x, Ax) \in G(A)$ by definition of $G(A)$.

Conversely, suppose $G(A)$ is closed in $X \times Y$. Let $\{x_n\} \subset D(A)$ be such that, as $n \to \infty$, $x_n \to x$ in X and $Ax_n \to y$ in Y. The sequence $\{(x_n, Ax_n)\} \subset G(A)$ converges in the norm of $X \times Y$ to (x, y). Since $G(A)$ is closed, $(x, y) \in G(A)$. By definition of $G(A)$ this means that $x \in D(A)$ and $y = Ax$. □

Theorem 4.157. *If A is an invertible closed linear operator, then A^{-1} is also closed.*

Proof. We can obtain $G(A^{-1})$ from the graph of $G(A)$ by the simple rearrangement $(x, Ax) \mapsto (Ax, x)$. Hence $G(A^{-1})$ is closed in $Y \times X$. □

We can now formulate Banach's *closed graph theorem.*

Theorem 4.158. *Let X and Y be Banach spaces. If A is a closed linear operator having $D(A) = X$, then A is continuous on X.*

See [32] for a proof. In applications the following simple consequence of the theorem can establish continuity of an operator.

Corollary 4.159. *Let X and Y be Banach spaces. If a closed linear operator A from X to Y is one-to-one and onto, then A^{-1} is continuous on Y.*

Proof. The operator A^{-1} is closed by Theorem 4.157 and continuous by Theorem 4.158. □

4.18 On the Sobolev Imbedding Theorem

The most important result obtained by S.L. Sobolev is the imbedding theorem. It gives some properties of the elements of Sobolev spaces and, in particular, relates them to continuously differentiable functions. An example of an imbedding can be seen from the estimate

$$\|f(\mathbf{x})\|_{W^{l,q}(\Omega)} \le m_{qp} \|f(\mathbf{x})\|_{W^{l,p}(\Omega)}, \qquad q < p, \tag{4.102}$$

which can be shown for any $f \in W^{l,p}(\Omega)$ to hold with a constant m_{qp} that depends on q, p, and Ω only. Note that for $q < p$ we have

$$f(\mathbf{x}) \in W^{l,p}(\Omega) \implies f(\mathbf{x}) \in W^{l,q}(\Omega);$$

hence the Sobolev space $W^{l,p}(\Omega)$ is a subset of the Sobolev space $W^{l,q}(\Omega)$:

$$W^{l,p}(\Omega) \subseteq W^{l,q}(\Omega), \qquad q < p.$$

But the estimate (4.102) gives us more than just this subset inclusion. We met inclusions of this type when considering the $L^p(\Omega)$ spaces. We called them imbeddings. Now we provide a general definition of this term.

Definition 4.160. The *operator of imbedding* from X to Y is the one-to-one correspondence between a space X and a subspace Y of a space Z under which we identify elements $x \in X$ with elements $y \in Y$ in such a way that the correspondence is linear. If, besides, the correspondence is continuous so that

$$\|y\|_Y \leq m \|x\|_X$$

for some constant m that does not depend on x, then we call it the continuous operator of imbedding. We sometimes employ the notation

$$X \hookrightarrow Y,$$

to indicate the existence of an imbedding from X to Y.

Some words of explanation are in order here. The reader should note that the formal definition of a continuous imbedding operator does not differ from that of a continuous linear operator. However, the term "imbedding" is reserved for situations in which we *identify* an element in X with its image in Y, and thereby effectively consider the "same element" as a member of two different spaces. (In this way an imbedding operator acts somewhat like the identity operator that serves to map elements of a space into themselves; the difference is that in the case of an identity operator the domain and range must be the same space.) The degree to which one may take literally the "identification" process between elements of X and their images in Y depends on the specific type of imbedding under consideration. In some instances the elements of X and Y are of the same basic nature (e.g., both are ordinary functions); in other instances this is not the case (e.g., the elements of Y may be functions while the elements of X are equivalence classes of Cauchy sequences of functions). Note, however, that even when the elements of Y and X are of the same nature, the norms associated with the spaces Y and X may be very different. Finally, we remark that there are imbedding operators that are *compact* and not merely *continuous*. We shall state this when it applies, but shall relegate coverage of compact operators to a later section of this chapter.

Example 4.161. Show that ℓ^q is continuously imbedded into ℓ^p if $p > q$.

Solution. The first step is to show that the norms $\|\cdot\|_p$ and $\|\cdot\|_q$ of the spaces ℓ^p and ℓ^q satisfy $\|\mathbf{x}\|_p \le \|\mathbf{x}\|_q$ whenever $p \ge q$ (Exercise 4.11). This gives the subset inclusion $\ell^q \subseteq \ell^p$ whenever $p \ge q$, and also shows that $\ell^q \hookrightarrow \ell^p$ with a constant $m = 1$ in the imbedding inequality. $\square$

Remark 4.162. There is only a limited analogy between the sequence space ℓ^p and the Lebesgue space $L^p(\Omega)$ for a bounded domain Ω. In the latter space an application of Hölder's inequality gives

$$\int_\Omega |f(\mathbf{x})|\, d\Omega \le \left(\int_\Omega 1^q\, d\Omega\right)^{1/q} \left(\int_\Omega |f(\mathbf{x})|^p\, d\Omega\right)^{1/p} = (\operatorname{mes}\Omega)^{1/q} \|f\|_p$$

but a similar application in ℓ^p would give

$$\sum_{k=1}^{\infty} |x_k| \le \left(\sum_{k=1}^{\infty} 1^q\right)^{1/q} \left(\sum_{k=1}^{\infty} |x_k|^p\right)^{1/p} = \infty \cdot \|\mathbf{x}\|_p .$$

Put simply, when we consider $L^p(\Omega)$ with bounded Ω, the "bad points" are those where some function f becomes infinite. Larger values of p make such behavior worse because $|f(\mathbf{x})|^p > |f(\mathbf{x})|^q$ for $p > q$. On the other hand, a sequence $\mathbf{x} \in \ell^p$ has terms x_k that satisfy $|x_k| < 1$ for sufficiently large k. In this case $|x_k|^p < |x_k|^q$ for $p > q$, so larger values of p aid in convergence. This is why for $p > q$ we have $L^p(\Omega) \hookrightarrow L^q(\Omega)$ (again, for bounded Ω) but $\ell^q \hookrightarrow \ell^p$. $\square$

Returning to Sobolev spaces, we see that the space $W^{l,p}(\Omega)$ is continuously imbedded into the space $W^{l,q}(\Omega)$ when $q < p$, and we write

$$W^{l,p}(\Omega) \hookrightarrow W^{l,q}(\Omega), \qquad q < p.$$

We are also interested in continuous imbeddings from Sobolev spaces into the spaces of continuously differentiable functions. To obtain a relevant example of an *imbedding theorem* let us consider the simple Sobolev space $W^{1,1}(0,1)$, the norm of which is

$$\|f(x)\|_{1,1} = \int_0^1 (|f(x)| + |f'(x)|)\, dx. \tag{4.103}$$

So $W^{1,1}(0,1)$ is the completion with respect to the norm (4.103) of the set of all functions that are continuously differentiable on $[0,1]$. Let $f(x)$ be continuously differentiable on $[0,1]$. Then for any $x, y \in [0,1]$ we have

$$f(x) - f(y) = \int_y^x f'(t)\, dt$$

and so

$$|f(x)| \le |f(y)| + \left| \int_y^x f'(t)\,dt \right| \le |f(y)| + \int_0^1 |f'(t)|\,dt.$$

Integrating this in y over $[0,1]$ we get

$$\int_0^1 |f(x)|\,dy \le \int_0^1 |f(y)|\,dy + \int_0^1 \int_0^1 |f'(t)|\,dt\,dy$$

or

$$\max_{x\in[0,1]} |f(x)| \le \int_0^1 |f(y)|\,dy + \int_0^1 |f'(t)|\,dt = \|f(x)\|_{1,1}\,. \tag{4.104}$$

Now let $F(x)$ be an equivalence class from $W^{1,1}(0,1)$. A representative of $F(x)$ is a Cauchy sequence $\{f_n(x)\}$ of continuously differentiable functions, and we have

$$\max_{x\in[0,1]} |f_{n+m}(x) - f_n(x)| \le \|f_{n+m}(x) - f_n(x)\|_{1,1}\,;$$

it follows that $\{f_n(x)\}$ is a Cauchy sequence in $C(0,1)$ as well, and thus has a limit that is continuous on $[0,1]$. From (4.104) it also follows that this limiting function does not depend on the choice of representative sequence of the element of $W^{1,1}(0,1)$. Hence we have a correspondence that is clearly linear, under which to an element $F(x) \in W^{1,1}(0,1)$ there corresponds a unique element $f(x) \in C(0,1)$ such that

$$\|f(x)\|_{C(0,1)} \le \|F(x)\|_{1,1}\,.$$

We identify this limit element with F, and call F by the name of this limit element. (We can really regard F as this element f if f is continuously differentiable on $[0,1]$ so there is a *stationary* representative sequence $(f, f, f, \ldots)$ from F.) In short, we have

$$W^{1,1}(0,1) \hookrightarrow C(0,1). \tag{4.105}$$

Similar results for $W^{l,p}(\Omega)$, where Ω is a compact subset of $\mathbb{R}^n$, are called *Sobolev imbedding theorems*. We shall state one such theorem next. We assume that Ω satisfies the *cone condition*: there is a finite circular cone in $\mathbb{R}^n$ that can touch any point of $\partial\Omega$ with its vertex while lying fully inside Ω (i.e., translations and rotations of the cone are allowed, but not changes in cone angle or height).

Theorem 4.163. *Let Ω_r be an r-dimensional piecewise smooth hypersurface in Ω. The imbedding*

$$W^{m,p}(\Omega) \hookrightarrow L^q(\Omega_r)$$

is continuous if one of the following conditions holds:

(i) $n > mp$, $r > n - mp$, $q \leq pr/(n - mp)$;
(ii) $n = mp$, *q is finite with* $q \geq 1$.

It is compact if

(i) $n > mp$, $r > n - mp$, $q < pr/(n - mp)$ *or*
(ii) $n = mp$ *and q is finite with* $q \geq 1$.

If $n < mp$ *then*

$$W^{m,p}(\Omega) \hookrightarrow C^{(k)}(\Omega)$$

for integers k such that $k \leq (mp - n)/p$, *and the imbedding is continuous. It is compact if* $k < (mp - n)/p$.

Although this theorem is appealing because of its generality, we shall employ only special cases involving $W^{1,2}(\Omega)$ and $W^{2,2}(\Omega)$. The following applies to equilibrium problems for membranes and two-dimensional elastic bodies:

Theorem 4.164. *Let γ be a piecewise differentiable curve in a compact set $\Omega \subset \mathbb{R}^2$. For any finite $q \geq 1$, there are compact (hence continuous) imbeddings*

$$W^{1,2}(\Omega) \hookrightarrow L^q(\Omega), \qquad W^{1,2}(\Omega) \hookrightarrow L^q(\gamma).$$

For use with problems of equilibrium of plates and shells, we have

Theorem 4.165. *Let Ω be a compact subset of $\mathbb{R}^2$. Then there is a continuous imbedding*

$$W^{2,2}(\Omega) \hookrightarrow C(\Omega).$$

For the first derivatives, the imbedding operators to $L^q(\Omega)$ and $L^q(\gamma)$ are compact for any finite $q \geq 1$.

The next result is used for problems of equilibrium of three-dimensional elastic bodies and dynamic problems for membranes and two-dimensional elastic bodies.

Theorem 4.166. *Let γ be a piecewise smooth surface in a compact set $\Omega \subset \mathbb{R}^3$. The imbeddings*

$$W^{1,2}(\Omega) \hookrightarrow L^q(\Omega), \qquad 1 \leq q \leq 6,$$
$$W^{1,2}(\Omega) \hookrightarrow L^p(\gamma), \qquad 1 \leq p \leq 4,$$

are continuous. They are compact if $1 \le q < 6$ *or* $1 \le p < 4$, *respectively.*

4.19 Some Energy Spaces in Mechanics

One may use various norms to distinguish between different states of mechanical objects. To characterize force magnitudes, for example, norms of the type (4.6) are appropriate. If the field is continuous, (4.4) is appropriate. The same can be said for fields of displacements, strains, and stresses. However, there is one important characteristic of a body: its energy due to deformation. It is sensible to try to use this quantity when we characterize the state of a body. We would like to consider this possibility in more detail. The most convenient fact is that the energy spaces we shall introduce are subspaces of Sobolev spaces, and thus we can use Sobolev's imbedding theorem to characterize the parameters of corresponding boundary value problems. Of course, it is possible to use Sobolev spaces directly for this, but energy spaces have many advantages. First, they can be closely customized to the nature of the problem, permitting a better use of mechanical intuition. Second, the energy norms and corresponding inner products permit a proper and direct use of such fundamental properties as mutual orthogonality of eigensolutions; these properties form the basis for solution by Fourier's technique.

Rod under tension

We begin with a simple problem that could be solved by direct integration. It describes the equilibrium of a rod stretched by a distributed load (Fig. 4.1). The double strain energy of a rod of length l is

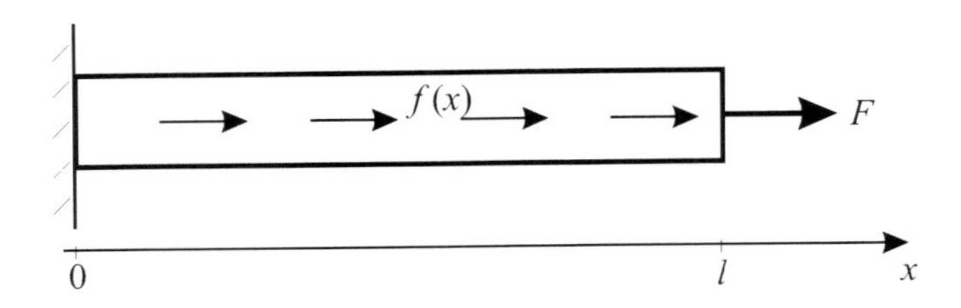

Fig. 4.1 Rod under distributed longitudinal load $f(x)$ and a point force F.

$$2\mathcal{E}(u) = \int_0^l ES(x)u'^2(x)\,dx$$

where the constant E is Young's modulus, $S(x)$ is the area of the cross-section with $0 < S_0 \le S(x) \le S_1$, and $u(x)$ is the displacement of the cross-section of the rod at point x in the longitudinal direction. Suppose the end at $x = 0$ is fixed:

$$u(0) = 0. \tag{4.106}$$

The strain energy generates a functional in two variables that can be considered as an *energy inner product*:

$$(u, v)_R = \int_0^l ES(x)u'(x)v'(x)\,dx. \tag{4.107}$$

The inner product has a clear mechanical meaning: it is the work of internal forces corresponding to the state of the rod $u(x)$ on the admissible displacement field $v(x)$. (Recall that the terms "admissible" and "virtual" are synonymous.) Considering (4.107) on the set C_{Rc} of all continuously differentiable functions on $[0, l]$ satisfying (4.106), the reader can verify that it really is an inner product. (Here the subscript pattern "Rc" reminds us we are dealing with a *clamped rod*: a rod fixed in space. Later, "Rf" will denote a *free rod*.) Let us demonstrate that on C_{Rc} the *energy norm*

$$\|u\|_R = (u, u)_R^{1/2} = \left(\int_0^l ES(x)u'^2(x)\,dx \right)^{1/2}$$

induced by (4.107) is equivalent to the norm of the Sobolev space $W^{1,2}(0, l)$, which is

$$\|u\|_{1,2} = \left(\int_0^l \left[u^2(x) + u'^2(x) \right] dx \right)^{1/2}.$$

We must show that there are positive constants m, M such that for any $u(x) \in C_{Rc}$ we have

$$m \|u\|_R \le \|u\|_{1,2} \le M \|u\|_R.$$

The left-hand inequality is a consequence of

$$\begin{aligned} \|u(x)\|_R^2 &= \int_0^l ES(x)u'^2(x)\,dx \\ &\le ES_1 \int_0^l \left(u^2(x) + u'^2(x) \right) dx \\ &= ES_1 \|u(x)\|_{1,2}^2 . \end{aligned}$$

To prove the right-hand inequality we begin with the identity

$$u(x) = \int_0^x u'(t)\,dt.$$

Squaring and then integrating over $[0, l]$ we get

$$\int_0^l u^2(x)\,dx = \int_0^l \left(\int_0^x u'(t)\,dt\right)^2 dx.$$

Applying the Hölder inequality we have

$$\begin{aligned} \int_0^l u^2(x)\,dx &= \int_0^l \left(\int_0^x 1 \cdot u'(t)\,dt\right)^2 dx \\ &\le \int_0^l \left(\int_0^x 1^2\,dt \int_0^x u'^2(t)\,dt\right) dx \\ &\le l^2 \int_0^l u'^2(x)\,dx, \end{aligned} \tag{4.108}$$

from which the needed fact follows immediately.

Applying the completion procedure in the set C_{Rc} with respect to the norms $\|\cdot\|_R$ and $\|\cdot\|_{1,2}$, we get spaces that contain the same elements and have equivalent norms, so they are considered as the same space. Let us denote this *energy space* by $\mathcal{E}_{Rc}$ and use the Sobolev imbedding theorem. Now

(1) $\mathcal{E}_{Rc}$ is a subspace $W^{1,2}(0, l)$,
(2) $W^{1,2}(0, l)$ is continuously imbedded into $W^{1,1}(0, l)$, and
(3) each element of $W^{1,1}(0, l)$ corresponds to a continuous function.

That is,

$$W^{1,2}(0, l) \hookrightarrow W^{1,1}(0, l) \hookrightarrow C(0, l).$$

Hence to each element of $\mathcal{E}_{Rc}$ there corresponds a continuous function. Clearly all these continuous functions satisfy (4.106). We shall identify them with the corresponding elements of $\mathcal{E}_{Rc}$, and in this sense say that the elements of $\mathcal{E}_{Rc}$ are continuous functions.

Free rod

In the same manner we can consider the energy space for a rod having both ends free of geometrical restriction. Since longitudinal motions are unrestricted by boundary conditions, when we try to use the energy inner product (4.107), we find that there are nontrivial displacements for which the induced energy norm is zero. One such state of the rod is $u(x) = c$. To show that there are no other states with zero strain energy, we derive an inequality to use in place of (4.108). We take the identity

$$u(x) = u(y) + \int_y^x u'(t)\,dt,$$

integrate with respect to y over $[0, l]$ to get

$$lu(x) = \int_0^l u(y)\,dy + \int_0^l \int_y^x u'(t)\,dt\,dy,$$

then take the absolute value of both sides and estimate the right side as in § 4.5:

$$l|u(x)| = \left| \int_0^l u(y)\,dy + \int_0^l \int_y^x u'(t)\,dt\,dy \right| \le \left| \int_0^l u(y)\,dy \right| + l \int_0^l |u'(t)|\,dt. \tag{4.109}$$

Let C_{Rf} be the set of functions $u(x)$ that are continuously differentiable on $[0, l]$ and satisfy

$$\int_0^l u(y)\,dy = 0. \tag{4.110}$$

Note that by subtracting the right constant c from a given function $u(x)$, corresponding to a free motion of the rod through the distance c, we can make the new displacement field satisfy (4.110). From (4.109) we have three consequences:

$$\int_0^l |u(x)|\,dx \le \left| \int_0^l u(x)\,dx \right| + l \int_0^l |u'(x)|\,dx, \tag{4.111}$$

$$l \max_{x\in[0,l]} |u(x)| \le \left| \int_0^l u(x)\,dx \right| + l \int_0^l |u'(x)|\,dx, \tag{4.112}$$

and

$$l \int_0^l |u(x)|^2\,dx \le 2 \left\{ \left| \int_0^l u(t)\,dt \right|^2 + l^3 \int_0^l |u'^2(y)|\,dy \right\}. \tag{4.113}$$

(cf., Exercise 4.62). From (4.111) it follows that the right side can serve as an equivalent norm in the space $W^{1,1}(0, l)$. Result (4.112) states that on the subspace of $W^{1,1}(0, l)$ that is the completion of C_{Rf} with respect to the norm of $W^{1,1}(0,l)$, we get the continuous imbedding of its elements into $C(0,l)$ and, moreover, the corresponding continuous functions satisfy (4.110). Finally, (4.113) implies that the completion of C_{Rf} with respect to the energy norm $\|\cdot\|_R$ is a subspace of $W^{1,2}(0,l)$, whose norm is equivalent to $\|\cdot\|_R$. This is one way of using the energy norm to circumvent the difficulty with free motions.

Another way is to introduce a factor space of continuously differentiable functions with respect to all constant functions. This means we declare that the union of all the constant functions is the zero element of the new space. Between this factor set and C_{Rf} there is a one-to-one correspondence preserving the energy distances between corresponding elements. So completion in both cases gives the same result from the standpoint of isometry, and hence the two approaches are equivalent.

Cantilever beam

The equilibrium of a flexible elastic beam (Fig. 4.2) is governed by

$$(EI\, y''(x))'' = f(x), \quad x \in [0,l], \tag{4.114}$$

where E, I are given characteristics of the beam, $y = y(x)$ is the transverse displacement, and $f = f(x)$ is the transverse load. If E and I are piecewise

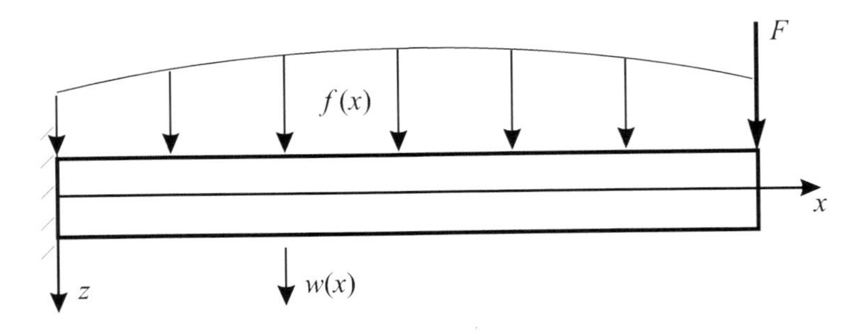

Fig. 4.2 Beam under load $f(x)$ and a point force F acting at the end.

continuous functions of x, then it is natural to assume that

$$0 < c_0 \le EI \le c_1, \tag{4.115}$$

where c_0 and c_1 are constants. We consider a cantilever beam for which

$$y(0) = 0 = y'(0). \tag{4.116}$$

So its left end is clamped and its right end is free from geometrical restrictions. In dimensionless variables, the strain energy is

$$E_B = \frac{1}{2}\int_0^l EI\, y''^2(x)\, dx. \tag{4.117}$$

On the subset C_B of those $C^{(2)}(0,l)$ functions satisfying (4.116), the energy expression suggests the metric

$$d(y,z) = \left(\int_0^l EI\,[y''(x) - z''(x)]^2\, dx\right)^{1/2} \tag{4.118}$$

(the reader should verify that all three metric axioms hold). This metric is induced by the energy norm

$$\|y\|_B = \left(\int_0^l EI\, y''^2(x)\, dx\right)^{1/2}, \tag{4.119}$$

which is in turn induced by the energy inner product

$$(y,z)_B = \int_0^l EI\, y''(x) z''(x)\, dx. \tag{4.120}$$

Completing C_B with respect to the norm $\|\cdot\|_B$, we obtain a Hilbert space denoted $\mathcal{E}_{Bc}$. By (4.115), the norm on $\mathcal{E}_{Bc}$ is equivalent to the auxiliary norm

$$\|y\|_2 = \left(\int_0^l y''^2(x)\, dx\right)^{1/2} \tag{4.121}$$

which we will use to study the properties of $y \in \mathcal{E}_{Bc}$. First let us mention that if $y \in C_B$ then $y' \in C_{Rc}$ and y' must satisfy (4.108):

$$\int_0^l y'^2(x)\, dx \le l^2 \int_0^l y''^2(x)\, dx.$$

In addition we have

$$\int_0^l y^2(x)\, dx \le l^2 \int_0^l y'^2(x)\, dx,$$

and thus for any smooth representer of the space $\mathcal{E}_{Bc}$ we have

$$\int_0^l y^2(x)\, dx + \int_0^l y'^2(x)\, dx \le c \int_0^l y''^2(x)\, dx \le c_2 \int_0^l EI\, y''^2(x)\, dx.$$

Therefore on $\mathcal{E}_{Bc}$ the energy norm is equivalent to the norm

$$\|y\|_{2,2}^2 = \int_0^l \left[y''^2(x) + y'^2(x) + y^2(x)\right] dx$$

of the Sobolev space $W^{2,2}(0,l)$. This means that on $\mathcal{E}_{Bc}$ we can use Sobolev's imbedding theorem for $W^{2,2}(0,l)$. Each element of $\mathcal{E}_{Bc}$ is thereby identified with a continuously differentiable function; in other words, $\mathcal{E}_{Bc}$ imbeds continuously into $C^{(1)}(0,l)$.

Free beam

In the absence of geometric constraints on the ends of the beam, the same functional $\|y\|_B$ satisfies all the norm axioms except one: the equation $\|y\|_B = 0$ has a nonzero solution of the form $y = a + bx$ where a and b are constants. Mechanically, this function is a rigid-body displacement of the beam. Recalling what we did with (4.108), we can use (4.113) to show that any function from $C^{(2)}(0,l)$ satisfies

$$l\int_0^l y'^2(x)\,dx \le 2\left(\int_0^l y'(x)\,dx\right)^2 + 2l^3\int_0^l y''^2(x)\,dx$$

and

$$l\int_0^l y^2(x)\,dx \le 2\left(\int_0^l y(x)\,dx\right)^2 + 2l^3\int_0^l y'^2(x)\,dx.$$

Hence

$$\int_0^l \left(y^2(x) + y'^2(x)\right) dx \le c_3\left[\left(\int_0^l y(x)\,dx\right)^2 + \right.$$
$$\left. + \left(\int_0^l y'(x)\,dx\right)^2 + \int_0^l y''^2(x)\,dx\right], \tag{4.122}$$

which means that the expression

$$\|y\|_2 = \left[\left(\int_0^l y(x)\,dx\right)^2 + \left(\int_0^l y'(x)\,dx\right)^2 + \int_0^l EI\,y''^2(x)\,dx\right]^{1/2} \tag{4.123}$$

is a norm equivalent to the norm of $W^{2,2}(0,l)$. To construct the energy space $\mathcal{E}_{Bf}$ for a free beam, we can use this fact in two ways, as was done for a stretched rod. First we can take a base set C_{Bf} consisting of smooth functions y for which

$$\int_0^l y(x)\,dx = 0 = \int_0^l y'(x)\,dx. \tag{4.124}$$

Indeed, to any smooth function $y = y(x)$ there corresponds a unique function satisfying (4.124), obtained by proper choice of the constants a and b in the expression $y(x) - a - bx$. This does not alter the stress distribution in the beam; it merely fixes the beam in space. Then (4.123) implies that on the set of functions from $C^{(2)}(0,l)$ satisfying (4.124) the norm $\|y\|_B$ is equivalent to the norm of $W^{2,2}(0,l)$, and thus after completion we can use the Sobolev imbedding theorem for $W^{2,2}(0,l)$. Any representative sequence of $\mathcal{E}_{Bf}$ has a continuous function as its limit; moreover, the sequence of first derivatives also converges to a continuous function. For the limit functions, (4.124) holds as well.

Alternatively we can employ a factor space, declaring that the zero element of the energy space is the set of all linear polynomials that are infinitesimal rigid motions of the beam, $a + bx$. In this case among all the representers of an element there is only one that satisfies (4.124), and thus we get an isometric one-to-one correspondence between the elements of the two versions of the energy space and can carry interpretations of results for one version over to the other.

Remark 4.167. In order to construct the energy space for an elastic beam subjected to normal and longitudinal loads, we can consider pairs of displacements (u, w) and combine the energy functionals, norms, and inner products for a rod and a beam. □

Membrane with clamped edge

The equilibrium of a clamped membrane (Fig. 4.3) occupying a domain $\Omega \subset \mathbb{R}^2$ is described by the equations

$$a\,\Delta u = -f, \qquad u\Big|_{\partial\Omega} = 0,$$

which together make up the Dirichlet problem for Laplace's equation. Here $u = u(x, y)$ is the transverse displacement of the membrane and $f = f(x, y)$ is the external load. The parameter a relates to the tension in the membrane. The potential energy of the membrane is

$$\mathcal{E}_M(u) = \frac{a}{2} \iint_\Omega \left[\left(\frac{\partial u}{\partial x} \right)^2 + \left(\frac{\partial u}{\partial y} \right)^2 \right] dx\, dy.$$

By a proper choice of dimensionless variables in what follows, we will put $a = 1$. A metric corresponding to this energy on the set of functions $u(x, y)$ from $C^{(1)}(\Omega)$ that satisfy the boundary condition

$$u(x, y)\Big|_{\partial\Omega} = 0 \tag{4.125}$$

is

$$d(u, v) = \left\{ \iint_\Omega \left[\left(\frac{\partial u}{\partial x} - \frac{\partial v}{\partial x} \right)^2 + \left(\frac{\partial u}{\partial y} - \frac{\partial v}{\partial y} \right)^2 \right] dx\, dy \right\}^{1/2}. \tag{4.126}$$

The resulting metric space is appropriate as a starting point for investigating the corresponding boundary value problem.

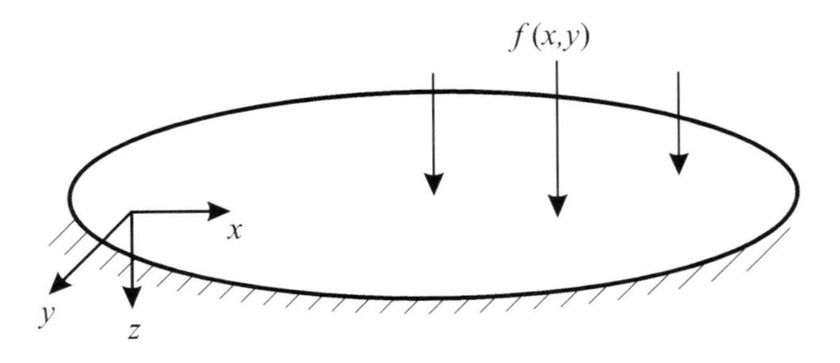

Fig. 4.3 Membrane clamped along the edge.

The subset C_{Mc} of $C^{(1)}(\Omega)$ consisting of all functions satisfying (4.125) with the metric (4.126) is an incomplete metric space. If we define an inner product

$$(u, v)_M = \iint_\Omega \left(\frac{\partial u}{\partial x}\frac{\partial v}{\partial x} + \frac{\partial u}{\partial y}\frac{\partial v}{\partial y} \right) dx\, dy$$

consistent with (4.126) we get an inner product space. Its completion in the metric (4.126) is the energy space for the clamped membrane, denoted $\mathcal{E}_{Mc}$. This is a real Hilbert space.

What can we say about the elements of $\mathcal{E}_{Mc}$? It is obvious that the sequences of first derivatives $\{\partial u_n/\partial x\}$, $\{\partial u_n/\partial y\}$, of a representative sequence $\{u_n\}$ are Cauchy sequences in the norm on $L^2(\Omega)$: i.e., if

$$d(u_m, u_n) = \left\{ \iint_\Omega \left[\left(\frac{\partial u_m}{\partial x} - \frac{\partial u_n}{\partial x} \right)^2 + \left(\frac{\partial u_m}{\partial y} - \frac{\partial u_n}{\partial y} \right)^2 \right] dx\, dy \right\}^{1/2}$$
$$\to 0 \quad \text{as } m, n \to \infty,$$

then

$$\left\{ \iint_\Omega \left(\frac{\partial u_m}{\partial x} - \frac{\partial u_n}{\partial x} \right)^2 dx\, dy \right\}^{1/2} = \left\| \frac{\partial u_m}{\partial x} - \frac{\partial u_n}{\partial x} \right\|_{L^2(\Omega)}$$
$$\to 0 \quad \text{as } m, n \to \infty,$$

and similarly for $\{\partial u_n/\partial y\}$. It takes more work to say something about $\{u_n\}$ itself; we need the *Friedrichs inequality.*

The Friedrichs inequality states that if a continuously differentiable function $u = u(x, y)$ has compact support in Ω, then there is a constant $C > 0$, depending on Ω only, such that

$$\iint_\Omega |u|^2\, d\Omega \leq C \iint_\Omega |\nabla u|^2\, d\Omega. \tag{4.127}$$

To prove this it is convenient to first suppose Ω is the square $|x| < a$, $|y| < a$. Since

$$u(x, y) = u(-a, y) + \int_{-a}^{x} \frac{\partial u(\xi, y)}{\partial \xi}\, d\xi$$

and $u(-a, y) = 0$, we have

$$\iint_\Omega |u(x, y)|^2\, d\Omega = \int_{-a}^{a} \int_{-a}^{a} \left| \int_{-a}^{x} \frac{\partial u(\xi, y)}{\partial \xi}\, d\xi \right|^2 dx\, dy.$$

Then

$$\begin{aligned}
\iint_\Omega |u(x, y)|^2\, d\Omega &= \int_{-a}^{a} \int_{-a}^{a} \left| \int_{-a}^{x} 1 \cdot \frac{\partial u(\xi, y)}{\partial \xi}\, d\xi \right|^2 dx\, dy \\
&\leq \int_{-a}^{a} \int_{-a}^{a} \int_{-a}^{x} 1^2\, d\xi \int_{-a}^{x} \left| \frac{\partial u(\xi, y)}{\partial \xi} \right|^2 d\xi\, dx\, dy \\
&\leq \int_{-a}^{a} \int_{-a}^{a} \int_{-a}^{a} 1^2\, d\xi \int_{-a}^{a} \left| \frac{\partial u(\xi, y)}{\partial \xi} \right|^2 d\xi\, dx\, dy \\
&= \int_{-a}^{a} 1^2\, d\xi \int_{-a}^{a} dx \int_{-a}^{a} \int_{-a}^{a} \left| \frac{\partial u(\xi, y)}{\partial \xi} \right|^2 d\xi\, dy \\
&= 4a^2 \int_{-a}^{a} \int_{-a}^{a} \left| \frac{\partial u(\xi, y)}{\partial \xi} \right|^2 d\xi\, dy,
\end{aligned}$$

hence

$$\iint_{\Omega} |u|^2 \, d\Omega \le 4a^2 \int_{-a}^{a} \int_{-a}^{a} \left| \frac{\partial u(x,y)}{\partial x} \right|^2 dx \, dy = 4a^2 \iint_{\Omega} \left| \frac{\partial u}{\partial x} \right|^2 d\Omega.$$

By the same reasoning, an analogous inequality holds with $\partial u/\partial y$ on the right side. Adding these two inequalities we obtain

$$\iint_{\Omega} |u|^2 \, d\Omega \le C \iint_{\Omega} \left(\left| \frac{\partial u}{\partial x} \right|^2 + \left| \frac{\partial u}{\partial y} \right|^2 \right) d\Omega$$

where $C = 2a^2$. If Ω is not square, we can enclose it in a square $\tilde{\Omega}$ and extend the function u onto the set $\tilde{\Omega}$ by setting $u \equiv 0$ on $\tilde{\Omega} - \Omega$ to obtain a new function $\tilde{u}$; in this case

$$\iint_{\tilde{\Omega}} |\tilde{u}|^2 \, d\tilde{\Omega} \le C \iint_{\tilde{\Omega}} \left(\left| \frac{\partial \tilde{u}}{\partial x} \right|^2 + \left| \frac{\partial \tilde{u}}{\partial y} \right|^2 \right) d\tilde{\Omega}$$

follows. (Note that the extension $\tilde{u}$ may have a discontinuous derivative on $\partial\Omega$; however, the presence of such a discontinuity does not invalidate any of the steps above when $\partial\Omega$ is sufficiently smooth.) The constant C depends only on a, hence only on Ω (which dictates the choice of a).

Above we observed that if $\{u_n\}$ is a representative of an element of $\mathcal{E}_{Mc}$, then $\{\partial u_n/\partial x\}$ and $\{\partial u_n/\partial y\}$ are Cauchy sequences in the norm of $L^2(\Omega)$. The Friedrichs inequality applied to $u = u_n(x,y)$ shows that $\{u_n\}$ is also a Cauchy sequence in the norm of $L^2(\Omega)$. Hence to each $U(x,y) \in \mathcal{E}_{Mc}$ having a representative sequence $\{u_n\}$, there correspond elements in $L^2(\Omega)$ having $\{u_n\}$, $\{\partial u_n/\partial x\}$ and $\{\partial u_n/\partial y\}$ as representatives. We denote these elements of $L^2(\Omega)$ by $U(x,y)$, $\partial U(x,y)/\partial x$, and $\partial U(x,y)/\partial y$, respectively. The elements $\partial U/\partial x$ and $\partial U/\partial y$ are assigned interpretations as generalized derivatives of the element U later on. However, we need a result for the elements of the completed energy space. Passage to the limit in (4.127) gives

$$\iint_{\Omega} U^2 \, dx \, dy \le C \iint_{\Omega} \left[\left(\frac{\partial U}{\partial x} \right)^2 + \left(\frac{\partial U}{\partial y} \right)^2 \right] dx \, dy \tag{4.128}$$

for any $U \in \mathcal{E}_{Mc}$ and a constant C independent of U.

Inequality (4.128) also means that in $\mathcal{E}_{Mc}$ the energy norm is equivalent to the norm of $W^{1,2}(\Omega)$, and thus for the space $\mathcal{E}_{Mc}$ there holds an imbedding result in the form of Theorem 4.164.

Free membrane

In the absence of geometrical constraints, a membrane is subject to uniform "rigid-body" displacements. These differ from the motions of an actual rigid body because the membrane model reflects only certain features of the real object that we regard as a membrane. To characterize the state of a free membrane, we choose the energy functional and hence the metric (4.126) or, equivalently, the norm

$$\|u\|_M = \left\{ \iint_\Omega \left[\left(\frac{\partial u}{\partial x} \right)^2 + \left(\frac{\partial u}{\partial y} \right)^2 \right] d\Omega \right\}^{1/2} . \tag{4.129}$$

This is not a norm on the function space $C^{(1)}(\Omega)$, where Ω is compact, as the equation $\|u\|_M = 0$ has a nonzero solution $u = c =$ constant. Physically, we cannot distinguish between two membrane states differing only in position by the constant c. This constant displacement is the only type of rigid motion permitted by the membrane model under consideration. Our method of circumventing the existence of rigid motions is similar to that used above for free rods and beams. It is based on *Poincaré's inequality*. This extends inequality (4.113) to a two-dimensional domain (in fact, to any compact n-dimensional domain with piecewise smooth boundary):

$$\iint_\Omega u^2 \, d\Omega \le C \left\{ \left(\iint_\Omega u \, d\Omega \right)^2 + \iint_\Omega \left[\left(\frac{\partial u}{\partial x} \right)^2 + \left(\frac{\partial u}{\partial y} \right)^2 \right] d\Omega \right\}, \tag{4.130}$$

with a constant C that does not depend on u. Although the proof for a rectangular domain is similar to that for (4.113), it is lengthy — even more so for a general compact domain with piecewise smooth boundary. The interested reader can refer to [6]. Inequality (4.130) implies that on functions from $C^{(1)}(\Omega)$ satisfying

$$\iint_\Omega u(x, y) \, d\Omega = 0 \tag{4.131}$$

the energy norm $\|u\|_M$ is equivalent to the norm of $W^{1,2}(\Omega)$. Thus, defining the energy space $\mathcal{E}_{Mf}$ as the completion of functions from $C^{(1)}(\Omega)$ satisfying (4.131) with respect to the norm (4.129), we get a subspace of $W^{1,2}(\Omega)$ and can use the Sobolev imbedding theorem for the elements of this energy space. Alternatively, we can collect all the constants into a single element and declare this as the zero element of the energy space. In this case the energy space is a factor space of $W^{1,2}(\Omega)$ with respect to the set of all the constant functions on Ω. Since there is one-to-one isometry between these two versions of the energy space, we can use either of them in what follows.

Elastic body

The internal energy of an elastic body occupying a three-dimensional bounded connected volume V is given by

$$\frac{1}{2}\int_V \sum_{ijkl=1}^{3} c^{ijkl} e_{kl} e_{ij}\, dV$$

where c^{ijkl} are the components of the tensor of elastic moduli and e_{ij} are the components of the tensor of small strains. From now on we shall omit the summation symbol when we meet a repeated index in an expression; this is called Einstein's rule for repeated indices. The components of the strain tensor relate to the components of the displacement vector $\mathbf{u} = (u_1, u_2, u_3)$ given in Cartesian coordinates according to

$$e_{ij} = e_{ij}(\mathbf{u}) = \frac{1}{2}\left(u_{i,j} + u_{j,i}\right),$$

where the indices after a comma mean differentiation with respect to the corresponding coordinates:

$$u_{i,j} = \frac{\partial u_i}{\partial x_j}.$$

We suppose that the elastic moduli have the usual properties of symmetry established in the theory of elasticity, and in addition possess the property providing positiveness of the functional of inner energy:

$$c^{ijkl} e_{kl} e_{ij} \geq c_0 e_{mn} e_{mn}$$

for any symmetric tensor with components e_{mn}. Here c_0 is a positive constant.

By symmetry of the c^{ijkl}, we can introduce the following symmetric bilinear functional as a candidate for an inner product:

$$(\mathbf{u}, \mathbf{v})_E = \iiint_V c^{ijkl} e_{kl}(\mathbf{u}) e_{ij}(\mathbf{v})\, dV.$$

Linearity in $\mathbf{u}$ and $\mathbf{v}$ is evident, as is the symmetry property

$$(\mathbf{u}, \mathbf{v})_E = (\mathbf{v}, \mathbf{u})_E.$$

It remains to check the first inner product axiom. By the properties of the elastic moduli we get

$$(\mathbf{u}, \mathbf{u})_E = \iiint_V c^{ijkl} e_{kl}(\mathbf{u}) e_{ij}(\mathbf{u})\, dV \geq 0.$$

If $(\mathbf{u}, \mathbf{u})_E = 0$ and the components of $\mathbf{u}$ are continuously differentiable, then $e_{ij}(\mathbf{u}) = 0$ for all i, j. The theory of elasticity states that this $\mathbf{u}$ describes an infinitesimal rigid-body motion:

$$\mathbf{u} = \mathbf{a} + \mathbf{b} \times \mathbf{r},$$

where $\mathbf{a}$ and $\mathbf{b}$ are constant vectors. If some part of the boundary of the body is fixed, then this provides that $\mathbf{u} = \mathbf{0}$. The needed demonstration is complete.

We consider the case in which the entire boundary is clamped:

$$\mathbf{u}\big|_{\partial\Omega} = \mathbf{0}. \tag{4.132}$$

As a base space we take the set C_{Ec} of all vector functions satisfying (4.132) whose components belong to $C^{(2)}(V)$. Denote by $\mathcal{E}_{Ec}$ the energy space of an elastic body with clamped boundary: i.e., the completion of C_{Ec} with respect to the induced norm $\|\mathbf{u}\|_E = (\mathbf{u}, \mathbf{u})_E^{1/2}$. We will study the properties of this Hilbert space.

Theorem 4.168. *The space $\mathcal{E}_{Ec}$ is a subspace of the space of three-dimensional vector functions, each Cartesian component of which belongs to $W^{1,2}(\Omega)$ (the latter space we shall denote by $(W^{1,2}(\Omega))^3$).*

The proof is based on the *Korn inequality*, which in this case can be written as

$$\iiint_V \left(|\mathbf{u}(\mathbf{x})|^2 + |\nabla \mathbf{u}(\mathbf{x})|^2\right) dV \le m \iiint_V e_{ij}(\mathbf{u}(\mathbf{x})) e_{ij}(\mathbf{u}(\mathbf{x}))\, dV. \tag{4.133}$$

We will prove (4.133) for the two-dimensional case in which the functions possess all second continuous derivatives on a compact domain S and vanish on the boundary ∂S. The space variables are x, y. The proof is shorter than that for the three-dimensional case, but contains all the necessary ideas. We rewrite (4.133) for the two-dimensional case in a modified form:

$$\iint_S \left(u^2 + v^2 + u_x^2 + u_y^2 + v_x^2 + v_y^2\right) dx\, dy$$
$$\le m \iint_S \left(u_x^2 + \frac{1}{2}(u_y + v_x)^2 + v_y^2\right) dx\, dy. \tag{4.134}$$

Here u, v are the components of vector function $\mathbf{u}$ that vanish on ∂S:

$$u|_{\partial S} = 0, \qquad v|_{\partial S} = 0, \tag{4.135}$$

and subscripts x, y mean partial derivatives with respect to the corresponding variables. Note the difference between the terms with derivatives of the norm of $\left(W^{1,2}(S)\right)^2$ and the right side of (4.134): the latter does not contain the squared derivatives u_y and v_x but their sum.

Let us prove (4.134). By Friedrichs' inequality it suffices to show that there is a constant $m_1 > 0$ such that

$$\iint_S \left(u_x^2 + \frac{1}{2}(u_y + v_x)^2 + v_y^2\right) dx\, dy \geq m_1 \iint_S \left(u_x^2 + u_y^2 + v_x^2 + v_y^2\right) dx\, dy. \tag{4.136}$$

Let us transform the intermediate term in the left side of (4.136):

$$\begin{aligned}\iint_S (u_y + v_x)^2\, dx\, dy &= \iint_S \left(u_y^2 + 2u_y v_x + v_x^2\right) dx\, dy \\ &= \iint_S \left(u_y^2 + 2u_x v_y + v_x^2\right) dx\, dy,\end{aligned}$$

where we integrated by parts with regard for (4.132), so we have

$$\begin{aligned}&\iint_S \left(u_x^2 + \frac{1}{2}(u_y + v_x)^2 + v_y^2\right) dx\, dy \\ &\quad= \iint_S \left(u_x^2 + \frac{1}{2}u_y^2 + \frac{1}{2}v_x^2 + v_y^2 + u_x v_y\right) dx\, dy \\ &\quad\geq \iint_S \left(u_x^2 + \frac{1}{2}u_y^2 + \frac{1}{2}v_x^2 + v_y^2 - \frac{1}{2}\left(u_x^2 + v_y^2\right)\right) dx\, dy \\ &\quad\geq \frac{1}{2}\iint_S \left(u_x^2 + u_y^2 + v_x^2 + v_y^2\right) dx\, dy.\end{aligned}$$

This completes the proof of the Korn inequality.

We recommend that the reader tackle the proof for a three-dimensional body. We will not prove Korn's inequality for a body with free boundary (i.e., when there are no boundary conditions for vector functions); the proof is technically much more complex and reader is referred to specialized books [20; 7]. We note that the form of this inequality is the same if we impose the two conditions

$$\iiint_V \mathbf{u}(\mathbf{x})\, dV = \mathbf{0}, \qquad \iiint_V \mathbf{r} \times \mathbf{u}(\mathbf{x})\, dV = \mathbf{0},$$

on each element of the space. These are four scalar conditions in the two-dimensional case and six conditions in the three-dimensional case, which coincides with the number of degrees of freedom of a rigid body.

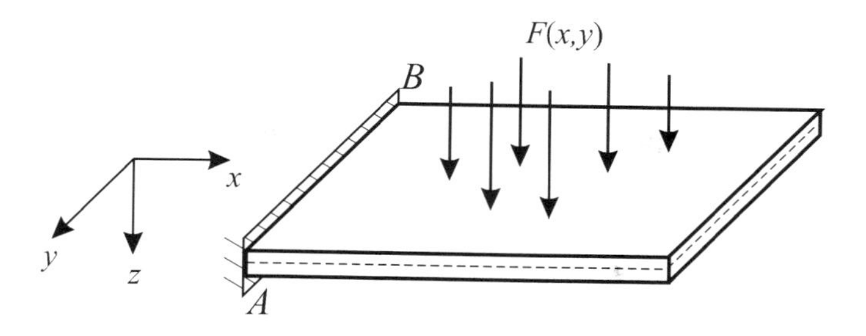

Fig. 4.4 A portion of a plate under a distributed load $F(x, y)$. The plate is clamped along AB.

Plate

The equilibrium of a linear plate (Fig. 4.4) is described by

$$D\Delta^2 w = F$$

where $w = w(x, y)$ is the transverse displacement of the midsurface of the plate, D is the plate rigidity, μ is Poisson's ratio, $0 < \mu < 1/2$, and $F = F(x, y)$ is a transverse load. The elastic energy of the plate referred to a compact domain Ω in $\mathbb{R}^2$ is

$$\frac{D}{2} \iint_\Omega \left(w_{xx} \left(w_{xx} + \mu w_{yy} \right) + 2(1 - \mu) w_{xy}^2 + w_{yy} \left(w_{yy} + \mu w_{xx} \right) \right) d\Omega$$

where subscripts x and y denote partial derivatives $\partial/\partial x$ and $\partial/\partial y$, respectively. Using dimensionless variables, for the role of a norm we will try the functional $\|\cdot\|_P$ where

$$\|w\|_P^2 = \iint_\Omega \left(w_{xx} \left(w_{xx} + \mu w_{yy} \right) + 2(1 - \mu) w_{xy}^2 + w_{yy} \left(w_{yy} + \mu w_{xx} \right) \right) d\Omega. \tag{4.137}$$

The associated inner product is

$$(u, v)_P = \iint_\Omega \left(u_{xx} \left(v_{xx} + \mu v_{yy} \right) + 2(1 - \mu) u_{xy} v_{xy} + u_{yy} \left(v_{yy} + \mu v_{xx} \right) \right) d\Omega.$$

Elementary calculations show that in $C^{(2)}(\Omega)$ the equation $\|w\|_P = 0$ has solutions only of the form $w = a + bx + cy$ with constants a, b, c. If the edge of the plate is hard-clamped, i.e.,

$$w\Big|_{\partial\Omega} = 0 = \frac{\partial w}{\partial n}\Big|_{\partial\Omega}, \tag{4.138}$$

then $\|w\|_P$ is a norm on the set C_P of functions in $C^{(2)}(\Omega)$ that satisfy (4.138). We will show that the completion $\mathcal{E}_{Pc}$ of C_P with respect to $\|\cdot\|_P$

is a subspace of $W^{2,2}(\Omega)$ (in fact, this is still the case if the plate is fixed only at three non-collinear points).

On C_P, the energy norm is equivalent to the norm

$$\|w\|_{2,2} = \left(\iint_\Omega \left(w_{xx}^2 + 2w_{xy}^2 + w_{yy}^2\right) d\Omega\right)^{1/2}$$

and so in the discussion we can use this norm. Next, if $w \in C_P$ then w_x and w_y are continuously differentiable on Ω, and (4.138) implies that on the boundary $w_x = 0 = w_y$. Thus we can apply Friedrichs' inequality, getting

$$\iint_\Omega w_x^2 \, d\Omega \le c \iint_\Omega \left(w_{xx}^2 + w_{xy}^2\right) d\Omega$$

and

$$\iint_\Omega w_y^2 \, d\Omega \le c \iint_\Omega \left(w_{yx}^2 + w_{yy}^2\right) d\Omega.$$

Combining this with Friedrichs' inequality for w we obtain

$$\begin{aligned}\iint_\Omega \left(w^2 + w_x^2 + w_y^2\right) d\Omega &\le c_1 \iint_\Omega \left(w_{xx}^2 + 2w_{xy}^2 + w_{yy}^2\right) d\Omega \\ &\le c_2 \iint_\Omega \left(w_{xx}\left(w_{xx} + \mu w_{yy}\right) + 2(1-\mu)w_{xy}^2 + w_{yy}\left(w_{yy} + \mu w_{xx}\right)\right) d\Omega.\end{aligned}$$

Together with a trivial inequality

$$\begin{aligned}&\iint_\Omega \left(w_{xx}\left(w_{xx} + \mu w_{yy}\right) + 2(1-\mu)w_{xy}^2 + w_{yy}\left(w_{yy} + \mu w_{xx}\right)\right) d\Omega \\ &\quad \le c_3 \iint_\Omega \left(w^2 + w_x^2 + w_y^2 + w_{xx}^2 + 2w_{xy}^2 + w_{yy}^2\right) d\Omega\end{aligned}$$

this proves that on C_P the energy norm is equivalent to the norm of $W^{2,2}(\Omega)$. Hence $\mathcal{E}_{Pc}$, which is the completion of C_P with respect to the energy norm (4.137), is a subspace of $W^{2,2}(\Omega)$. When dealing with $\mathcal{E}_{Pc}$, we can use Sobolev's imbedding theorem for the elements of $W^{2,2}(\Omega)$.

In the absence of geometrical constraints, rigid motions of the form $w = a + bx + cy$ are possible. To handle this we appeal to Poincaré's inequality for w_x and w_y,

$$\iint_\Omega w_x^2 \, d\Omega \le c_4 \left\{\left(\iint_\Omega w_x \, d\Omega\right)^2 + \iint_\Omega \left(w_{xx}^2 + w_{xy}^2\right) d\Omega\right\}$$

and

$$\iint_\Omega w_y^2 \, d\Omega \le c_4 \left\{\left(\iint_\Omega w_y \, d\Omega\right)^2 + \iint_\Omega \left(w_{yx}^2 + w_{yy}^2\right) d\Omega\right\}.$$

Together with Poincaré's inequality for w these give

$$\begin{aligned}
&\iint_\Omega \left(w^2 + w_x^2 + w_y^2\right) d\Omega \\
&\leq c_5 \left\{ \left(\iint_\Omega w \, d\Omega \right)^2 + \left(\iint_\Omega w_x \, d\Omega \right)^2 + \left(\iint_\Omega w_y \, d\Omega \right)^2 \right. \\
&\quad \left. + \iint_\Omega \left(w_{xx}^2 + 2w_{xy}^2 + w_{yy}^2\right) d\Omega \right\} \\
&\leq c_6 \left\{ \left(\iint_\Omega w \, d\Omega \right)^2 + \left(\iint_\Omega w_x \, d\Omega \right)^2 + \left(\iint_\Omega w_y \, d\Omega \right)^2 \right. \\
&\quad \left. + \iint_\Omega \left(w_{xx}\left(w_{xx} + \mu w_{yy}\right) + 2(1-\mu)w_{xy}^2 + w_{yy}\left(w_{yy} + \mu w_{xx}\right)\right) d\Omega \right\}.
\end{aligned}$$

Any given function from $C^{(2)}(\Omega)$ can be adjusted by subtracting a term $a + bx + cy$ to make

$$\iint_\Omega w \, d\Omega = 0, \quad \iint_\Omega w_x \, d\Omega = 0, \quad \iint_\Omega w_y \, d\Omega = 0, \tag{4.139}$$

and for such functions we get the inequality

$$\begin{aligned}
&\iint_\Omega \left(w^2 + w_x^2 + w_y^2\right) d\Omega \\
&\quad \leq c_6 \iint_\Omega \left(w_{xx}\left(w_{xx} + \mu w_{yy}\right) + 2(1-\mu)w_{xy}^2 + w_{yy}\left(w_{yy} + \mu w_{xx}\right)\right) d\Omega.
\end{aligned}$$

Hence the completion $\mathcal{E}_{Pf}$ of the set of functions from $C^{(2)}(\Omega)$ satisfying (4.139), with respect to the energy norm $\|\cdot\|_P$, is a closed subspace of $W^{2,2}(\Omega)$ whose norm is equivalent to $\|\cdot\|_P$. Note that $\mathcal{E}_{Pf}$ is a Hilbert space. Alternatively, $\mathcal{E}_{Pf}$ could be constructed as the factor space of $W^{2,2}(\Omega)$ with respect to the set of all linear polynomials $a + bx + cy$. These two versions of $\mathcal{E}_{Pf}$ are in one-to-one isometric correspondence, permitting the use of Sobolev's imbedding theorem for $W^{2,2}(\Omega)$ in either case.

4.20 Introduction to Spectral Concepts

The equation

$$A\mathbf{x} = \lambda \mathbf{x} \tag{4.140}$$

plays an important role in the theory of an $n \times n$ matrix A. Any number λ that satisfies (4.140) for some nonzero vector $\mathbf{x}$ is an *eigenvalue* of A, and

$\mathbf{x}$ is a corresponding *eigenvector*. Another form for (4.140) is, of course,

$$(A - \lambda I)\mathbf{x} = 0$$

where I is the $n \times n$ identity matrix. This equation is related to an inhomogeneous equation that corresponds to most mechanical problems involving periodic forced oscillations of finitely many oscillators:

$$(A - \lambda I)\mathbf{x} = \mathbf{b}. \tag{4.141}$$

If λ is not an eigenvalue of A, this equation is solvable for any $\mathbf{b}$. The eigenvalues of A correspond to the frequencies of external forces that put the system into the resonance state when the amplitude of vibrations grows without bound.

But relations such as (4.141) also occur outside matrix theory. Equations of the form

$$(A - \lambda I)x = b, \tag{4.142}$$

where A is a more general operator, arise naturally in continuum physics. Usually we get an equation of this form when studying the oscillations of a medium. Then A is a differential or integral operator acting on the set of admissible functions that represent distributions of displacement, strain, stress, heat, etc. This operator is linear. By properly defining the set of admissible functions x and loading terms b (note that b may represent actual mechanical loads in some problems, but may represent sources, say of heat, in other problems) we get an operator equation. If $b = 0$ we then have the problem of finding nontrivial solutions to the homogeneous equation. These are called *eigensolutions*. The terminology of matrix theory is retained in this case. These eigensolutions, as for a finite system of oscillators, represent eigen-oscillations of elastic bodies or fields. Even when they do not represent oscillations of the system, they still participate in the Fourier method of separation of variables to solve the problem and, in any case, provide an understanding of how the system functions. Note that unlike the situation for a matrix equation, where we seek solutions in the space $\mathbb{R}^n$ for which all norms are equivalent, the choice of admissible sets for continuum problems creates a new situation: with a proper choice of the solution space, we can gain or lose eigensolutions. A physical understanding of the corresponding processes may indicate which spaces are "correct."

The simple relation between the existence of solution for an inhomogeneous matrix equation and λ being or not being an eigenvalue may fail for continuum problems. There are situations in which λ is not an eigenvalue of the corresponding operator equation, so there are no eigenvectors of the operator A, but there is no solution to (4.142) that depends continuously on changes in b. The collection of "trouble spots" for λ in the complex plane (including the eigenvalues) is known as the *spectrum* of the operator A. We give a formal definition of this concept next, as well as a classification of the points of the spectrum.

Definition 4.169. Let A be a linear operator having domain and range in a complex normed space X. For a complex parameter λ, denote by A_λ the operator

$$A_\lambda = A - \lambda I \tag{4.143}$$

where I is the identity operator on X. The *resolvent set* of A is the set $\rho(A)$ of all $\lambda \in \mathbb{C}$ for which the range of A_λ is dense in X and for which A_λ has a bounded inverse. For any $\lambda \in \rho(A)$, we call A_λ^{-1} the *resolvent* of A at λ and write

$$R(\lambda; A) = (A - \lambda I)^{-1}. \tag{4.144}$$

The complement of $\rho(A)$ in $\mathbb{C}$ is the *spectrum* of A, denoted $\sigma(A)$.

Any value $\lambda \in \rho(A)$ is known as a *regular value* of A. Any $\lambda \in \sigma(A)$ is a *spectral value* of A. The spectrum of any operator A is naturally partitioned into three disjoint subsets:

(1) $P_\sigma(A)$, the *point spectrum* of A, is the set of all spectral values for which the resolvent $R(\lambda; A)$ does not exist. Its elements are the *eigenvalues* of A.
(2) $C_\sigma(A)$, the *continuous spectrum* of A, is the set of all spectral values for which $R(\lambda; A)$ exists on a dense subset of X but is not a bounded operator.
(3) $R_\sigma(A)$, the *residual spectrum* of A, is the set of all spectral values for which $R(\lambda; A)$ exists but with a domain that is not dense in X.

So

$$\sigma(A) = P_\sigma(A) \cup C_\sigma(A) \cup R_\sigma(A) \tag{4.145}$$

(we shall see that some of the sets on the right may be empty). The use of the term "eigenvalue" for an element $\lambda \in P_\sigma(A)$ may be justified as follows. We have $\lambda \in P_\sigma(A)$ if and only if the linear operator $A - \lambda I$ is not one-to-one, which is true if and only if its null space does not consist only of the zero vector. In other words, we can have $\lambda \in P_\sigma(A)$ if and only if the equation

$$(A - \lambda I)x = 0 \tag{4.146}$$

has a nontrivial solution x. Such an element x would be, of course, an eigenvector of A corresponding to the eigenvalue λ.

Example 4.170. Let $X = \ell^1$, and let A from X to X be given by

$$A\mathbf{x} = \left(\frac{\xi_1}{1}, \frac{\xi_2}{2}, \frac{\xi_3}{3}, \ldots\right)$$

for $\mathbf{x} = (\xi_1, \xi_2, \xi_3, \ldots) \in \ell^1$. Find $P_\sigma(A)$.

Solution. We have

$$(A - \lambda I)\mathbf{x} = \left(\left(\frac{1}{1} - \lambda\right)\xi_1, \left(\frac{1}{2} - \lambda\right)\xi_2, \left(\frac{1}{3} - \lambda\right)\xi_3, \ldots\right).$$

$A - \lambda I$ is not one-to-one if and only if λ is such that $\frac{1}{k} - \lambda = 0$ for some $k = 1, 2, 3, \ldots$. Hence $P_\sigma(A) = \left\{1, \frac{1}{2}, \frac{1}{3}, \ldots\right\}$. □

Example 4.171. Show that if A is a bounded linear operator and λ is an eigenvalue of A, then $|\lambda| \leq \|A\|$.

Solution. For some nonzero vector v we have $Av = \lambda v$, hence $|\lambda| \, \|v\| = \|Av\| \leq \|A\| \, \|v\|$. □

For a bounded operator we can display an important part of the resolvent set immediately.

Theorem 4.172. *Let A be a bounded linear operator on a Banach space X. All the $\lambda \in \mathbb{C}$ such that $\|A\| < |\lambda|$ are points of the resolvent set of operator A, that is $(A - \lambda I)^{-1}$ is a bounded linear operator on X. Moreover, there holds*

$$(A - \lambda I)^{-1} = -\frac{1}{\lambda} \sum_{k=0}^{\infty} \frac{1}{\lambda^k} A^k. \tag{4.147}$$

The series on the right is called the ***Neumann series*** *for A.*

Proof. Thus A is a bounded linear operator on a Banach space X. Let us take a value $\lambda \in \mathbb{C}$ and consider solving

$$Ax - \lambda x = y \tag{4.148}$$

for $x \in X$ when $y \in X$ is given. We rewrite this as

$$x = -\frac{1}{\lambda}y + \frac{1}{\lambda}Ax,$$

define the right member as the mapping $F(x) = -\lambda^{-1}y + \lambda^{-1}Ax$, and check whether F can be a contraction. We have

$$\|F(x_1) - F(x_2)\| = |\lambda|^{-1}\|Ax_1 - Ax_2\| \le |\lambda|^{-1}\|A\|\|x_1 - x_2\|,$$

hence F is a contraction whenever $|\lambda| > \|A\|$. If this condition holds we can use the iteration scheme

$$x_{j+1} = -\frac{1}{\lambda}y + \frac{1}{\lambda}Ax_j, \quad j = 0, 1, 2, \ldots$$

to solve (4.148). Starting with $x_0 = -y/\lambda$, we may generate a sequence of iterates:

$$\begin{aligned}
x_0 &= -\frac{1}{\lambda}y \\
x_1 &= -\frac{1}{\lambda}y + \frac{1}{\lambda}Ax_0 = -\frac{1}{\lambda}y - \frac{1}{\lambda^2}Ay \\
x_2 &= -\frac{1}{\lambda}y + \frac{1}{\lambda}Ax_1 = -\frac{1}{\lambda}y - \frac{1}{\lambda^2}Ay - \frac{1}{\lambda^3}A^2y \\
&\vdots \\
x_n &= -\frac{1}{\lambda}\sum_{k=0}^{n}\frac{1}{\lambda^k}A^k y.
\end{aligned}$$

These iterates converge to the unique solution

$$x = -\frac{1}{\lambda}\sum_{k=0}^{\infty}\frac{1}{\lambda^k}A^k y.$$

So the operator given by the absolutely convergent series

$$-\frac{1}{\lambda}\sum_{k=0}^{\infty}\frac{1}{\lambda^k}A^k,$$

is the inverse of the operator $A - \lambda I$. We can also check this statement explicitly. To see that it is a right inverse, we write

$$
\begin{aligned}
(A-\lambda I)\left(-\frac{1}{\lambda}\sum_{k=0}^{\infty}\frac{1}{\lambda^k}A^k\right) &= \left(I-\frac{1}{\lambda}A\right)\left(I+\sum_{k=1}^{\infty}\frac{1}{\lambda^k}A^k\right) \\
&= I-\left(\frac{1}{\lambda}A+\frac{1}{\lambda}A\sum_{k=1}^{\infty}\frac{1}{\lambda^k}A^k\right)+\sum_{k=1}^{\infty}\frac{1}{\lambda^k}A^k \\
&= I-\sum_{k=1}^{\infty}\frac{1}{\lambda^k}A^k+\sum_{k=1}^{\infty}\frac{1}{\lambda^k}A^k \\
&= I.
\end{aligned}
$$

Verification that it is a left inverse is similar. □

By this theorem the set

$$\{\lambda \in \mathbb{C}\colon |\lambda| > \|A\|\}$$

does not contain any points of the spectrum of A, which is another solution of Example 4.171.

Certain kinds of operators have simple and convenient spectral properties. We will need the following results.

Lemma 4.173. *Let A be a self-adjoint continuous linear operator A acting in a Hilbert space H. Then*

(i) the functional (Ax, x) is real valued;

(ii) the eigenvalues of A are real;

(iii) if x_1, x_2 are two eigenvectors corresponding to distinct eigenvalues λ_1, λ_2, then $(x_1, x_2) = 0$ and $(Ax_1, x_2) = 0$.

Proof. To prove item (i) we merely write

$$(Ax, x) = (x, Ax) = \overline{(Ax, x)}.$$

If $Ax = \lambda x$ then $(Ax, x) = \lambda(x, x)$, hence λ is real. This proves (ii). Now suppose $Ax_1 = \lambda_1 x_1$ and $Ax_2 = \lambda_2 x_2$ where $\lambda_2 \neq \lambda_1$. Forming inner products with x_2 and x_1 respectively, we obtain

$$\lambda_1(x_1, x_2) = (Ax_1, x_2), \qquad \lambda_2(x_1, x_2) = (x_1, Ax_2) = (Ax_1, x_2);$$

subtracting these we find $(\lambda_2 - \lambda_1)(x_1, x_2) = 0$, hence $(x_1, x_2) = 0$. Returning to $\lambda_1(x_1, x_2) = (Ax_1, x_2)$, we have $(Ax_1, x_2) = 0$. This proves (iii). □

4.21 The Fredholm Theory in Hilbert Spaces

It is common to seek solutions $\mathbf{x}$ of the following algebraic problem in $\mathbb{R}^n$:

$$A\mathbf{x} - \lambda\mathbf{x} = \mathbf{b}, \tag{4.149}$$

where A is an $n \times n$ matrix. When $\mathbf{b} = 0$, this is an eigenvalue problem for the matrix A. It is well known that if λ is not an eigenvalue of A, then (4.149) is solvable for any $\mathbf{b}$. There are no more than n eigenvalues of A. If λ is an eigenvalue of A, then (4.149) is solvable only for some set of values $\mathbf{b}$ that are orthogonal to all the eigenvectors of the conjugate-transpose matrix A^* that correspond to $\bar{\lambda}$, an eigenvalue of A^*. So to an eigenvalue λ_0 of A there corresponds an eigenvalue $\bar{\lambda}_0$ of A^*; moreover, the dimensions of the subspaces of the corresponding eigenvectors of A and A^* are the same. Furthermore, the situation for the solvability of the dual equation

$$A^*\mathbf{x} - \lambda\mathbf{x} = \mathbf{b}^*$$

is symmetric to the problem involving the operator A.

This was extended by Ivar Fredholm to the theory of certain integral equations, now known as *Fredholm equations of the second kind*:

$$\lambda u(\mathbf{x}) - \int_\Omega K(\mathbf{x},\mathbf{y})u(\mathbf{y})\,d\Omega_{\mathbf{y}} = f(\mathbf{x}).$$

If the operator is compact, this equation inherits nearly all the qualitative features of (4.149) except the number of possible eigenvalues: it may be countable, but the only possible accumulation point is zero. Fredholm's theory was later extended to Banach spaces [25; 26].

We present a particular case of this theory in a Hilbert space H, which will suffice to treat the eigenfrequency problems for bounded elastic objects like membranes, plates, shells, or elastic bodies. We recall that the Fredholm integral operator is compact in L^2. Thus we consider the following equation in H:

$$Ax - \lambda x = b,$$

with given $b \in H$. We suppose A is a compact linear operator in H. Let us exhibit the required notation. A^* is the adjoint to A, satisfying

$$(Ax, y) = (x, A^*y).$$

Correspondingly we introduce the equation

$$A^*x - \lambda x = b^*.$$

We denote by $N(\lambda)$ the subspace of H spanned by the eigenvectors of A corresponding to a given eigenvalue λ. With the exception of the zero element, each member of this subspace is an eigenvector of A. Indeed any finite linear combination of $x_1, \ldots, x_m \in N(\lambda)$ also belongs to $N(\lambda)$:

$$A\left(\sum_{i=1}^{m} \alpha_i x_i\right) = \sum_{i=1}^{m} \alpha_i A x_i = \sum_{i=1}^{m} \alpha_i \lambda x_i = \lambda\left(\sum_{i=1}^{m} \alpha_i x_i\right).$$

Note that $N(\lambda)$ contains all the eigenvectors corresponding to λ, along with the zero element of H.[5] We denote by $M(\lambda)$ the orthogonal complement of $N(\lambda)$ in H. The corresponding sets for A^* are denoted by $N^*(\lambda)$ and $M^*(\lambda)$. Let us state the facts of the Fredholm–Riesz–Schauder theory as

Theorem 4.174. *Let A be a compact linear operator in a Hilbert space H. Then*

(1) the spectrum of A consists only of eigenvalues, and thus the remaining points of the complex plane are all regular points of A;

(2) to any nonzero eigenvalue λ of A there corresponds a finite number of linearly independent eigenvectors (i.e., $N(\lambda)$ is finite dimensional);

(3) the only possible point of accumulation of the eigenvalues of A in the complex plane is zero;

(4) if λ is an eigenvalue of A then $\bar{\lambda}$ is an eigenvalue of A^* and vice versa, and the equation

$$Ax - \lambda x = b$$

is solvable if and only if b is orthogonal to the set $N^*(\bar{\lambda})$;

(5) the dimensions of $N(\lambda)$ and $N^*(\bar{\lambda})$ are equal;

(6) A^* is a compact linear operator, and thus

(6a) its spectrum consists only of eigenvalues with zero as the only possible point of accumulation of the eigenvalues;

(6b) to each eigenvalue there corresponds a space of eigenvectors $N^*(\lambda)$ that is finite dimensional;

(6c) the equation

$$A^*x - \lambda x = b^*$$

is solvable if and only if b^* is orthogonal to the subspace $N(\bar{\lambda})$.

[5] An alternative definition of $N(\lambda)$ is as the null space of the operator $A - \lambda I$, i.e., as the set of all $x \in H$ that satisfy $(A - \lambda I)x = 0$.

The proof will be formulated as a collection of lemmas. We begin with statement (2).

Lemma 4.175. *If λ is any nonzero eigenvalue of A, then $N(\lambda)$ is a closed, finite dimensional subspace of H.*

Proof. To see that $N(\lambda)$ is closed we use the continuity of A. Let x_* be a limit point of $N(\lambda)$. There is a sequence $\{x_n\} \subset N(\lambda)$ such that $x_n \to x_*$ in H. For each n we have $Ax_n = \lambda x_n$, and passage to the limit as $n \to \infty$ gives $Ax_* = \lambda x_*$. Hence $x_* \in N(\lambda)$. We next show that $N(\lambda)$ is finite dimensional. We recall Theorem 4.52 which states that any closed and bounded set is compact only in a finite dimensional Hilbert space. So let S be an arbitrary closed and bounded subset of $N(\lambda)$, and choose any sequence $\{x_k\} \subset S$. By compactness of A and the equality $x_k = \lambda^{-1}Ax_k$, we see that $\{x_k\}$ has a Cauchy subsequence. Hence S is precompact. But S is also a closed subset of a complete space H, hence it contains the limits of its Cauchy sequences. We conclude that S is compact, as desired. □

Remark 4.176. Here we do not consider the eigenvalue $\lambda = 0$, as it corresponds to the infinite eigenfrequency of a body. Its properties differ from those of the other eigenvalues. Take, for example, a one dimensional operator A of the form $Ax = F(x)x_0$ where x_0 is fixed and $F(x)$ is a continuous linear functional. Then by the equation $Ax = \lambda x$, the eigenvalues corresponding to $\lambda = 0$ are those elements x that satisfy $F(x)x_0 = 0$. By Theorem 4.100 we can express $F(x) = (x, f)$ for some fixed $f \in H$, hence any vector x that is orthogonal to f belongs to $N(0)$. A stronger example is afforded when A is the zero operator, which is of course compact. In this case the equation $Ax = \lambda x$ becomes $\lambda x = 0$, and with $\lambda = 0$ this holds for any $x \in H$. In this case $N(0) = H$. So $\lambda = 0$ was by necessity excluded from statement (2). In statement (3) we see that $\lambda = 0$ is the only possible accumulation point for the set of all eigenvalues. □

Statement (3) will be proved as Lemma 4.178. In preparation for this we establish some notation along with an auxiliary result. Let $\lambda_1, \ldots, \lambda_k$ be eigenvalues of A. We denote by

$$N(\lambda_1)\dot{+}\cdots\dot{+}N(\lambda_k)$$

the space spanned by the union of the eigenvectors that generate the individual eigenspaces $N(\lambda_1), \ldots, N(\lambda_k)$. Use of the notation for direct sum is justified by the next result which shows, in particular, that eigenspaces corresponding to distinct eigenvalues can intersect only in the zero vector.

Lemma 4.177. *Assume $S_i = \{x_1^{(i)}, x_2^{(i)}, \ldots, x_{n_i}^{(i)}\}$ is a linearly independent system of elements in $N(\lambda_i)$ for each $i = 1, \ldots, k$. Then the union $\cup_{i=1}^k S_i$ is linearly independent. If S_i is a basis of $N(\lambda_i)$ for each i, then $\cup_{i=1}^k S_i$ is a basis of $N(\lambda_1)\dot{+}\cdots\dot{+}N(\lambda_k)$.*

Proof. The proof is by induction. We want to show that under the hypothesis of the lemma $\cup_{i=1}^k S_i$ is linearly independent in $N(\lambda_1)\dot{+}\cdots\dot{+}N(\lambda_k)$ for each positive integer k. For $k = 1$ the statement holds trivially. Suppose it holds for $k = n$. Let us take the eigenvalue-eigenvector pairings

$$(\lambda_i, x_p^{(i)}), \quad p = 1, \ldots, n_i, \qquad i = 1, \ldots, n,$$

and renumber everything so that these same pairings are denoted as (λ_j, x_j), $j = 1, \ldots, r$. By assumption then,

$$\sum_{j=1}^{r} \alpha_j x_j = 0 \implies \alpha_j = 0 \text{ for } j = 1, \ldots, r. \tag{4.150}$$

We must show that the statement holds for $k = n + 1$. Appending S_{n+1} to $\cup_{i=1}^n S_i$, we assume that

$$\sum_{j=1}^{r+s} c_j x_j = 0 \tag{4.151}$$

and attempt to draw a conclusion regarding the c_j (here s is new notation for the number of elements in S_{n+1}). An application of A to both sides allows us to write

$$\frac{1}{\lambda_{n+1}} \sum_{j=1}^{r+s} c_j \lambda_j x_j = 0$$

and upon subtraction from the previous equation we obtain

$$\sum_{j=1}^{r+s} c_j \left(1 - \frac{\lambda_j}{\lambda_{n+1}}\right) x_j = \sum_{j=1}^{r} c_j \left(1 - \frac{\lambda_j}{\lambda_{n+1}}\right) x_j = 0.$$

We have $c_j = 0$ for $j = 1, \ldots, r$ by (4.150). Substitution into (4.151) gives

$$\sum_{j=r+1}^{r+s} c_j x_j = 0;$$

but the eigenvectors participating in this sum are all associated with λ_{n+1} and are linearly independent by assumption. Hence $c_j = 0$ for $j = r + 1, \ldots, r + s$.

The second statement of the lemma follows from the fact that the dimension of the direct sum $N(\lambda_1)\dot{+}\cdots\dot{+}N(\lambda_k)$ is less than or equal to the sum of the dimensions of the constituent eigenspaces $N(\lambda_i)$. Since we do have $n_1 + \cdots + n_k$ linearly independent vectors in the direct sum, we have found a basis. □

Lemma 4.178. *The only possible point of accumulation of the eigenvalues of A in the complex plane is $\lambda = 0$.*

Proof. Suppose λ_0 is a limit point of the set of eigenvalues of A, and $|\lambda_0| > 0$. There is a sequence $\{\lambda_n\}$ of distinct eigenvalues of A such that $\lambda_n \to \lambda_0$. For each λ_n take an eigenvector x_n, and denote by H_n the subspace spanned by $\{x_1, \ldots, x_n\}$. Thus $H_n \subseteq H_{n+1}$ for each n. Let $y_1 = x_1/\|x_1\|$. Successively, we can construct another sequence $\{y_n\}$, $n > 1$, as follows. By Lemma 4.177 we have $H_n \neq H_{n+1}$, so for each n there exists $y_{n+1} \in H_{n+1}$ such that $\|y_{n+1}\| = 1$ and y_{n+1} is orthogonal to H_n. Indeed, we use the orthogonal decomposition theorem to decompose H_{n+1} into H_n and another nonempty subspace orthogonal to H_n, from which we choose a normalized element. Now consider the sequence $\{y_n/\lambda_n\}$; because it is bounded in H, its image $\{A(y_n/\lambda_n)\}$ contains a Cauchy subsequence. We begin to seek a contradiction to this last statement by writing

$$A\left(\frac{y_{n+m}}{\lambda_{n+m}}\right) - A\left(\frac{y_n}{\lambda_n}\right) = y_{n+m} - \left(y_{n+m} - \frac{1}{\lambda_{n+m}}Ay_{n+m} + \frac{1}{\lambda_n}Ay_n\right) \tag{4.152}$$

for $m \geq 1$. On the right the first term y_{n+m} belongs to H_{n+m}; the second (parenthetical) term belongs to H_{n+m-1} because we can write $y_{n+m} = \sum_{k=1}^{n+m} c_k x_k$ and have

$$\begin{aligned} y_{n+m} - \frac{1}{\lambda_{n+m}}Ay_{n+m} &= \sum_{k=1}^{n+m} c_k x_k - \frac{1}{\lambda_{n+m}}A\left(\sum_{k=1}^{n+m} c_k x_k\right) \\ &= \sum_{k=1}^{n+m-1} c_k \left(1 - \frac{\lambda_k}{\lambda_{n+m}}\right) x_k \in H_{n+m-1} \end{aligned}$$

along with the fact that $\lambda_n^{-1}Ay_n \in H_n \subseteq H_{n+m-1}$. As the two terms on the right side of (4.152) are orthogonal, the Pythagorean theorem yields

$$\begin{aligned} &\left\|A\left(\frac{y_{n+m}}{\lambda_{n+m}}\right) - A\left(\frac{y_n}{\lambda_n}\right)\right\|^2 \\ &= \|y_{n+m}\|^2 + \left\|y_{n+m} - \frac{1}{\lambda_{n+m}}Ay_{n+m} + \frac{1}{\lambda_n}Ay_n\right\|^2 \geq 1, \end{aligned}$$

for any n and $m \geq 1$. Therefore $\{A(y_n/\lambda_n)\}$ cannot contain a Cauchy subsequence. □

Let us proceed to

Lemma 4.179. *Let λ be fixed. There are positive constants m_1 and m_2 such that*

$$m_1 \|x\| \leq \|Ax - \lambda x\| \leq m_2 \|x\| \tag{4.153}$$

for all $x \in M(\lambda)$.

Proof. We have

$$\|Ax - \lambda x\| \leq \|Ax\| + \|\lambda x\| \leq (\|A\| + |\lambda|) \|x\|,$$

thus establishing the inequality on the right. Suppose the inequality on the left is false. Then there is a sequence $\{x_n\} \subset M(\lambda)$ such that $\|x_n\| = 1$ and $\|Ax_n - \lambda x_n\| \to 0$ as $n \to \infty$. Because A is compact, $\{Ax_n\}$ has a Cauchy subsequence. By the equality

$$\lambda x_n = Ax_n - (Ax_n - \lambda x_n)$$

$\{x_n\}$ also has a Cauchy subsequence which we again denote as $\{x_n\}$. By completeness of $M(\lambda)$ we have $x_n \to x_0$ for some $x_0 \in M(\lambda)$. Continuity of A gives $Ax_n \to Ax_0$, and from

$$0 = \lim_{n \to \infty} \|Ax_n - \lambda x_n\| = \|Ax_0 - \lambda x_0\|$$

we get $Ax_0 = \lambda x_0$. This means that $x_0 \in N(\lambda)$. Thus we have $\|x_0\| = 1$, $x_0 \in N(\lambda)$, and $x_0 \in M(\lambda)$; this is impossible since the spaces $N(\lambda)$ and $M(\lambda)$ intersect only in the zero element. □

Lemma 4.179 shows that on $M(\lambda)$ we can impose a norm

$$\|x\|_1 = \|Ax - \lambda x\|$$

which is equivalent to the norm of H. The associated inner product is given by

$$(x, y)_1 = (Ax - \lambda x, Ay - \lambda y).$$

Similarly, on $M^*(\bar{\lambda})$ the norm $\|A^*x - \bar{\lambda}x\|$ is equivalent to the norm of H.

Lemma 4.180. *The equation*

$$Ax - \lambda x = b \tag{4.154}$$

is solvable if and only if b is orthogonal to each vector in $N^*(\bar{\lambda})$; equivalently,

$$R(A - \lambda I) = M^*(\bar{\lambda}). \tag{4.155}$$

Similarly, the equation

$$A^*x - \bar{\lambda}x = b^* \tag{4.156}$$

is solvable if and only if b^* is orthogonal to each vector in $N(\lambda)$; equivalently,

$$R(A^* - \bar{\lambda}I) = M(\lambda). \tag{4.157}$$

Proof. Suppose (4.154) is solvable with solution x_0. If $y \in N^*(\bar{\lambda})$ is arbitrary, then

$$(b, y) = (Ax_0 - \lambda x_0, y) = (x_0, A^*y - \bar{\lambda}y) = (x_0, 0) = 0.$$

Conversely, suppose $b \in M^*(\bar{\lambda})$. The functional (x, b) is linear and continuous on H (and so on $M^*(\bar{\lambda})$), hence by Theorem 4.100 can be represented on $M^*(\bar{\lambda})$ using $(\cdot, \cdot)_1$ as

$$(x, b) = (x, \tilde{b})_1 = (A^*x - \bar{\lambda}x, A^*\tilde{b} - \bar{\lambda}\tilde{b})$$

for some $\tilde{b} \in M^*(\bar{\lambda})$. This equality, being valid for $x \in M^*(\bar{\lambda})$, holds for all $x \in H$ too; indeed bearing $x = x_1 + x_2$, $x_1 \in N^*(\bar{\lambda})$, $x_2 \in M^*(\bar{\lambda})$, we have

$$A^*x - \bar{\lambda}x = A^*x_1 - \bar{\lambda}x_1 + A^*x_2 - \bar{\lambda}x_2 = A^*x_2 - \bar{\lambda}x_2$$

and so, for all $x \in H$,

$$(A^*x - \bar{\lambda}x, A^*\tilde{b} - \bar{\lambda}\tilde{b}) = (A^*x_2 - \bar{\lambda}x_2, A^*\tilde{b} - \bar{\lambda}\tilde{b}) = (x_2, \tilde{b})_1 = (x_2, b) = (x, b)$$

since $(x_1, b) = 0$. Denoting $A^*\tilde{b} - \bar{\lambda}\tilde{b}$ by g we have

$$(A^*x - \bar{\lambda}x, g) = (x, Ag - \lambda g) = (x, b) \text{ for all } x \in H,$$

hence $Ag - \lambda g = b$ and g satisfies (4.154). The rest of the lemma is proved analogously. □

By this lemma we have partially addressed part (4) of Theorem 4.174.

Lemma 4.181. *If N_n is the null space of $(A - \lambda I)^n$, then*

(i) N_n is a finite dimensional subspace of H;

(ii) $N_n \subseteq N_{n+1}$ for all $n = 1, 2, \ldots$;

(iii) there exists k such that $N_n = N_k$ for all $n > k$.

Proof.

(i) Writing $(A - \lambda I)^n x = 0$ as

$$(\lambda^n I - n\lambda^{n-1} A + \cdots)x = 0,$$

the sum of the terms beginning with the second is a compact operator $(-B)$ so denoting $\lambda^n = \gamma$ we get an eigenvalue problem $(B - \gamma I)x = 0$ with compact B and so N_n is finite dimensional.

(ii) If $(A - \lambda I)^n x = 0$, then $(A - \lambda I)^{n+1} x = 0$.

(iii) First we show that if $N_{k+1} = N_k$ for some k then $N_{k+m} = N_k$ for $m = 1, 2, 3, \ldots$. Consider the case $m = 2$. By part (ii) we know that $N_k \subseteq N_{k+2}$. Conversely

$$\begin{aligned} x_0 \in N_{k+2} &\implies 0 = (A - \lambda I)^{k+2} x_0 = (A - \lambda I)^{k+1}((A - \lambda I)x_0) \\ &\implies (A - \lambda I)x_0 \in N_{k+1} = N_k \\ &\implies (A - \lambda I)^{k+1} x_0 = 0 \\ &\implies x_0 \in N_{k+1} = N_k, \end{aligned}$$

so $N_{k+2} \subseteq N_k$. Hence $N_{k+2} = N_k$. Now we have $N_{k+1} = N_{k+2}$, and so by the previous argument we get $N_{k+1} = N_{k+3}$, hence $N_{k+3} = N_k$, and so on.

Now suppose there is no k such that $N_k = N_{k+1}$. Then there is a sequence $\{x_n\}$ such that $x_n \in N_n$, $\|x_n\| = 1$, and x_n is orthogonal to N_{n-1}. Since A is compact the sequence $\{Ax_n\}$ must contain a convergent subsequence. But

$$Ax_{n+m} - Ax_n = \lambda x_{n+m} + (Ax_{n+m} - \lambda x_{n+m} - Ax_n)$$

where on the right the first term belongs to N_{n+m} and the second (parenthetical) term belongs to N_{n+m-1}. (To see the latter note that $Ax_n \in N_n$ since

$$(A - \lambda I)^n A x_n = A(A - \lambda I)^n x_n = 0,$$

and $(A - \lambda I)^{n+m-1}(Ax_{n+m} - \lambda x_{n+m}) = (A - \lambda I)^{n+m} x_{n+m} = 0$.) By orthogonality of these two terms we have

$$\|Ax_{n+m} - Ax_n\|^2 = \|\lambda x_{n+m}\|^2 + \|Ax_{n+m} - \lambda x_{n+m} - Ax_n\|^2 \geq |\lambda|^2.$$

Since $\lambda \neq 0$ we have a contradiction. □

Lemma 4.182. *We have $R(A - \lambda I) = H$ if and only if $N(\lambda) = \{0\}$.*

Proof. Let $R(A - \lambda I) = H$ and suppose $N(\lambda) \neq \{0\}$. Take a nonzero $x_0 \in N(\lambda)$. Since $R(A - \lambda I) = H$ we can solve successively the equations in the following infinite system:

$$(A - \lambda I)x_1 = x_0; \quad (A - \lambda I)x_2 = x_1; \quad \cdots \quad (A - \lambda I)x_{n+1} = x_n; \quad \cdots$$

The sequence of solutions $\{x_n\}$ has the property that

$$(A - \lambda I)^n x_n = x_0 \neq 0 \qquad \text{but} \qquad (A - \lambda I)^{n+1} x_n = (A - \lambda I)x_0 = 0.$$

In the terminology of Lemma 4.181, these imply that $x_n \notin N_n$ but $x_n \in N_{n+1}$. So there is no finite k such that $N_{k+1} = N_k$, and this contradicts part (iii) of Lemma 4.181.

Conversely let $N(\lambda) = \{0\}$. Then $M(\lambda) = H$ hence by (4.157) we have $R(A^* - \bar{\lambda} I) = H$. By the proof of the converse given above, $N^*(\bar{\lambda}) = \{0\}$ and thus $M^*(\bar{\lambda}) = H$. The proof is completed by reference to (4.155). □

We can now establish part (1) of Theorem 4.174:

Lemma 4.183. *The spectrum of a compact linear operator A consists only of eigenvalues.*

Proof. Suppose λ is not an eigenvalue of A. Then $N(\lambda)$ contains only the zero vector, hence $M(\lambda) = H$ and (4.153) applies for all $x \in H$. This means, in conjunction with Theorem 4.78, that the operator $(A - \lambda I)^{-1}$ is bounded on the range of $A - \lambda I$, which is H by Lemma 4.182. Hence λ is a regular point of the spectrum of A. □

We continue to part (4) of Theorem 4.174:

Lemma 4.184. *If λ is an eigenvalue of A, then $\bar{\lambda}$ is an eigenvalue of A^*.*

Proof. Suppose λ is an eigenvalue of A but $\bar{\lambda}$ is not an eigenvalue of A^*. Then $N^*(\bar{\lambda}) = \{0\}$ and thus $M^*(\bar{\lambda}) = H$. By equation (4.155) we have $R(A - \lambda I) = H$ hence $N(\lambda) = \{0\}$ by Lemma 4.182. This is impossible since an eigenvalue must correspond to at least one eigenvector. □

Finally, part (5) of Theorem 4.174 is established as

Lemma 4.185. *The spaces $N(\lambda)$ and $N^*(\bar{\lambda})$ have the same dimension.*

Proof. Let the dimensions of $N(\lambda)$ and $N^*(\bar{\lambda})$ be n and m, respectively, and suppose that $n < m$. Choose orthonormal bases $\{x_1, \ldots, x_n\}$ and

$\{y_1, \ldots, y_m\}$ of $N(\lambda)$ and $N^*(\bar{\lambda})$, respectively. Let us define an auxiliary operator Q by

$$Qx = (A - \lambda I)x + \sum_{k=1}^{n}(x, x_k)y_k \equiv (C - \lambda I)x,$$

where C is a compact linear operator as the sum of the compact operator A and a finite dimensional operator.

First we show that the null space of Q cannot contain nonzero elements. Indeed if $Qx_0 = 0$ then

$$(A - \lambda I)x_0 + \sum_{k=1}^{n}(x_0, x_k)y_k = 0.$$

Because $R(A - \lambda I) = M^*(\bar{\lambda})$ and $M^*(\bar{\lambda})$ is orthogonal to $N^*(\bar{\lambda})$, the terms $(A - \lambda I)x_0 \in M^*(\bar{\lambda})$ and $\sum_{k=1}^{n}(x_0, x_k)y_k \in N^*(\bar{\lambda})$ must separately equal zero; furthermore, since $\{y_k\}$ is a basis we have $(x_0, x_k) = 0$ for $k = 1, \ldots, n$. From $(A - \lambda I)x_0 = 0$ it follows that $x_0 \in N(\lambda)$; because x_0 is orthogonal to all basis elements of $N(\lambda)$, we have $x_0 = 0$.

By Lemma 4.182 we have $R(Q) = H$ and thus the equation $Qx = y_{n+1}$ has a solution x_0. But

$$\begin{aligned}
1 &= (y_{n+1}, y_{n+1}) \\
&= (y_{n+1}, Qx_0) \\
&= (y_{n+1}, (A - \lambda I)x_0) + \left(y_{n+1}, \sum_{k=1}^{n}(x_0, x_k)y_k\right) \\
&= ((A^* - \bar{\lambda}I)y_{n+1}, x_0) \\
&= 0,
\end{aligned}$$

a contradiction. Hence $n \geq m$. But A is adjoint to A^* and by the proof above we have $m \geq n$, so $m = n$. □

The proof of Theorem 4.174 is now complete.

4.22 Exercises

4.1 Show that a set in a metric space is closed if and only if it contains the limits of all its convergent sequences. That is, S is closed in X if and only if for any sequence $\{x_n\} \subset S$ such that $x_n \to x$ in X, we have $x \in S$.

4.2 Show that the following sets are closed in any metric space X: (a) any closed ball, (b) the empty set $\emptyset$, (c) X itself, (d) the intersection of any number of closed sets, (e) the union of any finite number of closed sets.

4.3 Suppose a complete metric space X contains a sequence of closed balls $\{B(x_n, r_n)\}_{n=1}^{\infty}$ such that $B(x_{n+1}, r_{n+1}) \subseteq B(x_n, r_n)$ for each n, and such that the radii $r_n \to 0$. Show that there is a unique point $x \in X$ such that $x \in \bigcap_{n=1}^{\infty} B(x_n, r_n)$.

4.4 Verify that if U is a closed linear subspace of a normed space X, then X/U is a normed linear space under the norm $\|\cdot\|_{X/U}$ given by

$$\|x + U\|_{X/U} = \inf_{u \in U} \|x + u\|_X .$$

Prove that if U is a closed subspace of a Banach space X, then X/U is also a Banach space.

4.5 Let M be a closed subspace of a separable normed space X. Show that X/M is separable.

4.6 Show that on $X \times Y$ the norm

$$\|(x, y)\|_2 = \|x\|_X + \|y\|_Y$$

is equivalent to the norm (4.22). When X and Y are Hilbert spaces, the norm in question defines an inner product on $X \times Y$.

4.7 Let A be a continuous linear operator from X to Y, where X and Y are Banach spaces. Let M be a closed subspace of X that lies within the kernel of A (i.e., if $x \in M$ then $Ax = 0$). Show that A induces an operator from X/M to Y that is also continuous.

4.8 Prove that a bounded set in a normed space is precompact if and only if it is finite dimensional,

4.9 Let A be a compact linear operator acting in a Banach space X, and let M be a closed subspace of X that lies within the kernel of A. Demonstrate that A induces a compact linear operator from X/M to X.

4.10 Prove that in the space ℓ^p the norm of the space c is not equivalent to the norm of ℓ^p for $1 \le p < \infty$.

4.11 (a) Show that ℓ^2 is not finite dimensional. (b) The space ℓ^∞ of *uniformly bounded sequences* is the set of all $\mathbf{x}$ having $\|\mathbf{x}\|_\infty < \infty$ where

$$\|\mathbf{x}\|_\infty = \sup_{k \ge 1} |x_k|.$$

Show that we may regard $\|\cdot\|_\infty$ as a limiting case of $\|\cdot\|_p$ as $p \to \infty$. (c) Show that if $p \le q$, then $\|\mathbf{x}\|_q \le \|\mathbf{x}\|_p$ for $\mathbf{x} \in \ell^p$. Note that this constitutes an imbedding theorem. (d) Show that $\ell^1 \subseteq \ell^p \subseteq \ell^q \subseteq \ell^\infty$ whenever $q \ge p \ge 1$. (e) Extend this string of inclusions to $\ell^1 \subseteq \ell^p \subseteq \ell^q \subseteq c_0 \subseteq c \subseteq \ell^\infty$ for $1 \le p \le q$. (f) Prove that for any $p \in [1, \infty]$, the normed space ℓ^p is a Banach space. (g) Show that the spaces ℓ^p, $1 \le p < \infty$, are separable. (h) Show that ℓ^∞ is not separable. (i) Show that c_0 is separable.

4.12 The distance function $d(x, y) = |x^3 - y^3|$ is imposed on the set of all real numbers $\mathbb{R}$ to form a metric space. Verify the metric axioms for $d(x, y)$. Show that the resulting space is complete.

4.13 Show that if A is a bounded linear operator then $\|A\|$ is given by the following alternative expressions:

$$\|A\| = \sup_{\|x\|=1} \|Ax\| = \sup_{\|x\| \ne 0} \frac{\|Ax\|}{\|x\|}.$$

Note that we also have

$$\|A\| = \sup_{\|x\| \le 1} \|Ax\| = \sup_{\|x\| < 1} \|Ax\|.$$

4.14 For each energy space studied in this chapter, write out (a) the parallelogram equality, and (b) the expression for the orthogonality of two elements.

4.15 Prove that a system of vectors in a Hilbert space is linearly independent if and only if its Gram determinant does not vanish.

4.16 Show that convergence $\|A_n - A\| \to 0$ in operator norm, that is in $L(X, Y)$ where X is normed and Y is a Banach space, implies uniform convergence $A_n x \to Ax$ on any bounded subset $S \subset D(A)$.

4.17 As in $\mathbb{R}^n$, an operator acting in a space of infinite dimensional vectors can be represented by a matrix A, but one having infinitely many elements $(A_{ij})_{i,j=1}^\infty$. Find the conditions for continuity of the operator acting in (a) the space c, and (b) the space ℓ^p.

4.18 Let $\{g_n\}$ be an orthonormal sequence in a Hilbert space H, and let $\{c_n\} \in \ell^2$. Show that the series $\sum_{n=0}^\infty c_n g_n$ converges in H.

4.19 Derive the differentiation formula

$$\frac{d}{dt}(u(t), v(t)) = \left(\frac{du(t)}{dt}, v(t) \right) + \left(u(t), \frac{dv(t)}{dt} \right).$$

4.20 Show that if $\{x_n\}$ converges weakly to x in a Hilbert space H, then

$$\|x\| \le \liminf_{n \to \infty} \|x_n\|.$$

4.21 An operator A from a normed space V to a normed space W is *densely* *defined* if $D(A)$ is dense in V. Assume W is a Banach space, and show that if A is bounded, linear, and densely defined, then A has a unique bounded linear extension to V. Also show that $\|A_e\| = \|A\|$ where A_e is the extension of A.

4.22 Show that in a finite dimensional space weak convergence implies strong convergence.

4.23 Suppose A and its inverse are both bounded linear operators defined on a normed space X. The *condition number* of A is defined by $\text{cond}(A) = \|A\| \, \|A^{-1}\|$. (a) Show that $\text{cond}(A) \geq 1$. (b) Consider the operator equation $Ax = y$. Given y, let $\hat{x}$ be an approximate solution; denote the "error" by $\varepsilon = x - \hat{x}$ and the "discrepancy" by $r = y - A\hat{x}$. Show that

$$\frac{1}{\text{cond}(A)} \frac{\|r\|}{\|y\|} \leq \frac{\|\varepsilon\|}{\|x\|} \leq \text{cond}(A) \frac{\|r\|}{\|y\|}.$$

4.24 Let T from X to X be a compact operator on an infinite dimensional normed space X. Show that if T has an inverse defined on all of X, then this inverse cannot be bounded.

4.25 (a) Show that every metric space isometry is continuous and one-to-one. (b) Prove that a linear operator $A\colon X \to Y$ between normed spaces is an isometry if and only if $\|Ax\| = \|x\|$ for all $x \in X$. (Notes: (1) We have $\|A\| = 1$ if $X \neq \{0\}$. (2) If A is also an isomorphism between the linear spaces X and Y, then A is called an *isometric isomorphism*.)

4.26 Let $\{g_k\}$ be an orthonormal system in a Hilbert space H. Show that if Parseval's equality

$$\sum_{k=1}^{\infty} |(f, g_k)|^2 = \|f\|^2$$

holds for all $f \in H$, then $\{g_k\}$ is a basis of H.

4.27 Show that the operator d/dx is bounded from $C^{(1)}(-\infty, \infty)$ to $C(-\infty, \infty)$.

4.28 Show that the set of all functions $f(x)$ bounded on $[0, 1]$ and equipped with the norm

$$\|f(x)\| = \sup_{x \in [0,1]} |f(x)|$$

is not separable.

4.29 Show that if X is a normed space and Y is a Banach space then $L(X, Y)$ is a Banach space.

4.30 Assume that X and Y are Banach spaces, $A \in L(X, Y)$ is continuously invertible, and $B \in L(X, Y)$ is such that $\|B\| < \|A^{-1}\|^{-1}$. Then $A + B$ has an inverse $(A + B)^{-1} \in L(Y, X)$ and

$$\|(A + B)^{-1}\| \leq (\|A^{-1}\|^{-1} - \|B\|)^{-1}.$$

4.31 Verify the condition stated for equality to hold in (4.49).

4.32 Show that the functional $\|x\|^2 - Fx$ in a real normed space X is bounded from below if F is a linear continuous functional in X.

4.33 A subset S of a normed space X is said to be *open* if its complement $X \setminus S$ is a closed set. (a) Show that S is open if and only if every point of S is the center of an open ball contained entirely within S. Hence this statement is an equivalent definition of an open set. (b) Show that any open ball is an open set. (c) Show that an operator $f\colon X \to Y$ is continuous if and only if the inverse image of every open set in Y is open in X.

4.34 Give an example of a function that is discontinuous everywhere on its domain of definition.

4.35 Show that under the condition

$$\left(\int_0^1 \int_0^1 |k(s,t)|^2 \, ds \, dt\right)^{1/2} < \infty$$

the Fredholm integral operator A defined by

$$Au = \int_0^1 k(s,t)u(t) \, dt$$

is a continuous operator on $L^2(0,1)$.

4.36 Calculate the norm of the *forward shift operator* S on ℓ^2, defined by

$$S\mathbf{x} = S(x_1, x_2, x_3, \ldots) = (0, x_1, x_2, \ldots).$$

4.37 Consider the operator

$$(Ax)(t) = \int_0^t x^2(s) \, ds$$

acting in $C(0,1)$. Find a closed ball, centered at the origin, on which A is a contraction.

4.38 Consider the subspace S of ℓ^∞ that consists of all sequences $\mathbf{x} = (\xi_i)$ having at most *finite* numbers of nonzero components. Show that S is *not* a Banach space.

4.39 Let A be a bounded linear operator on a Banach space X. Show that if $\|A\| < 1$ then

$$\|(A - I)^{-1}\| \leq \frac{1}{1 - \|A\|}.$$

4.40 Show that if X and Y are Banach spaces, then so is the product space $X \times Y$ under the norm $\|(x, y)\| = \max\{\|x\|_X, \|y\|_Y\}$.

4.41 Show that if $x_n \to x$ then $y_n \equiv \frac{1}{n}\sum_{i=1}^n x_i \to x$.

4.42 We have observed that equivalent norms have the same convergence properties. Prove the converse of this statement.

4.43 Show that if $\{x_n\}$ is a Cauchy sequence in a normed space, then the sequence of norms $\{\|x_n\|\}$ converges. (Note that this implies that every Cauchy sequence is bounded.)

4.44 Show that if a metric space X has a dense subspace that is separable, then X is also separable.

4.45 Show that a normed space is complete if and only if every absolutely convergent series converges to an element of the space.

4.46 The operator A given by $A\mathbf{x} = (2^{-1}x_1, 2^{-2}x_2, 2^{-3}x_3, \cdots)$ acts in ℓ^2. Show that A is compact.

4.47 Show that the number $\lambda = 0$ belongs to the residual spectrum of the forward-shift operator $A\mathbf{x} = A(x_1, x_2, \ldots) = (0, x_1, x_2, \ldots)$ defined on ℓ^2.

4.48 A sequence of infinite dimensional vectors $\{\mathbf{x}_k\}$ is defined as follows:

$$\mathbf{x}_k = (\underbrace{1, \ldots, 1}_{\substack{\text{first } k \\ \text{positions}}}, 0, 0, \ldots), \qquad k = 1, 2, 3, \ldots.$$

Show that $\{\mathbf{x}_k\}$ is not weakly convergent in ℓ^2.

4.49 Prove that the sequence $\{\sin kx\}$ is weakly convergent to zero in $L^2(0, \pi)$. Then show that it contains no weakly convergent subsequence (and therefore is not weakly compact) in $W^{1,2}(0, \pi)$.

4.50 Use the Hölder inequality to place a bound on the norm of the imbedding operator from $L^p(\Omega)$ into $L^q(\Omega)$, $p \geq q$. Assume Ω is a compact domain in $\mathbb{R}^n$.

4.51 Show that if A is a compact linear operator acting in a Hilbert space H, and $\{x_n\}$ is an orthonormal sequence in H, then $Ax_n \to 0$ as $n \to \infty$.

4.52 Let Ω be a compact set in $\mathbb{R}^n$. Show that the imbedding $C^{(n)}(\Omega) \hookrightarrow C(\Omega)$ is continuous and compact for $n \geq 1$.

4.53 Suppose a and b are finite. Let P_n be the space consisting of all polynomials on $[a, b]$ having order up to n, supplied with the norm of $C(a, b)$. Describe the space that results when we apply the completion theorem to P_n.

4.54 Show that weak convergence is equivalent to strong convergence in a finite dimensional Hilbert space.

4.55 Use the orthogonal decomposition theorem to show that a closed subspace of a Hilbert space is weakly closed.

4.56 Let a sequence $\{x_n\}$ in a Hilbert space have the following property: any subsequence of $\{x_n\}$ contains a sub-subsequence that converges to x_0 (the same x_0 for any subsequence). Prove that whole sequence $\{x_n\}$ converges to x_0.

4.57 Let a sequence $\{x_n\}$ in a Hilbert space have the following property: any subsequence of $\{x_n\}$ contains a sub-subsequence that converges weakly to x_0 (the same x_0 for any subsequence). Prove that the whole sequence $\{x_n\}$ converges weakly to x_0.

4.58 Let S and T be subsets of a metric space. Show that (a) if S is closed and T is open, then $S \setminus T$ is closed, and (b) if S is open and T is closed, then $S \setminus T$ is open.

4.59 Show that if a system is complete in a set S that is dense in a Hilbert space H, then it is complete in H.

4.60 A function f satisfies a Lipschitz condition with constant L if it satisfies the inequality $|f(\mathbf{x}) - f(\mathbf{y})| \leq L|\mathbf{x} - \mathbf{y}|$. Let S be a uniformly bounded collection of functions given on a compact set $\Omega \subset \mathbb{R}^n$ and satisfying a Lipschitz condition on Ω with the same constant L. Show that S is precompact in $C(\Omega)$.

4.61 Let A be a closed linear operator from a normed space X to a normed space Y. Show that A maps compact sets into closed sets.

4.62 Derive inequality (4.113).

4.63 A beam is hinged at the point $x = 0$. Mechanically this means that the beam can rotate about its end at $x = 0$. Using the expression for the norm related to the free beam (4.123), write out the corresponding energy norm related to the equilibrium problem for this restricted beam.

Chapter 5

Applications of Functional Analysis in Mechanics

In Chapter 1 we studied the calculus of variations. As a rule each variational problem was assumed to have a solution. But Perron's paradox (page 17) suggests a great deal of caution when assuming the existence of an object while investigating its properties. The study of variational problems from the viewpoint of solvability is difficult, even for those problems that seem well posed. In the nonlinear elasticity of bodies under dead external load, for example, the existence of a minimizer of total potential energy is in general not shown. Fortunately, a class of variational problems corresponds to linear boundary value problems for which the problem of existence is solved completely. We shall use mechanical terminology for these problems; in fact, however, some are quite general and can describe objects from fields such as electrodynamics and biology.

5.1 Some Mechanics Problems from the Standpoint of the Calculus of Variations; the Virtual Work Principle

We have considered the equilibrium problem for a membrane. Historically, the membrane was investigated via Poisson's equation

$$-\Delta u(x,y) = f(x,y) \tag{5.1}$$

on a two-dimensional bounded domain Ω. If the edge $\partial\Omega$ is fixed (Fig. 4.3) in a form described by a given function $a(s)$, then the boundary condition

$$u\big|_{\partial\Omega} = a(s) \tag{5.2}$$

and (5.1) constitute a boundary value problem. Using this, we can derive the total potential energy functional whose minimum points are given by (5.1)–(5.2). Let $\mathcal{D}$ be a set of test functions that are infinitely differentiable on Ω and zero in some neighborhood of $\partial\Omega$. In what follows we consider only simple domains (as typically encountered in applications), so let Ω be bounded with $\partial\Omega$ piecewise smooth. Multiply (5.1) by $\varphi(x,y) \in \mathcal{D}$ and integrate over Ω:

$$-\iint_\Omega \Delta u(x,y)\varphi(x,y)\,dx\,dy = \iint_\Omega f(x,y)\varphi(x,y)\,dx\,dy. \tag{5.3}$$

Integration by parts on the left gives

$$\iint_\Omega \left(\frac{\partial u}{\partial x}\frac{\partial \varphi}{\partial x} + \frac{\partial u}{\partial y}\frac{\partial \varphi}{\partial y}\right) dx\,dy = \iint_\Omega f(x,y)\varphi(x,y)\,dx\,dy \tag{5.4}$$

since $\varphi(x,y) \equiv 0$ for $(x,y) \in \partial\Omega$. If we wish to regard $\varphi(x,y)$ in (5.4) as a variation of the solution $u(x,y)$, then the left side is the first variation of the integral

$$\frac{1}{2}\iint_\Omega \left[\left(\frac{\partial u}{\partial x}\right)^2 + \left(\frac{\partial u}{\partial y}\right)^2\right] dx\,dy.$$

This is the strain energy of the membrane, aside from a constant factor that characterizes the membrane and which may be regarded as absorbed into the given load f as a type of normalization factor. The integral on the right in (5.4) is linear in φ and can be considered as the first variation of the functional

$$\iint_\Omega f(x,y)u(x,y)\,dx\,dy,$$

which is the work of external forces on the displacement field $u(x,y)$. Hence (5.4) states that the first variation of the functional

$$\frac{1}{2}\iint_\Omega \left[\left(\frac{\partial u}{\partial x}\right)^2 + \left(\frac{\partial u}{\partial y}\right)^2\right] dx\,dy - \iint_\Omega f(x,y)u(x,y)\,dx\,dy$$

is zero. This functional, encountered in Chapter 1, expresses the total energy of the membrane: i.e., the sum of the internal energy and the potential energy due to the work of external forces. An extremal describes the equilibrium state of the membrane. Lagrange's theorem in classical mechanics states that the total potential energy of a particle system is minimized in a stable equilibrium state of the system. Of course, the membrane does not obey classical mechanics: it is an object of a different nature. However, Lagrange's theorem extends to this case.

We may replace (5.2) by other boundary conditions known in membrane theory, for example Neumann's condition

$$\left.\frac{\partial u}{\partial n}\right|_{\partial\Omega} = g(s). \tag{5.5}$$

This time, in repeating the steps that lead to (5.4), we need not take $\varphi \in \mathcal{D}$. Equation (5.3) still holds if φ is only sufficiently smooth, but Green's formula yields an additional term:

$$\iint_\Omega \left(\frac{\partial u}{\partial x}\frac{\partial \varphi}{\partial x} + \frac{\partial u}{\partial y}\frac{\partial \varphi}{\partial y}\right) dx\,dy - \int_{\partial\Omega} \frac{\partial u}{\partial n}\varphi(s)\,ds = \iint_\Omega f(x,y)\varphi(x,y)\,dx\,dy, \tag{5.6}$$

where $\varphi(s)$ denotes the values of $\varphi(x,y)$ on $\partial\Omega$. By (5.5) we have

$$\iint_\Omega \left(\frac{\partial u}{\partial x}\frac{\partial \varphi}{\partial x} + \frac{\partial u}{\partial y}\frac{\partial \varphi}{\partial y}\right) dx\,dy = \iint_\Omega f(x,y)\varphi(x,y)\,dx\,dy + \int_{\partial\Omega} g(s)\varphi(s)\,ds. \tag{5.7}$$

The last integral in (5.7) looks like the work of the force $g(s)$ acting through a displacement φ on the membrane edge, so Neumann's condition actually specifies a force distribution $g(s)$ over the edge.

In this problem statement we neglect inertia; we regard the membrane as a body having zero mass. If external forces that are not self-balanced act on a body free from geometrical restrictions, mechanical considerations show that the equilibrium problem is not solvable: the body should move as a whole and, having zero mass, should undergo infinite acceleration. So the self-balance condition is necessary for such problems. In the present instance the only kind of free motion as a whole is $u(x,y) = c$, as the inner energy is constant only for such displacements. By linearity we can put $u(x,y) = 1$. Therefore on this displacement the work of external forces must be zero:

$$\iint_\Omega f(x,y)\,dx\,dy + \int_{\partial\Omega} g(s)\,ds = 0. \tag{5.8}$$

If the external forces act only on the edge so that $f(x,y) = 0$, then

$$\int_{\partial\Omega} g(s)\,ds = 0. \tag{5.9}$$

This is the well-known solvability condition for the Neumann problem. Mechanically, the external forces must be self-balanced, which is expressed by equation (5.8).

In classical mechanics the self-balance condition consists of six equations: the three projections each of the resultant force and moment onto the frame axes are all zero. The membrane model fails to satisfy all the conditions, which is typical of the approximate particular models of continuum mechanics. In linear elasticity the self-balance condition appears exactly as in classical mechanics.

Equation (5.7) can be taken as a formulation of the virtual work principle. The left and right sides can be called the *work of internal forces* and the *work of external forces*, respectively. Then (5.7) states a fundamental physical law called the *virtual work principle*, which is

> *On any admissible displacements, the total work of internal and external forces of the system in equilibrium is zero.*

In this case, the equation can be obtained as the first variation of the total potential energy functional. Hence we can start with the principle of minimum potential energy. Although this principle cannot be used for body-force systems where the potential of external forces does not exist, the virtual work principle remains valid. Continuum mechanics treats the virtual work principle as independent and relates it to the variational principles of mechanics. Thus the variational part of mechanics contains not only problems of minimizing certain functionals, but also the theory encompassing equations which, like (5.7), contain admissible fields of displacements, strains, or stresses. The portion of continuum mechanics known as "the variational problems of mechanics" is not completely a subset of the classical calculus of variations. A mechanicist may regard as variational anything that involves integro-differential equations containing some virtual variables and from which, using the main lemma of the calculus of variations, it is possible to derive relations such as equilibrium or constitutive equations.

Finally, note that in deriving (5.7) we used a set of smooth admissible variations φ of a solution; we do so even if we seek a solution with singularities. If we begin with the principle of minimum potential energy, it is reasonable to consider all the functions for which the terms of the principle make sense; moreover, there is no reason why admissible variations should be smoother than the solution. This remark will lead us to the generalized setup of some boundary value problems in mechanics.

Many problems involving elastic objects (strings, beams, shells, two- and three-dimensional elastic bodies, etc.) can be described by a total potential energy functional whose first variation yields the equilibrium equations for the object. It has the structure $\mathcal{E} - V$ where $\mathcal{E}$ is the strain energy and V is the work of external forces.[1] Minimization of the energy functional entails setting its first variation to zero on all admissible variations of the corresponding solutions. The resulting integral equations, expressing the equality of the sum of the work of internal and external forces on admissible variations to zero, also express the virtual work principle for the corresponding problems. It is more generally applicable than the minimum potential energy principle.

We now list the total potential energy $\mathcal{E} - V$ and the equation of the virtual work principle for some objects of interest.

1. *Rod (Fig. 4.1):*

$$\mathcal{E} - V = \frac{1}{2} \int_0^l ES(x) u'^2(x)\, dx - \int_0^l f(x) u(x)\, dx - Fu(l) \tag{5.10}$$

and the equation of the virtual work principle is

$$\int_0^l ES(x) u'(x) v'(x)\, dx = \int_0^l f(x) v(x)\, dx + Fv(l), \tag{5.11}$$

where $f(x)$ is a force tangential to the axis, F is a stretching force at the free end, and u is the tangential displacement of points on the neutral axis.

2. *Beam (Fig. 4.2):*

$$\mathcal{E} - V = \frac{1}{2} \int_0^l EI(x) w''^2(x)\, dx - \int_0^l f(x) w(x)\, dx - Fw(l) \tag{5.12}$$

and the equation of the virtual work principle is

$$\int_0^l EI(x) w''(x) v''(x)\, dx = \int_0^l f(x) v(x)\, dx + Fv(l), \tag{5.13}$$

where w is the transverse displacement of the neutral axis, $f(x)$ is the transverse distributed force, and F is the transverse force on the end.

3. *Plate (Fig. 4.4):*

$$\mathcal{E} - V = \frac{D}{2} \iint_\Omega \left(w_{xx}^2 + w_{yy}^2 + 2\nu w_{xx} w_{yy} + 2(1-\nu) w_{xy}^2 \right) d\Omega - \iint_\Omega Fw\, d\Omega \tag{5.14}$$

[1] In the case of potential forces V is the potential of the forces and, by analogy with elementary physics terminology for gravitational forces, the expression $-V$ can be called the potential energy of the force field.

and the equation of the virtual work principle is

$$D \iint_{\Omega} \left(w_{xx} v_{xx} + w_{yy} v_{yy} + \nu \left(w_{xx} v_{yy} + w_{yy} v_{xx} \right) + 2(1 - \nu) w_{xy} v_{xy} \right) d\Omega$$
$$= \iint_{\Omega} F v \, d\Omega, \tag{5.15}$$

where D is the plate rigidity, ν is Poisson's ratio, and $w = w(x, y)$ is the deflection at point (x, y) of the domain S occupied by the mid-surface.

4. *Three-dimensional linearly elastic body:*

$$\mathcal{E} - V = \frac{1}{2} \iiint_V c^{ijkl} e_{kl}(\mathbf{u}) e_{ij}(\mathbf{u}) \, dV - \iiint_V \mathbf{F} \cdot \mathbf{u} \, dV - \iint_{\partial V_1} \mathbf{f} \cdot \mathbf{u} \, dS, \tag{5.16}$$

and the equation of the virtual work principle is

$$\iiint_V c^{ijkl} e_{kl}(\mathbf{u}) e_{ij}(\mathbf{v}) \, dV = \iiint_V \mathbf{F} \cdot \mathbf{v} \, dV + \iint_{\partial V_1} \mathbf{f} \cdot \mathbf{v} \, dS, \tag{5.17}$$

where $\mathbf{F}$ represents volume external forces and $\mathbf{f}$ forces acting over some portion of the boundary ∂V_1.

Relations (5.10)–(5.17) will permit us to study generalized solutions of these mechanics problems.

5.2 Generalized Solution of the Equilibrium Problem for a Clamped Rod with Springs

To discuss generalized setups while avoiding too much repetition, we consider an equilibrium problem for a rod with a longitudinal distributed load $f = f(x)$ (not shown in Fig. 5.1) and n point forces F_k acting at the points a_k. In addition, two identical springs having rigidity k are attached at the point c at angles φ.

This problem is normally solved by splitting the rod into sections at the points a_k and c. We take the variational approach. The total potential energy of the system is

$$\mathcal{E} - V = \frac{1}{2} \int_0^l ES(x) u'^2(x) \, dx + 2 \cdot \frac{1}{2} k z^2 - \int_0^l f(x) u(x) \, dx - \sum_{k=1}^{n} F_k u(a_k),$$

where $\frac{1}{2} k z^2$ is the elastic energy of a spring suffering extension z. The

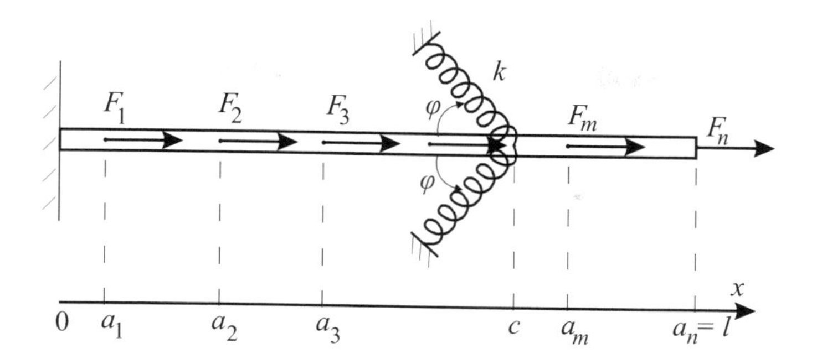

Fig. 5.1 Coupled system of rod and springs under load.

compatibility condition for small deformation is $z\cos\varphi = u(c)$, hence

$$\mathcal{E} - V = \frac{1}{2}\int_0^l ES(x)u'^2(x)\,dx + \frac{k}{\cos\varphi}u^2(c) - \int_0^l f(x)u(x)\,dx - \sum_{k=1}^n F_k u(a_k). \tag{5.18}$$

The left end of the rod is clamped:

$$u(0) = 0. \tag{5.19}$$

The force conditions will follow from the variational setup as natural conditions. Setting the first variation of $\mathcal{E} - V$ to zero, we get

$$\int_0^l ES(x)u'(x)v'(x)\,dx + 2\frac{k}{\cos\varphi}u(c)v(c) - \int_0^l f(x)v(x)\,dx - \sum_{k=1}^n F_k v(a_k) = 0, \tag{5.20}$$

where v is an admissible displacement satisfying $v(0) = 0$. To construct the energy space for the system, we start with the energy inner product

$$(u, v)_S = \int_0^l ES(x)u'(x)v'(x)\,dx + 2\frac{k}{\cos\varphi}u(c)v(c) \tag{5.21}$$

defined on the subset of $C^2(0, l)$ consisting of functions that vanish at $x = 0$. Equilibrium equation (5.20) becomes

$$(u, v)_S = W(v), \tag{5.22}$$

where

$$W(v) = \int_0^l f(x)v(x)\,dx + \sum_{k=1}^n F_k v(a_k) \tag{5.23}$$

is the work of the load on an admissible displacement v. The completion of this set with respect to the norm induced by the inner product $(\cdot,\cdot)_S$ is the energy space $\mathcal{E}_S$. By (4.105) we have $W^{1,1}(0,l) \hookrightarrow C(0,l)$. Since $W^{1,2}(0,l) \hookrightarrow C(0,l)$, the norms $\|\cdot\|_S$ and $\|\cdot\|_{W^{1,2}(0,l)}$ are equivalent on $\mathcal{E}_S$. We define a generalized solution of the equilibrium problem as a function $u \in \mathcal{E}_S$ satisfying (5.22) for all $v \in \mathcal{E}_S$. This is meaningful provided each term in (5.22) has meaning in $\mathcal{E}_S$. Suppose $f \in L(0,l)$. Clearly W is a linear functional in $\mathcal{E}_S$. It is also bounded, because

$$\begin{aligned}|W(v)| &= \left| \int_0^l f(x)v(x)\,dx + \sum_{k=1}^n F_k v(a_k)\right| \\ &\le \left(\int_0^l |f(x)|\,dx + \sum_{k=1}^n |F_k|\right) \max_{x\in[0,l]} |v(x)| \\ &\le c\,\|v\|_S\end{aligned}$$

with a constant c independent of $v \in \mathcal{E}_S$. Applying Theorem 4.100 in $\mathcal{E}_S$ to $W(v)$, we ascertain the existence of a unique $u_0 \in E_S$ such that (5.22) becomes

$$(u,v)_S = (u_0,v)_S \tag{5.24}$$

for all $v \in \mathcal{E}_S$. The unique solution is $u = u_0 \in \mathcal{E}_S$. We have proved existence and uniqueness of a generalized solution for the rod–spring system.

The reader may ask why such complicated analysis is warranted for a problem that could be solved by splitting into subdomains. First, it is advantageous to demonstrate the variational approach on an easy example. Second, the form (5.22) of the equilibrium equation is precisely the form used to introduce the finite element method for this system. Convergence is established using (5.22) in the energy space $\mathcal{E}_S$.

Finally, let us discuss one peculiar detail. If we try to solve the rod–spring equilibrium problem as a boundary value problem for the equation on the whole interval $(0,l)$, we obtain the relatively simple equation

$$(ES(x)u'(x))' = f(x) + \sum_{k=1}^n F_k \delta(x - a_k), \tag{5.25}$$

where $\delta(x)$ is the Dirac delta. Clearly this equation cannot have a classical solution with two continuous derivatives. Nonetheless, engineers have long dealt with δ-functions as *point forces* and have found practical ways to overcome the difficulties inherent in their use.

5.3 Equilibrium Problem for a Clamped Membrane and its Generalized Solution

We saw that the equilibrium of a membrane with fixed edge can be formulated as the problem of minimizing the functional

$$E_M(u) = \frac{1}{2}\iint_\Omega \left[\left(\frac{\partial u}{\partial x}\right)^2 + \left(\frac{\partial u}{\partial y}\right)^2\right] dx\,dy - \iint_\Omega f(x,y)u(x,y)\,dx\,dy. \tag{5.26}$$

Suppose

$$u\big|_{\partial\Omega} = 0. \tag{5.27}$$

In § 4.19 we constructed the Hilbert space $\mathcal{E}_{Mc}$ with inner product

$$(u,v)_M = \iint_\Omega \left(\frac{\partial u}{\partial x}\frac{\partial v}{\partial x} + \frac{\partial u}{\partial y}\frac{\partial v}{\partial y}\right) dx\,dy. \tag{5.28}$$

The first term in (5.26) can be written as

$$\frac{1}{2}(u,u)_M = \frac{1}{2}\left\|u\right\|_M^2 . \tag{5.29}$$

The second term,

$$W(u) = \int_\Omega f(x,y)u(x,y)\,dx\,dy, \tag{5.30}$$

is a linear functional in u. If $f \in L^p(\Omega)$ for some $p > 1$, then Hölder's inequality gives

$$\begin{aligned} |W(u)| &= \left|\iint_\Omega f(x,y)u(x,y)\,dx\,dy\right| \\ &\le \left(\iint_\Omega |f(x,y)|^p\,dx\,dy\right)^{1/p}\left(\iint_\Omega |u(x,y)|^q\,dx\,dy\right)^{1/q} \end{aligned}$$

with $q = p/(p-1)$. On the energy space, by equivalence of the norms $\|\cdot\|_M$ and $\|\cdot\|_{W^{1,2}(\Omega)}$ and Theorem 4.164, we have

$$\left(\iint_\Omega |u(x,y)|^q\,dx\,dy\right)^{1/q} \le m\left\|u\right\|_M$$

so

$$|W(u)| \le m\left(\iint_\Omega |f(x,y)|^p\,dx\,dy\right)^{1/p}\left\|u\right\|_M = m_1\left\|u\right\|_M .$$

Hence $W(u)$ is also continuous. By Theorem 4.100 there is a unique $u_0 \in \mathcal{E}_{Mc}$ such that

$$W(u) = (u, u_0)_M. \tag{5.31}$$

So the energy functional for a membrane with clamped edge can be represented in the energy space as

$$E_M(u) = \frac{1}{2} \|u\|_M^2 - (u, u_0)_M. \tag{5.32}$$

Let us consider the problem of minimizing $E_M(u)$ in $\mathcal{E}_{Mc}$.

Theorem 5.1. *In the energy space $\mathcal{E}_{Mc}$ the functional $E_M(u)$ attains its minimum at $u = u_0$, and the minimizer is unique.*

Proof. We have

$$\begin{aligned}
2E_M(u) &= \|u\|_M^2 - 2(u, u_0)_M \\
&= (u, u)_M - 2(u, u_0)_M + (u_0, u_0)_M - (u_0, u_0)_M \\
&= (u - u_0, u - u_0)_M - (u_0, u_0)_M \\
&= \|u - u_0\|_M^2 - \|u_0\|_M^2
\end{aligned}$$

so that

$$\min E_M(u) = -\frac{1}{2} \|u_0\|_M^2 .$$

Uniqueness of the minimizer u_0 is evident. □

Let us return to (5.26). The equation for the minimizer coincides with setting the first variation of $E_M(u)$ to zero:

$$\iint_\Omega \left(\frac{\partial u_0}{\partial x} \frac{\partial v}{\partial x} + \frac{\partial u_0}{\partial y} \frac{\partial v}{\partial y} \right) dx\, dy = \iint_\Omega f(x, y) v(x, y)\, dx\, dy. \tag{5.33}$$

It is usually said that (5.33) defines a generalized solution $u_0 \in \mathcal{E}_{Mc}$ to Poisson's equation $\Delta u = -f$ with boundary condition (5.27) if u_0 satisfies (5.33) for any $v \in \mathcal{E}_{Mc}$. This is often called the energy (or weak) solution. Note that (5.33) expresses the virtual work principle for a membrane with clamped edge.

5.4 Equilibrium of a Free Membrane

For the Neumann problem, the equilibrium equation (virtual work principle) is (see (5.7))

$$\iint_\Omega \left(\frac{\partial u}{\partial x}\frac{\partial \varphi}{\partial x} + \frac{\partial u}{\partial y}\frac{\partial \varphi}{\partial y}\right) dx\,dy = \iint_\Omega f(x,y)\varphi(x,y)\,dx\,dy + \int_{\partial\Omega} g(s)\varphi(s)\,ds. \tag{5.34}$$

The corresponding total potential energy functional is evidently

$$\begin{aligned} E_{M1}(u) = {} & \frac{1}{2}\iint_\Omega \left[\left(\frac{\partial u}{\partial x}\right)^2 + \left(\frac{\partial u}{\partial y}\right)^2\right] dx\,dy \\ & - \iint_\Omega f(x,y)u(x,y)\,dx\,dy - \int_{\partial\Omega} g(s)u(s)\,ds. \end{aligned} \tag{5.35}$$

Equation (5.34) is then the equality of the first variation of $E_{M1}(u)$ to zero, as follows from general considerations in the calculus of variations. Again, we put the equilibrium problem for a membrane with given edge forces $g(s)$ as a minimization problem for the energy functional $E_{M1}(u)$ on an energy space. We have the option to use a factor space $\mathcal{E}_{Mf}$ (see § 4.19), or its isometric variant where we take the balanced elements satisfying

$$\iint_\Omega u(x,y)\,dx\,dy = 0. \tag{5.36}$$

On the latter the problem of minimizing the energy is well defined if

$$f(x,y) \in L^{p_1}(\Omega), \qquad g(s) \in L^{p_2}(\partial\Omega), \tag{5.37}$$

for some $p_1, p_2 > 1$. But on the factor space the energy functional is not well defined if the forces are not self-balanced with

$$\iint_\Omega f(x,y)\,dx\,dy + \int_{\partial\Omega} g(s)\,ds = 0. \tag{5.38}$$

If (5.38) is not fulfilled, then for different representatives of zero, $u(x,y) = c$, the energy functional $E_{M1}(u)$ takes different values, which is impossible when we seek the minimum of the energy functional. This is a consequence of the fact that in this model we neglect the inertia properties of the membrane. Thus, considering the equilibrium problem on the factor space $E_{M1}(u)$, we get an additional necessary condition (5.38) of self-balance for the external forces. This condition does not arise when we adopt the second variant of the energy space, because (5.36) is an artificial geometric constraint that was absent from the initial problem statement and has been imposed as an auxiliary restriction. Although we do not need (5.38) when considering the problem in this way, we should nonetheless retain it since it is required by the initial setup.

Under the restriction (5.38) we can consider the equilibrium problem for a free membrane as the minimization problem for (5.35) on the space $\mathcal{E}_{Mf}$ of "usual" functions satisfying (5.36). Condition (5.37) is sufficient for $E_{M1}(u)$ to be well defined on $\mathcal{E}_{Mf}$. We need only show that the functional representing the work of external forces is well defined in this space. Hölder's inequality gives

$$\begin{aligned}
|W(u)| &= \left| \iint_\Omega f(x,y)u(x,y)\,dx\,dy + \int_{\partial\Omega} g(s)u(s)\,ds \right| \\
&\le \left(\iint_\Omega |f(x,y)|^{p_1}\,d\Omega \right)^{1/p_1} \left(\iint_\Omega |u(x,y)|^{q_1}\,d\Omega \right)^{1/q_1} \\
&\quad + \left(\int_{\partial\Omega} |g(s)|^{p_2}\,ds \right)^{1/p_2} \left(\int_{\partial\Omega} |u(s)|^{q_2}\,ds \right)^{1/q_2} \\
&\le m \left(\|f\|_{L^{p_1}(\Omega)} + \|g\|_{L^{p_2}(\partial\Omega)} \right) \|u\|_M
\end{aligned} \tag{5.39}$$

where $q_1 = p_1/(p_1-1)$, $q_2 = p_2/(p_2-1)$, and $\|\cdot\|_M$ is defined by (5.28). In the last transformation we used imbedding Theorem 4.164. Thus $W(u)$ is well defined on $\mathcal{E}_{Mf}$. Linearity of $W(u)$ is evident, and (5.39) guarantees continuity. Hence by Theorem 4.100

$$W(u) = (u, u_0)_M$$

where $u_0 \in \mathcal{E}_{Mf}$ is uniquely defined by the external forces f, g. Hence the minimization problem for $E_{M1}(u)$ can be reformulated as the minimization problem for

$$E_{M1}(u) = \frac{1}{2}\|u\|_M^2 - (u, u_0)_M. \tag{5.40}$$

There is formally no difference between the functionals (5.40) and (5.32), so we merely reformulate the results of § 5.3 for this problem as

Theorem 5.2. *Let (5.37) and (5.38) hold. In the energy space $\mathcal{E}_{Mf}$ the functional $E_{M1}(u)$ attains its minimum at $u = u_0$, and the minimizer is unique.*

The minimizer is a generalized solution of the equilibrium problem for a membrane with free edge. All the linear equilibrium problems we consider will reduce to minimization problems for quadratic functionals of the form

$$E(u) = \frac{1}{2}\|u\|^2 - W(u) \tag{5.41}$$

where $W(u)$ is a linear continuous functional. The proof of Theorem 5.1 does not depend on the nature of the space in which it is done, so we can immediately formulate

Theorem 5.3. *Let $W(u)$ be a linear continuous functional acting in a Hilbert space H. Then the minimization problem for* (5.41) *has a unique solution $u_0 \in H$ defined by the Riesz representation theorem: $W(u) = (u, u_0)$.*

Applications of this theorem appear in the next section.

5.5 Some Other Equilibrium Problems of Linear Mechanics

The mechanics problems for which we presented the energy functional and the virtual work principle in § 5.1 ((5.10)–(5.17)) all share the form (5.41) where the linear functional $W(u)$ is the potential of external forces (or, what amounts to the same thing, the work of external forces) on the displacement field u. Theorem 5.3 asserts the generalized solvability of a corresponding boundary value problem and the uniqueness of its generalized solution if $W(u)$ is continuous. To study continuity of $W(u)$, we shall use Sobolev's imbedding theorem and the fact that the corresponding energy space is a subspace of a Sobolev space $W^{l,2}(\Omega)$. The results are analogous to Theorem 5.2 and are left to the reader. We show only the restrictions on external forces to provide continuity of the potential of external forces as a functional in the energy space.

Rod

See (5.10) and (5.11). Here $u(0) = 0$ and

$$W(u) = \int_0^l f(x)u(x)\,dx + Fu(l). \tag{5.42}$$

In this case $u(x)$ is continuous on $[0, l]$ (i.e., each representative Cauchy sequence for an element of an energy space converges to a continuous function) and so if

$$f(x) \in L(0, l) \tag{5.43}$$

then

$$|W(u)| = \left| \int_0^l f(x)u(x)\,dx + Fu(l) \right|$$
$$\leq \left(\int_0^l |f(x)|\,dx + |F| \right) \max_{x\in[0,l]} |u(x)|$$
$$\leq m \left(\int_0^l |f(x)|\,dx + |F| \right) \|u\|_R$$

where

$$\|u\|_R = \left(\int_0^l ES(x)u'^2(x)\,dx \right)^{1/2}.$$

Beam

See (5.12) and (5.13). Now we can consider various boundary conditions. For clamped edges we formulate

$$w(0) = 0 = w'(0), \qquad w(l) = 0 = w'(l), \tag{5.44}$$

and the energy space for a bent beam with the norm

$$\|w\|_B = \left(\int_0^l EI(x)w''^2(x)\,dx \right)^{1/2} \tag{5.45}$$

is a subspace of $W^{2,2}(0,l)$ in which functions and their derivatives are continuous on $[0,l]$ and the corresponding operator of imbedding into the space of continuously differentiable functions is continuous. A sufficient condition for the potential of external forces

$$W(w) = \int_0^l f(x)w(x)\,dx + Fw(l)$$

is of the same type as for a stretched rod,

$$f(x) \in L(0,l), \tag{5.46}$$

and the proof is the same. However, in this case it is possible to include in the potential expression, and hence in the setup, the point external torques and transverse forces that are common in the strength of materials (represented by δ-functions). The proof is practically the same.

As to other variants of boundary conditions for a bent beam, the difference comes when the beam can move as a rigid whole. Then the situation is similar to that for a free membrane. A rigid motion of a free beam (i.e., a function w for which $\|w\|_B = 0$) has the form $w = a + bx$. Different boundary conditions can restrict the constants a and b (above they are zero). If the beam can move as a rigid body, the self-balance condition on external forces appears:

$$\int_0^l f(x)(a+bx)\,dx + F(a+bl) = 0 \tag{5.47}$$

for all admissible a, b. If the only geometrical boundary constraint is $w(0) = 0$, then

$$\int_0^l xf(x)\,dx + lF = 0. \tag{5.48}$$

Plate

It is possible to consider various boundary conditions. When the edge of the plate is clamped,

$$w\Big|_{\partial\Omega} = 0 = \frac{\partial w}{\partial n}\Big|_{\partial\Omega}. \tag{5.49}$$

The norm of the corresponding energy space $\mathcal{E}_{Pc}$, which is

$$\|w\|_P = \left(\iint_\Omega \left(w_{xx}^2 + w_{yy}^2 + 2\nu w_{xx}w_{yy} + 2(1-\nu)w_{xy}^2\right) d\Omega\right)^{1/2} \tag{5.50}$$

as shown in Chapter 4, is equivalent to the norm of $W^{2,2}(\Omega)$ when Ω is compact in $\mathbb{R}^2$. In this case $\mathcal{E}_{Pc}$ imbeds continuously into $C(\Omega)$. When both distributed and lumped forces are present, the potential of external forces

$$W(w) = \iint_\Omega F(x,y)w(x,y)\,d\Omega + \sum_{k=1}^N F_k w(x_k, y_k)$$

is a linear continuous functional in $\mathcal{E}_{Pc}$ provided that

$$F(x,y) \in L(\Omega). \tag{5.51}$$

Indeed,

$$\begin{aligned}
|W(w)| &= \left| \iint_\Omega F(x,y)w(x,y)\,d\Omega + \sum_{k=1}^{N} F_k w(x_k, y_k) \right| \\
&\le \left(\iint_\Omega |F(x,y)|\,dx\,dy + \sum_{k=1}^{N} |F_k| \right) \max_{x,y\in\Omega} w(x,y) \\
&\le m \left(\iint_\Omega |F(x,y)|\,dx\,dy + \sum_{k=1}^{N} |F_k| \right) \|w\|_P \\
&= m_1 \|w\|_P .
\end{aligned}$$

In this case there is a unique generalized solution to the equilibrium problem for the plate with clamped edge.

If the plate edge is free from geometrical constraints, there appear motions of the plate as a rigid whole that satisfy

$$\|w\|_P = 0. \tag{5.52}$$

The corresponding rigid motions are

$$w = ax + by + c \tag{5.53}$$

where a, b, c are constants. As in the theory of the free membrane, the condition of self-balance of the external forces appears:

$$W(ax+by+c) = \iint_\Omega F(x,y)(ax+by+c)\,d\Omega + \sum_{k=1}^{N} F_k(ax_k + by_k + c) = 0. \tag{5.54}$$

This holds for all a, b, c, so it represents three equations for the external forces that express equality to zero of the resultant force and resultant moments with respect to the coordinate axes (the reader should write them out and verify this). Condition (5.54) must be added to (5.51) as a necessary condition for solvability of the problem.

If there are other geometrical constraints on a plate, then the appearance of the self-balance condition depends on whether the constraints leave some freedom of movement. Fixation at three noncollinear points will prevent rigid motions. Rigid motions do arise if only a straight segment in the mid-surface is fixed, since the plate can rotate about this axis. In this case a self-balance condition appears.

Elastic body

When the boundary of the body is clamped, the energy norm

$$\|\mathbf{u}\|_E = \left(\iiint_V c^{ijkl} e_{kl}(\mathbf{u}) e_{ij}(\mathbf{u})\, dV\right)^{1/2} \tag{5.55}$$

is equivalent to the norm of the Sobolev space $\left(W^{1,2}(V)\right)^k$ provided V is compact in $\mathbb{R}^k$, $k = 2, 3$. In the two-dimensional case the imbedding result is exactly as for the membrane, and thus a sufficient condition for generalized solvability is that the Cartesian components of the vector of external forces belong to some $L^p(S)$ with $p > 1$. Mathematical physicists prefer "if and only if" conditions for solvability and have introduced the so-called negative Sobolev spaces. In terms of these the forces are completely characterized; but in a practical sense this condition gives us no more than if we simply say "the corresponding functional must be continuous in the space," so sufficient conditions are preferable in practice.

For a three-dimensional elastic body, the imbedding of $W^{1,2}(V)$, when V is compact, is a continuous operator to $L^p(V)$, $1 \le p \le 6$, and to $L^q(S)$, $1 \le q \le 4$, where S is a piecewise smooth surface in Ω. Conditions sufficient for generalized solvability of the equilibrium problem for a body with clamped boundary are

$$\begin{aligned} \mathbf{F} &\in (L^p(V))^3, \quad p \ge 6/5, \\ \mathbf{f} &\in (L^q(\partial V))^3, \quad q \ge 4/3. \end{aligned}$$

Indeed

$$\begin{aligned} |W(\mathbf{u})| &= \left| \iiint_V \mathbf{F}\cdot\mathbf{u}\, dV + \iint_{\partial V} \mathbf{f}\cdot\mathbf{u}\, dS \right| \\ &\le \left(\iiint_V |\mathbf{F}|^{6/5}\, dV\right)^{5/6} \left(\iiint_V |\mathbf{u}|^6\, dV\right)^{1/6} \\ &\quad + \left(\iint_{\partial V} |\mathbf{f}|^{4/3}\, dS\right)^{3/4} \left(\iint_{\partial V} |\mathbf{u}|^4\, dS\right)^{1/4} \\ &\le m \left[\left(\iiint_V |\mathbf{F}|^{6/5}\, dV\right)^{5/6} + \left(\iint_{\partial V} |\mathbf{f}|^{4/3}\, dS\right)^{3/4}\right] \|\mathbf{u}\|_E \end{aligned}$$

where we have used Hölder's inequality and the equivalence of the energy and Sobolev norms.

When we consider the equilibrium of a body free from geometrical constraints, rigid-body motions arise:

$$\mathbf{u} = \mathbf{a} + \mathbf{b} \times \mathbf{r} \tag{5.56}$$

(recall that these satisfy $\|\mathbf{u}\|_E = 0$), which imply that for a body free of geometrical constraints the forces must be self-balanced with

$$\iiint_V \mathbf{F} \cdot (\mathbf{a} + \mathbf{b} \times \mathbf{r})\, dV + \iint_{\partial V_1} \mathbf{f} \cdot (\mathbf{a} + \mathbf{b} \times \mathbf{r})\, dS = \mathbf{0}. \tag{5.57}$$

This equation must hold for all $\mathbf{a}$ and $\mathbf{b}$, giving six equations which are precisely the conditions of self-balance in classical mechanics: the resultant force and the resultant moments vanish.

In the case of mixed boundary conditions, if the body can move as a rigid whole, we must retain a subset of the self-balance conditions for the load. If the body can rotate about a fixed point, for example, the resultant moment with respect to the fixed point must vanish.

The one-dimensional problems and the plate problem allow us to formulate boundary conditions at a point, and the corresponding boundary value problems in their generalized setups are well posed. Note that this is not the case for the membrane or elastic body.

When the problem involves elastic support such as a Winkler foundation, or some interaction of elements with different models as would be the case with a three-dimensional elastic body coupled to a plate, the variational statement includes the sum of internal energies of all system elements. It is necessary to add some geometrical conditions of compatibility between the displacement fields of the bodies involved. The energy norm must contain all the internal energy functionals for the bodies (nonnegative quadratic terms) and sometimes the energy space is quite strange from the standpoint of classical Sobolev space theory. For such "coupled" models, we impose explicit geometrical constraints on interaction of the coupled elements, but not conditions for the stress terms: stress conditions on the interface are derived somewhat like natural boundary conditions. This prevents crude errors that commonly appear in the setup of similar problems, i.e., when someone tries to write down force balance equations for the interface elements in cases where the models approximate real stresses differently.

Nonhomogeneous geometrical boundary conditions

Homogeneous boundary conditions of the form $u|_{\partial\Omega} = 0$ provide linearity of the corresponding energy space. There are two ways to approach

$$u\big|_{\partial\Omega} = a(s) \tag{5.58}$$

where $a(s)$ is given. One is to minimize over a closed cone of all elements satisfying (5.58), as is done in variational inequalities. The other is traditional in mathematical physics: assume the existence of an element with some differential properties that satisfies (5.58), and seek a solution as a sum of this element and another element subject to homogeneous boundary conditions. The treatment of the membrane is typical. First we suppose there is an element $u^*(x, y) \in W^{1,2}(\Omega)$ (as usual we speak of functions with the understanding that such elements actually belong to the completion of the set of continuously differentiable functions) and seek the minimum point u of the energy functional

$$E_M(u) = \frac{1}{2}\iint_\Omega \left[\left(\frac{\partial u}{\partial x}\right)^2 + \left(\frac{\partial u}{\partial y}\right)^2\right] dx\, dy - \iint_\Omega f(x, y)u(x, y)\, dx\, dy$$

in the form

$$u(x, y) = u^*(x, y) + v(x, y) \tag{5.59}$$

where $v(x, y) \in \mathcal{E}_{Mc}$. So v satisfies the homogeneous condition $v|_{\partial\Omega} = 0$. Redenoting v by u, we get the following variational problem in $\mathcal{E}_{Mc}$:

$$\frac{1}{2}\iint_\Omega \left[\left(\frac{\partial(u + u^*)}{\partial x}\right)^2 + \left(\frac{\partial(u + u^*)}{\partial y}\right)^2\right] dx\, dy \\ - \iint_\Omega f(x, y)[u(x, y) + u^*(x, y)]\, dx\, dy \to \min\,.$$

Setting the first variation to zero, we get

$$\iint_\Omega \left(\frac{\partial u}{\partial x}\frac{\partial \varphi}{\partial x} + \frac{\partial u}{\partial y}\frac{\partial \varphi}{\partial y}\right) dx\, dy = \iint_\Omega f(x, y)\varphi(x, y)\, dx\, dy \\ - \iint_\Omega \left(\frac{\partial u^*}{\partial x}\frac{\partial \varphi}{\partial x} + \frac{\partial u^*}{\partial y}\frac{\partial \varphi}{\partial y}\right) dx\, dy. \tag{5.60}$$

A generalized solution of the equilibrium problem for a membrane with given edge displacement is an element $u(x, y) \in \mathcal{E}_{Mc}$ that satisfies (5.60) for any $\varphi(x, y) \in \mathcal{E}_{Mc}$. The first term on the right appeared in the problem with the homogeneous boundary condition. The second term is a bounded linear functional in φ because

$$\left| \iint_\Omega \left(\frac{\partial u^*}{\partial x}\frac{\partial \varphi}{\partial x} + \frac{\partial u^*}{\partial y}\frac{\partial \varphi}{\partial y} \right) dx\,dy \right| \le \left\| \frac{\partial u^*}{\partial x} \right\|_{L^2(\Omega)} \left\| \frac{\partial \varphi}{\partial x} \right\|_{L^2(\Omega)} + \left\| \frac{\partial u^*}{\partial y} \right\|_{L^2(\Omega)} \left\| \frac{\partial \varphi}{\partial y} \right\|_{L^2(\Omega)}$$
$$\le m \left\| u^* \right\|_{W^{1,2}(\Omega)} \left\| \varphi \right\|_M .$$

By Theorem 5.3, there is a unique generalized solution. The following question remains. Redenote the homogeneous part of the solution by u_1. Suppose we choose another fixed function u^{**} that takes the same boundary values, and find the homogeneous part of the solution denoted u_2. Do we have uniqueness in the sense that $u_1 + u^* = u_2 + u^{**}$? Denote $u_{21} = u_2 - u_1$ and subtract the equation for u_1 from the equation for u_2 with the same admissible variation φ. We have

$$\iint_\Omega \left(\frac{\partial u_{21}}{\partial x}\frac{\partial \varphi}{\partial x} + \frac{\partial u_{21}}{\partial y}\frac{\partial \varphi}{\partial y} \right) dx\,dy$$
$$= \iint_\Omega \left(\frac{\partial (u^{**} - u^*)}{\partial x}\frac{\partial \varphi}{\partial x} + \frac{\partial (u^{**} - u^*)}{\partial y}\frac{\partial \varphi}{\partial y} \right) dx\,dy.$$

But the difference $u^{**} - u^*$ belongs to $\mathcal{E}_{Mc}$ (why?), and since φ is an arbitrary element of $\mathcal{E}_{Mc}$ we have $u_{21} = u^{**} - u^*$. This completes the proof.

The general theory of Sobolev spaces is concerned with *trace theorems*. These deal with the question of which conditions must be stipulated on the boundary values in order to insure the existence of an element of a Sobolev space taking them as boundary conditions. The theorems require some smoothness of the domain boundary and are not convenient for practical verification; however, they provide "if and only if" conditions for existence of a continuation of the boundary functions as a function inside the domain, in such a way that the corresponding operator of continuation is continuous. Hence there arise Sobolev spaces $W^{l,p}(\Omega)$ with fractional parameters l.

Finally, we note that the study of generalized solutions is usually the first step in the study of the smoothness properties of solutions (see [27]). The birth of functional analysis was signaled when in this way Hilbert justified the Dirichlet principle (i.e., the same principle of minimum potential energy) for the solution of Laplace's equation with given boundary data, and showed that there exists an analytical solution of the latter under some restrictions on the given boundary function and the boundary itself. However, there is an important case for which practitioners find precisely the generalized solution. This is discussed in § 5.6.

5.6 The Ritz and Bubnov–Galerkin Methods

All problems in the linear mechanics of solids that we wish to consider have the form

$$E(u) = \frac{1}{2}\|u\|^2 - W(u) \to \min_H \tag{5.61}$$

where H is a Hilbert (energy) space and $W(u)$ is a linear continuous functional on H. By Theorem 4.100 this reduces to the problem

$$E(u) = \frac{1}{2}\|u\|^2 - (u, u_0) \to \min_H \tag{5.62}$$

with a given $u_0 \in H$. At first glance (5.62) seems trivial: the solution is u_0. However, u_0 is determined only theoretically; the term (u, u_0) stands in place of a functional W, and the role of (5.62) is simply to clarify some intermediate steps.

Ritz was the first to think, in practical terms, of the possibility of finding a minimizer, not on the whole space H but on some of its subspaces. In Ritz's time all calculations were done manually, so it was essential to find methods requiring as few steps as possible. Thus it was necessary (and still is, despite the capabilities of computers) to find a subspace having minimal dimension but capable of yielding a good approximation.[2] The finite dimensional subspace was constructed by the choice of basis elements $e_1, e_2, \ldots, e_n$. They should be linearly independent which, according to linear algebra, means that the Gram determinant

$$\begin{vmatrix} (e_1, e_1) & (e_1, e_2) & \cdots & (e_1, e_n) \\ (e_2, e_1) & (e_2, e_2) & \cdots & (e_2, e_n) \\ \vdots & \vdots & \ddots & \vdots \\ (e_n, e_1) & (e_n, e_2) & \cdots & (e_n, e_n) \end{vmatrix} \neq 0. \tag{5.63}$$

We also assume the set $e_1, e_2, \ldots, e_n, \ldots$ is complete in H; i.e., any element of H can be approximated to within given accuracy by a finite linear combination of elements from the set. Denote by H_n the space spanned by

[2]The approximate models of mechanics, like the theories of shells and plates, aims to reduce the full dimensionality of the problem. They reduce the dimensionality of the space coordinates for thin-walled structures from three to two dimensions by introducing some hypotheses on the form of deformation or the order of some strain components. The Ritz method also does this, but more directly: it reduces the possible forms of deformation of a body to forms expected to approximate the real ones more or less accurately.

$e_1, e_2, \ldots, e_n$. We call

$$E(u) = \frac{1}{2} \|u\|^2 - (u, u_0) \to \min_{H_n} \tag{5.64}$$

the Ritz method for the solution of (5.62).

Let us denote the minimizer of the problem by

$$u_n = \sum_{k=1}^{n} c_k e_k \tag{5.65}$$

where the c_k are constants. The equality to zero of the first variation of this functional for all admissible variations $v \in H_n$ is

$$(u_n, v) - (u_0, v) = 0. \tag{5.66}$$

Since $e_1, e_2, \ldots, e_n$ is a basis of H_n, the last equation is equivalent to the n simultaneous equations

$$\left(\sum_{k=1}^{n} c_k e_k, e_m \right) = (u_0, e_m), \qquad m = 1, \ldots, n. \tag{5.67}$$

called the Ritz system of the nth approximation step. The system can be rewritten as

$$\begin{aligned}
(e_1, e_1)c_1 + (e_2, e_1)c_2 + \cdots + (e_n, e_1)c_n &= W(e_1), \\
(e_1, e_2)c_1 + (e_2, e_2)c_2 + \cdots + (e_n, e_2)c_n &= W(e_2), \\
&\vdots \\
(e_1, e_n)c_1 + (e_2, e_n)c_2 + \cdots + (e_n, e_n)c_n &= W(e_n).
\end{aligned} \tag{5.68}$$

On the right side of (5.67) we have some given numbers. It is necessary to find the unknown c_k.

Theorem 5.4. *The system of simultaneous equations of the nth approximation has a unique solution $u_n = \sum_{k=1}^{n} c_k e_k$. The sequence $\{u_n\}$ converges strongly to the solution of the problem* (5.62).

Proof. The principal determinant of this system is the transposed Gram determinant. Hence by (5.63) the system (5.68) has a unique solution. Let us return to (5.66), which we rewrite as

$$(u_n - u_0, v) = 0 \quad \text{for all } v \in H_n.$$

This means $u_n - u_0$ is orthogonal to H_n, i.e., u_n is the orthogonal projection of u_0 onto H_n. Besides, it is easily seen from (5.67) that if $e_1, \ldots, e_n$ is an orthonormal basis of H_n, then (5.67) defines the Fourier coefficients $c_k = (u_0, e_k)$ of the solution u_0. Hence by Bessel's inequality

$$\|u_n\| \leq \|u_0\| .$$

Even if $e_1, \ldots, e_n$ is not an orthonormal basis of H_n, we can still construct an equivalent orthonormal basis of H_n which consequently defines an orthonormal basis of H. Thus the Fourier expansion of u_0 lies in the space spanned by this basis, and the Ritz approximation u_n coincides with the first n terms of that expansion. By the general Fourier theory, $\{u_n\}$ converges strongly to u_0 in H. □

We have noted that mechanics problems with free boundaries may be treated theoretically in factor spaces and in spaces of balanced functions. For numerical calculation by Ritz's method, only the balanced function spaces are appropriate. Were we to work in the corresponding factor spaces, the solution would contain the same undetermined constants of rigid motions, hence the corresponding determinant would be zero. Because of rounding errors and other numerical uncertainties, the system of Ritz's method (and any other numerical method) can lose the compatibility present in the initial setup. These issues do not occur with the energy space of balanced functions.

Ritz's method is the basis for various versions of the finite element methods. We should also note that the equations of Ritz method could be obtained from (5.66), which are not so elementary for certain problems where we cannot practically represent the work functional as an inner product. In this case the n elements v that define the linear algebraic system need not be e_k; they could be any other n linearly independent elements of the space which, for all n, constitute a complete system. This constitutes *Galerkin's method* (or the Bubnov–Galerkin method; cf., § 1.4).

5.7 The Hamilton–Ostrogradski Principle and Generalized Setup of Dynamical Problems in Classical Mechanics

One of the main variational principles of classical dynamics, the *Hamilton–Ostrogradski principle*, is not minimal. It asserts that the real motion of a system of material points, described by generalized coordinates

$$\mathbf{q}(t) = (q_1(t), \ldots, q_n(t))$$

and under the influence of potential forces, occurs in such manner that among all the motions from the initial position $\mathbf{q}_0$ taken at time t_0 to the final position $\mathbf{q}_1$ taken at time t_1, the real motion yields an extremal for the *action* functional

$$\int_{t_0}^{t_1} L(\mathbf{q}, \dot{\mathbf{q}}, t)\, dt. \tag{5.69}$$

Here an overdot denotes differentiation with respect to time t. The *kinetic potential* L is given by

$$L = K - E \tag{5.70}$$

where K and E are the kinetic and potential energies, respectively, of the system. The first variation of (5.69) is

$$\delta \int_{t_0}^{t_1} L(\mathbf{q}, \dot{\mathbf{q}}, t)\, dt = \int_{t_0}^{t_1} \sum_{i=1}^{n} \left(\frac{\partial L(\mathbf{q}, \dot{\mathbf{q}}, t)}{\partial q_i} \delta q_i + \frac{\partial L(\mathbf{q}, \dot{\mathbf{q}}, t)}{\partial \dot{q}_i} \delta \dot{q}_i \right) dt \tag{5.71}$$

where all variations δq_i of the generalized coordinates are considered as independent functions (cf., Chapter 1), and $\delta q_i(t_0) = 0 = \delta q_i(t_1)$ for $i = 1, 2, \ldots, n$. Setting the first variation to zero, we obtain *Lagrange's equations of motion*

$$\frac{d}{dt} \frac{\partial L}{\partial \dot{q}_i} - \frac{\partial L}{\partial q_i} = 0 \tag{5.72}$$

which form the basis of Lagrangian mechanics. In general the action does not attain a minimum or maximum. Normally for Lagrange's equations (if not in Hamiltonian form) a Cauchy problem is formulated in which equations (5.72) are supplemented with initial data

$$\mathbf{q}(t_0) = \mathbf{q}_0, \qquad \dot{\mathbf{q}}(t_0) = \mathbf{q}_{01}. \tag{5.73}$$

If we regard (5.71) as a generalized setup for some problem for (5.72), we see that (5.71) with the boundary conditions $\mathbf{q}(t_0) = \mathbf{q}_0$, $\mathbf{q}(t_1) = \mathbf{q}_1$, $\delta\mathbf{q}(t_0) = 0 = \delta\mathbf{q}(t_1)$ is formulated for a boundary value problem. How do we reformulate (5.71) and requirements on $q_i(t)$ to get a generalized setup for the Cauchy problem (5.72)–(5.73)? We would like to do this because the same operation will be performed when we transition from equilibrium problems to dynamical problems in solid mechanics. Let us take a special class D_1 of variations $\delta\mathbf{q}(t)$ that are continuously differentiable with $\delta\mathbf{q}(t_1) = \mathbf{0}$. Take $\delta\mathbf{q}(t) \in D_1$, multiply (5.72) by $\delta q_i(t)$, sum over i, and integrate over $[t_0, t_1]$:

$$\int_{t_0}^{t_1} \sum_{i=1}^{n} \left(\frac{d}{dt} \frac{\partial L}{\partial \dot{q}_i} - \frac{\partial L}{\partial q_i} \right) \delta q_i\, dt = 0. \tag{5.74}$$

Integration by parts (the operation inverse to the standard one done in the calculus of variations) gives

$$\int_{t_0}^{t_1} \sum_{i=1}^{n} \left(\frac{\partial L}{\partial \dot{q}_i} \delta \dot{q}_i + \frac{\partial L}{\partial q_i} \delta q_i \right) dt - \sum_{i=1}^{n} \left(\frac{\partial L}{\partial \dot{q}_i} \delta q_i \right) \Big|_{t=t_0} = 0. \tag{5.75}$$

In the second sum, the terms given at t_0, there stand values (5.73) so they do not contain q_i; the integrand involves only $q_i(t)$ and $\dot{q}_i(t)$ whereas (5.72) contains second derivatives of $q_i(t)$. Thus the requirements for $q_i(t)$ in (5.75) are less than in (5.72), and it is sensible to formulate a generalized setup of the Cauchy problem using (5.75) because in (5.75) we need not appoint values for $\mathbf{q}$ and $\dot{\mathbf{q}}$ at the instant t_1 in advance. It is clear that from (5.75), using the standard procedure of the calculus of variations, we can obtain (5.72) if we require (5.75) to hold for any $\delta \mathbf{q}(t) \in D_1$.

Next we must define a space in which to seek a solution. Usually the norm of this space would involve time integration, and this means we cannot stipulate on a generalized solution the point condition $\dot{\mathbf{q}}(t_0) = \mathbf{q}_1$, it comes into the definition through the second sum term of (5.75). The first initial condition $\mathbf{q}(t_0) = \mathbf{q}_0$ could be stipulated separately. We do not formulate exact statements here because, first of all, the form of the norm depends on the form of L and the statements would depend on this. More importantly, the generalized setup is avoided in classical mechanics. We have engaged in these considerations only to prepare ourselves for the more complex problems of continuum mechanics, for which all the pertinent details will be repeated.

5.8 Generalized Setup of Dynamic Problem for Membrane

In continuum mechanics the Hamilton–Ostrogradski principle can also be put in the form (5.69)–(5.70):

$$\delta \int_{t_0}^{t_1} L \, dt = 0, \qquad L = K - E, \tag{5.76}$$

where for each of the objects we have considered in equilibrium — beam, membrane, plate, elastic body — E is the energy functional we used (the difference between the elastic energy of an object and the potential of external forces acting on the object); here the state of the body at t_0 and t_1

must coincide with the real states of the body. The kinetic energy is

$$K = \frac{1}{2} \int_S \rho \dot{\mathbf{u}}^2 \, dS \tag{5.77}$$

where S is the domain taken by the object in a coordinate frame and ρ is the specific density of the material. For example, in the case of a three-dimensional elastic body the equation of the Hamilton–Ostrogradski principle looks like

$$\delta \int_{t_0}^{t_1} \left\{ \frac{1}{2} \iiint_V \rho \dot{\mathbf{u}}^2 \, dV - \left[\frac{1}{2} \iiint_V c^{ijkl} e_{kl}(\mathbf{u}) e_{ij}(\mathbf{u}) \, dV \right.\right.$$
$$\left.\left. - \left(\iiint_V \mathbf{F} \cdot \mathbf{u} \, dV + \iint_{\partial V} \mathbf{f} \cdot \mathbf{u} \, dS \right) \right] \right\} dt = 0$$

for any admissible variation of displacement vector $\delta \mathbf{u}$. Here $\mathbf{u}$ must satisfy the geometrical boundary conditions of the problem, $\delta \mathbf{u} = \delta \mathbf{u}(\mathbf{x}, t)$ the homogeneous geometrical boundary conditions and, besides,

$$\delta \mathbf{u}(\mathbf{x}, t_0) = \mathbf{0} = \delta \mathbf{u}(\mathbf{x}, t_1).$$

So this formulation corresponds to a boundary value problem as if the values of $\mathbf{u}(\mathbf{x}, t)$ are given at $t = t_0$ and $t = t_1$.

Now we would like to derive a generalized setup of the Cauchy problem for the dynamic problems under consideration. It is clear that the corresponding energy spaces should include the terms with integrals for the kinetic energy and, besides, if we would like to use Hilbert space tools, temporal integration should appear in the norm. The form of the integrand of the E term in the action remains the same, so we need only consider the kinetic energy term. We begin with the universal equation that is the virtual work principle in statics. To simplify the calculations we consider a membrane; the remaining problems are treated similarly. We combine the virtual work principle with *d'Alembert's principle*, which asserts that the system of external forces can be balanced by the inertia forces. For a membrane the work of external forces complemented by the inertia forces on a virtual displacement $v(\mathbf{x}, t)$ is

$$\iint_\Omega [f(\mathbf{x}, t) - \rho \ddot{u}(\mathbf{x}, t)] \, v(\mathbf{x}, t) \, d\Omega, \qquad d\Omega = dx \, dy.$$

Thus, for a membrane with clamped edge, the virtual work principle gives

$$\iint_\Omega \left(\frac{\partial u}{\partial x} \frac{\partial v}{\partial x} + \frac{\partial u}{\partial y} \frac{\partial v}{\partial y} \right) d\Omega = \iint_\Omega [f(\mathbf{x}, t) - \rho \ddot{u}(\mathbf{x}, t)] \, v(\mathbf{x}, t) \, d\Omega. \tag{5.78}$$

Of course we could begin with the differential equations of motion and obtain the same result step by step, but we take a shorter route. We suppose all functions are smooth enough to permit the required transformations and that the virtual displacement v satisfies

$$v(\mathbf{x}, T) = 0. \tag{5.79}$$

Let us integrate (5.78) over time and integrate by parts in the last term:

$$\int_0^T \iint_\Omega \left(\frac{\partial u}{\partial x}\frac{\partial v}{\partial x} + \frac{\partial u}{\partial y}\frac{\partial v}{\partial y} \right) d\Omega\, dt = \int_0^T \iint_\Omega f(\mathbf{x}, t) v(\mathbf{x}, t)\, d\Omega\, dt$$
$$+ \int_0^T \iint_\Omega \rho \dot{u}(\mathbf{x}, t) \dot{v}(\mathbf{x}, t)\, d\Omega\, dt + \iint_\Omega \rho u_1^*(\mathbf{x}) v(\mathbf{x}, 0)\, d\Omega. \tag{5.80}$$

Here $u_1^*(\mathbf{x})$ is an initial condition for $u(\mathbf{x}, t)$:

$$u(\mathbf{x}, t)\big|_{t=t_0} = u_0^*(\mathbf{x}), \qquad \dot{u}(\mathbf{x}, t)\big|_{t=t_0} = u_1^*(\mathbf{x}). \tag{5.81}$$

We shall use (5.80) for the generalized setup of the dynamic problem for a membrane. The first step is to define proper function spaces.

An energy space for a clamped membrane (dynamic case)

Without loss of generality we can set $t_0 = 0$ and denote $t_1 = T$. It is clear that the expression for an inner product in this space should include some terms from (5.80). Let it be given by

$$(u, v)_{[a,b]} = \int_a^b \iint_\Omega \rho \dot{u}(\mathbf{x}, t) \dot{v}(\mathbf{x}, t)\, d\Omega\, dt + \int_a^b \iint_\Omega \left(\frac{\partial u}{\partial x}\frac{\partial v}{\partial x} + \frac{\partial u}{\partial y}\frac{\partial v}{\partial y} \right) d\Omega\, dt. \tag{5.82}$$

The energy space $\mathcal{E}_{Mc}(a, b)$ is the completion of the set of twice continuously differentiable functions that satisfy the boundary condition

$$u|_{\partial\Omega} = 0, \tag{5.83}$$

with respect to the norm $\|u\| = (u, u)_{[a,b]}^{1/2}$. Denote $Q_{a,b} = \Omega \times [a, b]$.

Lemma 5.5. *$\mathcal{E}_{Mc}(a, b)$ is a closed subspace of $W^{1,2}(Q_{a,b})$. The norm of $\mathcal{E}_{Mc}(a, b)$ is equivalent to the norm of $W^{1,2}(Q_{a,b})$.*

Proof. It suffices to prove the last statement of the lemma for twice differentiable functions satisfying (5.83). The inequality

$$(u, u)_{[a,b]} \le M \|u\|^2_{W^{1,2}(Q_{a,b})}$$

is evident. Let us show that the reverse inequality with a positive constant m holds as well. From the Friedrichs inequality it follows that

$$\int_a^b \|u\|^2_{W^{1,2}(\Omega)}\, dt \leq m \int_a^b \iint_\Omega \left[\left(\frac{\partial u}{\partial x}\right)^2 + \left(\frac{\partial u}{\partial y}\right)^2 \right] d\Omega\, dt.$$

Adding to both sides the term

$$\int_a^b \iint_\Omega \rho \dot{u}^2(\mathbf{x}, t)\, d\Omega\, dt$$

after easy transformations, we get the needed inequality. □

By Sobolev's imbedding theorem, from Lemma 5.5 it follows that $\mathcal{E}_{Mc}(a, b)$ imbeds continuously into $L^6(Q(a, b))$ and at any fixed $t \in [a, b]$ into $L^4(\Omega)$, so we can pose an initial condition for u to satisfy in the sense of $L^4(\Omega)$. However we now demonstrate a general result that shows the meaning in which we can state the initial condition.

Let H be a separable Hilbert space. Consider the set of functions of the parameter $t \in [a, b]$ that take values in H. In what follows $H = L^2(\Omega)$. The theory of such functions is quite similar to the usual theory of functions in one variable. In particular, we can define the space $C(H; a, b)$ of all functions continuous on $[a, b]$ and taking values in H. Its properties are the same as those of $C(a, b)$: for separable H, it is a separable Banach space with the norm of an element $x(t)$ given by

$$\|x\|_{C(H;a,b)} = \max_{t \in [a,b]} \|x(t)\|_H \,.$$

For functions with values in H we can introduce the notion of derivative as

$$x'(t) = \lim_{\Delta t \to 0} \frac{x(t + \Delta t) - x(t)}{\Delta t},$$

as well as derivatives of higher order. The definite Riemann integral

$$\int_c^d x(t)\, dt$$

is the limit of Riemann sums that must not depend on the manner in which $[c, d]$ is partitioned. Analogous to the spaces $C^{(k)}(a, b)$, for functions with values in H we can introduce spaces $C^{(k)}(H; a, b)$ (we leave this to the reader). Finally we can employ an analogue of $L^2(a, b)$, denoted by $L^2(H; a, b)$. This is a Hilbert space with an inner product

$$(x, y)_{L^2(H;a,b)} = \int_a^b (x(t), y(t))_H \, dt, \tag{5.84}$$

and is the completion of $C(H;a,b)$ in the norm induced by (5.84). Note that $L^2(L^2(\Omega);a,b)$ is $L^2(Q_{a,b})$. Quite similarly, we can introduce a Sobolev space $W^{1,2}(H;a,b)$ as the completion of $C^{(1)}(H;a,b)$ with respect to the norm induced by

$$(x,y)_{W^{1,2}(H;a,b)} = \int_a^b \{(x(t),y(t))_H + (x'(t),y'(t))_H\}\,dt. \tag{5.85}$$

Lemma 5.6. *$W^{1,2}(H;a,b)$ is continuously imbedded into $C(H;a,b)$.*

The proof mimics that of the similar result for $W^{1,2}(a,b)$, so we leave it to the reader. Lemma 5.6 states that we can formulate the initial condition for $u(\mathbf{x},t)$ at a fixed t in the sense of $L^2(\Omega)$ since the element of $\mathcal{E}_{Mc}(a,b)$, by the form of the norm, belongs to $W^{1,2}(L^2(\Omega);a,b)$ as well. However, to pose the initial boundary value problem we need a stronger result. This is a particular imbedding theorem in a Sobolev space that is useful for hyperbolic boundary value problems.

Lemma 5.7. *If $\{u_n\}$ converges weakly to u_0 in $\mathcal{E}_{Mc}(a,b)$, then it also converges to u_0 uniformly with respect to t in the norm of $C(L^2(\Omega);a,b)$.*

Proof. By equivalence on $\mathcal{E}_{Mc}(a,b)$ of the norm of $\mathcal{E}_{Mc}(a,b)$ to the norm of $W^{1,2}(Q_{a,b})$, and Sobolev's imbedding theorem, we state that

$$\|u_n\|_{[a,b]} \le m \tag{5.86}$$

and that

$$\|u_n - u_0\|_{L^2(Q_{a,b})} \to 0 \quad \text{as } n \to \infty. \tag{5.87}$$

So u_n converges to u_0 strongly in $L^2(Q_{a,b})$. Now we need a special bound for an element of $W^{1,2}(L^2(\Omega);a,b)$, into which $W^{1,2}(Q_{a,b})$ imbeds continuously. We derive the estimate for elements that are smooth in time t, and then extend to all the elements. Let $c \in [a,b)$ and $\Delta > 0$ be such that $c+\Delta \in [a,b)$. Let $t,s \in [c,c+\Delta]$. The simple identity

$$v(\mathbf{x},t) = v(\mathbf{x},s) + \int_s^t \frac{\partial v(\mathbf{x},\theta)}{\partial\theta}\,d\theta$$

gives

$$\begin{aligned}\iint_\Omega v^2(\mathbf{x},t)\,d\Omega &= \iint_\Omega \left(v(\mathbf{x},s) + \int_s^t \frac{\partial v(\mathbf{x},\theta)}{\partial\theta}\,d\theta\right)^2 d\Omega \\ &\le 2\iint_\Omega v^2(\mathbf{x},s)\,d\Omega + 2\iint_\Omega \left(\int_s^t \frac{\partial v(\mathbf{x},\theta)}{\partial\theta}\,d\theta\right)^2 d\Omega.\end{aligned}$$

Let us integrate this with respect to s over $[c, c+\Delta]$. Dividing through by Δ we get

$$\iint_\Omega v^2(\mathbf{x},t)\,d\Omega \le \frac{2}{\Delta}\int_c^{c+\Delta}\iint_\Omega v^2(\mathbf{x},s)\,d\Omega\,ds + \frac{2}{\Delta}\int_c^{c+\Delta}\iint_\Omega\left(\int_s^t 1\cdot\frac{\partial v(\mathbf{x},\theta)}{\partial\theta}\,d\theta\right)^2 d\Omega\,ds.$$

Applying Hölder's inequality to the last term on the right we have

$$\iint_\Omega v^2(\mathbf{x},t)\,d\Omega \le \frac{2}{\Delta}\int_{Q_{c,c+\Delta}} v^2(\mathbf{x},s)\,d\Omega\,ds + \frac{2}{\Delta}\int_c^{c+\Delta}\iint_\Omega\left(\int_s^t 1^2\,d\theta\int_s^t\left(\frac{\partial v(\mathbf{x},\theta)}{\partial\theta}\right)^2 d\theta\right) d\Omega\,ds.$$

Finally, direct integration in the last integral and simple estimates yield

$$\iint_\Omega v^2(\mathbf{x},t)\,d\Omega \le \frac{2}{\Delta}\int_{Q_{c,c+\Delta}} v^2(\mathbf{x},\theta)\,d\Omega\,d\theta + \Delta\int_{Q_{c,c+\Delta}}\left(\frac{\partial v(\mathbf{x},\theta)}{\partial\theta}\right)^2 d\Omega\,d\theta, \tag{5.88}$$

which is the basis for the proof of Lemma 5.7. By the completion procedure (5.88) extends to any element of $\mathcal{E}_{a,b}$. We write it out for $u_n - u_0$:

$$\iint_\Omega (u_n(\mathbf{x},t)-u_0(\mathbf{x},t))^2\,d\Omega \le \frac{2}{\Delta}\int_{Q_{c,c+\Delta}} (u_n(\mathbf{x},\theta)-u_0(\mathbf{x},\theta))^2\,d\Omega\,d\theta + \Delta\int_{Q_{c,c+\Delta}}\left(\frac{\partial\,(u_n(\mathbf{x},\theta)-u_0(\mathbf{x},\theta))}{\partial\theta}\right)^2 d\Omega\,d\theta. \tag{5.89}$$

Let $\varepsilon > 0$ be an arbitrarily small positive number. To prove the lemma it is enough to find a number N such that the right side of (5.89) is less than ε for any $t \in [c, c+\Delta]$. Let us put $\Delta = \varepsilon/2m$ where m is the constant from (5.86). Then the last integral is less than $\varepsilon/2$. By (5.87) we can find N such that

$$\frac{2}{\Delta}\int_{Q_{c,c+\Delta}} (u_n(\mathbf{x},t)-u_0(\mathbf{x},t))^2\,d\Omega\,dt \le \frac{\varepsilon}{2}$$

independent of $t \in [c, c+\Delta]$. Since this is independent of $c \in [a,b]$ we establish the result for all $t \in [a,b]$. □

Generalized setup

Without loss of generality we consider the initial problem on $[0,T]$ for fixed but arbitrary T. In this case we use the energy space $\mathcal{E}_{Mc}(0,T)$. In addition, we must define a closed subspace D_0^T which is the completion of the subset of twice continuously differentiable functions satisfying (5.83) that vanish at $t = T$.

Definition 5.8. $u(\mathbf{x}, t) \in \mathcal{E}_{Mc}(0, T)$ is a *generalized solution* of the dynamical problem for a clamped membrane if it satisfies the equation

$$\int_0^T \iint_\Omega \left(\frac{\partial u}{\partial x}\frac{\partial v}{\partial x} + \frac{\partial u}{\partial y}\frac{\partial v}{\partial y} \right) d\Omega\, dt = \int_0^T \iint_\Omega f(\mathbf{x}, t) v(\mathbf{x}, t)\, d\Omega\, dt \\ + \int_0^T \iint_\Omega \rho \dot{u}(\mathbf{x}, t) \dot{v}(\mathbf{x}, t)\, d\Omega\, dt + \iint_\Omega \rho u_1^*(\mathbf{x}) v(\mathbf{x}, 0)\, d\Omega \tag{5.90}$$

with any $v(\mathbf{x}, t) \in D_0^T$ and the first initial condition

$$u(\mathbf{x}, t)\big|_{t=0} = u_0^*(\mathbf{x}) \tag{5.91}$$

in the sense of $L^2(\Omega)$, that is,

$$\iint_\Omega \left(u(\mathbf{x}, 0) - u_0^*(\mathbf{x}) \right)^2 d\Omega = 0. \tag{5.92}$$

Let us suppose that

(i) $u_0^*(\mathbf{x}) \in W^{1,2}(\Omega)$ and satisfies (5.83),
(ii) $u_1^*(\mathbf{x}) \in L^2(\Omega)$, and
(iii) $f(\mathbf{x}, t) \in L^2(Q_{0,T})$.

It is easy to show that under these restrictions all terms of (5.90) make sense. The goal is to prove the following.

Theorem 5.9. *Under restrictions (i)–(iii) there exists (in the sense of Definition 5.8) a unique generalized solution to the dynamical problem for a clamped membrane.*

The proof splits into several lemmas. First we construct an approximate method of solution for the problem under consideration, a variant of the Bubnov–Galerkin method called the Faedo–Galerkin method. Then we justify its convergence. Finally, we give an independent proof of uniqueness.

The Faedo–Galerkin method

Suppose there is a complete system of elements of $\mathcal{E}_{Mc}$, any finite set of which is a linearly independent system. In applications these are smooth functions except in the finite element method where they are piecewise smooth. Take the first n elements of the system. We can always "orthonormalize" the latter system with respect to the $L^2(\Omega)$ inner product:

$$\rho \iint_\Omega \varphi_i(\mathbf{x})\varphi_j(\mathbf{x})\, d\Omega = \delta_{ij} = \begin{cases} 1, & i = j, \\ 0, & i \neq j. \end{cases} \tag{5.93}$$

This is done only to simplify calculations (and to get the final equations in normal form); it is not necessary in principle. We seek the nth approximation to the solution in the form

$$u_n(\mathbf{x}, t) = \sum_{k=1}^{n} c_k(t)\varphi_k(\mathbf{x}) \tag{5.94}$$

where the $c_k(t)$ are time functions satisfying the following system of Faedo–Galerkin equations, which are implied by (5.78) in which we put u_n instead of u and consequently φ_i instead of v:

$$\iint_\Omega \left(\frac{\partial u_n}{\partial x}\frac{\partial \varphi_i}{\partial x} + \frac{\partial u_n}{\partial y}\frac{\partial \varphi_i}{\partial y} \right) d\Omega = \iint_\Omega \left[f(\mathbf{x}, t) - \rho \ddot{u}_n(\mathbf{x}, t) \right] \varphi_i(\mathbf{x})\, d\Omega \tag{5.95}$$

for $i = 1, \ldots, n$. These can be written as

$$\rho \iint_\Omega \ddot{u}_n(\mathbf{x}, t)\varphi_i(\mathbf{x})\, d\Omega = -(u_n, \varphi_i)_M + \iint_\Omega f(\mathbf{x}, t)\varphi_i(\mathbf{x})\, d\Omega, \quad i = 1, \ldots, n.$$

Finally, using (5.94) and (5.93), let us rewrite this as

$$\ddot{c}_i(t) = -\sum_{k=1}^{n} c_k(t)(\varphi_k, \varphi_i)_M + \iint_\Omega f(\mathbf{x}, t)\varphi_i(\mathbf{x})\, d\Omega, \quad i = 1, \ldots, n. \tag{5.96}$$

This is a system of simultaneous ordinary differential equations for which we must formulate initial conditions. The condition $\dot{u}(\mathbf{x}, t)|_{t=0} = u_1(\mathbf{x})$ and (5.93) imply

$$\dot{c}_i(0) = \rho^{-1/2} \iint_\Omega u_1^*(\mathbf{x})\varphi_i(\mathbf{x})\, d\Omega, \quad i = 1, \ldots, n. \tag{5.97}$$

From (5.91) we derive the following conditions for $c_i(0)$. Let us solve the problem

$$\left\| u_0^* - \sum_{k=1}^{n} a_k \varphi_k \right\|_M^2 \to \min_{a_1, \ldots, a_n}. \tag{5.98}$$

We know this is solvable; moreover, its solution $d_1, \ldots, d_n$ gives us $\sum_{k=1}^{n} d_k \varphi_k$, the orthogonal projection in $\mathcal{E}_{Mc}$ of u_0 onto the subspace spanned by $\varphi_1, \ldots, \varphi_n$. Thus the second set of initial conditions is

$$c_i(0) = d_i, \quad i = 1, \ldots, n. \tag{5.99}$$

So the setup of the nth approximation of the Faedo–Galerkin method consists of (5.96) supplemented with (5.97) and (5.99). We begin by establishing the properties of this Cauchy problem.

Unique solvability of the Cauchy problem for the nth approximation of the Faedo–Galerkin method

We would like to understand what we can say about the solution of the Cauchy problem (5.96), (5.97), (5.99). The simultaneous equations (5.96) are linear in the unknown $c_i(t)$. The load terms $\iint_\Omega f(\mathbf{x},t)\varphi_i(\mathbf{x})\,d\Omega$ belong to $L^2(0,T)$; indeed, by Schwarz's inequality

$$\begin{aligned}\int_0^T \left(\iint_\Omega f(\mathbf{x},t)\varphi_i(\mathbf{x})\,d\Omega\right)^2 dt \\ &\le \int_0^T \left(\iint_\Omega f^2(\mathbf{x},t)\,d\Omega\right)\left(\iint_\Omega \varphi_i^2(\mathbf{x})\,d\Omega\right) dt \\ &= \|\varphi_i\|^2_{L^2(\Omega)}\,\|f\|^2_{L^2(Q_{0,T})}\,.\end{aligned}$$

From general ODE theory the Cauchy problem (5.96), (5.97), (5.99) has a unique solution on $[0,T]$ with arbitrary T such that

$$c_i''(t) \in L^2(0,T) \tag{5.100}$$

and $c_i(t)$ and $c_i'(t)$ are continuous on $[0,T]$. This can be shown by the traditional way of proving such results, in which a Cauchy problem is transformed into a system of integral equations (by double integration of the equations in time taking into account the initial conditions). For the integral equations the existence of a unique continuous solution can be shown via Banach's contraction principle, and then time differentiation yields the remaining properties. Now we obtain the estimate of the solution that we need to prove the above theorem. The estimate for the solution $c_i(t)$, $i = 1, \ldots, n$, is

$$\max_{t\in[0,T]} \left(\sum_{k=1}^n (c_k'(t))^2 + \left\|\sum_{k=1}^n c_k(t)\varphi_k\right\|_M^2\right) \le m.$$

Indeed, let us multiply the ith equation in (5.96) by $c_i'(t)$ and sum over i:

$$\begin{aligned}\sum_{i=1}^n \ddot{c}_i(t)\dot{c}_i(t) &= -\sum_{i=1}^n\sum_{k=1}^n (c_k(t)\varphi_k, \dot{c}_i(t)\varphi_i)_M \\ &\quad + \sum_{i=1}^n \iint_\Omega f(\mathbf{x},t)\dot{c}_i(t)\varphi_i(\mathbf{x})\,d\Omega.\end{aligned} \tag{5.101}$$

The term on the left is

$$\sum_{i=1}^{n} \ddot{c}_i(t)\dot{c}_i(t) = \frac{1}{2}\frac{d}{dt}\sum_{i=1}^{n} \dot{c}_i(t)\dot{c}_i(t)$$

$$= \frac{1}{2}\rho\frac{d}{dt}\sum_{i,j=1}^{n} \dot{c}_i(t)\dot{c}_j(t)\iint_\Omega \varphi_i(\mathbf{x})\varphi_j(\mathbf{x})\,d\Omega$$

$$= \frac{d}{dt}\left(\frac{1}{2}\rho\iint_\Omega \dot{u}_n(\mathbf{x},t)\dot{u}_n(\mathbf{x},t)\,d\Omega\right).$$

Similarly

$$\sum_{i=1}^{n}\sum_{k=1}^{n} (c_k(t)\varphi_k, \dot{c}_i(t)\varphi_i)_M = \frac{1}{2}\frac{d}{dt}\left(u_n(\mathbf{x},t), u_n(\mathbf{x},t)\right)_M$$

and

$$\sum_{i=1}^{n}\iint_\Omega f(\mathbf{x},t)\dot{c}_i(t)\varphi_i(\mathbf{x})\,d\Omega = \iint_\Omega f(\mathbf{x},t)\dot{u}_n(\mathbf{x},t)\,d\Omega.$$

So (5.101) can be presented as

$$\frac{d}{dt}\left(\frac{1}{2}\rho\iint_\Omega \dot{u}(\mathbf{x},t)\dot{u}_n(\mathbf{x},t)\,d\Omega\right) + \frac{1}{2}\frac{d}{dt}\left(u_n(\mathbf{x},t), u_n(\mathbf{x},t)\right)_M$$
$$= \iint_\Omega f(\mathbf{x},t)\dot{u}_n(\mathbf{x},t)\,d\Omega,$$

or rewritten as

$$\frac{1}{2}\frac{d}{dt}\left(\rho\left\|\dot{u}_n(\mathbf{x},t)\right\|^2_{L^2(\Omega)} + \left\|u_n(\mathbf{x},t)\right\|^2_M\right) = \iint_\Omega f(\mathbf{x},t)\dot{u}_n(\mathbf{x},t)\,d\Omega.$$

Integrating over time t (renaming t by s) we have

$$\frac{1}{2}\int_0^t \frac{d}{ds}\left(\rho\left\|\dot{u}_n(\mathbf{x},s)\right\|^2_{L^2(\Omega)} + \left\|u_n(\mathbf{x},s)\right\|^2_M\right)ds$$
$$= \int_0^t\iint_\Omega f(\mathbf{x},s)\dot{u}_n(\mathbf{x},s)\,d\Omega\,ds$$

or

$$\frac{1}{2}\left(\rho\left\|\dot{u}_n(\mathbf{x},t)\right\|^2_{L^2(\Omega)} + \left\|u_n(\mathbf{x},t)\right\|^2_M\right)$$
$$= \frac{1}{2}\left(\rho\left\|\dot{u}_n(\mathbf{x},0)\right\|^2_{L^2(\Omega)} + \left\|u_n(\mathbf{x},0)\right\|^2_M\right)$$
$$+ \int_0^t\iint_\Omega f(\mathbf{x},s)\dot{u}_n(\mathbf{x},s)\,d\Omega\,ds.$$

Taking into account the way in which we derived the initial conditions for u_n, we have

$$\|\dot{u}_n(\mathbf{x},0)\|_{L^2(\Omega)} \leq \|u_1^*(\mathbf{x})\|_{L^2(\Omega)}, \qquad \|u_n(\mathbf{x},0)\|_M \leq \|u_0^*(\mathbf{x})\|_M.$$

We can then state that

$$\begin{aligned}
&\frac{1}{2}\left(\rho\,\|\dot{u}_n(\mathbf{x},t)\|_{L^2(\Omega)}^2 + \|u_n(\mathbf{x},t)\|_M^2\right) \\
&\quad \leq \frac{1}{2}\left(\rho\,\|u_1^*(\mathbf{x})\|_{L^2(\Omega)}^2 + \|u_0^*(\mathbf{x})\|_M^2\right) \\
&\qquad + \int_0^t \iint_\Omega f(\mathbf{x},s)\dot{u}_n(\mathbf{x},s)\,d\Omega\,ds.
\end{aligned}$$

The elementary inequality

$$|ab| \leq \frac{a^2}{2\varepsilon} + \frac{\varepsilon b^2}{2}$$

yields

$$\begin{aligned}
\left|\int_0^t \iint_\Omega f(\mathbf{x},s)\dot{u}_n(\mathbf{x},s)\,d\Omega\,ds\right| &\leq \frac{1}{2\varepsilon}\int_0^t \iint_\Omega f^2(\mathbf{x},s)\,d\Omega\,ds \\
&\quad + \frac{\varepsilon}{2}\int_0^t \iint_\Omega \dot{u}_n^2(\mathbf{x},s)\,d\Omega\,ds \\
&\leq \frac{1}{2\varepsilon}\int_0^T \iint_\Omega f^2(\mathbf{x},s)\,d\Omega\,ds \\
&\quad + \frac{\varepsilon T}{2}\max_{s\in[0,T]}\iint_\Omega \dot{u}_n^2(\mathbf{x},s)\,d\Omega
\end{aligned}$$

so

$$\begin{aligned}
\frac{1}{2}\left(\rho\,\|\dot{u}_n(\mathbf{x},t)\|_{L^2(\Omega)}^2 + \|u_n(\mathbf{x},t)\|_M^2\right) &\leq \frac{1}{2}\left(\rho\,\|u_1^*(\mathbf{x})\|_{L^2(\Omega)}^2 + \|u_0^*(\mathbf{x})\|_M^2\right) \\
&\quad + \frac{1}{2\varepsilon}\int_0^T \iint_\Omega f^2(\mathbf{x},s)\,d\Omega\,ds \\
&\quad + \frac{\varepsilon T}{2}\max_{s\in[0,T]}\iint_\Omega \dot{u}_n^2(\mathbf{x},s)\,d\Omega.
\end{aligned}$$

Putting $\varepsilon = \rho/(2T)$ and taking the maximum of the left side of the last inequality we get

$$\max_{t\in[0,T]} \frac{1}{2} \left(\rho \|\dot{u}_n(\mathbf{x},t)\|^2_{L^2(\Omega)} + \|u_n(\mathbf{x},t)\|^2_M\right)$$
$$\leq \frac{1}{2} \left(\rho \|u_1^*(\mathbf{x})\|^2_{L^2(\Omega)} + \|u_0^*(\mathbf{x})\|^2_M\right)$$
$$+ \frac{T}{\rho} \int_0^T \iint_\Omega f^2(\mathbf{x},s)\, d\Omega\, ds$$
$$+ \frac{\rho}{4} \max_{t\in[0,T]} \iint_\Omega \dot{u}_n^2(\mathbf{x},t)\, d\Omega$$

so

$$\max_{t\in[0,T]} \frac{1}{4} \left(\rho \|\dot{u}_n(\mathbf{x},t)\|^2_{L^2(\Omega)} + \|u_n(\mathbf{x},t)\|^2_M\right)$$
$$\leq \frac{1}{2} \left(\rho \|u_1^*(\mathbf{x})\|^2_{L^2(\Omega)} + \|u_0^*(\mathbf{x})\|^2_M\right)$$
$$+ \frac{T}{\rho} \int_0^T \iint_\Omega f^2(\mathbf{x},s)\, d\Omega\, ds.$$

This is the needed estimate, which can be written as

$$\max_{t\in[0,T]} \left(\rho \|\dot{u}_n(\mathbf{x},t)\|^2_{L^2(\Omega)} + \|u_n(\mathbf{x},t)\|^2_M\right) \leq m$$

where the constant m does not depend on the number n. In particular, from this follows the rougher estimate

$$\int_0^T \left(\rho \|u_n(\mathbf{x},t)\|^2_{L^2(\Omega)} + \|u_n(\mathbf{x},t)\|^2_M\right) dt \leq m_1$$

which can be written in terms of (5.82) as

$$(u_n, u_n)_{[0,T]} \leq m_1. \tag{5.102}$$

Convergence of the Faedo–Galerkin method

Now we show that there is a subsequence of $\{u_n(\mathbf{x},t)\}$ that converges to a generalized solution of the problem under consideration. By (5.102), $\{u_n\}$ has a subsequence that converges weakly to an element $u_0(\mathbf{x},t)$. We shall show that $u_0(\mathbf{x},t)$ is a generalized solution. By Lemma 5.6 we can consider it as a function continuous in t on $[0,T]$ with values in $L^2(\Omega)$. Let us renumber this subsequence, denoting it by $\{u_n\}$ (in fact, by the uniqueness theorem proved later, the whole sequence converges weakly so renumbering is not required; however, at this point we are not assured of uniqueness).

So, now we know that $u_n(\mathbf{x}, t)$ tends to $u_0(\mathbf{x}, t)$ weakly in $\mathcal{E}_{Mc}(0, T)$. First we show that u_0 satisfies (5.91). Indeed, by the method of constructing the Faedo–Galerkin approximations u_n, we see that $\{u_n(\mathbf{x}, 0)\}$ converges to the initial value $u_0^*(\mathbf{x})$ strongly in $W^{1,2}(\Omega)$ and thus in $L^2(\Omega)$. On the other hand, by Lemma 5.7, $\{u_n(\mathbf{x}, t)\}$ converges to $u_0(\mathbf{x}, t)$ in the norm of $C(L^2(\Omega); 0, T)$. Thus (5.91) holds for $u_0(\mathbf{x}, t)$. Let us verify that (5.90) with $u = u_0(\mathbf{x}, t)$ holds for any $v(\mathbf{x}, t) \in D_0^T$. First we reduce the set of admissible v to a subset of D_0^T defined as follows. Let

$$v_k(t, \mathbf{x}) = \sum_{k=1}^{n} d_k(t)\varphi_k(\mathbf{x}), \qquad k \leq n \tag{5.103}$$

where the $d_k(t)$ are continuously differentiable and $d_k(T) = 0$. Denote the set of all such finite sums by D_{0f}^T. This set is dense in D_0^T and thus, to complete the proof of Theorem 5.9, it is enough to demonstrate the validity of (5.90) for $u = u_0(\mathbf{x}, t)$ when $v \in D_{0f}^T$. Let us return to (5.95) for u_n:

$$\iint_\Omega \left(\frac{\partial u_n}{\partial x} \frac{\partial \varphi_i}{\partial x} + \frac{\partial u_n}{\partial y} \frac{\partial \varphi_i}{\partial y} \right) d\Omega = \iint_\Omega \left(f(\mathbf{x}, t) - \rho \ddot{u}_n(\mathbf{x}, t) \right) \varphi_i(\mathbf{x})\, d\Omega,$$
$$i = 1, \ldots, n.$$

Multiplying the ith equation by $d_i(t)$ and summing from $i = 1$ to k we get

$$\iint_\Omega \left(\frac{\partial u_n}{\partial x} \frac{\partial v_k}{\partial x} + \frac{\partial u_n}{\partial y} \frac{\partial v_k}{\partial y} \right) d\Omega = \iint_\Omega \left(f(\mathbf{x}, t) - \rho \ddot{u}_n(\mathbf{x}, t) \right) v_k(\mathbf{x}, t)\, d\Omega$$

for $k \leq n$. Let us integrate this with respect to t:

$$\int_0^T \iint_\Omega \left(\frac{\partial u_n}{\partial x} \frac{\partial v_k}{\partial x} + \frac{\partial u_n}{\partial y} \frac{\partial v_k}{\partial y} \right) d\Omega\, dt$$
$$= \int_0^T \iint_\Omega \left(f(\mathbf{x}, t) - \rho \ddot{u}_n(\mathbf{x}, t) \right) v_k(\mathbf{x}, t)\, d\Omega\, dt.$$

Integrating by parts in the last term we get

$$\int_0^T \iint_\Omega \left(\frac{\partial u_n}{\partial x} \frac{\partial v_k}{\partial x} + \frac{\partial u_n}{\partial y} \frac{\partial v_k}{\partial y} \right) d\Omega\, dt = \int_0^T \iint_\Omega f(\mathbf{x}, t) v_k(\mathbf{x}, t)\, d\Omega\, dt$$
$$+ \int_0^T \iint_\Omega \rho \dot{u}_n(\mathbf{x}, t) \dot{v}_k(\mathbf{x}, t)\, d\Omega\, dt + \iint_\Omega \rho \dot{u}_n(\mathbf{x}, 0) v_k(\mathbf{x}, 0)\, d\Omega.$$

Let us now fix $v_k(\mathbf{x}, t)$ and let $n \to \infty$. By the properties of u_n we have

$$\int_0^T \iint_\Omega \left(\frac{\partial u_0}{\partial x} \frac{\partial v_k}{\partial x} + \frac{\partial u_0}{\partial y} \frac{\partial v_k}{\partial y} \right) d\Omega\, dt = \int_0^T \iint_\Omega f(\mathbf{x}, t) v_k(\mathbf{x}, t)\, d\Omega\, dt$$
$$+ \int_0^T \iint_\Omega \rho \dot{u}_0(\mathbf{x}, t) \dot{v}_k(\mathbf{x}, t)\, d\Omega\, dt + \iint_\Omega \rho u_1^*(\mathbf{x}) \dot{v}_k(\mathbf{x}, 0)\, d\Omega,$$

as required by Definition 5.8.

Uniqueness of the generalized solution

Theorem 5.10. *A generalized solution of the dynamic problem for a membrane with clamped edge is unique.*

Proof. Suppose there are two generalized solutions u' and u''. Subtracting term by term the equations (5.90) for these solutions and introducing $u = u'' - u'$, we get

$$\int_0^T \iint_\Omega \rho \dot{u}(\mathbf{x}, t) \dot{v}(\mathbf{x}, t)\, d\Omega\, dt - \int_0^T \iint_\Omega \left(\frac{\partial u}{\partial x} \frac{\partial v}{\partial x} + \frac{\partial u}{\partial y} \frac{\partial v}{\partial y} \right) d\Omega\, dt = 0 \tag{5.104}$$

for any $v \in D_0^T$. Also,

$$u(\mathbf{x}, t)\big|_{t=0} = 0$$

holds in the sense of $L^2(\Omega)$. Let us define an auxiliary function

$$w(\mathbf{x}, t) = \begin{cases} \displaystyle\int_\tau^t u(\mathbf{x}, \vartheta)\, d\vartheta, & t \in [0, \tau], \\ 0, & t > \tau. \end{cases}$$

First we note that on $[0, \tau]$

$$\frac{\partial w(\mathbf{x}, t)}{\partial t} = u(\mathbf{x}, t).$$

This and other similar relations between w and u are established by simple differentiation of the representative functions of corresponding Cauchy sequences; then a limit passage justifies that they hold for the elements themselves. It is seen that $w(\mathbf{x}, t)$ belongs to D_0^T. Moreover, it has generalized derivatives $\partial^2 w/\partial t \partial x = \partial u/\partial x$, $\partial^2 w/\partial t \partial y = \partial u/\partial y$ in $L^2(Q_{0,\tau})$. Next, $\partial^2 w/\partial t^2 = \partial u/\partial t \in L^2(Q_{0,\tau})$. Finally, as follows from Lemma 5.6, w

and its first derivatives belong to $C(L^2(\Omega); 0, \tau)$ (the reader should verify this). Let us put $v = w$ in (5.104). This equality can be written as

$$\int_0^\tau \iint_\Omega \rho \frac{\partial u(\mathbf{x},t)}{\partial t} u(\mathbf{x},t)\, d\Omega\, dt - \int_0^\tau \iint_\Omega \left(\frac{\partial^2 w}{\partial x \partial t} \frac{\partial w}{\partial x} + \frac{\partial^2 w}{\partial y \partial t} \frac{\partial w}{\partial y} \right) d\Omega\, dt = 0,$$

and rewritten as

$$\frac{1}{2} \int_0^\tau \iint_\Omega \frac{\partial}{\partial t} \left\{ \rho u^2(\mathbf{x},t) - \left(\frac{\partial w}{\partial x} \right)^2 - \left(\frac{\partial w}{\partial y} \right)^2 \right\} d\Omega\, dt = 0.$$

Integrating over t we get

$$\iint_\Omega \left\{ \rho u^2(\mathbf{x},t) - \left(\frac{\partial w}{\partial x} \right)^2 - \left(\frac{\partial w}{\partial y} \right)^2 \right\} d\Omega \Bigg|_{t=0}^{t=\tau} = 0.$$

Using the initial condition for u and the definition of w we have

$$\iint_\Omega \rho u^2(\mathbf{x},\tau)\, d\Omega + \iint_\Omega \left\{ \left(\frac{\partial w(\mathbf{x},t)}{\partial x} \right)^2 + \left(\frac{\partial w(\mathbf{x},t)}{\partial y} \right)^2 \right\} d\Omega \Bigg|_{t=0} = 0.$$

Here all integrands are positive so

$$\iint_\Omega \rho u^2(\mathbf{x},\tau)\, d\Omega = 0.$$

Since τ is an arbitrary point of $[0, T]$ we have $u = 0$. □

Let us recall that because of uniqueness it can be shown (by way of contradiction) that the whole Faedo–Galerkin sequence of approximations $\{u_n\}$ converges weakly to the generalized solution of the problem under consideration in the energy space.

5.9 Other Dynamic Problems of Linear Mechanics

Let us briefly consider the changes needed to treat various other dynamical problems of mechanics.

We begin with a mixed problem for the membrane. If a portion of the edge is free from clamping and loading, how must the approach change? Only in the definition of the energy space. The removal of restrictions on the free part of the boundary simply requires a wider energy space; then everything carries through as before, and the same theorems are formally established.

When on some part Γ_1 of the edge a load $f(s,t)$ is given, the equation for generalized solution appears as follows:

$$\int_0^T \iint_\Omega \left(\frac{\partial u}{\partial x}\frac{\partial v}{\partial x} + \frac{\partial u}{\partial y}\frac{\partial v}{\partial y} \right) d\Omega\, dt = \int_0^T \iint_\Omega f(\mathbf{x},t) v(\mathbf{x},t)\, d\Omega\, dt$$
$$+ \int_0^T \int_{\Gamma_1} \varphi(s,t) v(s,t)\, ds\, dt + \int_0^T \iint_\Omega \rho \dot{u}(\mathbf{x},t)\dot{v}(\mathbf{x},t)\, d\Omega\, dt$$
$$+ \iint_\Omega \rho u_1^*(\mathbf{x}) v(\mathbf{x},0)\, d\Omega. \tag{5.105}$$

For solvability we also need

$$\varphi(s,t) \in W^{1,2}(L^2(\Gamma_1); 0, T). \tag{5.106}$$

Under this restriction it is possible to obtain an *a priori* estimate of the generalized solution, and thus to prove existence of a generalized solution. The formulation and uniqueness proof remain practically unchanged (except for the definition and notation for the energy space).

We shall not consider in detail all the other problems of dynamics for the objects we studied in statics. The introduction of the main equation of motion always repeats all the steps we performed for the membrane. The corresponding energy space formulation, in which the inner product is denoted by $(\cdot,\cdot)_\mathcal{E}$, yields

$$\int_0^T (\mathbf{u}(t), \mathbf{v}(t))_\mathcal{E}\, dt = \int_0^T \int_\Omega \mathbf{f}(\mathbf{x},t) \cdot \mathbf{v}(\mathbf{x},t)\, d\Omega\, dt$$
$$+ \int_0^T \int_{\Gamma_1} \boldsymbol{\varphi}(s,t) \cdot \mathbf{v}(s,t)\, ds\, dt + \int_0^T \int_\Omega \rho \dot{\mathbf{u}}(\mathbf{x},t) \cdot \dot{\mathbf{v}}(\mathbf{x},t)\, d\Omega\, dt$$
$$+ \int_\Omega \rho \mathbf{u}_1^*(\mathbf{x}) \cdot \mathbf{v}(\mathbf{x},0)\, d\Omega$$

which parallels (5.105) for the membrane. The reasoning leading to the main theorems remains the same, the differences residing only in the definitions of the appropriate energy spaces. The reader can formulate and prove the existence and uniqueness of generalized solutions for initial-boundary value problems in the theory of plates and for two- and three-dimensional elastic bodies.

5.10 The Fourier Method

A principal method of solving dynamics problems was developed by Fourier. The method facilitates the description of transient processes. Normally the class of loads considered analytically is not wide, and it is possible to find a partial solution that "removes" the effect of the load; it then remains to find how the behavior of a non-loaded object changes from some arbitrary initial state. For solution of the latter problem, Fourier proposed a method of separation of variables. As an example let us consider the dynamic problem for a string, described by

$$\frac{\partial^2 u}{\partial t^2} = \frac{\partial^2 u}{\partial x^2}, \qquad x \in [0, \pi] \tag{5.107}$$

with initial and boundary conditions

$$u(0,t) = 0 = u(\pi,t), \qquad u(x,0) = u_0(x), \qquad \frac{\partial u(x,0)}{\partial t} = u_1(x). \tag{5.108}$$

We seek a solution to (5.107) in the form $u(x,t) = T(t)v(x)$. From (5.107) we have

$$\frac{T''(t)}{T(t)} = \frac{X''(x)}{X(x)} = -\lambda^2.$$

The value λ can only be constant since each fraction of the equality depends on only one of the independent variables x or t. We seek nontrivial solutions of this form. The equation

$$X''(x) + \lambda^2 X(x) = 0 \tag{5.109}$$

with the necessary boundary conditions

$$X(0) = 0 = X(\pi) \tag{5.110}$$

has nontrivial solutions only when $\lambda = k$ with k a positive or negative integer; that is, $X_k(x) = c \sin kx$. There are no other nontrivial solutions to (5.109)–(5.110), which is typical of eigenvalue problems for distributed systems. Using this, we find an adjoint solution for the equation

$$T''(t) + k^2 T(t) = 0,$$

whose general solution is

$$T_k(t) = c_{k0} \cos kt + c_{k1} \sin kt.$$

Hence Fourier obtained a general solution to (5.107) as

$$\sum_{k=1}^{\infty} (c_{k0} \cos kt + c_{k1} \sin kt) \sin kx. \tag{5.111}$$

Finally, we can look for coefficients that satisfy (5.108). So a central role in Fourier theory is played by the eigenvalue problem, the problem of finding nontrivial solutions to a boundary value problem with a parameter, (5.109)–(5.110). A similar problem arises in all linear mechanical problems, and in a similar fashion. In fact, we could begin at once to seek a class of particular solutions of the form $e^{i\mu t}v(x)$ where $v(0) = 0 = v(\pi)$. Now we have the same eigenvalue problem for $v(x)$:

$$v''(x) + \mu^2 v(x) = 0.$$

Moreover, when we seek a general solution as a sum of particular real solutions, we come to the same expression (5.111). This can be said for any of the linear mechanical problems considered earlier. Thus in every case we come to a particular eigenvalue boundary value problem, then to the problem of finding the coefficients of the corresponding Fourier series of the type (5.111), and finally to the problem of convergence. This will be considered in detail in the next few sections.

5.11 An Eigenfrequency Boundary Value Problem Arising in Linear Mechanics

For each problem considered earlier, the dynamic equations with use of the D'Alembert principle have the form

$$(u, \eta)_{\mathcal{E}} = -\int_{\Omega} \rho \frac{\partial^2 u}{\partial t^2} \eta \, d\Omega \tag{5.112}$$

where $(\cdot, \cdot)_{\mathcal{E}}$ is a scalar product in the energy space and η is an admissible virtual displacement. To formulate an eigenfrequency problem, we put $u = e^{i\mu t} v(\mathbf{x})$ in (5.112) and obtain

$$(v, \eta)_{\mathcal{E}} = \rho\mu^2 \int_{\Omega} v\eta \, d\Omega. \tag{5.113}$$

For convenience let us take $\rho = 1$ (by choice of dimensional units). Since we now consider complex-valued u, we let η be complex as well. In this case (5.113) takes the form

$$(v, \eta)_{\mathcal{E}} = \mu^2 \int_{\Omega} v\overline{\eta} \, d\Omega. \tag{5.114}$$

Equation (5.114) defines the general form of the eigenfrequency problems for the elastic objects considered in this chapter.

Definition 5.11. If (5.114) has a nonzero solution v for some μ, then v is an *eigensolution* (*eigenvector*) and μ is the corresponding *eigenfrequency*. The value $\lambda = 1/\mu$ is the *eigenvalue* of the object.

Remark 5.12. We could arrive at the same eigenvalue problem by considering heat transfer described by

$$\frac{\partial T}{\partial t} = \Delta T$$

with zero temperature T on the boundary of the domain. If we seek a solution in the form $T(\mathbf{x}, t) = e^{-\mu t} v(\mathbf{x})$ in a generalized statement, we get the equation that coincides with (5.114) governing eigen-oscillations of a membrane taking the same domain in the plane; the only discrepancy is the form of the parameter in the equation: it is μ for heat transfer and μ^2 for the membrane. Next, introducing $\lambda = 1/\mu$ in the heat problem we get a parameter that is usually called the eigenvalue. However we will retain our terminology since it makes more mechanical sense. Next there is a discrepancy between our terminology and that which is common in textbooks on mathematical physics: we call eigenfrequencies the quantities that are called eigenvalues in mathematical physics; the reason is that in mathematical physics they normally consider the equation in $L^2(\Omega)$ so $A = \Delta$ is considered as an unbounded operator in $L^2(\Omega)$ and the terminology is borrowed from standard spectral theory. But in our approach this differential operator corresponds to the identity operator in an energy space. □

We formulated (5.114) in a complex energy space. The next lemma permits a return to real spaces.

Lemma 5.13. *All eigenfrequencies of the problem* (5.114) *are real.*

Proof. The result follows from the fact that $(v, v)_\mathcal{E}$ and $\int_\Omega v\overline{v}\, d\Omega$ are positive numbers for any v, hence so is $\mu^2 = (v, v)_\mathcal{E} / \int_\Omega v\overline{v}\, d\Omega$. □

Since (5.114) is linear in v, we can consider its real and imaginary parts separately, and hence consider it only in a real energy space. Thus the equation we shall study is formulated in a real energy space and the eigenfrequency problem is as follows.

Eigenvalue Problem. Find a nonzero u belonging to a real energy space $\mathcal{E}$ that satisfies the equation

$$(u, v)_\mathcal{E} = \mu^2 \int_\Omega uv\, d\Omega \tag{5.115}$$

for any $v \in \mathcal{E}$.

We require that $\mathcal{E}$ is a Hilbert space and that there is a constant $m > 0$ such that

$$\|u\|_{\mathcal{E}} \geq m \|u\|_{W^{1,2}(\Omega)} \tag{5.116}$$

for any $u \in \mathcal{E}$. All the energy spaces we introduced had this property; in the case of a three-dimensional elastic body, u is a vector function, and in the integral on the right of (5.115) uv must mean a dot product of the displacement vectors $\mathbf{u}$ and $\mathbf{v}$.

Let us transform (5.115) into an operator form using Theorem 4.100. At any fixed $u \in \mathcal{E}$, the integral $\int_\Omega uv\, d\Omega$ is a functional linear in v. Schwarz's inequality, Sobolev's imbedding theorem, and (5.116) give us

$$\begin{aligned} \left| \int_\Omega uv\, d\Omega \right| &\leq \|u\|_{L^2(\Omega)} \|v\|_{L^2(\Omega)} \\ &\leq m_1 \|u\|_{W^{1,2}(\Omega)} \|v\|_{W^{1,2}(\Omega)} \\ &\leq m_2 \|u\|_{\mathcal{E}} \|v\|_{\mathcal{E}}, \end{aligned} \tag{5.117}$$

which means this functional is continuous for $v \in \mathcal{E}$. Thus it can be represented as an inner product in $\mathcal{E}$:

$$\int_\Omega uv\, d\Omega = (w, v)_{\mathcal{E}}, \tag{5.118}$$

where $w \in \mathcal{E}$ is uniquely defined by u. (The second position of v in the inner product is unimportant by symmetry in the arguments.) Since to any $u \in \mathcal{E}$ there corresponds $w \in \mathcal{E}$, we have defined an operator A acting in $\mathcal{E}$:

$$w = Au. \tag{5.119}$$

With this notation (5.115) takes the form

$$(u, v)_{\mathcal{E}} = \mu^2 (Au, v)_{\mathcal{E}}. \tag{5.120}$$

Since $v \in \mathcal{E}$ is arbitrary we get

$$u = \mu^2 Au. \tag{5.121}$$

Although A has been introduced theoretically, we should be able to establish some of its properties through the defining equality

$$(Au, v)_{\mathcal{E}} = \int_\Omega uv\, d\Omega. \tag{5.122}$$

Let us begin.

Lemma 5.14. *The operator A is linear and continuous on $\mathcal{E}$.*

Proof. For linearity it suffices to establish that

$$A(\alpha_1 u_1 + \alpha_2 u_2) = \alpha_1 A u_1 + \alpha_2 A u_2 \tag{5.123}$$

for any real numbers α_i and elements $u_i \in \mathcal{E}$. By (5.122) we have

$$\begin{aligned}(A(\alpha_1 u_1 + \alpha_2 u_2), v)_{\mathcal{E}} &= \int_\Omega (\alpha_1 u_1 + \alpha_2 u_2) v \, d\Omega \\ &= \alpha_1 \int_\Omega u_1 v \, d\Omega + \alpha_2 \int_\Omega u_2 v \, d\Omega.\end{aligned}$$

On the other hand

$$(Au_i, v)_{\mathcal{E}} = \int_\Omega u_i v \, d\Omega, \qquad i = 1, 2,$$

and thus

$$(A(\alpha_1 u_1 + \alpha_2 u_2), v)_{\mathcal{E}} = \alpha_1 (Au_1, v)_{\mathcal{E}} + \alpha_2 (Au_2, v)_{\mathcal{E}} .$$

From this (5.123) follows by the arbitrariness of v. To prove continuity of A let us use (5.117), from which

$$|(Au, v)_{\mathcal{E}}| = \left| \int_\Omega uv \, d\Omega \right| \le m_2 \|u\|_{\mathcal{E}} \|v\|_{\mathcal{E}} .$$

Setting $v = Au$, we get for an arbitrary u

$$|(Au, Au)_{\mathcal{E}}| \le m_2 \|u\|_{\mathcal{E}} \|Au\|_{\mathcal{E}} .$$

It follows that

$$\|Au\|_{\mathcal{E}} \le m_2 \|u\|_{\mathcal{E}} ,$$

and this completes the proof. □

Definition 5.15. An operator B is *strictly positive* in a Hilbert space H if $(Bx, x) \ge 0$ for any $x \in H$, and from the equality $(Bx, x) = 0$ it follows that $x = 0$.

Lemma 5.16. *The operator A is strictly positive in $\mathcal{E}$.*

Proof. Clearly

$$(Au, u)_{\mathcal{E}} = \int_\Omega u^2 \, d\Omega \ge 0.$$

If $(Au, u)_{\mathcal{E}} = 0$, then $u = 0$ in $L^2(\Omega)$ and thus in $\mathcal{E}$. □

Lemma 5.17. *The operator A is self-adjoint.*

Proof. We use the symmetry in the arguments u, v in the definition (5.122) and continuity of A. Since

$$(Au, v)_{\mathcal{E}} = \int_{\Omega} uv \, d\Omega = \int_{\Omega} vu \, d\Omega = (Av, u)_{\mathcal{E}} = (u, Av)_{\mathcal{E}}$$

the proof is immediate. □

The last property we wish to establish is

Lemma 5.18. *The operator A is compact.*

Proof. It is enough to show that for any weakly Cauchy sequence $\{u_n\}$ the corresponding $\{Au_n\}$ is a strongly Cauchy sequence. Let $\{u_n\}$ be a weakly Cauchy sequence in $\mathcal{E}$. By (5.116) it is a weakly Cauchy sequence in $W^{1,2}(\Omega)$ and thus, by Sobolev's imbedding theorem, it is a strongly Cauchy sequence in $L^2(\Omega)$. Let us use an inequality following from (5.117),

$$\left| \int_{\Omega} uv \, d\Omega \right| \leq m_3 \left\| u \right\|_{L^2(\Omega)} \left\| v \right\|_{\mathcal{E}},$$

to write

$$\left| (A(u_n - u_m), v)_{\mathcal{E}} \right| = \left| \int_{\Omega} (u_n - u_m) v \, d\Omega \right| \leq m_3 \left\| u_n - u_m \right\|_{L^2(\Omega)} \left\| v \right\|_{\mathcal{E}}.$$

Putting $v = A(u_n - u_m)$ we get

$$\left| (A(u_n - u_m), A(u_n - u_m))_{\mathcal{E}} \right| \leq m_3 \left\| u_n - u_m \right\|_{L^2(\Omega)} \left\| A(u_n - u_m) \right\|_{\mathcal{E}}$$

so that

$$\left\| A(u_n - u_m) \right\|_{\mathcal{E}} \leq m_3 \left\| u_n - u_m \right\|_{L^2(\Omega)} \to 0 \quad \text{as } n, m \to \infty.$$

This completes the proof. □

5.12 The Spectral Theorem

The results of this section are general despite their formulation in energy spaces. They apply in any separable Hilbert space $\mathcal{E}$, whether or not the space pertains to a mechanical problem. We suppose A is a self-adjoint, strictly positive, compact operator acting in a real Hilbert space $\mathcal{E}$. The inner product in $\mathcal{E}$ is denoted $(u, v)_{\mathcal{E}}$. Because A is self-adjoint and strictly positive, the bilinear functional $(Au, v)_{\mathcal{E}}$ has all the properties of an inner product. Let us denote this inner product by

$$(u, v)_A = (Au, v)_{\mathcal{E}} \tag{5.124}$$

and its corresponding norm by $\|u\|_A = (u, u)_A^{1/2}$.

Since $\mathcal{E}$ is incomplete with respect to the new norm we can apply the completion theorem. The completion of $\mathcal{E}$ with respect to the norm $\|u\|_A$ is denoted by $\mathcal{E}_A$ and is called the energy space of the operator A. But, unlike the earlier energy spaces, this energy space for the problems under consideration does not relate to the system energy. Looking at the form of the inner product in $\mathcal{E}_A$ for A from the previous section, we see that it is an inner product in $L^2(\Omega)$. Moreover, from the general theory of the L^p spaces it is known that infinitely differentiable functions whose support is compact in Ω (so they are zero on the boundary of Ω) are dense in $L^2(\Omega)$. Hence the resulting space $\mathcal{E}_A$ for the problems of the previous section is $L^2(\Omega)$ (more precisely, the elements stand in one-to-one distance preserving correspondence). In what follows we need

Definition 5.19. A functional F is *weakly continuous* at a point u if for any sequence $\{u_n\}$ weakly convergent to u we have $F(u_n) \to F(u)$ as $n \to \infty$. A functional is weakly continuous on a domain M if it is weakly continuous at each point $u \in M$.

By definition a linear weakly continuous functional is continuous, and vice versa.

Lemma 5.20. *A functional $F(u)$, weakly continuous on the unit ball $\|u\|_{\mathcal{E}} \le 1$ of a Hilbert space $\mathcal{E}$, takes its minimal and maximal values on this ball.*

Proof. This is similar to a classical calculus theorem on the extremes of a continuous function given on a compact set. We prove the statement for maxima of F. The result for minima follows by consideration of $-F$. Let $\{u_n\}$ be a sequence in the unit ball, denoted by B, such that

$$F(u_n) \to \sup_{\|u\|_{\mathcal{E}} \le 1} F(u) \quad \text{as } n \to \infty.$$

Since $\{u_n\}$ lies in B it contains a weakly convergent subsequence $\{u_{n_k}\}$. Since B is weakly closed in $\mathcal{E}$ this subsequence has a weak limit u^* belonging to B. The value $F(u^*)$ is finite and since F is weakly continuous we have

$$F(u_{n_k}) \to F(u^*) = \sup_{\|u\|_{\mathcal{E}} \le 1} F(u),$$

so u^* is the needed point. □

Lemma 5.21. *Let A be a compact linear operator in a Hilbert space $\mathcal{E}$. Then $F(u) = (Au, u)_\mathcal{E}$ is a weakly continuous functional in $\mathcal{E}$.*

Proof. Let $\{u_n\}$ be weakly convergent to u. Consider

$$\begin{aligned}
|(Au_n, u_n)_\mathcal{E} - (Au, u)_\mathcal{E}| &= |(Au_n, u_n)_\mathcal{E} - (Au, u_n)_\mathcal{E} + (Au, u_n)_\mathcal{E} - (Au, u)_\mathcal{E}| \\
&\le |(Au_n, u_n)_\mathcal{E} - (Au, u_n)_\mathcal{E}| + |(Au, u_n)_\mathcal{E} - (Au, u)_\mathcal{E}| \\
&\le \|A(u_n - u)\|_\mathcal{E} \|u_n\|_\mathcal{E} + |(Au, u_n - u)_\mathcal{E}| \\
&\to 0 \quad \text{as } n \to \infty.
\end{aligned}$$

For the first term this happened since $\|u_n\|_\mathcal{E}$ is bounded and $A(u_n - u) \to 0$ strongly in $\mathcal{E}$. The second term tends to zero since it is a linear continuous functional in $u_n - u$. □

For a strictly positive operator all the eigenvalues are nonnegative (why?) so we will denote them as λ^2:

$$Ax = \lambda^2 x.$$

This is done to preserve the terminology of mechanics, where the corresponding value $\mu = 1/\lambda$ is called an eigenfrequency of the object. Let us formulate the main result of this section.

Theorem 5.22. *Let A be a self-adjoint, strictly positive, compact operator acting in a real separable Hilbert space. Then*

(i) A has a countable set of eigenfrequencies with no finite limit point;

(ii) to each eigenfrequency of A there corresponds a finite dimensional set of eigenvectors $\{\varphi_k\}$; we can choose eigenvectors constituting an orthonormal basis;

(iii) the union of all orthonormal bases $\{\varphi_k\}$ corresponding to the eigenfrequencies of A is orthonormal in $\mathcal{E}$;

(iv) the same union $\{\varphi_k\}$ is an orthogonal basis in $\mathcal{E}_A$;

(v) for any $u \in \mathcal{E}$ there holds

$$Au = \sum_{k=1}^{\infty} \lambda_k^2 (u, \varphi_k)_\mathcal{E} \varphi_k, \qquad A\varphi_k = \lambda_k^2 \varphi_k. \tag{5.125}$$

We subdivide the proof into Lemmas 5.20 through 5.27. Statements (i) and (ii) follow from the Fredholm–Riesz–Schauder theory for compact operators. Statement (iii) follows from the self-adjointness of A. So we know some properties of the eigenvalues of A, but it remains unknown whether the set of eigenvectors is nonempty. First we demonstrate the existence of such an eigenvector.

Lemma 5.23. *For a self-adjoint strictly positive compact linear operator A acting in $\mathcal{E}$,*

$$\lambda_1^2 = \sup_{\|u\|_{\mathcal{E}} \le 1} (Au, u)_{\mathcal{E}}$$

is an eigenvalue of A. It is also the largest eigenvalue of A, and the lowest eigenfrequency of A is $\mu_1 = 1/\lambda_1$.

Proof. If λ^2 is an eigenvalue then $Au = \lambda^2 u$ and it follows that $(Au, u)_{\mathcal{E}} = \lambda^2 \|u\|_{\mathcal{E}}^2$. So for $\|u\|_{\mathcal{E}} \le 1$ we have $(Au, u)_{\mathcal{E}} \le \lambda_1^2$, and thus all the eigenvalues are nonnegative and less than or equal to $\lambda_1^2 > 0$. Let us demonstrate that λ_1^2 is an eigenvalue of A. By Lemmas 5.20 and 5.21 we know that $\sup(Au, u)_{\mathcal{E}}$ is attained on some point φ_1 of the ball $\|u\|_{\mathcal{E}} \le 1$. Since the form $(Au, u)_{\mathcal{E}}$ is homogeneous in u, we know that φ_1 belongs to the unit sphere $\|u\|_{\mathcal{E}} = 1$:

$$\lambda_1^2 = (A\varphi_1, \varphi_1)_{\mathcal{E}}, \qquad \|\varphi_1\|_{\mathcal{E}} = 1.$$

We show that φ_1 is an eigenvector of A. It is clear that λ_1^2 can be defined as the maximum of the form $(Au, u)_{\mathcal{E}}$ on the unit sphere $\|u\|_{\mathcal{E}} = 1$. By homogeneity the same can be said about the functional

$$G(u) = \frac{(Au, u)_{\mathcal{E}}}{\|u\|_{\mathcal{E}}^2} = (Av, v)_{\mathcal{E}}, \qquad v = \frac{u}{\|u\|_{\mathcal{E}}}, \qquad \|v\|_{\mathcal{E}} = 1.$$

Thus $G(u)$ takes the same set of values as $(Au, u)_{\mathcal{E}}$ on the unit sphere $\|u\|_{\mathcal{E}} = 1$ and, moreover, it attains its maximal value equal to λ_1^2 at the same point φ_1. Consider $G(\varphi_1 + \alpha w)$ for a fixed $w \in \mathcal{E}$. This is a function continuously differentiable in α in some neighborhood of $\alpha = 0$, and attaining its maximum at $\alpha = 0$. Thus

$$\left. \frac{dG(\varphi_1 + \alpha w)}{d\alpha} \right|_{\alpha=0} = 0.$$

Calculating this we get

$$(A\varphi_1, w)_{\mathcal{E}} - \frac{(A\varphi_1, \varphi_1)_{\mathcal{E}}}{\|\varphi_1\|_{\mathcal{E}}} (\varphi_1, w)_{\mathcal{E}} = 0;$$

that is,

$$\left(A\varphi_1 - \lambda_1^2 \varphi_1, w\right)_{\mathcal{E}} = 0$$

for any $w \in \mathcal{E}$. So φ_1 is an eigenvector and λ_1^2 is an eigenvalue of A. $\square$

Next we describe a procedure for finding other eigenvectors and eigenvalues of A, using the established property that the set of all eigenvectors of A has an orthonormal basis. We know how to find the first eigenvector. For the rest we shall use a procedure having ith step as follows. Let $\varphi_1, \ldots, \varphi_n$ be mutually orthogonal eigenvectors determined by the procedure. Denote by $\mathcal{E}_{n\perp}$ the orthogonal complement in $\mathcal{E}$ of the subspace of $\mathcal{E}$ spanned by $\varphi_1, \ldots, \varphi_n$. Consider the operator A given on $\mathcal{E}_{n\perp}$. We can repeat the reasoning of Lemma 5.23 and find an eigenvalue denoted by λ_{n+1}^2 and an eigenvector φ_{n+1} of the restriction of A to $\mathcal{E}_{n\perp}$. So

$$\left(A\varphi_{n+1} - \lambda_{n+1}^2 \varphi_{n+1}, w\right)_{\mathcal{E}} = 0 \tag{5.126}$$

holds for any $w \in \mathcal{E}_{n\perp}$. We show that this holds for any $w \in \mathcal{E}$. By the orthogonal decomposition theorem, it is enough to prove that (5.126) holds when w is any of the previous eigenvectors $\varphi_1, \ldots, \varphi_n$. Since for any $i < n+1$

$$(\varphi_{n+1}, \varphi_i)_{\mathcal{E}} = 0 \text{ and } (A\varphi_{n+1}, \varphi_i)_{\mathcal{E}} = (\varphi_{n+1}, A\varphi_i)_{\mathcal{E}} = \lambda_i^2 (\varphi_{n+1}, \varphi_i)_{\mathcal{E}} = 0,$$

it follows that (5.126) holds for any $w \in \mathcal{E}$. Hence we really did obtain the next eigenpair.

Lemma 5.24. *For an infinite dimensional space $\mathcal{E}$, the eigenvalues of A are countable. The corresponding eigenfrequencies $\mu_i = 1/\lambda_i$, $\lambda_i > 0$, are such that $\mu_i \le \mu_{i+1} \to +\infty$ as $i \to \infty$.*

Proof. The above procedure can terminate only when we get some subspace $\mathcal{E}_{n\perp}$ on the unit ball of which $\sup(Au, u)_{\mathcal{E}} = 0$. But then $\mathcal{E}_{n\perp}$ contains only the zero element since A is strictly positive. So $\mathcal{E}$ is finite dimensional, a contradiction. The rest of the lemma follows from the method of constructing the eigenvalues. □

Lemma 5.25. *The set of all the constructed eigenvectors $\varphi_1, \ldots, \varphi_n, \ldots$ is an orthonormal basis of $\mathcal{E}$.*

Proof. Take any $u \in \mathcal{E}$ and consider the remainder of the Fourier series

$$u_n = u - \sum_{k=1}^{n} (u, \varphi_k)_{\mathcal{E}} \varphi_k.$$

We see that $(u_n, \varphi_k)_{\mathcal{E}} = 0$ for $k \le n$, and thus $u_n \in \mathcal{E}_{n\perp}$. From Fourier expansion theory we know that $\{\sum_{k=1}^{n} (u, \varphi_k)_{\mathcal{E}} \varphi_k\}$ is convergent, hence so is $\{u_n\}$. Suppose, contrary to the lemma statement, that the strong limit of $\{u_n\}$ is $u_0 \neq 0$. By the procedure for finding eigenvalues and the fact that u_n is in $\mathcal{E}_{n\perp}$, we have

$$\frac{(Au_n, u_n)_\mathcal{E}}{\|u_n\|_\mathcal{E}^2} \leq \lambda_{n+1}^2.$$

Passage to the limit in n implies

$$\frac{(Au_0, u_0)_\mathcal{E}}{\|u_0\|_\mathcal{E}^2} \leq 0.$$

Hence $u_0 = 0$, which completes the proof. □

Lemma 5.26. *For any $u \in \mathcal{E}$, relation* (5.125) *holds:*

$$Au = \sum_{k=1}^{\infty} \lambda_k^2 (u, \varphi_k)_\mathcal{E} \varphi_k, \qquad A\varphi_k = \lambda_k^2 \varphi_k.$$

Proof. The Fourier series

$$u = \sum_{k=1}^{\infty} (u, \varphi_k)_\mathcal{E} \varphi_k$$

is strongly convergent. Applying a compact (and hence continuous) operator A we get

$$Au = \sum_{k=1}^{\infty} (u, \varphi_k)_\mathcal{E} A\varphi_k = \sum_{k=1}^{\infty} \lambda_k^2 (u, \varphi_k)_\mathcal{E} \varphi_k,$$

as required. □

The last non-proven statement of the theorem follows from

Lemma 5.27. *The set $\psi_k = \varphi_k/\lambda_k$, $\lambda_k > 0$, $k = 1, 2, 3, \ldots$, is an orthonormal basis of $\mathcal{E}_A$.*

Proof. Mutual orthogonality of the ψ_k in $\mathcal{E}_A$ follows from

$$(\psi_i, \psi_j)_A = (A\psi_i, \psi_j)_\mathcal{E} = \left(\frac{1}{\lambda_i} A\varphi_i, \frac{\varphi_j}{\lambda_j}\right)_\mathcal{E} = \frac{\lambda_i^2}{\lambda_i \lambda_j} (\varphi_i, \varphi_j)_\mathcal{E}.$$

Hence the set is orthonormal as well. For the proof it is enough to demonstrate that Parseval's equality holds in $\mathcal{E}_A$ for any $u \in \mathcal{E}$:

$$\begin{aligned}(u,u)_A &= (Au,u)_\mathcal{E} = \left(\sum_{k=1}^{\infty}(u,\varphi_k)_\mathcal{E} A\varphi_k, u\right)_\mathcal{E} = \sum_{k=1}^{\infty}(u,\varphi_k)_\mathcal{E}(A\varphi_k,u)_\mathcal{E} \\ &= \sum_{k=1}^{\infty}\left(u, \frac{A\varphi_k}{\lambda_k^2}\right)_\mathcal{E}(A\varphi_k,u)_\mathcal{E} = \sum_{k=1}^{\infty}(u,A\psi_k)_\mathcal{E}(A\psi_k,u)_\mathcal{E} \\ &= \sum_{k=1}^{\infty}(u,\psi_k)_A^2.\end{aligned}$$

□

5.13 The Fourier Method, Continued

We have obtained general results on the structure of the spectrum and the properties of the eigenvalue problem for a strictly positive, self-adjoint, compact linear operator A. This eigenvalue problem includes all the eigenvalue problems of linear mechanics that we have considered.

In § 5.10 we began to study the Fourier method for dynamical linear problems. We sought a general solution of a general linear initial-boundary value problem for a body free from external load. However, the fact that the eigenvectors of A, satisfying

$$\lambda_k^2(\varphi_k, v)_\mathcal{E} = \int_\Omega \varphi_k(\mathbf{x})v(\mathbf{x})\,d\Omega,$$

constitute an orthogonal basis in $\mathcal{E}$ and $\mathcal{E}_A$ simultaneously, allows us to consider the problem for a loaded body as well. Here the Fourier method appears to relate to the Faedo–Galerkin method for a special basis, namely for the eigenvectors of the operator A which is now well defined by (5.118). Let us recall that for the basis

$$(\varphi_i,\varphi_j)_\mathcal{E} = \delta_{ij} = \begin{cases}1, & i=j,\\ 0, & i\neq j,\end{cases} \qquad \int_\Omega \varphi_i(\mathbf{x})\varphi_j(\mathbf{x})\,d\Omega = \lambda_i^2\delta_{ij}. \tag{5.127}$$

Let us review the general notation of this section. In $\mathcal{E}(0,T)$ an inner product is defined as

$$(u,v)_{[0,T]} = \int_0^T (u,v)_\mathcal{E}\,dt + \int_0^T\int_\Omega \dot{u}(\mathbf{x},t)\dot{v}(\mathbf{x},t)\,d\Omega\,dt \tag{5.128}$$

(changing the dimensions we put $\rho = 1$) and D_0^T denotes the subspace that is the completion of that subset of the base functions of $\mathcal{E}(0,T)$ which vanish at $t = T$. A generalized solution $u \in \mathcal{E}(0,T)$ is defined by

$$\int_0^T (u,v)_{\mathcal{E}}\,dt = \int_0^T \int_\Omega f(\mathbf{x},t)v(\mathbf{x},t)\,d\Omega\,dt + \int_0^T \int_\Omega \dot{u}(\mathbf{x},t)\dot{v}(\mathbf{x},t)\,d\Omega\,dt + \int_\Omega u_1^*(\mathbf{x})v(\mathbf{x},0)\,d\Omega \tag{5.129}$$

for any $v \in D_0^T$. Note that the initial condition for the first time derivative, that is u_1^*, is taken into account in (5.129); we do not require it to hold separately. Another initial condition

$$u(\mathbf{x},t)\big|_{t=0} = u_0^*(\mathbf{x}) \tag{5.130}$$

must be satisfied in the sense of $L^2(\Omega)$; see Definition 5.8. The boundary conditions are hidden in the definition of $\mathcal{E}$. We recall that we require $u_0^*(\mathbf{x}) \in \mathcal{E}$, $u_1^*(\mathbf{x}) \in \mathcal{E}_A$, and $f(\mathbf{x},t) \in L^2(\Omega \times [0,T])$. Now we return to the Faedo–Galerkin method with the basis elements φ_k, $k = 1,2,\ldots$, that are eigenvectors of A with the properties studied earlier. Let us seek the nth Faedo–Galerkin approximation

$$u_n = \sum_{k=1}^{n} c_k(t)\varphi_k$$

to the generalized solution given by the equations

$$\ddot{c}_i(t)\int_\Omega \varphi_i^2(\mathbf{x})\,d\Omega = -(\varphi_i,\varphi_i)_{\mathcal{E}}c_i(t) + \int_\Omega f(\mathbf{x},t)\varphi_i(\mathbf{x})\,d\Omega, \qquad i = 1,\ldots,n \tag{5.131}$$

or, because of (5.127),

$$\ddot{c}_i(t) + \mu_i^2 c_i(t) = f_i(t), \quad \mu_i = 1/\lambda_i, \qquad i = 1,\ldots,n \tag{5.132}$$

where

$$f_i(t) = \mu_i^2 \int_\Omega f(\mathbf{x},t)\varphi_i(\mathbf{x})\,d\Omega$$

and eigenfrequencies $\mu_i = 1/\lambda_i \to \infty$. We see that equations (5.132) are mutually independent. Let us derive the initial conditions for these equations. Denoting $c_i(0) = d_{0i}$, $\dot{c}_i(0) = d_{1i}$, and remembering that d_{0i} are defined by

$$\left\| u_0^* - \sum_{k=1}^{n} d_{0k}\varphi_k \right\|_{\mathcal{E}}^2 \to \min$$

we get

$$d_{0i}(\varphi_i,\varphi_i)_{\mathcal{E}} = (u_0^*,\varphi_i)_{\mathcal{E}}$$

so

$$c_i(0) = d_{0i} = (u_0^*, \varphi_i)_\mathcal{E} = (u_0^*, \mu_i A\varphi_i)_\mathcal{E} = \mu_i \int_\Omega u_0^*(\mathbf{x})\varphi_i(\mathbf{x})\, d\Omega. \tag{5.133}$$

Similarly, minimizing

$$\left\| u_1^* - \sum_{k=1}^n d_{1k}\varphi_k \right\|_A^2 \to \min$$

we obtain

$$d_{1i}(\varphi_i, \varphi_i)_A = (\varphi_i, u_1^*)_A$$

or

$$\dot{c}_i(0) = d_{1i} = \mu_i^2(\varphi_i, u_1^*)_A = \mu_i^2 \int_\Omega u_1^*(\mathbf{x})\varphi_i(\mathbf{x})\, d\Omega \tag{5.134}$$

so we see that the initial conditions are split as well. Because of the mutual orthogonality and basis properties of $\{\varphi_i\}$ in $\mathcal{E}$ and $\mathcal{E}_A$ we can rewrite the corresponding Parseval equalities

$$\sum_{i=1}^\infty d_{0i}^2 = \|u_0^*\|_\mathcal{E}^2 \tag{5.135}$$

and

$$\sum_{i=1}^\infty d_{1i}^2(\varphi_i, \varphi_i)_A = \sum_{i=1}^\infty d_{1i}^2\lambda_i^2 = \|u_1^*\|_A^2 . \tag{5.136}$$

The solution of the problem (5.132), (5.133), (5.134) is

$$c_i(t) = d_{0i}\cos(\mu_i t) + d_{1i}\sin(\mu_i t) + \frac{1}{\mu_i}\int_0^t f_i(\tau)\sin\mu_i(t-\tau)\, d\tau.$$

It is easily seen that $c_i(t)$ is continuously differentiable on $[0, T]$. Note that, unlike the case of a general complete system of basis elements, the coefficients of the Faedo–Galerkin method do not depend on the step number. Let us examine the behavior of the corresponding partial sums of the formal series

$$u(\mathbf{x}, t) = \sum_{i=1}^\infty \Bigg(d_{0i}\cos(\mu_i t) + d_{1i}\sin(\mu_i t) + \frac{1}{\mu_i}\int_0^t f_i(\tau)\sin\mu_i(t-\tau)\, d\tau \Bigg)\varphi_i(\mathbf{x}). \tag{5.137}$$

Let us note that the portion

$$u(\mathbf{x},t) = \sum_{i=1}^{\infty} \left(d_{0i}\cos(\mu_i t) + d_{1i}\sin(\mu_i t)\right)\varphi_i(\mathbf{x})$$

is a formal solution for the dynamical problem for a load-free elastic body by the Fourier method. From the above we know these partial sums converge weakly to a generalized solution of the dynamical problem. So in a certain way $u(\mathbf{x},t)$ given formally by (5.137) is this solution. We will establish the properties of the series (5.137) and hence those of the generalized solution.

Let us consider the convergence of series (5.137). Multiply the identity (5.131) termwise by $\dot{c}_i(t)$ and sum over i:

$$\sum_{i=1}^{n} \ddot{c}_i(t)\dot{c}_i(t)\int_\Omega \varphi_i^2(\mathbf{x})\,d\Omega + \sum_{i=1}^{n} c_i(t)\dot{c}_i(t)(\varphi_i,\varphi_i)_\mathcal{E}$$
$$= \sum_{i=1}^{n}\int_\Omega f(\mathbf{x},t)\dot{c}_i(t)\varphi_i(\mathbf{x})\,d\Omega$$

or

$$\frac{1}{2}\frac{d}{dt}\left(\sum_{i=1}^{n}\dot{c}_i^2(t)\int_\Omega \varphi_i^2(\mathbf{x})\,d\Omega + \sum_{i=1}^{n} c_i^2(t)(\varphi_i,\varphi_i)_\mathcal{E}\right)$$
$$= \int_\Omega f(\mathbf{x},t)\left(\sum_{i=1}^{n}\dot{c}_i(t)\varphi_i(\mathbf{x})\right)d\Omega.$$

We used this procedure in obtaining the estimate of the Faedo–Galerkin approximation. So redenoting t by τ and integrating the last equality in τ over $[0,t]$ we get

$$\frac{1}{2}\left(\sum_{i=1}^{n}\dot{c}_i^2(t)\int_\Omega \varphi_i^2(\mathbf{x})\,d\Omega + \sum_{i=1}^{n} c_i^2(t)(\varphi_i,\varphi_i)_\mathcal{E}\right)$$
$$= \frac{1}{2}\left(\sum_{i=1}^{n}\dot{c}_i^2(0)\int_\Omega \varphi_i^2(\mathbf{x})\,d\Omega + \sum_{i=1}^{n} c_i^2(0)(\varphi_i,\varphi_i)_E\right)$$
$$+ \int_0^t\int_\Omega f(\mathbf{x},\tau)\left(\sum_{i=1}^{n}\dot{c}_i(\tau)\varphi_i(\mathbf{x})\right)d\Omega\,d\tau$$

and therefore

$$\frac{1}{2}\left(\sum_{i=1}^{n}\dot{c}_i^2(t)\int_\Omega \varphi_i^2(\mathbf{x})\,d\Omega+\sum_{i=1}^{n}c_i^2(t)(\varphi_i,\varphi_i)_\varepsilon\right)$$
$$\le \frac{1}{2}\sum_{i=1}^{n}\left(d_{1i}^2\lambda_i^2+d_{0i}^2\right)+T\int_0^t\int_\Omega f^2(\mathbf{x},\tau)\,d\Omega\,d\tau$$
$$+\frac{1}{4T}\int_0^t\int_\Omega\left(\sum_{i=1}^{n}\dot{c}_i(\tau)\varphi_i(\mathbf{x})\right)^2 d\Omega\,d\tau$$
$$=\frac{1}{2}\sum_{i=1}^{n}\left(d_{1i}^2\lambda_i^2+d_{0i}^2\right)+T\int_0^t\int_\Omega f^2(\mathbf{x},\tau)\,d\Omega\,d\tau$$
$$+\frac{1}{4T}\int_0^t\sum_{i=1}^{n}\dot{c}_i^2(\tau)\left(\int_\Omega\varphi_i^2(\mathbf{x})\,d\Omega\right)^2 d\tau.$$

Here we used the elementary inequality

$$|ab|\le\frac{a^2}{4T}+Tb^2$$

and mutual orthogonality of the φ_i in $\mathcal{E}_A=L^2(\Omega)$. Taking maximum values on $[0,T]$ in the last inequalities we get

$$\frac{1}{2}\max_{t\in[0,T]}\left(\sum_{i=1}^{n}\dot{c}_i^2(t)\int_\Omega \varphi_i^2(\mathbf{x})\,d\Omega+\sum_{i=1}^{n}c_i^2(t)(\varphi_i,\varphi_i)_\varepsilon\right)$$
$$\le \frac{1}{2}\sum_{i=1}^{n}\left(d_{1i}^2\lambda_i^2+d_{0i}^2\right)+T\int_0^T\int_\Omega f^2(\mathbf{x},\tau)\,d\Omega\,d\tau$$
$$+\frac{1}{4T}T\max_{\tau\in[0,T]}\sum_{i=1}^{n}\dot{c}_i^2(\tau)\left(\int_\Omega\varphi_i^2(\mathbf{x})\,d\Omega\right)^2$$

so

$$\frac{1}{2}\max_{t\in[0,T]}\left(\frac{1}{2}\sum_{i=1}^{n}\dot{c}_i^2(t)\int_\Omega \varphi_i^2(\mathbf{x})\,d\Omega+\sum_{i=1}^{n}c_i^2(t)(\varphi_i,\varphi_i)_\varepsilon\right)$$
$$\le \frac{1}{2}\sum_{i=1}^{n}\left(d_{1i}^2\lambda_i^2+d_{0i}^2\right)+T\int_0^T\int_\Omega f^2(\mathbf{x},\tau)\,d\Omega\,d\tau. \tag{5.138}$$

The right side of (5.138), by (5.135) and (5.136), is bounded by some constant M independent of n. By orthogonality of the basis elements and the form of the norm of a partial sum for series (5.137), which is

$$u_n(\mathbf{x},t)=\sum_{i=1}^{n}c_i(t)\varphi_i(\mathbf{x}),$$

we know that $\{u_n\}$ converges in $C(\mathcal{E};0,T)$ and $\{du_n/dt\}$ converges in $C(\mathcal{E}_A;0,T) = C(L^2(\Omega);0,T)$. Thus the series (5.137), which is also a generalized solution to the problem under consideration, belongs to $C(\mathcal{E};0,T)$, whereas its time derivative $\partial u/\partial t$ belongs to $C(\mathcal{E}_A;0,T)$. Simultaneously we justified convergence of the Fourier method for a free-load dynamical problem for an elastic body. Assuming existence of time derivatives of the force term f, in the same way we can show that the solution has additional time derivatives. Moreover, for the free-load case we can show that the time derivative of any order of the solution is in $C(\mathcal{E}_A;0,T)$.

5.14 Equilibrium of a von Kármán Plate

So far we have considered only linear mechanical problems. Of course, these represent only simple approximations to natural processes: although some weakly nonlinear processes can be analyzed with sufficient accuracy via linear models, many important physical effects are inherently nonlinear. It is fortunate that the speed of machine computation has increased to the point where more realistic simulation has become possible. But the availability of numerical methods has also underscored the importance of analytical considerations. To work effectively we must know whether a solution exists and to which function space it belongs. We should also understand the differences between various solution methods and be prepared to place rigorous bounds on the error.

An important nonlinear problem, and one that can be regarded as a touchstone for many numerical methods, is the equilibrium problem for a thin elastic plate under transverse load q. The plate is described by two nonlinear equations,

$$D\Delta^2 w = [f,w] + q, \tag{5.139}$$

$$\Delta^2 f = -[w,w], \tag{5.140}$$

given over a two-dimensional region Ω representing the mid-surface of the plate. Here $w = w(x,y)$ is the transverse displacement of a point (x,y) of the mid-surface, $f = f(x,y)$ is the Airy stress function, D is the rigidity coefficient of the plate, and the notation $[u,v]$ is defined by

$$[u,v] = u_{xx}v_{xx} + u_{yy}v_{yy} - 2u_{xy}v_{xy} \tag{5.141}$$

where the subscripts x and y denote the partial derivatives $\partial/\partial x$ and $\partial/\partial y$, respectively. With suitably chosen dimensionless variables we can get $D =$

1. We shall consider the problem with the boundary conditions

$$w\big|_{\partial\Omega} = 0 = \frac{\partial w}{\partial n}\bigg|_{\partial\Omega} \tag{5.142}$$

and

$$f\big|_{\partial\Omega} = 0 = \frac{\partial f}{\partial n}\bigg|_{\partial\Omega}. \tag{5.143}$$

Conditions (5.142) mean that the edge of the plate is fixed against transverse displacement and rotation, and (5.143) means that the lateral boundary is not subjected to tangential load. In mechanics, (5.143) is derived for a simply connected domain. As usual we take Ω to be compact with a piecewise smooth boundary so that Sobolev's imbedding theorem for $W^{2,2}(\Omega)$ applies. If we neglect the term $[f, w]$ in (5.139), we get the linear equation of equilibrium of a plate under transverse load as considered in Chapter 4. We would like to apply the tools of generalized setup of mechanical problems. Let us begin with the pair of integro-differential equations

$$a(w, \zeta) = B(f, w, \zeta) + \iint_\Omega q\zeta \, d\Omega, \tag{5.144}$$

$$a(f, \eta) = -B(w, w, \eta), \tag{5.145}$$

where

$$a(u, v) = \iint_\Omega (u_{xx}(v_{xx} + \mu v_{yy}) + 2(1 - \mu)u_{xy}v_{xy} + u_{yy}(v_{yy} + \mu v_{xx}))\, d\Omega,$$

μ is Poisson's ratio for the material ($0 < \mu < 1/2$), and

$$B(u, v, \varphi) = \iint_\Omega ((u_{xy}v_y - u_{yy}v_x)\varphi_x + (u_{xy}v_x - u_{xx}v_y)\varphi_y)\, d\Omega.$$

From a variational perspective, (5.144)–(5.145) would appear to constitute the first variation of some functional; we could regard ζ and η as arbitrary admissible smooth variations of w and f. Because such a viewpoint would return us to (5.139)–(5.140), we could try (5.144)–(5.145) as equations appropriate for the generalized setup. Other forms of the bilinear functional $a(u, v)$ may also yield (5.139)–(5.140) as a consequence of the variational technique; however, for types of boundary conditions that differ from (5.142) this would lead to incorrect natural boundary conditions. If we wish to consider boundary conditions for f including tangential load, we must take a different form of the left side in (5.145) (see, for example, [31]). But for conditions (5.143) we can forget about the physical meaning of the Airy function and use the same form of $a(u, v)$ in the generalized equation. Hence we shall use (5.144)–(5.145) for the generalized setup of the equilibrium problem for von Kármán's plate. Experience with the linear equilibrium problem for a plate suggests that we exploit the form $a(u, v)$ as an inner product in "energy" spaces for w and f. This means, by the results for a linear plate, that the solution will be sought in the subspace of $W^{2,2}(\Omega)$ consisting of the functions satisfying the boundary conditions (5.142). We need to see whether the terms of (5.144)–(5.145) make sense when the functions included therein reside in the energy spaces (note that we now consider dimensionless versions of the equations). Of course, we suppose that q satisfies at least the same conditions as for the generalized setup of the corresponding linear plate problem. For definiteness, let $q \in L(\Omega)$. We will check that the other terms in the equations make sense. It is necessary to consider only the trilinear form $B(u, v, w)$. Apply Hölder's inequality for three functions to a typical term:

$$\left| \iint_\Omega u_{xx} v_y w_x \, d\Omega \right| \leq \left(\iint_\Omega u_{xx}^2 \, d\Omega \right)^{1/2} \left(\iint_\Omega v_y^4 \, d\Omega \right)^{1/4} \left(\iint_\Omega w_x^4 \, d\Omega \right)^{1/4}$$
$$\leq m \, \|u\|_P \, \|v\|_P \, \|w\|_P \, , \tag{5.146}$$

where we have used the fact that in $\mathcal{E}_{Pc}$ the norm

$$\|w\|_P = (a(w, w))^{1/2}$$

is equivalent to the norm of $W^{2,2}(\Omega)$ and elements of $W^{2,2}(\Omega)$ have the first derivatives belonging to $L^p(\Omega)$ with any finite $p > 1$, in particular for $p = 4$, which is necessary in Hölder's inequality. So all terms make sense in the energy space.

Definition 5.28. A *generalized solution* to the equilibrium problem is a pair w, f belonging to $\mathcal{E}_{Pc} \times \mathcal{E}_{Pc}$ and satisfying (5.144)–(5.145) for any $\zeta, \eta \in \mathcal{E}_{Pc}$.

Equation (5.145) is linear in f. Using this we will eliminate f from the explicit statement of the problem. The right side of (5.145) is linear in η; estimates of the type (5.146) give us

$$|B(w, w, \eta)| \leq m \, \|w\|_P^2 \, \|\eta\|_P \, . \tag{5.147}$$

This means $B(w, w, \eta)$ is continuous in η so we can apply Theorem 4.100 and state that for any fixed $w \in \mathcal{E}_{Pc}$

$$-B(w, w, \eta) = (C, \eta)_P = a(C, \eta). \tag{5.148}$$

This $C \in \mathcal{E}_{Pc}$, uniquely defined by w, is considered as the value of an operator in $\mathcal{E}_{Pc}$ at w: $C = C(w)$. Then (5.145) is rewritten as

$$a(f, \eta) = a(C(w), \eta) \tag{5.149}$$

and thus $f = C(w)$. We will make further use of this.

Let us call a *nonlinear* operator in a Hilbert space *completely continuous* if it takes any weakly Cauchy sequence into a strongly Cauchy sequence.

Lemma 5.29. *The operator $C(w)$ is completely continuous in $\mathcal{E}_{Pc}$.*

The proof is based on the following symmetry property of the trilinear form $B(u, v, w)$.

Lemma 5.30. *For any $u, v, w \in \mathcal{E}_{Pc}$,*

$$\begin{aligned} B(u, v, w) &= B(w, u, v) = B(v, w, u) = B(v, u, w) \\ &= B(w, v, u) = B(u, w, v). \end{aligned} \tag{5.150}$$

Proof. We introduced the energy spaces as completions of the sets of functions satisfying appropriate boundary conditions and having all the continuous derivatives (in this case up to second order) that are included in the energy expression for the body. However, the set of infinitely differentiable functions is dense in subspaces of $C^{(k)}(\Omega)$, and this means we can use it as a base to get a corresponding energy space (in other words, among representative Cauchy sequences of an element of an energy space there are those which consist of infinitely differentiable functions only). The validity of (5.150) is shown by direct integration by parts for functions u, v, w having all the third continuous derivatives (they cancel mutually after transformations). Taking then representative Cauchy sequences for elements $u, v, w \in \mathcal{E}_{Pc}$ that have infinitely differentiable members we get the needed property by the limit passage in the equalities (5.150) written for the members. Equation (5.147) justifies the limit passage. □

Proof. [Proof of Lemma 5.29] By (5.150) and (5.148), for any $\eta \in \mathcal{E}_{Pc}$ we have

$$(C(w), \eta)_P = -B(w, w, \eta) = -B(\eta, w, w). \tag{5.151}$$

Let $\{w_n\}$ be a weakly Cauchy sequence in $\mathcal{E}_{Pc}$ and thus $\|w_n\|_P < c_0$ with c_0 independent of n. We must show that $\{C(w_n)\}$ is a strongly Cauchy sequence. From (5.151) it follows that

$$|(C(w_{n+m}) - C(w_n), \eta)_P| = |B(\eta, w_{n+m}, w_{n+m}) - B(\eta, w_n, w_n)|. \tag{5.152}$$

Consider a typical pair of corresponding terms of the right side:

$$\begin{aligned}
&\left|\iint_\Omega \eta_{xx}(w_{n+m_y}w_{n+m_x} - w_{ny}w_{nx})\,d\Omega\right| \\
&= \left|\iint_\Omega \eta_{xx}(w_{n+m_y}w_{n+m_x} - w_{n+m_y}w_{nx} + w_{n+m_y}w_{nx} - w_{ny}w_{nx})\,d\Omega\right| \\
&\le \left|\iint_\Omega \eta_{xx}w_{n+m_y}(w_{n+m_x} - w_{nx})\,d\Omega\right| + \left|\iint_\Omega \eta_{xx}w_{nx}(w_{n+m_y} - w_{ny})\,d\Omega\right|.
\end{aligned}$$

Applying Hölder's inequality to each term on the right as in (5.146), we have

$$\begin{aligned}
&\left|\iint_\Omega \eta_{xx}(w_{n+m_y}w_{n+m_x} - w_{ny}w_{nx})\,d\Omega\right| \\
&\le \left(\iint_\Omega \eta_{xx}^2\,d\Omega\right)^{1/2}\left(\iint_\Omega w_{n+m_y}^4\,d\Omega\right)^{1/4}\left(\iint_\Omega (w_{n+m_x} - w_{nx})^4\,d\Omega\right)^{1/4} \\
&\quad + \left(\iint_\Omega \eta_{xx}^2\,d\Omega\right)^{1/2}\left(\iint_\Omega w_{nx}^4\,d\Omega\right)^{1/4}\left(\iint_\Omega (w_{n+m_y} - w_{ny})^4\,d\Omega\right)^{1/4} \\
&\le M\,\|\eta\|_P\,c_0\left(\left\|w_{n+m_x} - w_{nx}\right\|_{L^4(\Omega)} + \left\|w_{n+m_y} - w_{ny}\right\|_{L^4(\Omega)}\right)
\end{aligned}$$

with a constant M defined by the imbedding theorem for $\mathcal{E}_{Pc}$. Doing this for each corresponding pair on the right side of (5.152) we get

$$\begin{aligned}
&|(C(w_{n+m}) - C(w_n), \eta)_P| \\
&\le M_1\,\|\eta\|_P\left(\left\|w_{n+m_x} - w_{nx}\right\|_{L^4(\Omega)} + \left\|w_{n+m_y} - w_{ny}\right\|_{L^4(\Omega)}\right)
\end{aligned}$$

Putting $\eta = C(w_{n+m}) - C(w_n)$ we get

$$\begin{aligned}
&|(C(w_{n+m}) - C(w_n), C(w_{n+m}) - C(w_n))_P| \\
&\le M_1\,\|C(w_{n+m}) - C(w_n)\|_P \cdot \\
&\qquad \cdot\left(\left\|w_{n+m_x} - w_{nx}\right\|_{L^4(\Omega)} + \left\|w_{n+m_y} - w_{ny}\right\|_{L^4(\Omega)}\right)
\end{aligned}$$

or

$$\begin{aligned}
&\|C(w_{n+m}) - C(w_n)\|_P \\
&\le M_1\left(\left\|w_{n+m_x} - w_{nx}\right\|_{L^4(\Omega)} + \left\|w_{n+m_y} - w_{ny}\right\|_{L^4(\Omega)}\right). \qquad (5.153)
\end{aligned}$$

But by Sobolev's imbedding theorem for $W^{2,2}(\Omega)$, which also applies to its subspace $\mathcal{E}_{Pc}$, we know that for a sequence $\{w_n\}$ weakly convergent in $\mathcal{E}_{Pc}$,

$$\left\|w_{n+m_x} - w_{nx}\right\|_{L^4(\Omega)} + \left\|w_{n+m_y} - w_{ny}\right\|_{L^4(\Omega)} \to 0 \quad \text{as } n \to \infty.$$

This implies the needed statement of the lemma:

$$\|C(w_{n+m}) - C(w_n)\|_P \to 0 \quad \text{as } n \to \infty. \qquad \square$$

Now we return to the generalized setup and eliminate $f = C(w)$ from the statement. Then (5.144)–(5.145) reduce to the single equation

$$(w, \zeta)_P = B(C(w), w, \zeta) + \iint_\Omega q\zeta \, d\Omega. \tag{5.154}$$

Let us present (5.154) in operator form. Consider the right side of (5.154) as a functional in ζ at a fixed w. It is linear in ζ. Next we get

$$\begin{aligned}\left| B(C(w), w, \zeta) + \iint_\Omega q\zeta \, d\Omega \right| &\leq m_1 \|C(w)\|_P \|w\|_P \|\zeta\|_P \\ &\quad + \max_\Omega |\zeta| \iint_\Omega |q| \, d\Omega \\ &\leq m_2 \|\zeta\|_P\end{aligned}$$

where we have used a consequence of inequality (5.146), the inequality

$$\|C(w)\|_P \leq M_1 \|w\|_P^2 \tag{5.155}$$

that can be obtained in the same fashion as (5.153) with use of Sobolev's imbedding theorem in $W^{2,2}(\Omega)$. This means the right side of (5.154) is a continuous linear functional in $\zeta \in \mathcal{E}_{Pc}$. Applying Theorem 4.100 we get

$$B(C(w), w, \zeta) + \iint_\Omega q\zeta \, d\Omega = (G, \zeta)_P$$

where $G \in \mathcal{E}_{Pc}$ is uniquely defined by $w \in \mathcal{E}_{Pc}$. Thus G can be considered as the result of an operator $G = G(w)$ acting in $\mathcal{E}_{Pc}$. Then (5.154) becomes

$$(w, \zeta)_P = (G(w), \zeta)_P$$

and so, by the arbitrariness of $\zeta \in \mathcal{E}_{Pc}$, we get an operator equation

$$w = G(w) \tag{5.156}$$

where G is a nonlinear operator in $\mathcal{E}_{Pc}$.

Lemma 5.31. *The operator G is completely continuous in $\mathcal{E}_{Pc}$; that is, it takes any weakly Cauchy sequence into a strongly Cauchy sequence.*

The proof practically repeats all the steps of the proof of Lemma 5.29 (and in fact is easier since C is a completely continuous operator).

To use the calculus of variations we should present (5.156) as the equality of the first variation of some functional to zero. The appropriate functional is

$$F(w) = \frac{1}{2}a(w,w) + \frac{1}{4}a(C(w), C(w)) - \iint_\Omega qw\, d\Omega. \tag{5.157}$$

Let us introduce

Definition 5.32. Suppose a functional Φ at point x in a real Hilbert space H can be represented as

$$\Phi(x+y) - \Phi(x) = (K(x), y)_H + o(\|y\|_H) \tag{5.158}$$

for any y, $\|y\|_H \le \varepsilon$ with some small $\varepsilon > 0$. The correspondence from x to $K(x)$ is called the *gradient* of W and is denoted as $\operatorname{grad}\Phi(x) = K(x)$.

This is a way of representing the first variation of a functional in a real Hilbert space, which was the central point of Chapter 1. The main term in the representation can often be found by formal differentiation with respect to a parameter t:

$$(K(x), y)_H = \frac{d}{dt}\Phi(x+ty)\bigg|_{t=0}. \tag{5.159}$$

For example, the gradient of the functional $\frac{1}{2}\|x\|_H^2$ is the identity operator:

$$\frac{d}{dt}\left(\frac{1}{2}(x+ty, x+ty)\right)_H\bigg|_{t=0} = (x,y)_H.$$

The reader can check this by direct calculation according to Definition 5.32. As in Chapter 1, we have

Lemma 5.33. *Suppose a functional $\Phi(x)$ has at any point x of a real Hilbert space H a continuous gradient $K(x)$. If $\Phi(x)$ attains a minimum at x_0, then $K(x_0) = 0$.*

Proof. For any fixed y and small t, by (5.158) we have

$$0 \le \Phi(x_0 + ty) - \Phi(x_0) = t(K(x_0), y)_H + o(|t|).$$

From this inequality we conclude, as is standard reasoning in Chapter 1, that $(K(x_0), y)_H = 0$. Hence $K(x_0) = 0$ by the arbitrariness of y. □

Note that we derived a version of the Euler equation for an abstract functional. The points x at which $K(x) = 0$ are called *critical points* of Φ. Thus points of minimum of a smooth functional Φ are its critical points. Let us apply this to our equation.

Theorem 5.34. *Let $q \in L(\Omega)$. There exists a generalized solution $w_0, f_0 \in \mathcal{E}_{Pc}$ to the equilibrium problem for von Kármán's plate with boundary conditions (5.142)–(5.143). The element w_0 is a point of minimum of the functional $F(w)$ defined by (5.157).*

We present the proof as several lemmas.

Lemma 5.35. *At any $w \in \mathcal{E}_{Pc}$ we have* $\operatorname{grad} F(w) = w - G(w)$.

Proof. Let us consider $F(w + t\zeta)$ at any fixed $w, \zeta \in \mathcal{E}_{Pc}$. In t this is a simple polynomial so we can define $\operatorname{grad} F$ by (5.159). Consider

$$\begin{aligned}\left.\frac{d}{dt}F(w+t\zeta)\right|_{t=0} &= \frac{d}{dt}\left(\frac{1}{2}a(w+t\zeta, w+t\zeta)\right.\\ &\qquad \left.\left. + \frac{1}{4}a(C(w+t\zeta), C(w+t\zeta)) - \iint_\Omega q(w+t\zeta)\,d\Omega\right)\right|_{t=0}\\ &= a(w,w) + \frac{1}{2}a\left.\left(\frac{dC(w+t\zeta)}{dt}, C(w)\right)\right|_{t=0} - \iint_\Omega q\zeta\,d\Omega.\end{aligned} \tag{5.160}$$

From (5.151), using the symmetry of its right side in w, we have

$$a\left.\left(\frac{dC(w+t\zeta)}{dt}, \eta\right)\right|_{t=0} = -\frac{d}{dt}B(\eta, w+t\zeta, w+t\zeta)\bigg|_{t=0} = -2B(\eta, w, \zeta).$$

So

$$a\left.\left(\frac{dC(w+t\zeta)}{dt}, C(w)\right)\right|_{t=0} = -2B(C(w), w, \zeta).$$

Combining this with (5.160) we get

$$\left.\frac{d}{dt}F(w+t\zeta)\right|_{t=0} = a(w,\zeta) - B(C(w), w, \zeta) - \iint_\Omega q\zeta\,d\Omega = (w - G(w), \zeta)_P,$$

which completes the proof. □

From this and the above we get

Lemma 5.36. *Any critical point $w \in \mathcal{E}_{Pc}$ of the functional F given by (5.157) implies the pair $w, f = C(w)$ is a generalized solution of the problem under consideration.*

Now we show that there is a point at which $F(w)$ attains its minimum. First we note that this minimum point is in a ball centered at the origin and having radius defined only by the load q. This follows from

$$\begin{aligned} 2F(w) &\geq a(w,w) - 2\left|\iint_\Omega qw\,d\Omega\right| \\ &\geq \|w\|_P^2 - 2\max_\Omega |w| \iint_\Omega |q|\,d\Omega \\ &\geq \|w\|_P^2 - M_0\,\|w\|_P\,, \end{aligned} \tag{5.161}$$

where the constant M_0 is defined by the norm of q in $L(\Omega)$ and the norm of the imbedding operator from $\mathcal{E}_{Pc}$ to $C(\Omega)$. Since $F(0) = 0$ and outside the sphere $\|w\|_P = M_0 + 1$, we have $F(w) \geq M_0 + 1$ and thus

Lemma 5.37. *If there is a minimum point of the functional F, then it belongs to the ball $\|w\|_P < M_0 + 1$. Moreover, the functional F is growing in $\mathcal{E}_{Pc}$; i.e., $F(w) \to \infty$ as $\|w\|_P \to \infty$.*

The fact that F is a growing functional follows immediately from (5.161). Now we must prove that F attains its limit point.

Lemma 5.38. *The functional*

$$\Phi(w) = \frac{1}{4}a(C(w), C(w)) - \iint_\Omega qw\,d\Omega$$

is weakly continuous in $\mathcal{E}_{Pc}$, thus the functional $F(w)$ is represented as

$$F(w) = \frac{1}{2}\|w\|_P^2 + \Phi(w)$$

with a weakly continuous functional Φ.

Proof. Evident since $\int_\Omega qw\,d\Omega$ is a continuous linear functional and C is a completely continuous operator. □

The proof of Theorem 5.34 is completed by the following result due to Tsitlanadze:

Theorem 5.39. *Let $f(x)$ be a growing functional in a Hilbert space H that has the form*

$$f(x) = \|x\|_H^2 + \varphi(x)$$

where $\varphi(x)$ is a weakly continuous functional in H. Then

(i) *there is a point x_0 at which $f(x)$ attains its absolute minimum, i.e., $f(x_0) \leq f(x)$ for any $x \in H$;*

(ii) any sequence $\{x_n\}$ minimizing f, so that

$$\lim_{n\to\infty} f(x_n) = f(x_0),$$

contains a subsequence strongly convergent to x_0.

Proof. On any ball $\varphi(x)$ is bounded and thus $f(x)$ is bounded as well. Because $f(x)$ is growing we state that a possible minimum point is inside a closed ball B of a radius R. Let a be the infimum of values of $f(x)$. Then

$$\inf_{x\in H} f(x) = \inf_{\|x\|_H \le R} f(x) = a.$$

Take a minimizing sequence $\{x_n\}$ of f. We can consider it is inside B and thus contains a weakly convergent subsequence that we redenote by $\{x_n\}$ again. Without loss of generality, we can consider the sequence of norms of x_n to converge to b, such that $b \le R$. Since a closed ball centered at the origin is weakly closed we know that $\{x_n\}$ converges weakly to an element $x_0 \in B$. Now it is enough to show that $\{x_n\}$ converges strongly to x_0. We know that if for a weak Cauchy sequence the sequence of norms of the elements converges to the norm of the weak limit element, then it converges strongly. Thus we must show only that $\|x_0\|_H = b$. It is clear that

$$\|x_0\|_H \le b.$$

Indeed, because of weak convergence of $\{x_n\}$ to x_0 we have

$$\|x_0\|_H^2 = \lim_{n\to\infty} (x_n, x_0)_H \le \|x_0\|_H \lim_{n\to\infty} \|x_n\|_H = b\, \|x_0\|_H \,.$$

Next, because of weak continuity of φ we have

$$\lim_{n\to\infty} \varphi(x_n) = \varphi(x_0)$$

and thus

$$a = \lim_{n\to\infty} f(x_n) = \lim_{n\to\infty} \left(\|x_n\|_H^2 + \varphi(x_n)\right) = b^2 + \varphi(x_0).$$

But

$$f(x_0) = \|x_0\|_H^2 + \varphi(x_0) \ge a$$

so $\|x_0\|_H^2 \ge b^2$ which means that $\|x_0\|_H = b$. All statements of the theorem are proved. □

By this theorem the proof of Theorem 5.34 is also completed. Note that Theorem 5.39 prepared everything to formulate the theorem on convergence of the Ritz approximations to a generalized solution of the problem under consideration. We leave this to the reader.

5.15 A Unilateral Problem

Let us consider deformation of a membrane constrained by a surface beneath it. The membrane cannot penetrate the surface. Such a *unilateral problem* can be formulated as a problem involving a so-called variational inequality. By this approach we obtain problems with free boundaries; i.e., the boundary of the domain over which some equations are applicable is determined during solution, not in advance. Our previous use of the term "free" indicated a lack of geometrical constraints on the displacements. Now there is an obstacle, and the border of contact between this obstacle and the membrane is undetermined (free).

Consider a membrane under load f occupying a compact domain Ω with clamped edge. Beneath the membrane there is an obstacle described by a function $\varphi = \varphi(x, y)$. The obstacle is impenetrable so that

$$u(x, y) \geq \varphi(x, y) \tag{5.162}$$

for all $(x, y) \in \Omega$. Let the clamped edge of the membrane be described by

$$u\big|_{\partial\Omega} = a(s) \tag{5.163}$$

where for the sake of compatibility between the boundary condition and the obstacle

$$\varphi\big|_{\partial\Omega} \leq a(s). \tag{5.164}$$

If the membrane lays against the obstacle φ, it must take the form of the obstacle over a domain called a *coincidence set*. Mechanically it is clear that the membrane equation should not be applied over such a set (in fact it does hold but contains an unknown force reaction of the obstacle). We do not know beforehand how to determine a coincidence set, its border, or the conditions for a solution on the border.

Classical setup of the problem

Let us attempt to apply the calculus of variations. As we would like to obtain a classical statement of the problem, we suppose all the functions employed are sufficiently smooth. Since the mechanics of the problem ensures applicability of the minimum total energy principle, a solution minimizes the energy functional

$$F(u) = \frac{1}{2} \iint_\Omega \left[\left(\frac{\partial u}{\partial x}\right)^2 + \left(\frac{\partial u}{\partial y}\right)^2 \right] d\Omega - \iint_\Omega f u \, d\Omega$$

over the set of functions satisfying (5.162)–(5.163). Supposing there is a solution belonging to $C^{(2)}(\Omega)$, we will find equations for a minimizer over the subset of $C^{(2)}(\Omega)$ consisting of functions satisfying (5.162)–(5.163). We denote this subset by C_φ. Note that we must assume $\varphi \in C^{(2)}(\Omega)$ as well. Later we will "forget" this requirement. Hence we seek equations governing a minimizer $u \in C_\varphi$ of $F(u)$ over C_φ. It is clear that the set C_φ is convex in $C^{(2)}(\Omega)$, which means that if u_1 and u_2 belong to C_φ then for any $t \in [0,1]$ we have $(1-t)u_1 + tu_2 \in C_\varphi$. Let us take an arbitrary $v \in C_\varphi$. By convexity we see that $u + t(v-u) = (1-t)u + tv$ belongs to C_φ for any $t \in [0,1]$ as well. So by the principle of minimum total energy we have

$$F(u + t(v-u)) \geq F(u)$$

for any $v \in C_\varphi$ and $t \in [0,1]$. Remembering the notation

$$(u,v)_M = \iint_\Omega \left(\frac{\partial u}{\partial x}\frac{\partial v}{\partial x} + \frac{\partial u}{\partial y}\frac{\partial v}{\partial y} \right) d\Omega \tag{5.165}$$

we have

$$\frac{1}{2}(u + t(v-u), u + t(v-u))_M - \frac{1}{2}(u,u)_M - t\iint_\Omega f(v-u)\, d\Omega \geq 0$$

or

$$t\left[(u, v-u)_M - \iint_\Omega f(v-u)\, d\Omega\right] + \frac{1}{2}t^2(v-u, v-u)_M \geq 0 \tag{5.166}$$

for any $t \in [0,1]$. For a fixed v, the coefficient of t is nonnegative:

$$(u, v-u)_M - \iint_\Omega f(v-u)\, d\Omega \geq 0. \tag{5.167}$$

Otherwise, choosing sufficiently small t, we find that (5.166) is violated since t^2 tends to zero faster than t as $t \to 0$. Hence a minimizer u must satisfy (5.167) for any $v \in C_\varphi$. This is an example of a *variational inequality*. Denote $\eta = v - u$. It is clear that on the boundary

$$\eta\big|_{\partial\Omega} = 0. \tag{5.168}$$

Then (5.167) takes the form

$$(u,\eta)_M - \iint_\Omega f\eta\, d\Omega \geq 0. \tag{5.169}$$

The left side of (5.169) is the first variation of functional F with virtual displacement η. In the calculus of variations, from (5.169) we stated that the first variation is equal to zero for any η. This was done because η was sufficiently arbitrary; this time, however, we have $\eta \geq 0$ on the coincidence set for u, so we cannot use the trick involving a sign change on η to obtain an equality in (5.169). Let us derive the differential equations from (5.169). Traditional integration by parts with regard for (5.168) yields

$$\iint_{\Omega} (-\Delta u - f)\, \eta \, d\Omega \geq 0. \tag{5.170}$$

If we restrict the support of η to the coincidence set of u denoted by Ω_{φ}, all we obtain is

$$-\Delta u - f \geq 0$$

inside Ω_{φ}. This means that on Ω_{φ} there is a reaction of the supporting obstacle applied to the membrane. Recall that on the coincidence set we have $u = \varphi$. We consider u to be of the class of $C^{(2)}(\Omega)$, and hence on the boundary Γ_{φ} of Ω_{φ} all the first derivatives of u and φ are equal:

$$\nabla(u - \varphi)\big|_{\Gamma_{\varphi}} = 0.$$

From this we can determine the position of Γ_{φ}. Let us consider what happens outside the coincidence set Ω_{φ}. Here the only restriction for η is some smallness of its negative values. For sufficiently small η with compact support lying in $\Omega \backslash \Omega_{\varphi}$ we have equality to zero in (5.170). Thus the usual tools of the calculus of variations imply that in $\Omega \backslash \Omega_{\varphi}$ there holds the Poisson equation

$$\Delta u = -f \tag{5.171}$$

as expected from mechanical considerations. Let us summarize the setup:

$$\begin{aligned}
&\Delta u = -f \text{ on } \Omega \backslash \Omega_{\varphi}, \\
&\Delta u + f \leq 0, \; u = \varphi \text{ on } \Omega_{\varphi}, \\
&\nabla(u - \varphi) = 0 \text{ on } \Gamma_{\varphi}, \\
&u = a \text{ on } \partial\Omega.
\end{aligned}$$

The equation of equilibrium on Ω becomes

$$(\Delta u + f)(u - \varphi) = 0 \text{ in } \Omega.$$

Generalized setup

It is difficult to prove the existence of a classical solution to the above problem. When the coincidence set is complicated or the load is non-smooth, the energy approach to the solution is quite appropriate. For the problem setup we shall use an energy space whose elements are sets of equivalent Cauchy sequences, so we must explain the meaning of (5.162). We begin with the inequality $u(x,y) \geq 0$. We say that $u(x,y) \geq 0$, $u \in W^{1,2}(\Omega)$, if there is a representative Cauchy sequence of $u(x,y)$ such that each of its terms $u_n(x,y) \geq 0$. We say that $u(x,y) \geq \varphi(x,y)$ if $u(x,y) - \varphi(x,y) \geq 0$. If $\varphi(x,y) \in W^{1,2}(\Omega)$, then the set of functions $u(x,y) \geq \varphi(x,y)$ is closed in $W^{1,2}(\Omega)$ and in any closed subspace of this space. Let us assume that the obstacle function $\varphi(x,y) \in W^{1,2}(\Omega)$ and satisfies (5.164). Now we seek a minimizer $u = u(x,y) \in W^{1,2}(\Omega)$ of

$$F(u) = \frac{1}{2} \iint_\Omega \left[\left(\frac{\partial u}{\partial x} \right)^2 + \left(\frac{\partial u}{\partial y} \right)^2 \right] d\Omega - \iint_\Omega f u \, d\Omega$$

over a subset W_φ of elements of $W^{1,2}(\Omega)$ satisfying

$$u\big|_{\partial\Omega} = a(s), \qquad u(x,y) \geq \varphi(x,y).$$

This minimizer is called a generalized solution of the unilateral problem for the clamped membrane. We suppose $\varphi \in W^{1,2}(\Omega)$ and $f \in L^p(\Omega)$ for some $p > 1$. In this case the problem of minimization of $F(u)$ over W_φ is well defined. As before we find that a minimizer $u \in W_\varphi$ satisfies the variational inequality (5.167) for all $v \in W_\varphi$. We would like to reduce the problem to the case we have studied. Assume there is an element $g = g(x,y) \in W^{1,2}(\Omega)$ that satisfies the same boundary condition as a solution,

$$g(x,y)\big|_{\partial\Omega} = a(s), \tag{5.172}$$

and define another unknown function w by the equality

$$u = w + g.$$

From the properties of u it follows that

$$w(x,y)\big|_{\partial\Omega} = 0.$$

We see that $w \in W^{1,2}(\Omega)$ and thus $w \in \mathcal{E}_{Mc}$. To pose the setup in terms of w, it is clear that w should satisfy

$$w(x,y) \geq \varphi(x,y) - g(x,y). \tag{5.173}$$

Let $W_{\varphi-g}$ denote the subset of $\mathcal{E}_{Mc}$ consisting of elements satisfying (5.173). The functional $F(u)$ reduces to the functional

$$F_1(w) = \frac{1}{2} \iint_\Omega \left[\left(\frac{\partial(w+g)}{\partial x} \right)^2 + \left(\frac{\partial(w+g)}{\partial y} \right)^2 \right] d\Omega - \iint_\Omega f(w+g) \, d\Omega.$$

Since f and g are fixed, the problem of minimizing $F(u)$ becomes the problem of minimizing the functional

$$\Phi(w) = \frac{1}{2}\iint_\Omega \left[\left(\frac{\partial(w+g)}{\partial x}\right)^2 + \left(\frac{\partial(w+g)}{\partial y}\right)^2\right] d\Omega - \iint_\Omega fw\, d\Omega$$

over the set $W_{\varphi-g}$. Let us formulate the problem explicitly:

> Given $\varphi, g \in W^{1,2}(\Omega)$ such that (5.172) and (5.164) hold, find a minimizer of $\Phi(w)$ over $W_{\varphi-g}$.

Using (5.165) we can write

$$\Phi(w) = \frac{1}{2}(w+g, w+g)_M - \iint_\Omega fw\, d\Omega.$$

Let w^* be a minimizer of $\Phi(w)$ over $W_{\varphi-g}$. We repeat the reasoning that led to (5.167). Fixing an arbitrary $w \in W_{\varphi-g}$, we have

$$\Phi(w^* + t(w - w^*)) \geq \Phi(w^*)$$

for any $t \in [0, 1]$. For such t it follows that

$$\frac{1}{2}(w^* + t(w-w^*) + g, w^* + t(w-w^*) + g)_M$$
$$- \frac{1}{2}(w^* + g, w^* + g)_M - t\iint_\Omega f(w - w^*)\, d\Omega \geq 0$$

or

$$t\left\{(w^*, w - w^*)_M + (g, w - w^*)_M - \iint_\Omega f(w-w^*)\, d\Omega\right\}$$
$$+ \frac{1}{2}t^2(w - w^*, w - w^*)_M \geq 0.$$

Since this holds for any $t \in [0, 1]$ we conclude that the coefficient of t must be nonnegative:

$$(w^*, w - w^*)_M \geq \iint_\Omega f(w - w^*)\, d\Omega - (g, w - w^*)_M$$

for all $w \in W_{\varphi-g}$. This is a necessary condition for w^* to be a minimizer of $\Phi(w)$ over $W_{\varphi-g}$.

Theorem 5.40. *There exists a generalized solution to the unilateral problem for the membrane with clamped edge; it is the unique minimizer w^* of $\Phi(w)$ over $W_{\varphi-g}$.*

Proof. Let us show uniqueness of the minimizer w^*. Suppose to the contrary that there are two minimizers w_1^* and w_2^*. Then

$$(w_i^*, w - w_i^*)_M \geq \iint_\Omega f(w - w_i^*)\, d\Omega - (g, w - w_i^*)_M.$$

We put $w = w_2^*$ in the inequality for w_1^* and $w = w_1^*$ in the inequality for w_2^*. Adding the results we get

$$(w_1^* - w_2^*, w_2^* - w_1^*)_M \geq 0,$$

which is possible only when $w_1^* = w_2^*$ since $w_i^* \in \mathcal{E}_{Mc}$.

Now let us show existence of a minimizer of $\Phi(w)$. It is clear that $\Phi(w)$ is bounded from below on $W_{\varphi-g}$ (why?). Let $d = \inf \Phi(w)$ over $W_{\varphi-g}$, and let $\{w_n\}$ be a minimizing sequence for $\Phi(w)$ in $W_{\varphi-g}$:

$$\Phi(w_n) \to d \quad \text{as } n \to \infty.$$

Now we show that $\{w_n\}$ is a Cauchy sequence. Indeed, consider

$$\begin{aligned}(w_n - w_m, w_n - w_m)_M &= 2(w_n, w_n)_M + 2(w_m, w_m)_M \\ &\quad - 4\left(\frac{1}{2}(w_n + w_m), \frac{1}{2}(w_n + w_m)\right)_M. \end{aligned} \tag{5.174}$$

An elementary transformation shows that

$$\begin{aligned} 2(w_n, w_n)_M + 2(w_m, w_m)_M - 4\left(\frac{1}{2}(w_n + w_m), \frac{1}{2}(w_n + w_m)\right)_M \\ = 4\Phi(w_n) + 4\Phi(w_m) - 8\Phi\left(\frac{1}{2}(w_n + w_m)\right). \end{aligned} \tag{5.175}$$

Next $\Phi(w_n) = d + \varepsilon_n$ where $\varepsilon_n \to 0$ as $n \to \infty$. Because $W_{\varphi-g}$ is convex the element $\frac{1}{2}(w_n + w_m)$ belongs to $W_{\varphi-g}$, so $\Phi\left(\frac{1}{2}(w_n + w_m)\right) \geq d$, hence (5.174)–(5.175) imply

$$\begin{aligned}(w_n - w_m, w_n - w_m)_M &\leq 2(d + \varepsilon_n) + 2(d + \varepsilon_m) - 4d \\ &= 2(\varepsilon_n + \varepsilon_m) \to 0 \quad \text{as } n, m \to \infty.\end{aligned}$$

This completes the proof. □

We have proved solvability of a unilateral problem for a clamped membrane. Since all the problems we considered for plates, rods, and elastic bodies have the same structure, and since in the reasoning for the membrane we used only the structure of the energy functional, we can immediately reformulate unilateral problems for all the objects just mentioned (of course, for a three-dimensional body we can stipulate the unilateral condition only on the boundary). This work is left to the reader. The theory of unilateral problems and variational inequalities contains harder questions than the existence of an energy solution: it studies the problem of regularity of this solution, which is how the smoothness of solutions depends on the smoothness of the load. The interested reader may consult more advanced sources (e.g., [8; 11]).

5.16 Exercises

5.1 For all the bodies discussed in § 5.1 (except a stretched bar), write out the functional of total potential energy and the virtual work principle in the case when some part of the object (of its boundary for a three-dimensional body) is supported by a foundation of Winkler's type (i.e., when there is a contact force of supports whose amplitude is proportional to the corresponding displacements).

5.2 By analogy with in § 5.4, consider the generalized setup of the equilibrium problem for a membrane with mixed boundary conditions. Assume that on some part of the boundary $u = 0$, while on the rest there is a given force $g(s)$. Formulate the corresponding theorem on existence and uniqueness of solution in this setup.

5.3 Consider a beam under bending and stretching. Formulate the generalized setup for this problem, combining the setups for a stretched rod and bent beam. Formulate the corresponding existence-uniqueness theorem.

5.4 (a) Which terms are necessary to add to the equilibrium equation (5.13) to include a finite number external point couples and forces acting on the beam into the generalized setup? (b) Is it possible to the consider generalized setup when there is a countable set of point couples and forces?

5.5 For the structures from Exercises 2.1–2.8, use the virtual work principle and the results of solution of the Exercises to introduce the appropriate energy spaces and investigate their properties. Then formulate the generalized setup of the corresponding problems, and formulate and prove corresponding existence–uniqueness theorems for the generalized solutions.

5.6 For a free plate, consider a case when forces are given on the plate edge. Formulate the form of the potential and the conditions for solvability of the corresponding problem.

5.7 Using § 5.8 as an example, reproduce the form of the Hamilton–Ostrogradski principle for each type of object we considered.

5.8 Derive equations for solving the minimum problem (5.98).

5.9 Show that if $\mathcal{E}$ is not finite dimensional, then the norm $\|u\|_A$ of § 5.12 cannot be equivalent to the initial norm of the space $\mathcal{E}$ because A is compact.

5.10 Show that the set $\left\{\frac{\sqrt{2}}{\pi}\sin kx\right\}$, $k = 1, 2, \ldots$, is an orthonormal basis of $L^2[0, \pi]$.

5.11 Reformulate the statements of § 5.12 for each of the mechanics problems.

5.12 Suppose that in the conditions of Theorem 5.39 the minimum point is unique. Prove that any minimizing sequence strongly converges to the minimum point.

5.13 Referring to § 5.15, demonstrate uniqueness of solution to the problem under consideration in $W^{1,2}(\Omega)$, that $w^* + g$ does not depend on the choice of $g \in W^{1,2}(\Omega)$.

Appendix A

Hints for Selected Exercises

1.1.

(a)

$$\frac{y(x)}{\sqrt{1+y'^2(x)+y^2(x)}} - \left(\frac{y'(x)}{\sqrt{1+y'^2(x)+y^2(x)}}\right)' = 0 \text{ in } (0,1),$$

$$\left.\frac{y'(x)}{\sqrt{1+y'^2(x)+y^2(x)}}\right|_{x=0} = 0, \qquad \left.\frac{y'(x)}{\sqrt{1+y'^2(x)+y^2(x)}}\right|_{x=1} = 0.$$

(b)

$$(1+x^2)\,y - y'' = 0 \text{ in } (-1,1),$$
$$y'\big|_{x=-1} = 0, \quad y'\big|_{x=1} = 0.$$

(c)

$$2(1+2x^2)\,y + y'' = 0 \text{ in } (1,3),$$
$$y'\big|_{x=1} = 0, \quad y'\big|_{x=3} = 0.$$

(d)

$$(1+x^2)\,y + 7y'' = 0 \text{ in } (a,b),$$
$$y'\big|_{x=a} = 0, \quad y'\big|_{x=b} = 0.$$

(e) Denote the functional by $F(y)$. Suppose y is a minimizer and take an admissible variation φ. The function $F(y+t\varphi)$ of the real variable t takes its minimum value at $t=0$, so $dF(y+t\varphi)/dt|_{t=0} = 0$. This equation is

$$2\int_a^b \big[y'(x)\varphi'(x) + (1+x^6)\,y(x)\varphi(x)\big]\,dx + 2y(a)\,\varphi(a) = 0.$$

Canceling the factor of 2 and integrating by parts, we get

$$\int_a^b \big[-y''(x) + (1+x^6)\,y(x)\big]\varphi(x)\,dx + y'(x)\,\varphi(x)\Big|_{x=a}^{x=b} + y(a)\,\varphi(a) = 0.$$

Taking the set of φ that vanish at the endpoints, we get the Euler equation

$$(1+x^6)\,y - y'' = 0 \text{ in } (a,b)$$

and the integral vanishes for any admissible φ. Taking φ equal to zero at only one of the endpoints, and then the other, we get the natural conditions

$$(-y'+y)\big|_{x=a} = 0, \quad y'\big|_{x=b} = 0.$$

(f)

$$(x^2-9)\,y - (xy')' = 0 \text{ in } (1,3),$$
$$(-y'+5y)\big|_{x=1} = 0, \quad (3y'+y)\big|_{x=3} = 0.$$

(g)

$$\sqrt{x-a}\,y - 5y'' = 0 \text{ in } (a,b),$$
$$(-5y'+y)\big|_{x=a} = 0, \quad y'\big|_{x=b} = 0.$$

(h)

$$x^2\,y - y'' = 0 \text{ in } (1,2)\cup(2,4),$$
$$y'\big|_{x=1} = 0, \quad y'\big|_{x=2-0} - y'\big|_{x=2+0} + y\big|_{x=2} = 0, \quad y'\big|_{x=4} = 0.$$

(i)

$$x\,y - y'' = 0 \text{ in } (a,c)\cup(c,b),$$
$$y'\big|_{x=a} = 0, \quad y'\big|_{x=c-0} - y'\big|_{x=c+0} + 1/2 = 0, \quad y'\big|_{x=b} = 0.$$

(j)

$$4\,y^3 - 3y'' = 0 \text{ in } (0,\pi),$$
$$3y'\big|_{x=0} + \big[y(\pi)-y(0)\big] = 0, \quad 3y'\big|_{x=\pi} + \big[y(\pi)-y(0)\big] = 0.$$

(k)

$$y'^2 + \sin y - 2(y\,y')' = 0 \text{ in } (0,\pi),$$
$$(yy')\big|_{x=0} = 0, \quad (yy'+y)\big|_{x=\pi} = 0.$$

(l)

$$y - 2\left[\left(y'^2-1\right)y'\right]' = 0 \text{ in } (0,1),$$
$$\left(y'^2-1\right)y'\Big|_{x=0} = 0, \quad \left(y'^2-1\right)y'\Big|_{x=1} = 0.$$

1.2.

(a)

$$y'''' + 2y = 0 \text{ in } (0,1),$$
$$y''\big|_{x=0,1} = 0, \quad y'''\big|_{x=0,1} = 0.$$

(b)

$$y'''' - 2y'' = 0 \text{ in } (0,1),$$
$$y''\big|_{x=0,1} = 0, \quad (y' - y''')\big|_{x=0,1} = 0.$$

(c)

$$y'''' - y'' + y = 0 \text{ in } (0,1),$$
$$y''\big|_{x=0,1} = 0, \quad (y' - y''')\big|_{x=0,1} = 0.$$

(d)

$$-y^{(6)} - y^{(4)} + 2(1-x^2)y = 0 \text{ in } (0,1),$$
$$y'''\big|_{x=0,1} = 0, \quad (y'' + y^{(4)})\big|_{x=0,1} = 0, \quad (y^{(3)} + y^{(5)})\big|_{x=0,1} = 0.$$

(e)

$$y^{(8)} + y'' + 1 = 0 \text{ in } (a,b),$$
$$y^{(4)}\big|_{x=a,b} = 0, \quad y^{(5)}\big|_{x=a,b} = 0, \quad y^{(6)}\big|_{x=a,b} = 0, \quad (y^{(7)} + y')\big|_{x=a,b} = 0.$$

1.3.

(a)

$$-u_{xx} - 2u_{yy} + 3u - 1 = 0 \text{ in } S,$$
$$u_x\big|_{x=a,b} = 0, \quad u_y\big|_{y=c,d} = 0.$$

(b)

$$-u_{xx} - u_{yy} + u = 0 \text{ in } S,$$
$$u_x\big|_{x=a,b} = 0, \quad u_y\big|_{y=d} = 0, \quad (-u_y + u)\big|_{y=c} = 0.$$

(c)

$$u_{xx} + u_{yy} + u = 0 \text{ in } S,$$
$$(u_x + u)\big|_{x=b} = 0, \quad u_x\big|_{x=a} = 0, \quad u_y\big|_{y=c,d} = 0.$$

(d)

$$-u_{xx} + u_{yy} + 1 = 0 \text{ in } S,$$
$$u_x\big|_{x=a,b} = 0, \quad u_y\big|_{y=c,d} = 0.$$

(e)

$$\frac{\partial}{\partial x}(u_x)^{n-1} + \frac{\partial}{\partial y}(u_y)^{n-1} = 0 \text{ in } S,$$
$$u_x\big|_{x=a,b} = 0, \quad u_y\big|_{y=c,d} = 0.$$

(f)

$$u_{xx} \sin u_x + u_{yy} \sin u_y = 0 \text{ in } S,$$
$$\cos u_x\big|_{x=a,b} = 0, \quad \cos u_y\big|_{y=c,d} = 0.$$

(g)

$$u_{xx}\cos u_x + u_{yy}\cos u_y = 0 \text{ in } S,$$
$$\sin u_x\Big|_{x=a,b} = 0, \quad \sin u_y\Big|_{y=c,d} = 0.$$

(h)

$$\left(\frac{u_x}{\sqrt{1+(u_x)^2+(u_y)^2}}\right)_x + \left(\frac{u_y}{\sqrt{1+(u_x)^2+(u_y)^2}}\right)_y = 0 \text{ in } S,$$
$$\frac{u_x}{\sqrt{1+(u_x)^2+(u_y)^2}}\Bigg|_{x=0,1} = 0, \quad \frac{u_y}{\sqrt{1+(u_x)^2+(u_y)^2}}\Bigg|_{y=0,1} = 0.$$

(i)

$$p''(u_x)u_{xx} + q''(u_y)u_{yy} = 0 \text{ in } S,$$
$$p'(u_x)\Big|_{x=0,1} = 0, \quad q'(u_y)\Big|_{y=0,1} = 0.$$

(j)

$$\left(\left(1+(u_x)^2+(u_y)^2\right)^{n-1}u_x\right)_x + \left(\left(1+(u_x)^2+(u_y)^2\right)^{n-1}u_y\right)_y = 0 \text{ in } S,$$
$$\left(1+(u_x)^2+(u_y)^2\right)^{n-1}u_x\Big|_{x=0,1} = 0, \quad \left(1+(u_x)^2+(u_y)^2\right)^{n-1}u_y\Big|_{y=0,1} = 0.$$

1.4. We first show that the Euler equation for the simplest functional can be rewritten in the equivalent form

$$\frac{1}{y'}\left[\frac{d}{dx}\left(f - f_{y'}y'\right) - f_x\right] = 0.$$

Observe that if $f(x, y, y')$ does not depend explicitly on x, then one integration can be performed to give

$$f - f_{y'}y' = \text{constant}.$$

Indeed, multiplying and dividing the left member of the Euler equation by y', we have

$$\frac{1}{y'}\left[f_y y' - y'\frac{d}{dx}f_{y'}\right] = 0.$$

Adding and subtracting a couple of terms inside the brackets, we obtain

$$\frac{1}{y'}\left[f_x + f_y y' + f_{y'}y'' - f_{y'}y'' - y'\frac{d}{dx}f_{y'} - f_x\right] = 0.$$

But the first three terms inside the brackets add to produce df/dx (total derivative), and the next two terms add to produce $-d(f_{y'}y')/dx$ (product rule).

For the surface of revolution problem, the area functional is

$$\int_a^b 2\pi y\sqrt{1+(y')^2}\,dx.$$

Note that x does not appear explicitly; using the integrated version of Euler's equation, we get

$$2\pi y\sqrt{1+(y')^2} - 2\pi y\frac{y'}{\sqrt{1+(y')^2}}y' = \text{constant} = \alpha.$$

Divide through by 2π, then multiply through by $\sqrt{1+(y')^2}$ and simplify to get $y = \beta\sqrt{1+(y')^2}$ where $\beta = \alpha/2\pi$. Now solve for y' to obtain the separable ODE

$$y' = \sqrt{\frac{y^2}{\beta^2} - 1}.$$

The solution, obtained by direct integration, is

$$\beta\cosh^{-1}\left(\frac{y}{\beta}\right) = x + \gamma,$$

hence

$$y(x) = \beta\cosh\left(\frac{x+\gamma}{\beta}\right)$$

is the general form of the curve sought. The constants β, γ must be determined from the two endpoint conditions. We see that the minimal surface of revolution is a catenoid.

1.5. We need to find a smooth curve connecting the points (a, y_0) and (b, y_1), $a < b$. It is clear that for solvability of the problem it is necessary that $y_0 > y_1$.

First show that if f takes the general form

$$f(x, y, y') = p(y)\sqrt{1+(y')^2},$$

where $p(y)$ depends explicitly on y only, then

$$\int \frac{dy}{\sqrt{\frac{p^2(y)}{\alpha^2} - 1}} = x + \beta$$

where α and β are constants. The functional giving the time taken for the motion along a curve $y(x)$ is obtained by putting $p(y) = 1/\sqrt{2gy}$ where g is the acceleration due to gravity.

Using the specific form of p given and introducing a new constant $\gamma = 1/2\alpha^2 g$, we have

$$\int \frac{dy}{\sqrt{\frac{\gamma}{y} - 1}} = x + \beta.$$

The substitution $y = \gamma \sin^2\left(\frac{\theta}{2}\right)$ reduces this to

$$\frac{\gamma}{2} \int (1 - \cos\theta)\, d\theta = x + \beta$$

after the use of a couple of trig identities. Hence

$$x + \beta = \frac{\gamma}{2}(\theta - \sin\theta).$$

The other equation of the cycloid is

$$y = \gamma \sin^2\left(\frac{\theta}{2}\right) = \frac{\gamma}{2}(1 - \cos\theta).$$

Of course, the constants β and γ would be determined by given endpoint conditions.

1.6. The Euler equation $f_y - f_{y'x} - f_{y'y}y' - f_{y'y'}y'' = 0$ reduces to $f_{y'y'}y'' = 0$. This holds if $y'' = 0$ or $f_{y'y'} = 0$. The equation $y'' = 0$ is satisfied by any line of the form $y = c_1 x + c_2$. If the equation $f_{y'y'} = 0$ has a real root $y' = \gamma$, then $y = \gamma x + c_3$; this, however, merely gives a family of particular straight lines (all having the same slope γ). In any case, the extremals are all straight lines.

1.7. The average kinetic energy is given by

$$\frac{1}{T} \int_0^T \frac{1}{2} m x'^2(t)\, dt.$$

Since the integrand depends explicitly on x' only, the extremal is of the general form $x(t) = c_1 t + c_2$. Imposing the end conditions to find the constants c_1 and c_2 we obtain

$$x(t) = \frac{x_1 - x_0}{T} t + x_0.$$

The solution means the motion should be at constant speed. Any acceleration would increase the energy of the motion.

1.9. (a) Vanishing of the first variation requires that equation (1.57) hold. Let us review for a moment. We know that if we appoint a condition such as $y(a) = c_0$ then, since we need $\phi(a) = 0$ to keep the variations $y(x) + \varphi(x)$ admissible, we need $\varphi(a) = 0$ and equation (1.57) yields

$$f_{y'}(b, y(b), y'(b)) = 0.$$

This natural condition makes reference purely to b. Now consider the mixed condition given in the problem. To keep the variations $y(x) + \varphi(x)$ admissible we need $\varphi(a) + \varphi(b) = 0$ or $\varphi(a) = -\varphi(b)$. Equation (1.57) yields

$$f_{y'}(b, y(b), y'(b)) + f_{y'}(a, y(a), y'(a)) = 0.$$

This is the supplemental "natural" boundary condition. (b) To keep the variation admissible this time we need

$$\psi(y(a) + \varphi(a), y(b) + \varphi(b)) = 0.$$

As before, we're looking for a relation between $\phi(a)$ and $\phi(b)$ that we can substitute into (1.57). Restricting ourselves to infinitesimal variations $\phi(x)$, we use Taylor's formula in two variables to write, approximately,

$$\begin{aligned}\psi(y(a) + \varphi(a), y(b) + \varphi(b)) &= \psi(y(a), y(b)) \\ &\quad + \left(\varphi(a)\frac{\partial}{\partial\alpha} + \varphi(b)\frac{\partial}{\partial\beta}\right)\psi(\alpha, \beta)\bigg|_{\substack{\alpha=y(a)\\ \beta=y(b)}}.\end{aligned}$$

The first term on the right side is zero by the condition given in the problem. Therefore we need

$$\varphi(a)\frac{\partial\psi(\alpha,\beta)}{\partial\alpha}\bigg|_{\substack{\alpha=y(a)\\ \beta=y(b)}} + \varphi(b)\frac{\partial\psi(\alpha,\beta)}{\partial\beta}\bigg|_{\substack{\alpha=y(a)\\ \beta=y(b)}} = 0$$

or

$$\varphi(a) = K\varphi(b), \qquad K = -\frac{\left.\dfrac{\partial\psi(\alpha,\beta)}{\partial\beta}\right|_{\substack{\alpha=y(a)\\ \beta=y(b)}}}{\left.\dfrac{\partial\psi(\alpha,\beta)}{\partial\alpha}\right|_{\substack{\alpha=y(a)\\ \beta=y(b)}}}.$$

Equation (1.57) yields

$$f_{y'}(b, y(b), y'(b)) - Kf_{y'}(a, y(a), y'(a)) = 0$$

as the corresponding natural condition. In part (a) we had $\psi(\alpha, \beta) = \alpha + \beta - 1$, which gave us $K = -1$.

1.10. This is a mixed problem. However, the general solution of the Euler equation is the same as for the brachistochrone problem:

$$x + \beta = \frac{\gamma}{2}(\theta - \sin\theta), \qquad y = \frac{\gamma}{2}(1 - \cos\theta).$$

The condition at $x = a$ determines β. The condition at $x = b$ is the free-end condition $f_{y'}\big|_{x=b} = 0$. Here

$$f(x, y, y') = \frac{1}{\sqrt{2gy}}\sqrt{1 + (y')^2}$$

(again, the same as for the brachistochrone problem) so that

$$f_{y'} = \frac{y'}{\sqrt{2gy}\sqrt{1 + (y')^2}}.$$

Thus the condition at $x = b$ is $y'(b) = 0$; i.e., the required curve must "flatten out" at this endpoint.

1.11. Arc length on the cylinder is given by $(ds)^2 = (a\,d\phi)^2 + (dz)^2$. Parameterizing the desired curve as $\phi = \phi(t)$, $z = z(t)$, we seek to minimize the functional

$$\int_a^b [a^2(\phi')^2 + (z')^2]\,dt.$$

Each equation of the system (1.63) involves only the derivative of the dependent variable; hence the extremals are straight lines:

$$\phi(t) = c_1 t + c_2, \qquad z(t) = c_3 t + c_4.$$

Eliminating t we find $z(\phi) = \alpha\phi + \beta$, a family of helices on the cylinder.

1.12. Repetition of the steps leading to (1.50) gives the system

$$\int_a^b f_y\left(x, \sum_{i=0}^n c_i\varphi_i(x), \sum_{i=0}^n c_i\varphi_i'(x), \sum_{i=0}^n c_i\varphi_i''(x)\right)\varphi_k(x)\,dx$$
$$+ \int_a^b f_{y'}\left(x, \sum_{i=0}^n c_i\varphi_i(x), \sum_{i=0}^n c_i\varphi_i'(x), \sum_{i=0}^n c_i\varphi_i''(x)\right)\varphi_k'(x)\,dx$$
$$+ \int_a^b f_{y''}\left(x, \sum_{i=0}^n c_i\varphi_i(x), \sum_{i=0}^n c_i\varphi_i'(x), \sum_{i=0}^n c_i\varphi_i''(x)\right)\varphi_k''(x)\,dx = 0$$

for $k = 1, \ldots, n$.

1.13. Recall how the functional $\int_{x_0}^{x_1} f(x, y, y')\,dx$ was treated in § 1.11. Assume the endpoints x_0, x_1 change so we get arbitrary variations δx_0 and δx_1. In § 1.11 we used linear extrapolation for the function outside $[x_0, x_1]$. The approach here is similar, but we must take into account that the functional involves y''; hence we suppose that y' also has variations at the endpoints $\delta y_0'$ and $\delta y_0'$. The technique of linear extrapolation outside $[a, b]$ must also be applied to the derivative.

As a result, the extended curve $y = y(x)$ is determined by the endpoint coordinates $(x_0 + \delta x_0, y_0 + \delta y_0)$ and $(x_1 + \delta x_1, y_1 + \delta y_1)$ *and* the values of the first derivatives, which are $y_0' + \delta y_0'$ and $y_1' + \delta y_1'$ respectively.

Our problem is to derive the linear part of the increment for (1.166) when φ, φ', φ'' δx_0, δy_0, $\delta y_0'$, δx_1, δy_1, and $\delta y_1'$ have the same order of smallness; that is, to extract the part of the increment that is linear in each of these quantities. Denote

$$\varepsilon = \|\varphi\|_{C^{(2)}(x_0,x_1)} + |\delta x_0| + |\delta y_0| + |\delta y_0'| + |\delta x_1| + |\delta y_1| + |\delta y_1'|.$$

The increment is

$$\Delta F(y) = \int_{x_0+\delta x_0}^{x_1+\delta x_1} f(x, y+\varphi, y'+\varphi', y''+\varphi'')\,dx - \int_{x_0}^{x_1} f(x,y,y',y'')\,dx.$$

The first integral can be decomposed as

$$\int_{x_0+\delta x_0}^{x_1+\delta x_1} (\cdots)\,dx = \int_{x_0}^{x_1} (\cdots)\,dx + \int_{x_1}^{x_1+\delta x_1} (\cdots)\,dx - \int_{x_0}^{x_0+\delta x_0} (\cdots)\,dx.$$

Recall that the functions $y = y(x)$, $y' = y'(x)$, $\varphi' = \varphi'(x)$, and $\varphi = \varphi(x)$ are all linearly extrapolated outside $[x_0, x_1]$, preserving continuity of the functions and their first derivatives. Thus

$$\begin{aligned}\Delta F(y) = &\int_{x_0}^{x_1} [f(x, y+\varphi, y'+\varphi', y''+\varphi'') - f(x,y,y',y'')]\,dx\\ &+ \int_{x_1}^{x_1+\delta x_1} f(x, y+\varphi, y'+\varphi', y''+\varphi'')\,dx\\ &- \int_{x_0}^{x_0+\delta x_0} f(x, y+\varphi, y'+\varphi', y''+\varphi'')\,dx.\end{aligned}$$

The integral over $[x_0, x_1]$ can be transformed in the usual manner:

$$\begin{aligned}&\int_{x_0}^{x_1} [f(x, y+\varphi, y'+\varphi', y''+\varphi'') - f(x,y,y',y'')]\,dx\\ &\quad= \int_{x_0}^{x_1} \left[f_y(x,y,y',y'') - \frac{d}{dx} f_{y'}(x,y,y',y'') + \frac{d^2}{dx^2} f_{y''}(x,y,y',y'')\right] \varphi\,dx\\ &\quad+ f_{y''}(x, y(x), y'(x), y''(x))\varphi'(x)\Big|_{x=x_0}^{x=x_1}\\ &\quad+ \Big[f_{y'}(x, y(x), y'(x), y''(x))\\ &\qquad- \frac{d}{dx} f_{y''}(x, y(x), y'(x), y''(x))\Big]\,\varphi(x)\Big|_{x=x_0}^{x=x_1} + o(\varepsilon).\end{aligned}$$

As earlier, φ at the endpoints using δy_0 and δy_1 is

$$\varphi(x_1) = \delta y_1 - y'(x_1)\delta x_1 + o(\varepsilon), \qquad \varphi(x_0) = \delta y_0 - y'(x_0)\delta x_0 + o(\varepsilon). \tag{A.1}$$

For φ' at the endpoints we should use $\delta y_0'$ and $\delta y_1'$, obtaining

$$\varphi'(x_1) = \delta y_1' - y''(x_1)\delta x_1 + o(\varepsilon), \qquad \varphi'(x_0) = \delta y_0' - y''(x_0)\delta x_0 + o(\varepsilon). \tag{A.2}$$

Thus

$$\begin{aligned}
&\int_{x_0}^{x_1} [f(x, y+\varphi, y'+\varphi', y''+\varphi'') - f(x,y,y',y'')]\,dx \\
&\quad = \int_{x_0}^{x_1} \left[f_y(x,y,y',y'') - \frac{d}{dx} f_{y'}(x,y,y',y'') + \frac{d^2}{dx^2} f_{y''}(x,y,y',y'') \right] \varphi\,dx \\
&\qquad + f_{y''}(x, y(x), y'(x), y''(x))\delta y' \Big|_{x=x_0}^{x=x_1} \\
&\qquad + \left[f_{y'}(x, y(x), y'(x), y''(x)) - \frac{d}{dx} f_{y''}(x, y(x), y'(x), y''(x)) \right] \delta y \Big|_{x=x_0}^{x=x_1} \\
&\qquad - \left[y'' f_{y''} + y'\left(f_{y'} - \frac{d}{dx} f_{y''} \right) \right] \delta x \Big|_{x=x_0}^{x=x_1} + o(\varepsilon).
\end{aligned}$$

The two other terms for ΔF mimic the ones for the simplest functional:

$$\begin{aligned}
&\int_{x_1}^{x_1+\delta x_1} f(x, y+\varphi, y'+\varphi', y''+\varphi'')\,dx \\
&\qquad = f(x_1, y(x_1), y'(x_1), y''(x_1))\delta x_1 + o(\varepsilon)
\end{aligned}$$

and

$$\begin{aligned}
&\int_{x_0}^{x_0+\delta x_0} f(x, y+\varphi, y'+\varphi', y''+\varphi'')\,dx \\
&\qquad = f(x_0, y(x_0), y'(x_0), y''(x_0))\delta x_0 + o(\varepsilon).
\end{aligned}$$

Collecting terms and selecting the first-order terms, we get the general form of the first variation of the functional when the ends of the curve can move:

$$\begin{aligned}
\delta F &= \int_{x_0}^{x_1} \left(f_y - \frac{d}{dx} f_{y'} + \frac{d^2}{dx^2} f_{y''} \right) \varphi\,dx \\
&\quad + f_{y''}\delta y' \Big|_{x=x_0}^{x=x_1} + \left[f_{y'} - \frac{d}{dx} f_{y''} \right] \delta y \Big|_{x=x_0}^{x=x_1} \\
&\quad + \left[f - y'' f_{y''} + y'\left(f_{y'} - \frac{d}{dx} f_{y''} \right) \right] \delta x \Big|_{x=x_0}^{x=x_1}.
\end{aligned} \tag{A.3}$$

1.14. Split the functional into two parts,

$$F(y) = \int_a^b f\,dx = \int_a^c f\,dx + \int_c^b f\,dx = F_1(y) + F_2(y),$$

find the first derivative $\delta F = \delta F_1 + \delta F_2$, apply (A.3) to each F_k, and consider the equation $\delta F = 0$ using the continuity conditions, the arbitrariness of φ, and the endpoint variations. We have

$$\begin{aligned}\delta F_1 = &\int_a^c \left(f_y - \frac{d}{dx}f_{y'} + \frac{d^2}{dx^2}f_{y''}\right)\varphi\,dx\\ &+ f_{y''}\delta y'\Big|_a^c + \left[f_{y'} - \frac{d}{dx}f_{y''}\right]\delta y\Big|_{x=a}^{x=c-0}\\ &+ \left[f - y''f_{y''} + y'\big(f_{y'} - \frac{d}{dx}f_{y''}\big)\right]\delta x\Big|_{x=c-0},\end{aligned}$$

where for brevity we denote $\varphi(a) = \delta y|_{x=a}$ and $\varphi'(a) = \delta y'|_{x=a}$. Introducing similar notation at b, i.e., $\varphi(b) = \delta y|_{x=b}$ and $\varphi'(b) = \delta y'|_{x=b}$, we get

$$\begin{aligned}\delta F_2 = &\int_c^b \left(f_y - \frac{d}{dx}f_{y'} + \frac{d^2}{dx^2}f_{y''}\right)\varphi\,dx\\ &+ f_{y''}\delta y'\Big|_c^b + \left[f_{y'} - \frac{d}{dx}f_{y''}\right]\delta y\Big|_{x=c+0}^{x=b}\\ &- \left[f - y''f_{y''} + y'\big(f_{y'} - \frac{d}{dx}f_{y''}\big)\right]\delta x\Big|_{x=c+0}.\end{aligned}$$

Adding δF_k and setting $\delta F = 0$, we should step-by-step choose subsets for φ and the endpoint variations as was done to derive the Weierstrass conditions. In this way we find that

$$f_y - \frac{d}{dx}f_{y'} + \frac{d^2}{dx^2}f_{y''} = 0$$

holds on (a, c) and (c, b). At the endpoints a and b we get the natural conditions

$$f_{y''} = 0, \quad f_{y'} - \frac{d}{dx}f_{y''} = 0 \quad \text{for } x = a \text{ and } x = b.$$

Remembering that the variations at c from the left and right must match, we obtain

$$\delta x\Big|_{x=c-0} = \delta x\Big|_{x=c+0} = \delta x_c, \quad \delta y\Big|_{x=c-0} = \delta y\Big|_{x=c+0} = \delta y_c,$$

and

$$\delta y'\Big|_{x=c-0} = \delta y'\Big|_{x=c+0} = \delta y'_c.$$

Hence we have

$$
\begin{aligned}
0 = &\left[f_{y''}\big|_{x=c-0} - f_{y''}\big|_{x=c+0} \right] \delta y'_c \\
&+ \left[\left(f_{y'} - \frac{d}{dx} f_{y''} \right)\Big|_{x=c-0} - \left(f_{y'} - \frac{d}{dx} f_{y''} \right)\big|_{x=c+0} \right] \delta y_c \\
&+ \left[\left(f - y'' f_{y''} + y' \big(f_{y'} - \frac{d}{dx} f_{y''} \big) \right)\Big|_{x=c-0} \right. \\
&\left. - \left(f - y'' f_{y''} + y' \big(f_{y'} - \frac{d}{dx} f_{y''} \big) \right)\Big|_{x=c+0} \right] \delta x_c .
\end{aligned}
$$

Taking into account arbitrariness and independence of the variations at c, we get conditions analogous to the Weierstrass–Erdmann conditions:

$$
f_{y''}\big|_{x=c-0} = f_{y''}\big|_{x=c+0}, \tag{A.4}
$$

$$
\left((f_{y'} - \frac{d}{dx} f_{y''} \right)\Big|_{x=c-0} = \left(f_{y'} - \frac{d}{dx} f_{y''} \right)\big|_{x=c+0}, \tag{A.5}
$$

$$
\left(f - y'' f_{y''} + y' \big(f_{y'} - \frac{d}{dx} f_{y''} \big) \right)\Big|_{x=c-0} = \left(f - y'' f_{y''} + y' \big(f_{y'} - \frac{d}{dx} f_{y''} \big) \right)\Big|_{x=c+0} . \tag{A.6}
$$

In beam theory, $f = EI(y'')^2/2$, $M = EIy''$, and $Q = -EIy'''$. In terms of moment M and shear force Q, equations (A.4)–(A.6) are

$$
M\big|_{x=c-0} = M\big|_{x=c+0}, \quad Q\big|_{x=c-0} = Q\big|_{x=c+0}, \tag{A.7}
$$

$$
\left(f - y''M + y'Q \right)\Big|_{x=c-0} = \left(f - y''M + y'Q \right)\Big|_{x=c+0} . \tag{A.8}
$$

Equation (A.7) expresses equality of the moments and shear forces, while (A.8) can be related with the energy-release when the defect at $x = c$ moves; see [12].

Let us analyze the setup for finding an extremal. We have equations on two intervals. Each equation is of fourth order in general, which means we get eight independent constants in the solution if the equations are linear. The ninth unknown constant is c. Now let us count the boundary conditions. At a and b we have four equations for unknown constants; taken together with the three above equations at c, the total number is seven. Two more conditions require continuity of y and y' at c:

$$
y(c-0) = y(c+0), \qquad y'(c-0) = y'(c+0).
$$

1.15. All the equations are the same except (A.6).

1.16. The Euler equation is $D\Delta^2 w = F$, and the natural boundary conditions are

$$\nu\Delta w + (1-\nu)(w_{xx}n_x^2 + w_{yy}n_y^2 + 2w_{xy}n_x n_y)\Big|_{\partial S} = 0,$$

$$D[(w_{xx}+\nu w_{yy})_y n_x + (w_{yy}+\nu w_{xx})_y n_y + (1-\nu)(w_{xyy}n_x + w_{xxy}n_y)] + D(1-\nu)\frac{d}{ds}[(w_{yy}-w_{xx})n_x n_y + w_{xy}(n_x^2 - n_y^2)]\Big|_{\partial S} = f.$$

1.17. The natural boundary conditions are

$$D_1\left[\nu\Delta w + (1-\nu)(w_{xx}n_x^2 + w_{yy}n_y^2 + 2w_{xy}n_x n_y)\right]\Big|_{\Gamma_-} = D_2\left[\nu\Delta w + (1-\nu)(w_{xx}n_x^2 + w_{yy}n_y^2 + 2w_{xy}n_x n_y)\right]\Big|_{\Gamma_+},$$

$$D_1\left\{[(w_{xx}+\nu w_{yy})_y n_x + (w_{yy}+\nu w_{xx})_y n_y + (1-\nu)(w_{xyy}n_x + w_{xxy}n_y)] +(1-\nu)\frac{d}{ds}[(w_{yy}-w_{xx})n_x n_y + w_{xy}(n_x^2-n_y^2)]\right\}\Big|_{\Gamma_-}$$
$$= D_2\left\{[(w_{xx}+\nu w_{yy})_y n_x + (w_{yy}+\nu w_{xx})_y n_y + (1-\nu)(w_{xyy}n_x + w_{xxy}n_y)] +(1-\nu)\frac{d}{ds}[(w_{yy}-w_{xx})n_x n_y + w_{xy}(n_x^2-n_y^2)]\right\}\Big|_{\Gamma_+}.$$

Here $(\cdot)\big|_{\Gamma_\pm}$ denote one-sided limits calculated as the argument tends to Γ from different sides (cf., Fig. 1.5).

1.18. Use the solution of Example 1.28 (page 47) and the identity

$$\int_{\partial S} \frac{\partial u}{\partial s}\frac{\partial \delta u}{\partial s}\, ds = -\int_{\partial S} \frac{\partial^2 u}{\partial s^2}\delta u\, ds.$$

The answer is

$$\Delta u + f = 0, \qquad \left(\frac{\partial u}{\partial n} - \alpha\frac{\partial^2 u}{\partial s^2}\right)\Big|_{\partial S} = g.$$

1.19. Without loss of generality we may take S as the unit square $S = \{(x,y) \in [0,1]\times[0,1]\}$. On three sides $\Omega_1 = ([0,1]\times\{0\})\cup(\{0\}\times[0,1])\cup(\{1\}\times[0,1])$ of S take $\alpha = 0$ and on the fourth side let $\alpha < 0$. Now

$$2E(u) = \int_0^1\int_0^1 \left(u_x^2 + u_y^2\right)\, dx\, dy + \alpha\int_0^1 u_x^2\big|_{y=1}\, dx.$$

We show that $E(u)$ is unbounded from below. Indeed, let $u_{kn} = y^n \sin \pi k x$, where n, k are integers. We get

$$E(u_{kn}) = \frac{(\pi k)^2}{4}\left(\alpha + \frac{1}{2n+1}\right) + \frac{1}{4}\frac{n^2}{2n-1}.$$

For $\alpha < 0$ we find such an n^* that

$$\alpha + \frac{1}{2n^*+1} < 0.$$

Then $E(u_{kn^*}) \to -\infty$ as $k \to \infty$.

2.1. With the spring at c loaded with a point force $-P$, the energy functional becomes

$$\mathcal{E}(y) = \frac{1}{2}\int_0^a EI(y'')^2\,dx - \int_0^a qy\,dx + Py(a) + \frac{1}{2}ky^2(a) + \frac{1}{2}k_2y^2(c).$$

Point c splits the beam into two parts that must be joined by the continuity conditions $y(c-0) = y(c+0)$ and $y'(c-0) = y'(c+0)$. Admissible φ should satisfy the same condition: $\varphi(c-0) = \varphi(c+0)$, $\varphi'(c-0) = \varphi'(c+0)$. In the corresponding equilibrium equation, we first select a subset of admissible functions φ that are zero on $[c, a]$; in this way we find that $E(Iy'')'' - q = 0$ on $(0, c)$. Then, selecting φ equal to zero on $[0, c]$, we establish the same equilibrium equation on (c, a). Finally, using the equations $\varphi(c-0) = \varphi(c+0)$ and $\varphi'(c-0) = \varphi'(c+0)$, we get two additional conditions at point c. These form part of the natural conditions supplementing the conditions at point a in Example 2.1. So we have two linear equations, the general solution of each containing four indefinite constants. Taken together, the continuity conditions and boundary conditions at points 0, c, and a provide eight conditions sufficient to determine the eight constants of the solution uniquely.

2.2. The strain energy functional is

$$W = \frac{1}{2}\int_0^{2a} EIw''^2(x)\,dx + \frac{1}{2}kw^2(a) + \frac{1}{2}cw'^2(2a).$$

The work of external forces is

$$A = \int_0^{2a} q(x)w(x)\,dx.$$

The kinematic boundary conditions are $w(0) = 0 = w'(0)$. The coupling conditions of the springs with beams are taken into account in the energy equation. As usual, $\mathcal{E} = W - A$. When deriving the equations, we first take admissible virtual displacements to be zero on the right portion of the beam $[a, 2a]$, and then on the left portion. We get the equilibrium equation

$$EIw_2^{(4)}(x) = q(x) \quad \text{in } (0,a) \cup (a, 2a).$$

Next we derive the natural boundary conditions, taking into account the continuity of w and w' at $x = a$, i.e., $w(a-0) = w(a+0)$ and $w'(a-0) = w'(a+0)$.

2.3. Denote the rod/beam displacements as follows: (u_1, w_1) for BD, (u_2, w_2) for CD, and (u_3, w_3) for AD. Let BD have length a. The boundary conditions for the rod system are $u_1(0) = u_2(0) = u_3(0) = 0$ and $w_1(0) = w_2(0) = w_3(0) = 0$. For the beam problem these should be supplemented with $w_1'(0) = w_2'(0) = w_3'(0) = 0$.

The total potential energy functional for the rod system is $\mathcal{E}_R = e_R - A$ where the strain energy e_R is

$$e_R = \frac{1}{2}\int_{BD} ESu_1'^2\,dx + \frac{1}{2}\int_{CD} ESu_2'^2\,dx + \frac{1}{2}\int_{AD} ESu_3'^2\,dx.$$

For the beam system it is $\mathcal{E}_B = e_B - A$, where e_B includes e_R and three terms for bending energy as in the previous exercise. The work A of the external load has the same form for both systems:

$$A = \int_{AD} (q(x)u_3(x)\cos\alpha + q(x)w_3(x)\sin\alpha)\,dx + \int_{AC} F(x)u_2(x)\,dx.$$

However, for the rod system we should note that rod CD is not flexible; in the w-direction it rotates as a rigid body, and so we should express $w_3(x)$ in terms of the value of $w_3|_D$ at the extreme point. For the rod system this is

$$w_3(x) = \frac{x\sin\alpha}{a} w_3|_D.$$

To place kinematic restrictions at junction D for both systems, we should use (2.1) where for the pair BD–CD we should change (u, w) to (u_1, w_1) and (u_1, w_1) in (2.1) to (u_2, w_2). The angle in (2.1) is changed to $-(\pi/2 - \alpha)$. For the other coupled rod at D, the angle is $\pi/2 - \alpha$. For the beam system, these equations for the displacements must be supplemented with the condition of equality of the rotation angle of the beams at D: $w_1'|_D = w_2'|_D = w_3'|_D$. The restrictions for the real displacements also apply to the admissible virtual displacements. The remainder of the solution is similar to that of the previous exercise.

2.4. Here we can neglect the beam elongations and use only the beam model for each of the elements. This and the following beam problems present interesting boundary conditions that may appear artificial to someone inexperienced in the strength of materials.

We number the beams from beam AB as number 1 to CD as number 3. We number the normal displacements w_k similarly, describing the beams in clockwise fashion starting at point A. The total potential energy of the system is

$$\mathcal{E} = \sum_{i=1}^{3} \frac{1}{2} \int_0^{a_i} EIw_i''^2(x)\,dx - \int_0^{a_2} q(x)w_2(x)\,dx - Pw_3(a_3).$$

The kinematic (geometric) boundary conditions are $w_1(0) = 0$ and $w_1'(0) = 0$. Since we neglect the beam elongations, we should regard the beams as unchangeable along their length directions. This yields the following restrictions at the junction points:

$$w_2(0) = 0, \qquad w_1'(a_1) = w_2'(0),$$

and

$$w_2(a_2) = 0, \qquad w_3(0) = -w_1(a_1), \qquad w_2'(a_2) = w_3'(0).$$

The equilibrium equations for the beams are

$$EIw_1^{(4)}(x) = 0, \qquad EIw_2^{(4)}(x) = q(x), \qquad EIw_3^{(4)}(x) = 0.$$

The natural boundary conditions are the supplementary conditions of equilibrium for the points B, C, and D.

2.5. Let A and D be the initial points for the length parameters of the beams. Winkler's foundation is a simple model of a junction when the elastic deformation at each point does not depend on the deformation at other points: when the foundation thickness changes by $u(x)$, then its elastic reaction is $ku(x)$. Denoting the deflections of AB and DC by w_1 and w_2, respectively, we get $u(x) = w_2(x) - w_1(x)$. The total potential energy is

$$\mathcal{E} = \sum_{k+1}^{2} \frac{1}{2} \int_0^{a} EIw_k''^2(x)\,dx + \frac{k}{2} \int_0^{a} (w_2(x) - w_1(x))^2\,dx + \int_0^{a} q(x)w_2(x)\,dx.$$

The kinematic boundary conditions are

$$w_1(0) = 0 = w_1'(0), \qquad w_2(0) = 0 = w_2'(0).$$

Answer: The equilibrium equations are

$$EIw_1^{(4)}(x) - k(w_2(x) - w_1(x)) = 0,$$
$$EIw_2^{(4)}(x) + k(w_2(x) - w_1(x)) = -q(x).$$

The natural boundary conditions at B and C mean that the shear stresses and the moments are zero: $w_k''(a) = 0 = w'''(a)$.

2.6. This is a free system of four beams, none of which is clamped. For such problems there are always additional conditions for the load under which the problem has a solution. We number the beams starting with AB. The positive direction for the deflection w is "inward" for each beam. The total potential energy functional is

$$\mathcal{E} = \frac{1}{2}\sum_{k=1}^{4}\int_0^a \left(EIw_k''^2(x) - 2q_k(x)w_k(x)\right)dx + Pw_3\left(\frac{a}{2}\right).$$

The restrictions at the junction nodes are that the rotation angles of the beams at a node are equal:

$$w_1'(0) = w_4'(a), \qquad w_1'(a) = w_2'(0), \qquad w_2'(a) = w_3'(0), \qquad w_3'(a) = w_4'(0).$$

We should also relate the displacements at the nodes, considering the beams to be rigid along their axial directions:

$$\begin{aligned} w_4(a) &= -w_2(0), \quad & w_1(a) &= -w_3(0), \\ w_2(a) &= -w_4(0), \quad & w_3(a) &= -w_1(0). \end{aligned}$$

Finally, we should state that at the point where force P is applied, w and w' are continuous. This implies two more natural boundary equations.

The beam equilibrium equations are

$$EIw_k^{(4)}(x) = q_k(x),$$

which for $k = 1, 2, 4$ hold on $(0, a)$ and for $k = 3$ holds on $(0, a/2)$ and on $(a/2, a)$.

To obtain solvability conditions for the minimization problem for the load, note that the rigid displacements (i.e., when the structure moves as a rigid body) occur through parallel displacements: $w_1(x) = -w_3(x) = c_1$ and $w_2(x) = -w_4(x) = c_2$ with independent constants c_k. If we denote the tangential displacements of the beams by u_k (they are constant for each beam), then for rigid rotation we get the following displacements:

$$\begin{aligned} &w_1 = cx, \quad u_1 = 0; \qquad & &w_2 = cx, \quad u_2 = ca; \\ &w_3 = -ca + cx, \quad u_3 = ca; \qquad & &w_4 = -ca + cx, \quad u_4 = 0. \end{aligned}$$

These three displacements are admissible, since they satisfy the kinematics of the structure and the conditions shown above. When we substitute these into the energy, we see that the quadratic part of the energy is zero for each of them. Hence the work functional must vanish on these; otherwise, by selecting appropriate values for c_k we can get any negative value for the energy functional and so the minimum problem becomes senseless. Besides, the equilibrium equations (the first variation set to zero) for these three displacements give us the following equations:

$$\int_0^a q_1(x)\,dx + P - \int_0^a q_3\,dx = 0,$$
$$\int_0^a q_2\,dx - \int_0^a q_4(x)\,dx = 0,$$
$$\int_0^a xq_1(x)\,dx + \int_0^a xq_2(x)\,dx + \int_0^a (-a+x)q_3(x)\,dx$$
$$-P\frac{a}{2} + \int_0^a (-a+x)q_4(x)\,dx = 0.$$

It can be shown that there are no other linearly independent displacements of the structure for which the elastic energy is zero. Hence the above equations are necessary (and sufficient) for solvability of the equilibrium problem; they are the self-balance conditions for the load.

The natural boundary conditions can be found using the usual procedure of selecting special displacement fields.

2.7. See the solution to Example 2.2.

2.8. See the solution to Example 2.2.

2.9. (1) The first conservation law is the same as the answer to Example 2.14. Indeed, our functional is a particular case of the functional from that example. Considering the symmetry transformation

$$x \to x^* = x + \varepsilon, \qquad y \to y^* = y,$$

which corresponds to $\xi = 1$, $\phi = 0$, we get the conservation law

$$P' = 0, \quad P = f - y' f_{y'}.$$

(2) Considering the symmetry transformation

$$x \to x^* = x, \qquad y \to y^* = y + \varepsilon,$$

which corresponds to $\xi = 0$, $\phi = 1$, we get the conservation law

$$P' = 0, \quad P = f_{y'}.$$

Answer:

$$f - y' f_{y'} = \text{constant}, \quad f_{y'} = \text{constant}.$$

2.10. The first two conservation laws were established in Exercise 2.10. One more law can be obtained by considering the scaling transformation

$$x \to x^* = x + \varepsilon x, \qquad y \to y^* = y + \frac{1}{2}\varepsilon y,$$

to which there correspond functions $\xi = x$, $\phi = y/2$. For this, the conservation law is

$$\frac{1}{2}y'y'' - \frac{x}{2}y'^2 = \text{constant}$$

Answer:

$$-y'^2/2 = \text{constant}, \quad y' = \text{constant}, \quad y'y''/2 - xy'^2/2 = \text{constant}.$$

2.11. Let us write out the infinitesimal invariance condition (2.63) for the functional under consideration and find all the functions $\xi(x, y)$ and $\phi(x, y)$ that satisfy the condition. We have

$$\begin{aligned}
&\left[\xi\frac{\partial}{\partial x} + \phi\frac{\partial}{\partial y} + \left(\frac{d\phi}{dx} - y'\frac{d\xi}{dx}\right)\frac{\partial}{\partial y'} + \frac{d\xi}{dx}\right] f \\
&= \left(\frac{d\phi}{dx} - y'\frac{d\xi}{dx}\right) y' + \frac{1}{2}\frac{d\xi}{dx}y'^2 \\
&= \frac{d\phi}{dx}y' - \frac{1}{2}\frac{d\xi}{dx}y'^2 = \left(\frac{\partial\phi}{\partial x} + \frac{\partial\phi}{\partial y}y'\right) y' - \frac{1}{2}\left(\frac{\partial\xi}{\partial x} + \frac{\partial\xi}{\partial y}y'\right) y'^2 \\
&= \frac{\partial\phi}{\partial x}y' + \left(\frac{\partial\phi}{\partial y} - \frac{1}{2}\frac{\partial\xi}{\partial x}\right) y'^2 - \frac{1}{2}\frac{\partial\xi}{\partial y}y'^3 = 0.
\end{aligned}$$

The factors before the potentials of y' do not depend on y'. It follows that to get the expressions to be zero for any y', it is necessary that they be zero independently of y'. Thus, to define $\xi(x, y)$ and $\phi(x, y)$, we have three simultaneous equations:

$$\frac{\partial\phi}{\partial x} = 0, \qquad \frac{\partial\phi}{\partial y} - \frac{1}{2}\frac{\partial\xi}{\partial x} = 0, \qquad -\frac{1}{2}\frac{\partial\xi}{\partial y} = 0.$$

The first and third equations imply $\phi = \phi(y)$ and $\xi = \xi(x)$. The second one implies

$$\xi = 2C_2 x + C_1, \qquad \phi = C_2 y + C_3,$$

where C_1, C_2, and C_3 are integration constants. These functions $\xi(x, y)$ and $\phi(x, y)$ define the most general case when variational symmetry is possible for the functional. To the three constants there correspond three transformations and the conservation laws found in Exercises 2.9 and 2.10.

2.12. Now F^* has the form

$$F^*(y^*) = \int_{a^*}^{b^*} f\left(x^*, y^*, \frac{d^k y^*}{dx^{*k}}\right) dx^*.$$

A principal difficulty is the derivation of the formula for $d^k y^*/dx^{*k}$. We skip the details and present the formulas, omitting the second and higher orders of ε:

$$\frac{dy^*}{dx^*} = y' + \varepsilon\left(\frac{d\phi}{dx} - y'\frac{d\xi}{dx}\right) + O(\varepsilon^2),$$

$$\frac{d^2y^*}{dx^{*2}} = y'' + \varepsilon\left(-y''\frac{d\xi}{dx} + \frac{d}{dx}\left(\frac{d\phi}{dx} - y'\frac{d\xi}{dx}\right)\right) + O(\varepsilon^2)$$

$$= y'' + \varepsilon\left(\frac{d^2\phi}{dx^2} - 2y''\frac{d\xi}{dx} - y'\frac{d^2\xi}{dx^2}\right) + O(\varepsilon^2),$$

$$\frac{d^3y^*}{dx^{*3}} = y''' + \varepsilon\left(-y'''\frac{d\xi}{dx} + \frac{d}{dx}\left(-y''\frac{d\xi}{dx} + \frac{d}{dx}\left(\frac{d\phi}{dx} - y'\frac{d\xi}{dx}\right)\right)\right) + O(\varepsilon^2),$$

$$\dots$$

$$\frac{d^ky^*}{dx^{*k}} = y^{(k)} + \varepsilon\left(\frac{d^k\phi}{dx^k} - \frac{d\xi}{dx}\frac{d^ky}{dx^k} - \frac{d}{dx}\left(\frac{d\xi}{dx}\frac{d^{k-1}y}{dx^{k-1}}\right) - \dots\right.$$
$$\left. - \frac{d^{k-1}}{dx^{k-1}}\left(\frac{d\xi}{dx}\frac{dy}{dx}\right)\right) + O(\varepsilon^2)$$
$$= y^{(k)} + \varepsilon\left(\frac{d^k\phi}{dx^k} - \sum_{p=0}^{k-1}\frac{d^p}{dx^p}\left(\frac{d\xi}{dx}\frac{d^{k-p}y}{dx^{k-p}}\right)\right) + O(\varepsilon^2).$$

So for F^* we get

$$F^* = F + \varepsilon\int_a^b [f_x\xi + f_y\phi$$
$$+ f_{y^{(k)}}\left[\frac{d^k\phi}{dx^k} - \sum_{p=0}^{k-1}\frac{d^p}{dx^p}\left(\frac{d\xi}{dx}\frac{d^{k-p}y}{dx^{k-p}}\right)\right] + f\frac{d\xi}{dx}\Bigg]\,dx + O(\varepsilon^2),$$

which implies the variational symmetry condition

$$\left[\xi\frac{\partial}{\partial x} + \phi\frac{\partial}{\partial y} + \left[\frac{d^k\phi}{dx^k} - \sum_{p=0}^{k-1}\frac{d^p}{dx^p}\left(\frac{d\xi}{dx}\frac{d^{k-p}y}{dx^{k-p}}\right)\right]\frac{\partial}{\partial y^{(k)}} + \frac{d\xi}{dx}\right] f = 0.$$

2.13. We use the solution of Exercise 2.11. Let us write out the infinitesimal divergence invariance condition (2.94) for the functional under consideration and find all the functions $\xi(x,y)$, $\phi(x,y)$, and $K(x,y,y')$ that satisfy the condition. In this case, (2.94) takes the form

$$\left[\xi\frac{\partial}{\partial x} + \phi\frac{\partial}{\partial y} + \left(\frac{d\phi}{dx} - y'\frac{d\xi}{dx}\right)\frac{\partial}{\partial y'} + \frac{d\xi}{dx}\right] f - \frac{d}{dx}K$$
$$= \frac{\partial\phi}{\partial x}y' + \left(\frac{\partial\phi}{\partial y} - \frac{1}{2}\frac{\partial\xi}{\partial x}\right)y'^2 - \frac{1}{2}\frac{\partial\xi}{\partial y}y'^3 - \frac{\partial K}{\partial x} - \frac{\partial K}{\partial y}y' - \frac{\partial K}{\partial y'}y'' = 0.$$

As y satisfies the Euler–Lagrange equation, the multiplier of y'' is zero. So the new conservation law does not change whether K depends on y' or not. This means we can take K independent of y': $K = K(x, y)$. As in Exercise 2.11, the coefficients of the potentials of y' do not depend on y'. It follows that they must vanish. Thus, $\xi(x, y)$, $\phi(x, y)$, and $K(x, y, y')$ are defined by the four simultaneous equations

$$\frac{\partial K}{\partial x} = 0, \tag{A.9}$$

$$\frac{\partial \phi}{\partial x} - \frac{\partial K}{\partial y} = 0, \tag{A.10}$$

$$\frac{\partial \phi}{\partial y} - \frac{1}{2}\frac{\partial \xi}{\partial x} = 0, \tag{A.11}$$

$$-\frac{1}{2}\frac{\partial \xi}{\partial y} = 0. \tag{A.12}$$

From (A.9) and (A.12) it follows that $K = K(u)$ and $\xi = \xi(x)$, respectively.

Differentiate (A.10) with respect to y and (A.11) with respect to x, then subtract the results to get

$$\frac{\partial^2 K}{\partial y^2} - \frac{1}{2}\frac{\partial^2 \xi}{\partial x^2} = 0.$$

As K depends only on y and ξ depends only on x, this equation can hold only if both terms are constant:

$$\frac{\partial^2 K}{\partial y^2} = C_1 = \frac{1}{2}\frac{\partial^2 \xi}{\partial x^2}.$$

This gives us

$$K = \frac{1}{2}C_1 y^2 + C_2 y + C_0, \qquad \xi = C_1 x^2 + C_3 + C_4,$$

where the C_k are constants. We can put $C_0 = 0$, as K is defined up to a constant. Indeed, adding a constant to K does not change the conservation law. Next, from (A.10) we find that

$$\phi = (C_1 y + C_2)x + \phi_0(y).$$

The function $\phi_0(y)$ is defined by (A.11), which reduces to

$$\phi_0'(y) = \frac{1}{2}C_3$$

and yields $\phi_0(y) = C_4 y/2 + C_5$. Therefore

$$\phi = C_1 xy + C + 2x + \frac{1}{2}C_3 y + C_5.$$

Finally, the conservation law is given by

$$P - K = \phi y' - \frac{1}{2}\xi y'^2 - K = \text{constant}.$$

Because we have five independent integration constants, we have obtained five independent conservation laws. They are obtained, respectively, when we set one of the constants to 1 and the rest to zero. Three of these conservation laws were found in Exercise 2.11.

Exercise 2.11 showed that the functional has three variational symmetries only, to which there correspond the three conservation laws. This exercise shows that existence of divergence symmetry extends the number of conservation laws to five. So we obtained five conservation laws.

3.2. The result follows from differentiation of the equality

$$\mathbf{\Psi}(t) \cdot \mathbf{\Psi}^{-1}(t) = \mathbf{E}.$$

We have

$$\left(\mathbf{\Psi}(t) \cdot \mathbf{\Psi}^{-1}(t)\right)' = \mathbf{\Psi}'(t) \cdot \mathbf{\Psi}^{-1}(t) + \mathbf{\Psi}(t) \cdot \left(\mathbf{\Psi}^{-1}(t)\right)' = \mathbf{E}' = \mathbf{0},$$

hence

$$\mathbf{\Psi}(t) \cdot \left(\mathbf{\Psi}^{-1}(t)\right)' = -\mathbf{\Psi}'(t) \cdot \mathbf{\Psi}^{-1}(t)$$

and can premultiply both sides by $\mathbf{\Psi}^{-1}(t)$.

3.3. Use the linearity of the main part of the increment with respect to the increment of the control function.

3.4. Introduce an additional component y_{n+1} of the vector $\mathbf{y}$ by the equations $y'_{n+1}(t) = G(\mathbf{y}(t))$, $y_{n+1}(0) = 0$.

3.5. By Pontryagin's maximum principle we get that F take the values $+1$ or -1 for optimal solution. Solve the problems with this F and collect the whole solution using these solutions.

4.1. Assume S is closed in X. Let $\{x_n\} \subset S$ be convergent (in X) so that $x_n \to x$ for some $x \in X$. We want to show that $x \in S$. Let us suppose $x \notin S$ and seek a contradiction. Given any $\varepsilon > 0$ there exists x_k ($\neq x$) such that $d(x_k, x) < \varepsilon$ (by the assumed convergence), so x is a limit point of S. Therefore S fails to contain all its limit points, and by definition is not closed.

Conversely, assume S contains the limits of all its convergent sequences. Let y be a limit point of S. By virtue of this, construct a convergent sequence $y_n \subset S$ as follows: for each n, take a point $y_n \in S$ such that $d(y_n, y) < 1/n$. Then $y_n \to y$ (in X). By hypothesis then, $y \in S$. This shows that S contains all its limit points, hence S is closed by definition.

4.2. (a) Let $B(p, r)$ denote the closed ball centered at point p and having radius r, and let q be a limit point of $B(p, r)$. There is a sequence of points p_k in $B(p, r)$

such that $d(p_k, q) \to 0$ as $k \to \infty$. For each k we have

$$d(q,p) \le d(q,p_k) + d(p_k,p) \le d(q,p_k) + r,$$

hence as $k \to \infty$ we get $d(p,q) \le r$. This proves that $q \in B(p,r)$. (b) True vacuously. (c) Obvious. (d) Let $S = \cap_{i\in I} S_i$ be an intersection of closed sets S_i. If $S = \emptyset$ then it is closed by part (b). Otherwise let q be any limit point of S and choose a sequence $\{p_k\} \subset S$ such that $p_k \to q$. We have $\{p_k\} \subset S_i$ for each i, and each S_i is closed so that we must have $q \in S_i$ for each i. This means that $q \in \cap_{i\in I} S_i$. (e) We communicate the general idea by outlining the proof for a union of two sets. Let $S = A \cup B$ where A, B are closed. Choose a convergent sequence $\{x_n\} \subset S$ and call its limit x. There is a subsequence $\{x_{n_k}\}$ that consists of points belonging to one of the given sets. Without loss of generality suppose $\{x_{n_k}\} \subset A$. But $x_{n_k} \to x$, hence $x \in A$ since A is closed. Therefore $x \in S$.

4.3. It is clear that the sequence of centers $\{x_n\}$ is a Cauchy sequence. By completeness, $x_n \to x$ for some $x \in X$. For each n, the sequence $\{x_{n+p}\}_{p=1}^{\infty}$ lies in $B(x_n, r_n)$ and converges to x; since the ball is closed we have $x \in B(x_n, r_n)$. This proves existence of a point in the intersection of all the balls. If y is any other such point, then $d(y,x) \le d(y,x_n) + d(x_n,x) \le 2\varepsilon_n \to 0$ as $n \to \infty$. Hence $y = x$ and we have proved uniqueness.

4.4. Let us verify the norm properties for $\|\cdot\|_{X/U}$. Certainly we have $\|x+U\|_{X/U} \ge 0$. Recalling that the zero element of X/U is U, we have

$$\|0_{X/U}\|_{X/U} = \|0_X + U\|_{X/U} = \inf_{u\in U} \|0_X + u\|_X = 0$$

since $0_X \in U$. Conversely, if $\|x+U\|_{X/U} = 0$ then

$$\inf_{u\in U} \|x+u\|_X = 0,$$

hence for every $\varepsilon > 0$ there exists $u \in U$ such that $\|x+u\|_X < \varepsilon$. From this we can infer the existence of a sequence $\{u_k\} \subset U$ such that

$$\lim_{k\to\infty} \|x+u_k\|_X = 0.$$

But this implies $x + u_k \to 0$, or $u_k \to -x$. Since U is closed we have $-x \in U$, hence $x + U = U$. Next,

$$\begin{aligned} \|\alpha(x+U)\|_{X/U} &= \|\alpha x + U\|_{X/U} = \inf_{u\in U} \|\alpha x + u\|_X = |\alpha| \inf_{u\in U} \left\| x + \frac{1}{\alpha} u \right\|_X \\ &= |\alpha| \inf_{u\in U} \|x+u\|_X = |\alpha| \; \|x+U\|_{X/U} \,. \end{aligned}$$

Finally

$$\|(x+U)+(y+U)\|_{X/U} = \|(x+y)+U\|_{X/U} = \inf_{u\in U} \|(x+y)+u\|_X$$
$$= \inf_{u,u'\in U} \left\|(x+y)+u+u'\right\|_X$$
$$= \inf_{u,u'\in U} \left\|(x+u)+(y+u')\right\|_X$$

so that

$$\|(x+U)+(y+U)\|_{X/U} \le \inf_{u,u'\in U} (\|(x+u)\|_X + \left\|(y+u')\right\|_X)$$
$$= \inf_{u,u'\in U} \|(x+u)\|_X + \inf_{u,u'\in U} \left\|(y+u')\right\|_X$$
$$= \inf_{u\in U} \|(x+u)\|_X + \inf_{u'\in U} \left\|(y+u')\right\|_X$$
$$= \|(x+U)\|_{X/U} + \|(y+U)\|_{X/U},$$

and the triangle inequality holds.

Now suppose X is complete. Choose a Cauchy sequence $\{y_k+U\} \subset X/U$. A "diagonal sequence" argument may be used to extract a subsequence $\{x_k+U\}$ of $\{y_k+U\}$ such that

$$\|(x_2+U)-(x_1+U)\|_{X/U} < 1/2,$$
$$\|(x_3+U)-(x_2+U)\|_{X/U} < 1/2^2,$$
$$\vdots$$

i.e., such that

$$\|(x_{k+1}+U)-(x_k+U)\|_{X/U} = \|(x_{k+1}-x_k)+U\|_{X/U} < 1/2^k$$

for each k. Then by definition of $\|\cdot\|_{X/U}$ we can assert the existence of an element $u_k \in (x_{k+1}-x_k)+U$ having $\|u_k\|_X < 1/2^k$. Choose a sequence $\{z_k\} \subset X$ such that for each k

$$z_k \in x_k + U, \qquad z_{k+1} - z_k = u_k.$$

(We indicate how this is done; see [3] for a more formal argument. Choose $z_1 \in x_1 + U$. We now wish to choose z_2 so that $z_2 \in x_2 + U$ and $z_2 - z_1 = u_1$. Write

$$u_1 = x_2 - x_1 + v \quad \text{for some } v \in U$$

and also

$$z_1 = x_1 + w \quad \text{for some } w \in U.$$

Then $u_1 + x_1 = x_2 + v$; add w to both sides and let $v + w = w' \in U$ to get

$$z_1 + u_1 = x_2 + w'.$$

Hence define $z_2 = x_2 + w'$. Repeat this procedure to generate $z_3, z_4, \ldots$.) Then

$$\|z_{k+1} - z_k\|_X < 1/2^k.$$

If $m > n$ then

$$\begin{aligned} \|z_m - z_n\|_X &\leq \|z_m - z_{m-1}\|_X + \cdots + \|z_{n+1} - z_n\|_X \\ &< \frac{1}{2^{m-1}} + \cdots + \frac{1}{2^n} < \frac{1}{2^{n-1}} \end{aligned}$$

so $\{z_k\}$ is Cauchy in X. Since X is complete, $z_k \to z$ for some $z \in X$. By the way the z_k were defined we have $x_k + U = z_k + U$. Then

$$\begin{aligned} \|(x_k + U) - (z + U)\|_{X/U} &= \|(z_k + U) - (z + U)\|_{X/U} \\ &= \|(z_k - z) + U\|_{X/U} \\ &= \inf_{u \in U} \|(z_k - z) + u\|_X \\ &\leq \|z_k - z\|_X \to 0 \end{aligned}$$

so that $x_k + U \to z + U$. We have therefore shown that some subsequence of the Cauchy sequence $\{y_k + U\}$ has a limit.

4.5. Since X is separable it has a countable dense subset A. The set

$$S = \{[x] : x \in A\} \subseteq X/M$$

is evidently countable; let us show that it is also dense in X/M. Because the norm on X/M is given by

$$\|\,[x]\,\| = \inf_{m \in M} \|x + m\|,$$

the distance between any two of its elements $[x]$ and $[y]$ can be expressed as

$$\|\,[x] - [y]\,\| = \|\,[x - y]\,\| = \inf_{m \in M} \|(x - y) + m\|.$$

So let $[z] \in X/M$ and $\varepsilon > 0$ be given. We can find $w \in A$ such that $\|z - w\| < \varepsilon$.

Then the distance between $[z]$ and $[w]$ is given by

$$\begin{aligned}\inf_{m\in M}\|(z-w)+m\| &\le \inf_{m\in M}(\|z-w\|+\|m\|)\\ &= \|z-w\| + \inf_{m\in M}\|m\|\\ &= \|z-w\|\\ &< \varepsilon.\end{aligned}$$

The element $[w]$ belongs to S and lies within distance ε of $[z]$ in the space X/M.

4.7. Let us propose a linear mapping T: to each $[x] \in X/M$ there corresponds the image element $T([x]) = Ax_0$, where x_0 is that representative of $[x]$ which has minimum norm. (The existence of x_0 is guaranteed because M is closed.) We have

$$\|x_0\|_X = \|\,[x]\,\|_{X/M}\,,$$

so

$$\|T([x])\|_Y = \|Ax_0\|_Y \le c\,\|x_0\|_X = c\,\|\,[x]\,\|_{X/M}\,.$$

Therefore T is bounded.

4.9. Let T be defined by $T([x]) = A\bar{x}$, where $\bar{x}$ is the minimum-norm representative of $[x]$. Take a bounded sequence $\{\,[x]_n\,\}$ from X/M so that $\|\,[x]_n\,\|_{X/M} < R$ for some finite R. For each n, choose from $[x]_n$ the minimum-norm representative $\bar{x}_n$. We have $T([x]_n) = A\bar{x}_n$ for each n, and the sequence $\{\bar{x}_n\}$ is bounded (in X) because $\|\bar{x}_n\|_X = \|\,[x]_n\,\|_{X/M}$. By compactness of A, there is a subsequence $\{\bar{x}_{n_k}\}$ such that $\{A\bar{x}_{n_k}\}$ is a Cauchy sequence in X. Therefore $\{\,[x]_n\,\}$ contains a subsequence $\{\,[x]_{n_k}\,\}$ whose image under T is a Cauchy sequence in X.

4.11. (a) Let e_n denote the sequence with nth term 1 and remaining terms 0. Each $e_n \in \ell^2$, and any finite set $\{e_1, \ldots, e_N\}$ is linearly independent. (b) For any positive integer n we have

$$\lim_{p\to\infty}\left(\sum_{k=1}^{n}|x_k|^p\right)^{1/p} = \max_{1\le k\le n}|x_k| \le \sup_{k\ge 1}|x_k|$$

so that

$$\lim_{n\to\infty}\lim_{p\to\infty}\left(\sum_{k=1}^{n}|x_k|^p\right)^{1/p} = \lim_{p\to\infty}\left(\sum_{k=1}^{\infty}|x_k|^p\right)^{1/p} \le \sup_{k\ge 1}|x_k|.$$

But for each $k \ge 1$

$$|x_k| \le \lim_{p\to\infty}\left(\sum_{k=1}^{\infty}|x_k|^p\right)^{1/p}$$

so that

$$\sup_{k\geq 1}|x_k| \leq \lim_{p\to\infty}\left(\sum_{k=1}^{\infty}|x_k|^p\right)^{1/p}.$$

Hence

$$\lim_{p\to\infty}\left(\sum_{k=1}^{\infty}|x_k|^p\right)^{1/p} = \sup_{k\geq 1}|x_k|.$$

(c) For $\mathbf{x} = 0$ the inequality is obvious, hence we take $\|\mathbf{x}\|_p \neq 0$. Assume $q \geq p$. Note that $0 \leq a \leq 1$ implies $a^q \leq a^p$. If $0 \leq a_k \leq 1$ for each k then, we have

$$\sum_{k=1}^{n}(a_k)^q \leq \sum_{k=1}^{n}(a_k)^p.$$

Because

$$|x_k| = (|x_k|^p)^{1/p} \leq \left(\sum_{j=1}^{n}|x_j|^p\right)^{1/p} \leq \|\mathbf{x}\|_p, \tag{A.13}$$

we have $|x_k|/\|\mathbf{x}\|_p \leq 1$ for each k, and shall momentarily let $|x_k|/\|\mathbf{x}\|_p$ play the role of a_k above. Now

$$\frac{(\|\mathbf{x}\|_q)^q}{(\|\mathbf{x}\|_p)^q} = \sum_{k=1}^{n}\left(\frac{|x_k|}{\|\mathbf{x}\|_p}\right)^q \leq \sum_{k=1}^{n}\left(\frac{|x_k|}{\|\mathbf{x}\|_p}\right)^p = 1.$$

Hence $(\|\mathbf{x}\|_q)^q \leq (\|\mathbf{x}\|_p)^q$, and the desired inequality follows. (d) To see that $\ell^1 \subseteq \ell^p$, observe that

$$\sum_{k=1}^{\infty}|x_k|^p \leq \left(\sum_{k=1}^{\infty}|x_k|\right)^p = (\|\mathbf{x}\|_1)^p$$

so $\|\mathbf{x}\|_p \leq \|\mathbf{x}\|_1$. If $\mathbf{x} \in \ell^1$ then $\|\mathbf{x}\|_1 < \infty$, hence $\|\mathbf{x}\|_p < \infty$ so $\mathbf{x} \in \ell^p$. The inclusion $\ell^p \subseteq \ell^q$ follows from the inequality of part (c). Finally, we may take the supremum of (A.13) to obtain $\|\mathbf{x}\|_\infty \leq \|\mathbf{x}\|_p$. The inclusion $\ell^p \subseteq \ell^\infty$ follows. (e) Every summable sequence converges to zero, every sequence that converges to zero converges, and every convergent sequence is bounded. (f) Let $p < \infty$ and let $\{\mathbf{x}^n\}$ be a Cauchy sequence in ℓ^p. Each $\mathbf{x}^n = (x_1^n, x_2^n, \ldots, x_k^n, \ldots)$. Let $\varepsilon > 0$ be given and choose N such that whenever $m, n > N$,

$$(\|\mathbf{x}^m - \mathbf{x}^n\|_p)^p = \sum_{k=1}^{\infty}|x_k^m - x_k^n|^p < \varepsilon^p. \tag{A.14}$$

Suppose $m \geq n$ and fix $n > N$. By (A.13) we have for each k

$$|x_k^m - x_k^n| \leq \|\mathbf{x}^m - \mathbf{x}^n\|_p < \varepsilon;$$

hence, for each k the sequence $\{x_k^m\}$ is a Cauchy sequence in $\mathbb{R}$. By completeness of $\mathbb{R}$ we have $x_k^m \to x_k$, say. Now let $\mathbf{x} = (x_1, x_2, \ldots, x_k, \ldots)$. We will show that $\mathbf{x}^n \to \mathbf{x}$. By (A.14) for any finite j we have

$$\sum_{k=1}^{j} |x_k^m - x_k^n|^p < \varepsilon^p.$$

Hence

$$\lim_{m\to\infty} \sum_{k=1}^{j} |x_k^m - x_k^n|^p \le \varepsilon^p$$

which gives us

$$\sum_{k=1}^{j} |x_k - x_k^n|^p \le \varepsilon^p.$$

As $j \to \infty$ we therefore have

$$\sum_{k=1}^{\infty} |x_k - x_k^n|^p \le \varepsilon^p.$$

On one hand, this means that $\mathbf{x} \in \ell^p$. Indeed,

$$\|\mathbf{x}\|_p \le \|\mathbf{x} - \mathbf{x}^{N+1}\|_p + \|\mathbf{x}^{N+1}\|_p \le \varepsilon + \|\mathbf{x}^{N+1}\|_p < \infty.$$

On the other hand, this can be reworded: $\|\mathbf{x} - \mathbf{x}^n\|_p \le \varepsilon$ whenever $n > N$, hence $\mathbf{x}^n \to \mathbf{x}$ and so ℓ^p for $1 \le p < \infty$ is complete.

Now consider the case $p = \infty$. Let $\{\mathbf{x}^n\}$ be a Cauchy sequence in ℓ^∞. Each $\mathbf{x}^n = (x_k^n)_{k=1}^\infty$. Fix $\varepsilon > 0$ and choose N such that whenever $m, n > N$,

$$\sup_k |x_k^m - x_k^n| < \varepsilon.$$

Suppose $m \ge n$ and fix $n > N$. For each k

$$|x_k^m - x_k^n| < \varepsilon, \tag{A.15}$$

hence for each k the sequence $\{x_k^m\}$ is a Cauchy sequence of real numbers. By completeness of $\mathbb{R}$ we have $x_k^m \to x_k$, say. Now let $\mathbf{x} = (x_k)_{k=1}^\infty$ and show that $\mathbf{x}^n \to \mathbf{x}$. As $m \to \infty$ (A.15) gives

$$|x_k - x_k^n| \le \varepsilon$$

for each k. Hence

$$\sup_k |x_k - x_k^n| \le \varepsilon$$

for $n > N$, proving that $\mathbf{x}^n \to \mathbf{x}$. Since $\|\mathbf{x} - \mathbf{x}^n\|_\infty \le \varepsilon$ for $n > N$ we have

$$\|\mathbf{x}\|_\infty \le \|\mathbf{x} - \mathbf{x}^{N+1}\|_\infty + \|\mathbf{x}^{N+1}\|_\infty \le \varepsilon + \|\mathbf{x}^{N+1}\|_\infty,$$

hence $\mathbf{x} \in \ell^\infty$. (g) Let $\mathbf{x} = (\xi_1, \xi_2, \ldots) \in \ell^p$. Since $\sum_{k=1}^\infty |\xi_k|^p$ converges we can choose n large enough to make $\sum_{k=n+1}^\infty |\xi_k|^p$ as small as desired. Hence we can approximate $\mathbf{x}$ arbitrarily closely by an element $\mathbf{x}_n$ having the form

$$\mathbf{x}_n = (\xi_1, \xi_2, \ldots, \xi_n, 0, 0, 0, \ldots).$$

Furthermore each ξ_i may be approximated by a rational number r_i. The set S consisting of all elements of the form

$$\mathbf{y}_n = (r_1, r_2, \ldots, r_n, 0, 0, 0, \ldots)$$

is countable and dense in ℓ^p. More formally, let $\varepsilon > 0$ be given. Choose n so that $\sum_{k=n+1}^\infty |\xi_k|^p < \varepsilon^p/2$, then choose the r_i so that $|\xi_i - r_i| < \varepsilon/(2n)^{1/p}$ for each $i = 1, \ldots, n$. We have

$$\|\mathbf{x} - \mathbf{y}_n\|^p = \sum_{k=1}^n |\xi_k - r_k|^p + \sum_{k=n+1}^\infty |\xi_k|^p < n\frac{\varepsilon^p}{2n} + \frac{\varepsilon^p}{2} = \varepsilon^p$$

as desired. (h) Fix any countable subset $\{\mathbf{x}^{(n)}\}_{n=1}^\infty$ of ℓ^∞. Denote the components of $\mathbf{x}^{(n)}$ by

$$\mathbf{x}^{(n)} = (\xi_1^{(n)}, \xi_2^{(n)}, \xi_3^{(n)}, \ldots).$$

We now construct $\mathbf{z} \in \ell^\infty$ such that $\|\mathbf{z} - \mathbf{x}^{(n)}\|_\infty \ge 1$ for all n. Denoting

$$\mathbf{z} = (\zeta_1, \zeta_2, \zeta_3, \ldots)$$

we let

$$\zeta_k = \begin{cases} \xi_k^{(k)} + 1, & |\xi_k^{(k)}| \le 1, \\ 0, & |\xi_k^{(k)}| > 1 \end{cases}$$

for each $k = 1, 2, 3, \ldots$. Then

$$\|\mathbf{z} - \mathbf{x}^{(n)}\|_\infty = \sup_{m \ge 1} |\zeta_m - \xi_m^{(n)}| \ge |\zeta_n - \xi_n^{(n)}| \ge 1$$

as desired. (i) Let S be the set of all vectors whose components form rational sequences that converge to 0. This set is evidently countable. We show that it is dense in c_0. Given $\mathbf{x} = (\xi_1, \xi_2, \ldots) \in c_0$ and $\varepsilon > 0$, choose $\mathbf{y} = (r_1, r_2, \ldots) \in S$ such that $|\xi_i - r_i| < \varepsilon$ for all $i = 1, 2, \ldots$. Then $\|\mathbf{x} - \mathbf{y}\|_\infty = \sup_i |\xi_i - r_i| < \varepsilon$.

4.12. Let $\{x_n\}$ be a Cauchy sequence in $(\mathbb{R}, d)$. We first show that $\{x_n\}$ is a Cauchy sequence in $(\mathbb{R}, |\cdot|)$. We have

$$|x_n^3 - x_m^3| = \underbrace{|x_n - x_m|}_{\text{factor 1}} \underbrace{|x_n^2 + x_n x_m + x_m^2|}_{\text{factor 2}} \to 0 \quad \text{as } m, n \to \infty.$$

This implies that either factor 1 or factor 2 approaches zero, or both. However, if factor 2 approaches zero then $x_n \to 0$ as $n \to \infty$, and this in turn implies that factor 1 approaches zero. So factor 1 must approach zero in any case.

Next, by the known completeness of $(\mathbb{R}, |\cdot|)$, we can name a limit element $x \in \mathbb{R}$ for $\{x_n\}$.

Finally, we show that $x_n \to x$ in $(\mathbb{R}, d)$. This follows from the equality

$$|x_n^3 - x^3| = |x_n - x|\, |x_n^2 + x_n x + x^2|,$$

because the first factor on the right approaches zero and the second factor is bounded (since $\{x_n\}$ is bounded).

Note that here we have no inequality $|x^3 - y^3| < m|x - y|$ for all x, y in $\mathbb{R}$, but the notions of sequence convergence with both metrics are equivalent. This distinguishes the notion of equivalence of metrics from that of equivalence of norms.

4.13. Call

$$\alpha = \sup_{\|x\| \neq 0} \frac{\|Ax\|}{\|x\|}.$$

By linearity of A, α is also equal to the other expression given in the exercise. By definition of supremum we have two things:

(1) For every $\varepsilon > 0$ there exists some $x_0 \neq 0$ such that

$$\frac{\|Ax_0\|}{\|x_0\|} > \alpha - \varepsilon.$$

Equivalently, $\|Ax_0\| > (\alpha - \varepsilon)\, \|x_0\|$. This implies, by the definition of $\|A\|$, that

$$\alpha - \varepsilon < \|A\|\,.$$

So $\alpha < \|A\| + \varepsilon$, and since $\varepsilon > 0$ is arbitrary we have $\alpha \leq \|A\|$.

(2) For every $x \neq 0$ we have

$$\frac{\|Ax\|}{\|x\|} \leq \alpha.$$

So $\|Ax\| \leq \alpha\, \|x\|$ for $x \neq 0$; in fact, this obviously holds when $x = 0$ as well so it holds for all x. By definition of $\|A\|$ we have $\|A\| \leq \alpha$.

Combining the inequalities from parts 1 and 2 we obtain $\|A\| = \alpha$.

4.15. We can show that the f_i are linearly dependent if and only if the Gram determinant is zero. The proof can rest on the fact that a linear homogeneous system $Ax = 0$ has a nontrivial solution if and only if $\det A = 0$.

Assume linear dependence. Then $\sum_{i=1}^{n} \alpha_i f_i = 0$ for some α_i not all zero. Taking inner products of this equation with the f_i in succession, we get

$$\begin{aligned} \alpha_1(f_1, f_1) + \cdots + \alpha_n(f_1, f_n) &= 0, \\ &\vdots \\ \alpha_1(f_n, f_1) + \cdots + \alpha_n(f_n, f_n) &= 0, \end{aligned} \tag{A.16}$$

or

$$\begin{pmatrix} (f_1, f_1) & \cdots & (f_1, f_n) \\ \vdots & \ddots & \vdots \\ (f_n, f_1) & \cdots & (f_n, f_n) \end{pmatrix} \begin{pmatrix} \alpha_1 \\ \vdots \\ \alpha_n \end{pmatrix} = \begin{pmatrix} 0 \\ \vdots \\ 0 \end{pmatrix}.$$

A nontrivial solution for the vector (α) implies that the Gram determinant vanishes. Conversely, assume the determinant vanishes so that (A.16) holds for some nontrivial (α). Rewrite (A.16) as

$$\left(f_i \,, \sum_{j=1}^{n} \alpha_j f_j \right) = 0, \qquad i = 1, \ldots, n,$$

multiply by α_i to get

$$\left(\alpha_i f_i \,, \sum_{j=1}^{n} \alpha_j f_j \right) = 0, \qquad i = 1, \ldots, n,$$

and then sum over i to obtain

$$\left(\sum_{i=1}^{n} \alpha_i f_i \,, \sum_{j=1}^{n} \alpha_j f_j \right) = \left\| \sum_{i=1}^{n} \alpha_i f_i \right\|^2 = 0.$$

Hence $\sum_{i=1}^{n} \alpha_i f_i = 0$ for some scalars α_i that are not all zero.

4.16. The statement $\|A_n - A\| \to 0$ means that

$$\|(A_n - A)x\| \le c_n \|x\| \quad \text{where } c_n \to 0$$

and each c_n is independent of x. Since $\|x\| \le M$ for all $x \in S$, we have

$$\|A_n x - Ax\| \le c_n M.$$

But $c_n M \to 0$ together with $c_n \to 0$ when $n \to \infty$, thus $A_n x \to Ax$.

4.18. We have

$$\left\| \sum_{n=0}^{\infty} c_n g_n \right\|^2 = \left(\sum_{n=0}^{\infty} c_n g_n \, , \, \sum_{k=0}^{\infty} c_k g_k \right) = \sum_{n=0}^{\infty} |c_n|^2 < \infty.$$

4.19. Assume $u(t)$ and $v(t)$ are each differentiable at t. Form the difference quotient

$$\frac{(u(t+h), v(t+h)) - (u(t), v(t))}{h} = \frac{1}{h}(u(t+h), v(t+h)) - \frac{1}{h}(u(t), v(t))$$

and on the right side subtract and add the term

$$\frac{1}{h}(u(t), v(t+h))$$

to write the difference quotient as

$$\left(\frac{u(t+h) - u(t)}{h} \, , \, v(t+h) \right) + \left(u(t) \, , \, \frac{v(t+h) - v(t)}{h} \right).$$

Then let $h \to 0$.

4.20. We can use the Cauchy–Schwarz inequality to write

$$\|x_n\| \, \|x\| \geq |(x_n, x)|$$

for each n, hence

$$\liminf_{n \to \infty} \|x_n\| \, \|x\| \geq \liminf_{n \to \infty} |(x_n, x)| = \lim_{n \to \infty} |(x_n, x)| = |(x, x)| = \|x\|^2 \, .$$

So

$$\|x\| \liminf_{n \to \infty} \|x_n\| \geq \|x\|^2 \, .$$

For $x \neq 0$ we can divide through by $\|x\|$ to get the desired inequality. It holds trivially when $x = 0$.

4.21. Because A is densely defined, for each $x \in V$ there is a sequence $\{x_n\} \subset D(A)$ such that $x_n \to x$. Since this sequence converges it is a Cauchy sequence. Because A is bounded, $\{Ax_n\}$ is a Cauchy sequence in W, hence converges to some $w \in W$. Furthermore, w does not depend on the Cauchy sequence used. (That is, if $x_n \to x$ and $x'_n \to x$, and $Ax_n \to w$, then $Ax'_n \to w$. Indeed for each n we have,

$$0 \leq \|Ax_n - Ax'_n\| = \|Ax_n - Ax + Ax - Ax'_n\| \leq \|A\| \left(\|x_n - x\| + \|x - x'_n\| \right);$$

as $n \to \infty$ we have $\lim_{n\to\infty} \|Ax_n - Ax'_n\| = 0$ and by continuity of the norm we have the conclusion.) Thus we can define an extension A_e by

$$A_e x = \lim_{n\to\infty} Ax_n = w \quad \text{for any } x \in V.$$

Linearity is evident. Since

$$\|A_e x\| = \left\|\lim_{n\to\infty} Ax_n\right\| = \lim_{n\to\infty} \|Ax_n\| \le \lim_{n\to\infty} \|A\| \, \|x_n\| = \|A\| \, \|x\| \, ,$$

A_e is bounded with $\|A_e\| \le \|A\|$. The reverse inequality follows by noting that $Ax = A_e x$ whenever $x \in D(A)$. Finally, we prove uniqueness: if A'_e is another bounded (hence continuous) linear extension of A, then for any sequence $\{x_n\} \subset D(A)$ with $x_n \to x$ we have

$$A'_e x = \lim_{n\to\infty} A'_e x_n = \lim_{n\to\infty} Ax_n = A_e x,$$

which gives $A'_e = A_e$.

4.22. Suppose $v_k \rightharpoonup v$ in V where the dimension of V is n. Choose a basis $\{e_k\}$ of V and write

$$v_k = \sum_{j=1}^{n} \alpha_j^{(k)} e_j, \qquad v = \sum_{j=1}^{n} \alpha_j e_j.$$

For an arbitrary bounded linear functional f on V we have $f(v_k) \to f(v)$ as $k \to \infty$. For $i = 1, \ldots, n$, put f equal to f_i defined for any $x = \sum_{k=1}^{n} \xi_k e_k$ by $f_i(x) = \xi_i$. Then $f_i(v_k) = \alpha_i^{(k)} \to f_i(v) = \alpha_i$ as $k \to \infty$, and we have

$$\lim_{k\to\infty} \|v - v_k\| = \lim_{k\to\infty} \left\| \sum_{j=1}^{n} (\alpha_j^{(k)} - \alpha_j) e_j \right\| \le \lim_{k\to\infty} \sum_{j=1}^{n} |\alpha_j^{(k)} - \alpha_j| \, \|e_j\| = 0.$$

4.23. (a) From $x = AA^{-1}x$ we obtain $\|x\| \le \|A\| \, \|A^{-1}\| \, \|x\|$ and the result follows. (b) Using $x = A^{-1}y$ we have $A\varepsilon = r$, hence $\varepsilon = A^{-1}r$. The four inequalities

$$\|x\| \le \|A^{-1}\| \, \|y\| \, , \qquad \|r\| \le \|A\| \, \|\varepsilon\| \, ,$$
$$\|y\| \le \|A\| \, \|x\| \, , \qquad \|\varepsilon\| \le \|A^{-1}\| \, \|r\| \, ,$$

follow immediately and yield the desired result.

4.24. Let B be the unit ball in X. The image of the bounded set B under T is precompact; T^{-1} returns this image into B. But a continuous operator maps precompact sets into precompact sets, hence if T^{-1} were bounded then B would be precompact. Since X is infinite dimensional, this is impossible.

4.25. (a) Let $F\colon X \to Y$ be an isometry between metric spaces (X, d_X) and (Y, d_Y). Then, by the definition,

$$d_Y(F(x_2), F(x_1)) = d_X(x_2, x_1) \quad \text{for all } x_1, x_2 \in X.$$

Continuity is evident. To see that F is one-to-one, suppose $F(x_2) = F(x_1)$. Then $d_Y(F(x_2), F(x_1)) = 0 = d_X(x_2, x_1)$, so $x_2 = x_1$ by the metric axioms. (b) First suppose $\|Ax\| = \|x\|$ for all $x \in X$. Replacing x by $x_2 - x_1$ we have $\|Ax_2 - Ax_1\| = \|x_2 - x_1\|$ as required. Conversely suppose that $\|Ax_2 - Ax_1\| = \|x_2 - x_1\|$ for any pair $x_1, x_2 \in X$. Putting $x_1 = 0$ and $x_2 = x$ we have the desired conclusion.

4.26. Suppose Parseval's equality holds for all f in H. We fix f and use the equality, equation (4.89), and continuity to write

$$\begin{aligned} 0 &= \lim_{n\to\infty}\left(\|f\|^2 - \sum_{k=1}^{n} |(f, g_k)|^2\right) \\ &= \lim_{n\to\infty}\left\|f - \sum_{k=1}^{n}(f, g_k)g_k\right\|^2 \\ &= \left\|f - \sum_{k=1}^{\infty}(f, g_k)g_k\right\|^2. \end{aligned}$$

This shows that

$$f = \sum_{k=1}^{\infty} \alpha_k g_k \quad \text{where } \alpha_k = (f, g_k).$$

4.27. The inequality

$$\left\|\frac{df}{dx}\right\|_{C(-\infty,\infty)} \le \alpha \|f\|_{C^{(1)}(-\infty,\infty)},$$

i.e.

$$\sup\left|\frac{df(x)}{dx}\right| \le \alpha\left(\sup|f(x)| + \sup\left|\frac{df(x)}{dx}\right|\right)$$

obviously holds with $\alpha = 1$.

4.28. We construct a subset M of the space whose elements cannot be approximated by functions from a countable set. Let α be an arbitrary point of $[0, 1]$. Form M from functions defined as follows:

$$f_\alpha(x) = \begin{cases} 1, & x \ge \alpha, \\ 0, & x < \alpha. \end{cases}$$

The distance from $f_\alpha(x)$ to $f_\beta(x)$ is

$$\|f_\alpha(x) - f_\beta(x)\| = \sup_{x\in[0,1]} |f_\alpha(x) - f_\beta(x)| = 1 \text{ if } \alpha \neq \beta.$$

Take a ball B_α of radius 1/3 about $f_\alpha(x)$. If $\alpha \neq \beta$ then $B_\alpha \cap B_\beta$ is empty.

If a countable subset is dense in the space then each of the B_α must contain at least one element of this subset, but this contradicts Theorem 4.11 since the set of balls B_α is of equal power with the continuum.

4.29. Let $\{A_n\}$ be a Cauchy sequence in $L(X, Y)$, i.e.,

$$\|A_{n+m} - A_n\| \to 0 \quad \text{as } n \to \infty, \qquad m > 0.$$

We must show that there is a continuous linear operator A such that $A_n \to A$. For any $x \in X$, $\{A_n x\}$ is also a Cauchy sequence because

$$\|A_{n+m}x - A_n x\| \le \|A_{n+m} - A_n\| \, \|x\| \, ;$$

hence there is a $y \in Y$ such that $A_n x \to y$ since Y is a Banach space. For every $x \in X$ this defines a unique $y \in Y$, i.e., defines an operator A such that $y = Ax$. This operator is clearly linear. Since $\{A_n\}$ is a Cauchy sequence, the sequence of norms $\{\|A_n\|\}$ is bounded:

$$\|Ax\| = \lim_{n\to\infty} \|A_n x\| \le \limsup_{n\to\infty} \|A_n\| \, \|x\| \, .$$

That is, A is continuous.

4.30. We can see that the equation $(A + B)x = y$ has a solution for any $y \in Y$ by applying the contraction mapping theorem. Indeed, pre-multiplication by A^{-1} allows us to rewrite this equation as $x = Cx + x_0$ where $C = -A^{-1}B$ and $x_0 = A^{-1}y$. Defining $F(x) = Cx + x_0$, we see that $F(x)$ is a contraction mapping:

$$\|F(x) - F(y)\| = \|Cx - Cy\| \le \|C\| \, \|x - y\| \, , \qquad \|C\| \le \|A^{-1}\| \, \|B\| < 1.$$

Since the equation $x = F(x)$ has a unique solution $x^* \in X$, so does the original equation.

From $x = A^{-1}Ax$ it follows that $\|x\| \le \|A^{-1}\| \, \|Ax\|$, hence

$$\|Ax\| \ge \|A^{-1}\|^{-1} \, \|x\| \, .$$

So for any $y \in Y$ we can write

$$\|y\| = \|(A + B)x\| \ge \|Ax\| - \|Bx\| \ge \|A^{-1}\|^{-1} \, \|x\| - \|B\| \, \|x\|$$

and therefore

$$\|x\| \le (\|A^{-1}\|^{-1} - \|B\|)^{-1} \, \|y\| \, .$$

The desired inequality follows.

4.31. First assume that $y = \lambda x$ for some scalar λ. Then

$$|(x,y)| = |(x,\lambda x)| = |\bar{\lambda}||(x,x)| = |\lambda|\,\|x\|^2 = \|x\|\,\|\lambda x\| = \|x\|\,\|y\|\,,$$

hence equality holds. Conversely, assume equality holds in (4.49). Squaring both sides, we obtain the relation

$$(x,y)\overline{(x,y)} = \|x\|^2\,\|y\|^2\,.$$

Using this it is easily verified that

$$|(y,y)x - (x,y)y|^2 = ((y,y)x - (x,y)y\,,\,(y,y)x - (x,y)y) = 0,$$

hence $(y,y)x - (x,y)y = 0$.

4.32. As F is continuous, $|Fx| \leq \|F\|\,\|x\|$. Next,

$$\|x\|^2 - Fx \geq \|x\|^2 - \|F\|\,\|x\| = \|x\|\,(\|x\| - \|F\|).$$

It seen that if $\|x\| \geq \|F\|$ then $\|x\|^2 - Fx \geq 0$ and if $\|x\| \leq \|F\|$ then

$$\|x\|^2 - Fx \geq -Fx \geq -\|F\|\,\|x\| \geq -\|F\|^2\,.$$

4.33. (a) Let us denote $X \setminus S$ by S^c. First suppose that S is open. Let y be an arbitrary point of S. Assume to the contrary that every open ball centered at y contains a point of S^c. In particular, each such ball having radius $1/n$, $n = 1, 2, 3, \ldots$, contains some point $x_n \in S^c$. So there is a sequence $\{x_n\} \subset S^c$ such that $x_n \to y$. But S^c is closed so we must have $y \in S^c$, a contradiction. Conversely, suppose that every point of S is the center of some open ball contained entirely within S. Suppose to the contrary that S is not open. Then S^c is not closed, and there is a convergent sequence $\{z_n\} \subset S^c$ having a limit $y \in S$. This means there are points of $\{z_n\}$ that are arbitrarily close to y, so it is impossible to find a ball centered at y that is contained entirely within S. This contradiction completes the proof. (b) Take an open ball of radius r centered at x, and denote by U the complement of this ball. Now take any sequence $\{x_n\} \subset U$ such that $x_n \to x$. Since $\|x_n - x\| \geq r$ for each n, we have $\|x_0 - x\| \geq r$ by continuity of the norm. This shows that $x_0 \in U$, hence U is closed. So the original ball is open by definition. (c) Let f be continuous and let S be open in Y. The set $f^{-1}(S)$ is open if it is empty, so we suppose it to be nonempty. Choose any $x \in f^{-1}(S)$. Then $f(x) \in S$, and since S is open there is an open ball $B(f(x), \varepsilon)$ contained entirely in S. By continuity there exists a ball $B(x, \delta)$ whose image $f(B(x, \delta))$ is contained in $B(f(x), \varepsilon)$ and therefore in S. So $B(x, \delta)$ is contained in $f^{-1}(S)$. This shows that $f^{-1}(S)$ is open. Next let $f^{-1}(S)$ be open whenever S is open, and pick an arbitrary $x \in X$. The ball $B(f(x), \varepsilon)$ is open so its inverse image is open and contains x. Hence there is a ball $B(x, \delta)$ contained in this inverse image. We have $f(B(x, \delta))$ contained in $B(f(x), \varepsilon)$, so f is continuous at x.

4.34. The function

$$f(x) = \begin{cases} 1, & x \text{ rational}, \\ 0, & x \text{ irrational}, \end{cases}$$

can be defined on $\mathbb{R}$. Now for any real number x_0, whether rational or irrational, there are sequences tending to x_0 that consist of purely rational or purely irrational elements (i.e., both the rationals and the irrationals are dense in the reals). For one type of sequence the limit is 1 and for the other type the limit is zero. Thus at point x_0 there is no limit value and the function is not continuous by definition.

4.35. We can write

$$\begin{aligned} \|Au\|^2_{L^2(0,1)} &= \int_0^1 \left(\int_0^1 k(s,t)u(t)\,dt \right)^2 ds \\ &\le \int_0^1 \left(\int_0^1 |k(s,t)|^2\,dt \right) \left(\int_0^1 u^2(t)\,dt \right) ds \\ &= \left(\int_0^1 \int_0^1 |k(s,t)|^2\,dt\,ds \right) \int_0^1 u^2(t)\,dt \\ &= M^2 \|u\|^2_{L^2(0,1)} \end{aligned}$$

where

$$M = \left(\int_0^1 \int_0^1 |k(s,t)|^2\,ds\,dt \right)^{1/2}.$$

Therefore

$$\|Au\|_{L^2(0,1)} \le M \|u\|_{L^2(0,1)}$$

and we have $\|A\| \le M$.

4.36. Since $\|S\mathbf{x}\| = \|\mathbf{x}\|$, we have $\|S\| = 1$.

4.37. We have

$$\begin{aligned} \|Ax - Ay\| &= \max_{t\in[0,1]} \left| \int_0^t x^2(s)\,ds - \int_0^t y^2(s)\,ds \right| \\ &\le \max_{t\in[0,1]} \int_0^t |x(s)+y(s)| \cdot |x(s)-y(s)|\,ds \\ &\le \left(\max_{t\in[0,1]} |x(t)| + \max_{t\in[0,1]} |y(t)| \right) \cdot \max_{t\in[0,1]} |x(t)-y(t)| \cdot \max_{t\in[0,1]} \int_0^t ds \\ &= (\|x\| + \|y\|) \cdot \|x-y\|\,. \end{aligned}$$

On any ball of the form $\|x\| \le \frac{1}{2} - \varepsilon$ where $\varepsilon > 0$, we have $\|Ax - Ay\| \le q\,\|x-y\|$ where $q < 1$.

4.38. All elements of the form

$$\mathbf{x}_n = \left(1, \frac{1}{2}, \frac{1}{3}, \ldots, \frac{1}{n}, 0, 0, 0, \ldots\right)$$

belong to S. The sequence $\{\mathbf{x}_n\}$ is a Cauchy sequence because for $m \geq 1$ we have

$$\|\mathbf{x}_{n+m} - \mathbf{x}_n\| = \sup_{n+1 \leq k \leq n+m} \frac{1}{k} = \frac{1}{n+1} \to 0 \quad \text{as } n \to \infty.$$

However, the element $\lim_{n\to\infty} \mathbf{x}_n$ does not belong to S.

4.39. The Neumann series for $(A - I)^{-1}$ is

$$(A - I)^{-1} = -\sum_{k=0}^{\infty} A^k.$$

So

$$\|(A - I)^{-1}\| \leq \sum_{k=0}^{\infty} \|A^k\| \leq \sum_{k=0}^{\infty} \|A\|^k = \frac{1}{1 - \|A\|}.$$

4.40. The reader should verify that the norm axioms are satisfied for the norm in question. Then take a Cauchy sequence $\{(x_k, y_k)\} \subset X \times Y$ so that

$$\begin{aligned}\|(x_m, y_m) - (x_n, y_n)\|_{X\times Y} &= \|(x_m - x_n, y_m - y_n)\|_{X\times Y} \\ &= \max\{\|x_m - x_n\|_X, \|y_m - y_n\|_Y\} \\ &\to 0 \quad \text{as } m, n \to \infty.\end{aligned}$$

This implies that

$$\|x_m - x_n\|_X \to 0 \quad \text{and} \quad \|y_m - y_n\|_Y \to 0 \quad \text{as } m, n \to \infty.$$

So $\{x_k\}$ and $\{y_k\}$ are each Cauchy sequences in their respective spaces X, Y; by completeness of these spaces we have $x_k \to x$ and $y_k \to y$ for some $x \in X$ and $y \in Y$. Finally, we have $(x_k, y_k) \to (x, y)$ in the norm of $X \times Y$:

$$\begin{aligned}\|(x_k, y_k) - (x, y)\| &= \|(x_k - x, y_k - y)\| \\ &= \max\{\|x_k - x\|_X, \|y_k - y\|_Y\} \\ &\to 0 \quad \text{as } k \to \infty.\end{aligned}$$

4.41. We have

$$\|y_n - x\| = \left\| \frac{\sum_{i=1}^{n}(x_i - x)}{n} \right\| \leq \frac{1}{n} \sum_{i=1}^{n} \kappa_i \quad \text{where } \kappa_i \equiv \|x_i - x\|.$$

Then for any m between 1 and n we can write

$$\begin{aligned}
\|y_n - x\| &\le \frac{1}{n}\sum_{i=1}^{m} \kappa_i + \frac{1}{n}\sum_{i=m+1}^{n} \kappa_i \\
&\le \frac{1}{n}\left(m \cdot \max_{1\le i\le m} \kappa_i\right) + \left(\frac{n-m}{n}\right) \cdot \max_{m+1\le i\le n} \kappa_i \\
&\le \frac{1}{n}\left(m \cdot \max_{1\le i\le m} \kappa_i\right) + \max_{i\ge m+1} \kappa_i .
\end{aligned}$$

Let $\varepsilon > 0$ be given. Choose and fix m sufficiently large that the second term is less than $\varepsilon/2$. In the first term the quantity in parentheses is then fixed, and we can therefore choose $N > m$ so that the first term is less than $\varepsilon/2$ for $n > N$.

4.42. Assume $\|\cdot\|_1$ and $\|\cdot\|_2$ have the property that $\|x_n - x\|_1 \to 0$ if and only if $\|x_n - x\|_2 \to 0$. Now suppose to the contrary that there is *no* positive constant C such that $\|x\|_2 \le C\|x\|_1$ for all $x \in X$. Then for each positive integer n there exists $x_n \in X$ such that

$$\|x_n\|_2 > n\|x_n\|_1 .$$

Define

$$y_n = \frac{1}{\sqrt{n}} \frac{x_n}{\|x_n\|_1}.$$

Then

$$\|y_n\|_1 = \frac{1}{\sqrt{n}} \to 0 \quad \text{as } n \to \infty$$

while

$$\|y_n\|_2 = \frac{1}{\sqrt{n}} \frac{\|x_n\|_2}{\|x_n\|_1} > \frac{1}{\sqrt{n}} \cdot n = \sqrt{n} \to \infty \quad \text{as } n \to \infty.$$

This contradiction shows that the required constant C does exist. Interchange the norms to get the reverse inequality.

4.43. We have $|\,\|x_m\| - \|x_n\|\,| \le \|x_m - x_n\| \to 0$ as $m, n \to \infty$, hence the sequence of norms is a Cauchy sequence in $\mathbb{R}$.

4.44. Let U be a separable, dense subspace of X. We take a countable dense subset A of U and show that A is also dense in X. Let $x \in X$ and $\varepsilon > 0$ be given. Since U is dense in X there exists $x' \in U$ such that $d(x, x') < \varepsilon/2$. Since A is dense in U there exists $x'' \in A$ such that $d(x', x'') < \varepsilon/2$. So $d(x, x'') < \varepsilon$ as required.

4.45. Let X be a Banach space so that any Cauchy sequence in it has a limit. Now let $\sum_{k=1}^{\infty} x_k$ be an absolutely convergent series of elements $x_k \in X$. Denote by s_i the ith partial sum of this series. Now $\{s_i\}$ is a Cauchy sequence in X because for $m > n$ we have

$$\|s_m - s_n\| = \left\| \sum_{k=n+1}^{m} x_k \right\| \le \sum_{k=n+1}^{\infty} \|x_k\| \to 0 \quad \text{as } m, n \to \infty.$$

Therefore $s_i \to s$ for some $s \in X$ by completeness.

Conversely suppose every absolutely convergent series of elements taken from X is convergent. Let $\{x_k\}$ be any Cauchy sequence in X. For every positive integer k we can find $N = N(k)$ such that $\|x_m - x_n\| < 1/2^k$ whenever $m, n > N$; furthermore, we can choose each such N so that $N(k)$ is a strictly increasing function of k. The series $\sum_{k=1}^{\infty}[x_{N(k+1)} - x_{N(k)}]$ converges absolutely:

$$\sum_{k=1}^{\infty} \left\|x_{N(k+1)} - x_{N(k)}\right\| < \sum_{k=1}^{\infty} \frac{1}{2^k} = 1.$$

Hence it converges and by definition its sequence of partial sums

$$s_j = \sum_{k=1}^{j}[x_{N(k+1)} - x_{N(k)}] = x_{N(j+1)} - x_{N(1)}$$

converges. Let s be its limit. From the last equality we see that $\{x_{N(j)}\}$ also converges and its limit is $x = s + x_{N(1)}$. But if a subsequence of a Cauchy sequence has a limit the entire sequence converges to it.

4.46. It suffices to show that the image of the unit ball, i.e., the set of all vectors $\mathbf{x} \in \ell^2$ having

$$\|\mathbf{x}\|^2 = \sum_{k=1}^{\infty} |x_k|^2 \leq 1,$$

is precompact. We call this image S and show that it is totally bounded (cf., Definition 4.46). Let $\varepsilon > 0$ be given. Note that if $\mathbf{z} = A\mathbf{x}$ is any element of S, we have

$$\sum_{n=N+1}^{\infty} |z_n|^2 = \sum_{n=N+1}^{\infty} |2^{-n}x_n|^2 \leq 2^{-2(N+1)} \sum_{n=1}^{\infty} |x_n|^2 \leq 2^{-2(N+1)},$$

hence it is possible to choose $N = N(\varepsilon)$ such that

$$\sum_{n=N+1}^{\infty} |z_n|^2 < \varepsilon^2/2$$

for all $\mathbf{z} \in S$. Now consider the set M of all "reduced" elements of the form $(z_1, \ldots, z_N, 0, 0, 0, \ldots)$ derivable from the elements of S. It is clear that $M \subseteq S$, which is bounded. Besides, the N-tuples of $\mathbf{z}$ belong to a bounded set in the finite dimensional space $\mathbb{R}^N$ in which any bounded set is precompact. Hence there is a finite $\varepsilon^2/2$-net of N-tuples from which for an arbitrary $\mathbf{z}$ we select $(\zeta_1, \ldots, \zeta_N)$ so that

$$\sum_{n=1}^{N} |z_n - \zeta_n|^2 < \varepsilon^2/2.$$

Thus an element $\mathbf{z}^{\varepsilon} = (\zeta_1, \dots, \zeta_N, 0, 0, \dots) \in \ell^2$ is an element of a finite ε-net of S, since

$$\|\mathbf{z} - \mathbf{z}^{\varepsilon}\|_{\ell^2}^2 = \sum_{n=1}^{N} |z_n - \zeta_n|^2 + \sum_{n=N+1}^{\infty} |z_n|^2 < \varepsilon^2/2 + \varepsilon^2/2 = \varepsilon^2.$$

4.47. For $\lambda = 0$ the operator $A - \lambda I$ is the same as A, hence the corresponding resolvent operator is simply A^{-1}. This operator exists; it is the backward-shift operator and its domain is $R(A)$. But $R(A)$ is not dense in ℓ^2 so the conclusion follows.

4.48. The ℓ^2-norms of the sequence elements are given by

$$\|\mathbf{x}_k\|_{\ell^2} = \left(\sum_{i=1}^{k} 1^2 \right)^{1/2} = k^{1/2}.$$

We see that $\|\mathbf{x}_k\|_{\ell^2} \to \infty$ as $k \to \infty$. But ℓ^2 is a Hilbert space, and in a Hilbert space every weakly convergent sequence is bounded.

4.49. It is clear that the sequence $\{\sin kx\}$ converges weakly if and only if the normalized sequence $\{\sqrt{\frac{2}{\pi}} \sin kx\}$ converges weakly. The latter sequence is orthonormal in $L^2(0, \pi)$, and any orthonormal sequence converges weakly to zero. Indeed Bessel's inequality shows that for any orthonormal sequence $\{e_k\}$ and any element $x \in H$ we have

$$\sum_{k=1}^{\infty} |(x, e_k)|^2 < \infty, \qquad \text{hence } \lim_{k \to \infty} (x, e_k) = 0.$$

In the Sobolev space, on the other hand, we have

$$\begin{aligned} \left\| \sqrt{\frac{2}{\pi}} \sin kx \right\|_{W^{1,2}(0,\pi)} &= \left(\int_0^{\pi} \left[\frac{2}{\pi} \sin^2 kx + \frac{2k^2}{\pi} \cos^2 kx \right] dx \right)^{1/2} \\ &= \sqrt{1 + k^2} \to \infty \quad \text{as } k \to \infty. \end{aligned}$$

For any subsequence the norms tend to infinity as well. Since any weakly convergent sequence in a Hilbert space is bounded, no subsequence can be weakly convergent.

4.50. In the process of introducing Lebesgue integration we obtained the inequality

$$\|F(\mathbf{x})\|_q \le (\operatorname{mes} \Omega)^{\frac{1}{q} - \frac{1}{p}} \|F(\mathbf{x})\|_p, \qquad 1 \le q \le p.$$

So a bound on the norm is $(\operatorname{mes} \Omega)^{\frac{1}{q} - \frac{1}{p}}$. Taking $F = 1$ we see that it is not a simple bound but the norm of the operator.

4.51. Since $\{x_n\}$ is an orthonormal sequence, it converges weakly to zero. The image sequence $\{Ax_n\}$ converges strongly to zero by compactness of A.

4.52. The subset inclusion $C^{(n)}(\Omega) \subset C(\Omega)$ certainly holds, so the imbedding operator I exists. It is continuous because $\|f\|_{C(\Omega)} \leq \|f\|_{C^{(k)}(\Omega)}$, as is seen from the form of the norms on these spaces. We must still show that I is compact.

Take a bounded set $S \subset C^{(n)}(\Omega)$, $n \geq 1$. The image $I(S)$ is uniformly bounded (since it is bounded in the max norm of $C(\Omega)$). Furthermore, S is a bounded subset of $C^{(1)}(\Omega)$. This latter fact, along with the mean value theorem

$$f(\mathbf{y}) - f(\mathbf{x}) = \nabla f(\mathbf{z}) \cdot (\mathbf{y} - \mathbf{x})$$

implies equicontinuity of $I(S)$. (Here $\mathbf{z}$ is an intermediate point on a segment from $\mathbf{x}$ to $\mathbf{y}$.) So $I(S)$ is compact by Arzelà's theorem. Therefore I maps bounded sets into precompact sets as required.

4.53. The space of polynomials P_n is linear but not complete. Weierstrass' theorem states that any function from $C(a, b)$ can be arbitrarily approximated by polynomials with respect to the norm of $C(a, b)$. So for $f \in C(a, b)$ there is a sequence of polynomials that converges to f, and this is necessarily a Cauchy sequence in $C(a, b)$. Clearly the norm of this sequence as a representer of an element of the completion space is equal to the norm of f in $C(a, b)$. This means that the result of completing P_n in the norm of $C(a, b)$ is a space that stands in one-to-one correspondence with $C(a, b)$ and can be identified with $C(a, b)$.

4.54. We already know that strong convergence implies weak convergence, and this does not depend on the dimension of the space. Let H be an n-dimensional Hilbert space having an orthonormal basis $\{e_1, \ldots, e_n\}$, and suppose $\{x_k\}$ is a sequence of elements in H such that $x_k \rightharpoonup x$. Then

$$\|x_k\|^2 = \sum_{i=1}^{n} |\langle x_k, e_i \rangle|^2 \to \sum_{i=1}^{n} |\langle x, e_i \rangle|^2 = \|x\|^2 \quad \text{as } k \to \infty,$$

and we have $x_k \to x$ according to Theorem 4.117.

4.55. Let M be a closed subspace of a Hilbert space H. Suppose $\{x_n\} \subset M$ converges weakly to $x \in H$. This means that $(x_n, f) \to (x, f)$ for every $f \in H$. Decompose H as $M \oplus M_\perp$. For every $g \in M_\perp$ we have

$$(x, g) = \lim_{n \to \infty} (x_n, g) = 0,$$

so $x \perp M_\perp$. This means that $x \in M$.

4.56. Suppose to the contrary that $\{x_n\}$ does not converge to x_0. So for some $\varepsilon_0 > 0$ we cannot find an integer N such that $\|x_n - x_0\| < \varepsilon_0$ whenever $n > N$. Thus there is a subsequence $\{x_{n_k}\}$ such that $\|x_{n_k} - x_0\| \geq \varepsilon_0$. But this means $\{x_{n_k}\}$ does not contain a sub-subsequence that converges to x_0, which contradicts the condition of the exercise.

4.57. Suppose to the contrary that $\{x_n\}$ does not converge weakly to x_0. Then there is a linear continuous functional F such that Fx_0 is not the limit of $\{Fx_n\}$. So there is an $\varepsilon_0 > 0$ and a subsequence $\{x_{n_k}\}$ such that $|Fx_{n_k} - Fx_0| \geq \varepsilon_0$. Hence $\{x_{n_k}\}$ does not contain a sub-subsequence that converges weakly to x_0.

4.58. (a) Assume S is closed and T is open. Take a sequence $\{x_n\} \subset S \setminus T$ such that $x_n \to x$. Since $\{x_n\} \subset S$, we have $x \in S$. We claim that $x \notin T$. For if not, then x belongs to the open set T and is therefore the center of some small open ball that lies entirely in T — a contradiction. (b) Assume S is open and T is closed. Let $x \in S \setminus T$. Since $x \in S$ we know that x is the center of an open ball that lies entirely in S; we claim that the radius of this ball can be chosen so small that no points of T can belong to it. For if not, then for each n the ball $B(x, 1/n)$ contains a point $x_n \in T$, and the sequence $\{x_n\} \subset T$ is convergent to x. Since T is closed we must have $x \in T$. However, this contradicts the assumption that $x \in S \setminus T$.

4.59. For any element f and any $\varepsilon > 0$ we can find an element $f^* \in S$ such that $\|f - f^*\| < \varepsilon/2$. Next, we can approximate f^* with a finite linear sum of system elements up to accuracy $\varepsilon/2$: $\left\|f^* - \sum_k c_k e_k\right\| < \varepsilon/2$. So the same sum approximates f to within accuracy ε.

4.60. We can take $\delta = \varepsilon/L$ in the definition of equicontinuity. Since uniform boundedness is given in the problem statement, S satisfies the conditions of Arzelà's theorem.

4.61. Suppose S be a compact subset of X. Let $\{y_n\}$ be a convergent sequence in $A(S)$, with $y_n \to y$. We need to show that $y \in A(S)$. The inverse image of $\{y_n\}$ under A is a sequence in S, and contains a convergent subsequence whose limit belongs to S: $x_k \to x \in S$, say. Noting that $\{A(x_k)\}$ is a subsequence of $\{y_n\}$, we have $A(x_k) \to y$. By definition of closed operator it follows that $x \in D(A)$ and $y = Ax$. Since $x \in S$ we have $y \in A(S)$, as desired.

4.62. We begin with

$$l|u(x)| \leq \left| \int_0^l u(t)\, dt \right| + l \int_0^l |u'(y)|\, dy,$$

square both sides and use the elementary inequality $2|ab| \leq a^2 + b^2$ to get

$$l^2 |u(x)|^2 \leq 2 \left| \int_0^l u(t)\, dt \right|^2 + 2l^2 \left(\int_0^l |u'(y)|\, dy \right)^2,$$

then integrate this over x:

$$l^2 \int_0^l |u(x)|^2\, dx \leq 2l \left\{ \left| \int_0^l u(t)\, dt \right|^2 + l^2 \left(\int_0^l |u'(y)|\, dy \right)^2 \right\},$$

so

$$l \int_0^l |u(x)|^2\,dx \le 2\left\{\left|\int_0^l u(t)\,dt\right|^2 + l^2\left(\int_0^l |u'(y)|\,dy\right)^2\right\}.$$

Finally, because of

$$\begin{aligned}\left(\int_0^l |u'(y)|\,dy\right)^2 &= \left(\int_0^l 1\cdot|u'(y)|\,dy\right)^2\\ &\le \int_0^l 1^2\,dy \int_0^l |u'^2(y)|\,dy\\ &= l\int_0^l |u'^2(y)|\,dy\end{aligned}$$

we get

$$l \int_0^l |u(x)|^2\,dx \le 2\left\{\left|\int_0^l u(t)\,dt\right|^2 + l^3\int_0^l |u'^2(y)|\,dy\right\}.$$

4.63.

$$\|y\| = \left(\left(\int_0^l y'(x)\,dx\right)^2 + \int_0^l EI\,y''^2(x)\,dx\right)^{1/2}.$$

When treating the problem of solvability of the equilibrium problems for a structure that can move as a rigid body, we should exclude rigid motions. For the beam under consideration, this is done with two conditions. One is the boundary condition $y|_0 = 0$. The other is somewhat artificial; it fixes the free rotations: $\int_0^l y(x)\,dx = 0$. On the set of smooth functions satisfying these conditions, the energy norm takes the form

$$\|y\| = \left(\int_0^l EI\,y''^2(x)\,dx\right)^{1/2}.$$

Note. In the following hints, k (with subscripts) denotes Winkler's coefficient, Ω_1, V_1 are subdomains, and γ is a sufficiently smooth curve (may be a part of the boundary).

5.1.

(1) *Membrane.* Total potential energy:

$$\frac{1}{2}\iint_\Omega \left[\left(\frac{\partial u}{\partial x}\right)^2 + \left(\frac{\partial u}{\partial y}\right)^2\right] dx\,dy + \frac{1}{2}\iint_{\Omega_1} k\,(u(x,y))^2\,dx\,dy$$
$$+\frac{1}{2}\int_\gamma k_1\,(u(x,y))^2\,ds - \iint_\Omega f(x,y)u(x,y)\,dx\,dy.$$

Virtual work principle:

$$\iint_\Omega \left(\frac{\partial u}{\partial x}\frac{\partial \varphi}{\partial x} + \frac{\partial u}{\partial y}\frac{\partial \varphi}{\partial y} \right) dx\, dy + \iint_{\Omega_1} k u(x,y)\varphi(x,y)\, dx\, dy$$
$$+ \int_\gamma k_1 u(x,y)\varphi(x,y)\, ds = \iint_\Omega f(x,y)\varphi(x,y)\, dx\, dy + \int_{\partial\Omega} g(s)\varphi(s)\, ds.$$

(2) *Stretched rod.* Here the notion of Winkler foundation makes no sense, because only longitudinal displacements are taken into account. However, we can suppose that at a point x_0 there is attached a linear spring with coefficient k, acting along the rod (which is analogous to Winkler's foundation). In that case we have the following. Total potential energy:

$$\frac{1}{2}\int_0^l ES(x){u'}^2(x)\, dx + \frac{1}{2}\left(ku(x_0)\right)^2 - \int_0^l f(x)u(x)\, dx - Fu(l).$$

Virtual work principle:

$$\int_0^l ES(x)u'(x)v'(x)\, dx + ku(x_0)v(x_0) = \int_0^l f(x)v(x)\, dx + Fv(l).$$

(Consider the case of several springs along the rod as well.)

(3) *Bent beam.* Total potential energy:

$$\frac{1}{2}\int_0^l EI(x){w''}^2(x)\, dx + \frac{1}{2}\int_a^b kw^2(x)\, dx + \frac{1}{2}k_1 w^2(x_0)\, dx$$
$$- \int_0^l f(x)w(x)\, dx - Fw(l).$$

Virtual work principle:

$$\int_0^l EI(x)w''(x)v''(x)\, dx + \int_a^b kw(x)v(x)\, dx + k_1 w(x_0)v(x_0)$$
$$= \int_0^l f(x)v(x)\, dx + Fv(l).$$

Here the region of the foundation is $[a,b]$, $0 \le a < b \le l$. We added a spring with coefficient k_1 at point x_0.

(4) *Plate.* Total potential energy:

$$\frac{D}{2}\iint_\Omega \left(w_{xx}^2 + w_{yy}^2 + 2\nu w_{xx}w_{yy} + 2(1-\nu)w_{xy}^2\right) d\Omega$$
$$+ \frac{1}{2}\iint_{\Omega_1} kw^2\, d\Omega + \frac{1}{2}\int_\gamma k_1 w^2\, ds - \iint_\Omega Fw\, d\Omega.$$

Virtual work principle:

$$D \iint_{\Omega} (w_{xx}v_{xx} + w_{yy}v_{yy} + \nu(w_{xx}v_{yy} + w_{yy}v_{xx}) + 2(1-\nu)w_{xy}v_{xy})\, d\Omega$$
$$+ \iint_{\Omega_1} kwv\, d\Omega + \int_{\gamma} k_1 wv\, ds = \iint_{\Omega} Fv\, d\Omega.$$

(5) *3D linearly elastic body.* Total potential energy:

$$\frac{1}{2}\iiint_V c^{ijkl} e_{kl}(\mathbf{u}) e_{ij}(\mathbf{u})\, dV + \frac{1}{2}\iint_{\partial V_2} k(\mathbf{u}\cdot\mathbf{n})^2\, dS$$
$$- \iiint_V \mathbf{F}\cdot\mathbf{u}\, dV - \iint_{\partial V_1} \mathbf{f}\cdot\mathbf{u}\, dS,$$

where $\mathbf{n}$ is the unit outward normal to the boundary. Virtual work principle:

$$\iiint_V c^{ijkl} e_{kl}(\mathbf{u}) e_{ij}(\mathbf{v})\, dV + \iint_{\partial V_2} k(\mathbf{u}\cdot\mathbf{n})(\mathbf{v}\cdot\mathbf{n})\, dS$$
$$= \iiint_V \mathbf{F}\cdot\mathbf{v}\, dV + \iint_{\partial V_1} \mathbf{f}\cdot\mathbf{v}\, dS.$$

5.2. For this case the equation of the virtual work principle takes the form

$$\iint_{\Omega} \left(\frac{\partial u}{\partial x}\frac{\partial \varphi}{\partial x} + \frac{\partial u}{\partial y}\frac{\partial \varphi}{\partial y}\right) dx\, dy = \iint_{\Omega} f(x,y)\varphi(x,y)\, dx\, dy + \int_{\partial\Omega_2} g(s)\varphi(s)\, ds.$$

It is valid for all functions $\varphi(x,y) \in C^1(\overline{\Omega})$ such that $\varphi(x,y)|_{\partial\Omega_1} = 0$, when $u = u_0(x,y)$ is a sufficiently smooth solution of the problem under consideration so it satisfies $u(x,y)|_{\partial\Omega_1} = 0$. If $\partial\Omega_1 \cup \partial\Omega_2$ does not cover $\partial\Omega$, this means that on $\Omega \setminus (\partial\Omega_1 \cup \partial\Omega_2)$ there is given zero load and so here $\partial u/\partial n = 0$.

Now the energy inner product takes the same form as for the above considered problems for a membrane $(u,v)_M$, but the energy space $\mathcal{E}_{Mm}$ is the completion of the set of functions $u \in C^1(\overline{\Omega})$ satisfying $u(x,y)|_{\partial\Omega_1} = 0$. On $\mathcal{E}_{Mm}$ the norm induced by the inner product is equivalent to the norm of $W^{1,2}(\Omega)$.

The generalized setup of the problem under consideration is defined by the above equation of the VWP, so $u \in \mathcal{E}_{Mm}$ is a generalized solution if this equation is valid for all $\varphi(x,y) \in \mathcal{E}_{Mm}$.

The minimum problem now takes on the form

$$E_{Mm}(u) = \frac{1}{2}\|u\|_M^2 - \Phi(u),$$

where

$$\Phi(u) = \iint_{\Omega} f(x,y)u(x,y)\, dx\, dy + \int_{\partial\Omega_2} g(s)u(s)\, ds.$$

If

$$f(x, y) \in L^{p_1}(\Omega), \qquad g(s) \in L^{p_2}(\partial\Omega_2), \tag{A.17}$$

then $\Phi(u)$ is a linear continuous functional in $\mathcal{E}_{Mm}$. The existence/uniqueness theorem is as follows:

Let (A.17) be valid. In the energy space $\mathcal{E}_{Mm}$ the functional $E_{Mm}(u)$ attains its minimum at $u = u_0$ and the minimizer satisfying the equation of the VWP is unique.

5.3. The total potential energy is now

$$\begin{aligned} E_{BR}(\mathbf{u}) &= \frac{1}{2}\int_0^l ES(x)u'^2(x)\,dx + \frac{1}{2}\int_0^l EI(x)w''^2(x)\,dx \\ &\quad - \int_0^l f(x)u(x)\,dx - Fu(l) - \int_0^l q(x)w(x)\,dx - Qw(l), \end{aligned} \tag{A.18}$$

where $q(x)$ is the distributed normal load and Q is the transverse force on the end.

The equation of the VWP is

$$\begin{aligned} &\int_0^l ES(x)u'(x)v'(x)\,dx + \int_0^l EI(x)w''(x)\varphi''(x)\,dx \\ &\quad = \int_0^l f(x)v(x)\,dx + Fv(l) + \int_0^l q(x)\varphi(x)\,dx + Q\varphi(l). \end{aligned} \tag{A.19}$$

Now the energy inner product for pairs $\mathbf{u}_i = (u_i, w_i)$ takes the form

$$(\mathbf{u}_1, \mathbf{u}_2)_{BR} = \int_0^l ES(x)u_1'(x)u_2'(x)\,dx + \int_0^l EI(x)w_1''(x)w_2''(x)\,dx.$$

With the boundary conditions $u(0) = 0$ and $w(0) = 0$, $w'(0) = 0$, construct the energy space $\mathcal{E}_{BR}$. On $\mathcal{E}_{BR}$ its induced norm is equivalent to the norm of $W^{1,2}(0,l) \times W^{2,2}(0,l)$. The total energy functional now takes the form

$$E_{BR}(\mathbf{u}) = \frac{1}{2}\|u\|_{BR}^2 - \Phi_{BR}(u)$$

with

$$\Phi_{BR}(u) = \int_0^l f(x)u(x)\,dx + Fu(l) + \int_0^l q(x)w(x)\,dx + Qw(l).$$

If $f(x) \in L(0,l)$ and $q(x) \in L(0,l)$ the functional $\Phi_{BR}(u)$ is linear and continuous in $\mathcal{E}_{BR}$ and this is enough to state that the total energy functional $E_{BR}(\mathbf{u})$ attains its minimum $\mathbf{u}_0$ in $\mathcal{E}_{BR}$ that is unique. This minimum is a generalized solution to the combined problem under consideration.

5.4. (a) The VWP takes the form

$$\int_0^l EI(x)w''(x)v''(x)\,dx = \int_0^l f(x)v(x)\,dx + \sum_k F_k v(x_k) + \sum_j M_j v'(x_j) + Fv(l),$$

where point force F_k acts at point x_k and point couple M_j acts at point x_j. *Remark:* This is meaningful because the energy space imbeds continuously to the space $C^{(1)}(0,l)$. For membranes and three-dimensional elastic bodies in the energy setup, point forces are impossible. For a plate we can consider a generalized setup with external point forces acting on the plate. (b) The generalized setup for countable sets of external point forces and couples is possible when the series $\sum_k F_k$ and $\sum_j M_j$ are absolutely convergent and the the beam ends are clamped, since the corresponding part of the work of external forces $\sum_k F_k v(x_k) + \sum_j M_j v'(x_j)$ is a linear continuous functional in the energy space:

$$\left| \sum_k F_k v(x_k) + \sum_j M_j v'(x_j) \right| \le \max_{[0,l]} |v(x)| \sum_k |F_k| + \max_{[0,l]} |v'(x)| \sum_j |F_j| \le m \|u\|_B .$$

5.5. The energy spaces for the problems are some subspaces of corresponding combinations of Sobolev spaces.

5.6. The functional $\Phi(w)$ (the potential) takes the form

$$\Phi(w) = \iint_\Omega F(x,y)w(x,y)\,d\Omega + \int_{\partial\Omega} f(s)w(x,y)\,ds + \sum_{k=1}^N F_k w(x_k,y_k).$$

The (self-balance) condition for solvability of the problem is

$$\Phi(ax+by+c) = \iint_\Omega F(x,y)(ax+by+c)\,d\Omega + \int_{\partial\Omega} f(s)(ax+by+c)\,ds + \sum_{k=1}^N F_k(ax_k+by_k+c) = 0 \quad \text{for all constants } a,b,c.$$

5.7. Use the following forms of the kinetic energy functionals.

Rod:

$$K = \int_0^l \rho \left(\frac{\partial u}{\partial t}\right)^2 dx.$$

Beam:

$$K = \int_0^l \rho \left(\frac{\partial w}{\partial t} \right)^2 dx.$$

Plate:

$$K = \iint_\Omega \rho \left(\frac{\partial w}{\partial t} \right)^2 dx\, dy.$$

5.8. It is necessary to solve the following simultaneous algebraic equations with respect to $a_1, \ldots, a_n$:

$$\sum_{k=1}^{n} a_k (\varphi_k, \varphi_j)_M = (u_0^*, \varphi_j)_M, \qquad j = 1, \ldots, n.$$

5.9. For an infinite dimensional space $\mathcal{E}$ the inequality $\|u\|_A \geq m \|u\|_\mathcal{E}$ with constant $m > 0$ independent of u is impossible. Indeed, take an orthonormal sequence $\{e_n\}$ in $\mathcal{E}$, so $\|e_n\|_\mathcal{E} = 1$. This sequence converges to zero weakly and thus, because A is compact, we get $\|Ae_n\|_\mathcal{E} \to 0$. Then $\|e_n\|_A^2 = (Ae_n, e_n)_\mathcal{E} \to 0$ as well.

5.10. This set is the set of eigenfunctions of the eigenvalue problem

$$u'' + \lambda^2 u = 0, \qquad u(0) = 0 = u(\pi).$$

What is the energy space for this problem where the set is an orthogonal basis?

5.11. We recall only that for each of our problems the operator A is defined by the following equalities (and the Riesz representation theorem).

Beam:

$$(Aw, v)_B = \int_0^l \rho w(x) v(x)\, dx.$$

Plate:

$$(Aw, v)_P = \iint_\Omega \rho w(x, y) v(x, y)\, d\Omega.$$

Three-dimensional elastic body:

$$(A\mathbf{u}, \mathbf{v})_E = \iiint_V \rho \mathbf{u} \cdot \mathbf{v}\, dV.$$

These operators have all the properties needed in Theorem 5.22, and so the theorem can be formulated for each of the problems without change.

5.12. Suppose there is a minimizing sequence $\{x_n\}$ that does not strongly converge to x_0. This means that there is $\varepsilon > 0$ and a subsequence $\{x_{n_k}\}$ such that $\|x_0 - x_{n_k}\|_H > \varepsilon$. But $\{x_{n_k}^*\}$ is a minimizing sequence as well, and so it contains a subsequence that strongly converges to a minimizer (by the theorem). By uniqueness this minimizer is x_0, which contradicts the above inequality.

5.13. Suppose that for g_1 and g_2 we get solutions $w_1^* + g_1$ and $w_2^* + g_2$. Then $(g_2 - g_1)|_{\partial\Omega} = 0$. Consider the "difference" of the corresponding equations. We come to the same problem for $w_3 = w_2 - w_1$ with $f = 0$ and the function $(g_1 - g_2)$ taken as g. This problem, by the theorem, has a unique solution w_3^*. By the structure of the equation of the problem it is evident that $w_3^* = g_1 - g_2$, and so $w_1^* + g_1 = w_2^* + g_2$.

Bibliography

Adams, R.A. (1975). *Sobolev Spaces* (Academic Press)., New York, 1975.

Antman, S.S. (1996). *Nonlinear Problems of Elasticity* (Springer).

Bachman, G., and Narici, L. (1966). *Functional Analysis* (Academic Press).

Chróścielewski, J., Makowski, J., and Pietraszkiewicz, W. (2004). *Statics and Dynamics of Multifold Shells: Nonlinear Theory and Finite Element Method (in Polish)* (Wydawnictwo IPPT PAN, Warszawa).

Ciarlet, P.G. (1988–2000). *Mathematical Elasticity* (North Holland).

Courant, R., and Hilbert, D. (1989) *Methods of Mathematical Physics, Volume 1* (Wiley)

Fichera, G. (1972). Existence theorems in elasticity (XIII.15), and Boundary value problems of elasticity with unilateral constraints (YII.8, XIII.15, XIII.6), in Handbuch der Physik YIa/2, C. Truesdell, ed. (Springer).

Friedman, A. (1982). *Variational Principles and Free-Boundary Problems* (Wiley).

Gelfand, I.M., and Fomin, S.V. (1963). *Calculus of Variations* (Prentice–Hall).

Hardy, G.H., Littlewood, J.E., and Pólya, G. (1952). *Inequalities* (Cambridge University Press).

Kinderlehrer, D., and Stampacchia, G. (1980). *An Introduction to Variational Inequalities and their Applications* (Academic Press).

Kienzler, R. J., and Herrmann, G. (2000). *Mechanics in Material Space with Applications to Defect and Fracture Mechanics* (Springer).

Lebedev, L.P., Vorovich, I.I., and Gladwell, G.M.L. (1996). *Functional Analysis: Applications in Mechanics and Inverse Problems* (Kluwer).

Lebedev, L.P., and Vorovich, I.I. (2002). *Functional Analysis in Mechanics* (Springer).

Lebedev, L.P., and Cloud, M.J. (2009). *Introduction to Mathematical Elasticity* (World Scientific).

Lebedev, L.P., Cloud, M.J., and Eremeyev, V.A. (2010). *Tensor Analysis With Applications in Mechanics* (World Scientific).

Libai, A. and Simmonds, J.G. (2005). *The Nonlinear Theory of Elastic Shells* (Cambridge University Press).

Love, A.E.H. (2011). *A Treatise on the Mathematical Theory of Elasticity* (Dover).

Maugin, G.A. (2011). *Configurational Forces: Thermomechanics, Physics, Mathematics, and Numerics* (CRC Press).
Mikhlin, S.G. (1965). *The Problem of Minimum of a Quadratic Functional* (Holden–Day).
Mura, T., and Koya, T. (1992). *Variational Methods in Mechanics* (Oxford University Press).
Novozhilov, V.V., Chernykh, K.Ph. and Mikhailovskiy E.M. (1991). *Linear Theory of Thin Shells (in Russian)* (Politechnika, Leningrad).
Olver, P.J. (1993). *Applications of Lie Groups to Differential Equations* (Springer).
Pinch, E. (1993). *Optimal Control and the Calculus of Variations* (Oxford University Press).
Riesz, F. (1918). Über lineare Funktionalgleichungen. Acta Math. 41 71–98.
Schauder, J. (1930). Über lineare, volstetige Funktionaloperationen. Stud. Math. 2, 1–6.
Sobolev, S.L. (1951). *Some Applications of Functional Analysis to Mathematical Physics* (LGU).
Svetlitsky, V.A. (2000). *Statics of Rods* (Springer).
Timoshenko, S. (1983). *Strength of Materials, Parts 1 and 2* (Krieger).
Timoshenko, S. (1970). *Theory of Elasticity* (McGraw-Hill).
Vorovich, I.I. (1999). *Nonlinear Theory of Shallow Shells* (Springer).
Yosida, K. (1965). *Functional Analysis* (Springer).

Index

absolute convergence, 230
action, 382
action functional, 114
adjoint operator, 299
admissible function, 15
admissible state, 8
approximation, 277
Arzelà's theorem, 255
autonomic system, 164

Banach space, 229
basis, 285
basis dyads, 179
basis functions, 24
beam problem, 40
Bessel's inequality, 288
Bessel–Hagen extension, 147
bounded set, 228
brachistochrone, 96
Bubnov–Galerkin method, 26

calculus of variations, 1
  broken extremals, 80
  first variation, 54, 72
  fundamental lemma of, 18
  isoperimetric problems, 65
  more general functionals, 34, 43, 49
  movable ends of extremals, 76
  natural boundary conditions, 31
  Ritz's method, 23
  simplest problem of, 15
  sufficient conditions, 85
  variational derivative, 61
catenary, 71
Cauchy sequence, 227
  strong, 291
  weak, 291
central field, 91
clamped rod, 321
closed ball, 221
closed graph theorem, 315
closed operator, 311
closed set, 231
closed system, 289
coincidence set, 425
compact operator, 304
compact set, 252
complete metric space, 229
complete system, 25, 285
completeness, 229
completion, 238
completion theorem, 238
cone, 234
cone condition, 318
conjugate equation, 173
conjugate point, 90
conservation law, 127
continuous spectrum, 339
continuum, 224
contraction mapping, 270
control variables, 161
control vector, 161
convergence, 221
  absolute, 230

weak, 292
convex set, 278
correspondence, 264
countable set, 223
critical point, 421

dense set, 224
diagonal sequence, 257
dimension, 223
divergence symmetry, 148
domain, 265
drilling moment, 126
dual problem, 69
dual space, 284
dyad, 180
dynamical system, 190

eigensolution, 338, 401
eigenvalue, 339, 401
eigenvector, 338, 401
elastic moduli, 109
energy inner product, 321, 325
energy norm, 321, 325
energy space, 322
equicontinuity, 255
equivalent Cauchy sequences, 238
Euler equation, 20
  integrated form, 22
  invariance of, 23
Euler–Lagrange equation, 37
extension, 312
extrema, 15
extremal, 20

field
  central, 91
  of extremals, 91
  proper, 91
finite ε-net, 251
finite dimensional operator, 305
first differential, 2
  of a function, 57
first integral, 128
first variation, 18, 58, 60, 64
fixed point, 270
flux, 134
Fourier coefficients, 287
Fourier series, 287
Fréchet derivative, 60
Fréchet differential, 73
free rod, 321
Friedrichs inequality, 329
function, 264
functional, 1, 2, 7, 15, 265
  continuous, 16
  linear, 72
  local minimum, 17
  stationary, 20, 64
fundamental lemma, 18
fundamental solution, 170, 187

Gâteaux derivative, 60
Galerkin's method, 26
Galilean boost, 150
generalized derivative, 249
global minimum, 1, 4
gradient, 185, 421
Gram determinant, 287
Gram–Schmidt procedure, 286
graph, 314

Hölder inequality, 243, 247
Hamilton form, 173
Hamilton's principle, 113, 119
Hamilton's variational principle
  elasticity, 114
  plate, 119
Hamilton–Ostrogradski principle, 381
hanging chain problem, 69
Hausdorff criterion, 251
Hessian, 5
Hilbert cube, 252
Hilbert space, 263
Hilbert's invariant integral, 93
Hooke's law, 109

imbedding, 316
increment, 64
infinitesimal divergence invariance, 147
infinitesimal invariance under transformation, 131

infinitesimal transformation, 131
inner product, 260
inner product space, 261
invariance
  infinitesimal, 131
  infinitesimal divergence, 147
inverse operator, 267
isometry, 238
isoperimetric problem, 8, 65

kernel, 301
Korn inequality, 333
Kronecker delta, 39

Lagrange formula, 2
Lagrange multipliers, 6
Lagrange variational principle
  elasticity, 111
  plate, 117
Lagrangian function, 6
Lamé's constants, 109
Lebesgue integral, 246
Legendre's condition, 89
limit, 221
limit point, 231
linear operator, 265
linear space, 218
Lipschitz condition, 358
local minimum, 1, 4, 17
  strict, 17

map, 7, 264
mapping, 264
mathematical programming, 7
mean value theorem, 61
metric, 220
  natural, 220
metric space, 220
  complete, 229
minimizer
  sufficient conditions, 90
minimum
  global, 1, 4
  local, 1, 4
Minkowski inequality, 233, 243
mixed problem, 110

natural boundary conditions, 31
needle-shaped increment, 166
neighborhood, 17
Neumann series, 340
neutral action method, 152
Noether's theorem, 134, 143
norm(s), 72, 218
  equivalent, 222
  of functional, 72
  operator, 266
  product, 237, 357
normed space, 218
null Lagrangian, 148, 151
null space, 268

open ball, 221
open set, 356
operator(s), 7, 264
  adjoint, 299
  bounded, 266
  bounded below, 268
  closed, 311
  compact, 304
  condition number of, 355
  densely defined, 355
  extension of, 312
  forward shift, 356
  Fredholm integral, 356
  graph of, 314
  imbedding, 316
  inverse, 267
  linear, 265
  resolvent, 339
  self-adjoint, 300
  strictly positive, 403
  weakly continuous, 302
order of smallness
  O, 55
  o, 55
orthogonal decomposition, 281
orthogonal subspaces, 281
orthogonality, 280
orthonormal system, 286

parallelogram equality, 262
Parseval's equality, 288

Peano's form, 3, 56
pencil, 91
Perron's paradox, 21
Poincaré inequality, 331
point spectrum, 339
Pontryagin's function, 173, 196
Pontryagin's principle, 174
pre-Hilbert space, 261
precompact set, 250
product norm, 237, 357
proper field, 91
Pythagorean theorem, 263

quotient space, 353

range, 265
regular value, 339
Reissner–Mindlin plate, 115
representative Cauchy sequence, 238
residual, 27
residual spectrum, 339
resolvent operator, 339
resolvent set, 339
Riesz lemma, 254
Riesz representation theorem, 282
Ritz approximation, 25
Ritz method, 23

Schauder basis, 285
Schwarz inequality, 261
second variation, 87
self-adjoint operator, 300
separability, 225
sequence
  Cauchy, 227
  convergent, 221
sequence spaces, 233
set(s)
  bounded, 228
  closed, 231
  compact, 252
  convex, 278
  countable, 223
  open, 356
  precompact, 250
  totally bounded, 251
  weakly closed, 296
  weakly compact, 296
  weakly precompact, 296
Sobolev imbedding theorem, 318
Sobolev norm, 237
Sobolev space, 249
space(s)
  Banach, 229
  Hilbert, 263
  inner product, 261
  metric, 220
  normed, 218
  sequence, 233
  Sobolev, 249
  strictly normed, 278
spectral value, 339
spectrum, 339
sphere, 221
stationarity, 20, 64
stationary equivalence class, 238
stationary sequence, 238
strain energy function, 110
strain tensor, 109
stress tensor, 108
strict local minimum, 17
strictly normed space, 278
strictly positive operator, 403
strong derivative, 249
successive approximations, 270
sufficiently smooth function, 15
support, 61

Taylor's formula, 3
Taylor's theorem, 56
tensor product, 179
total derivative, 45
totally bounded set, 251
transformation, 264
transversality, 78

unilateral problem, 425
unit ball, 304
unit tensor, 182

variational derivative, 62, 64, 168
variational inequality, 426

variational problem, 15
variational symmetry, 132
virtual displacement, 57
virtual state, 8
virtual work principle, 113, 362
  plate, 118
von Mises truss, 155

wave equation, 146
weak Cauchy sequence, 291
weak completeness, 295
weak continuity, 302
weak convergence, 292
weak derivative, 249
weakly closed set, 296
weakly compact set, 296
weakly precompact set, 296
Weierstrass approximation theorem, 226
Weierstrass conditions, 93
Weierstrass excess function, 93
Weierstrass–Erdmann conditions, 80